Post-Transcriptional Gene Regulation in Human Disease

Translational Epigenetics Series

Trygve Tollefsbol, Ph.D., D.O., Distinguished Professor - Series Editor
Professor of Biology, University of Alabama at Birmingham, and Senior Scientist, Comprehensive Cancer Center, Comprehensive Center for Healthy, Birmingham, AL, United States
Aging, Comprehensive Diabetes Center and Nutrition Obesity Research Center, Birmingham, AL, United States
Director, Cell Senescence Culture Facility, Birmingham, AL, United States

Transgenerational Epigenetics
Edited by Trygve O. Tollefsbol, 2014

Personalized Epigenetics
Edited by Trygve O. Tollefsbol, 2015

Epigenetic Technological Applications
Edited by Y. George Zheng, 2015

Epigenetic Cancer Therapy
Edited by Steven G. Gray, 2015

DNA Methylation and Complex Human Disease
By Michel Neidhart, 2015

Epigenomics in Health and Disease
Edited by Mario F. Fraga and Agustin F. F Fernández, 2015

Epigenetic Gene Expression and Regulation
Edited by Suming Huang, Michael Litt and C. Ann Blakey, 2015

Epigenetic Biomarkers and Diagnostics
Edited by Jose Luis García-Giménez, 2015

Drug Discovery in Cancer Epigenetics
Edited by Gerda Egger and Paola Barbara Arimondo, 2015

Medical Epigenetics
Edited by Trygve O. Tollefsbol, 2016

Chromatin Signaling and Diseases
Edited by Olivier Binda and Martin Fernandez-Zapico, 2016

Genome Stability
Edited by Igor Kovalchuk and Olga Kovalchuk, 2016

Chromatin Regulation and Dynamics
Edited by Anita Göndör, 2016

Neuropsychiatric Disorders and Epigenetics
Edited by Dag H. Yasui, Jacob Peedicayil and Dennis R. Grayson, 2016

Polycomb Group Proteins
Edited by Vincenzo Pirrotta, 2016

Epigenetics and Systems Biology
Edited by Leonie Ringrose, 2017

Cancer and Noncoding RNAs
Edited by Jayprokas Chakrabarti and Sanga Mitra, 2017

Nuclear Architecture and Dynamics
Edited by Christophe Lavelle and Jean-Marc Victor, 2017

Epigenetic Mechanisms in Cancer
Edited by Sabita Saldanha, 2017

Epigenetics of Aging and Longevity
Edited by Alexey Moskalev and Alexander M. Vaiserman, 2017

The Epigenetics of Autoimmunity
Edited by Rongxin Zhang, 2018

Epigenetics in Human Disease, Second Edition
Edited by Trygve O. Tollefsbol, 2018

Epigenetics of Chronic Pain
Edited by Guang Bai and Ke Ren, 2018

Epigenetics of Cancer Prevention
Edited by Anupam Bishayee and Deepak Bhatia, 2018

Computational Epigenetics and Diseases
Edited by Loo Keat Wei, 2019

Pharmacoepigenetics
Edited by Ramón Cacabelos, 2019

Epigenetics and Regeneration
Edited by Daniela Palacios, 2019

Chromatin Signaling and Neurological Disorders
Edited by Olivier Binda, 2019

Transgenerational Epigenetics, Second Edition
Edited by Trygve Tollefsbol, 2019

Nutritional Epigenomics
Edited by Bradley Ferguson, 2019

Prognostic Epigenetics
Edited by Shilpy Sharma, 2019

Epigenetics of the Immune System
Edited by Dieter Kabelitz, 2020

Stem Cell Epigenetics
Edited by Eran Meshorer and Giuseppe Testa, 2020

Epigenetics Methods
Edited by Trygve Tollefsbol, 2020

Histone Modifications in Therapy
Edited by Pedro Castelo-Branco and Carmen Jeronimo, 2020

Environmental Epigenetics in Toxicology and Public Health
Edited by Rebecca Fry, 2020

Genome Stability
Edited by Igor Kovalchuk and Olga Kovalchuk, 2021

Twin and Family Studies of Epigenetics
Edited by Shuai Li and John Hopper, 2021

Epigenetics and Metabolomics
Edited by Paban K. Agrawala and Poonam Rana, 2021

Medical Epigenetics, Second Edition
Edited by Trygve Tollefsbol, 2021

Epigenetics in Precision Medicine
Edited by José Luis García-Giménez, 2022

Epigenetics of Stress and Stress Disorders
Edited by Nagy A. Youssef, 2022

Translational Epigenetics

Post-Transcriptional Gene Regulation in Human Disease

Volume 32

Editors

Buddhi Prakash Jain

Department of Zoology, Mahatma Gandhi Central University, Motihari, Bihar, India

Shyamal K. Goswami

School of Life Sciences, Jawaharlal Nehru University, New Delhi, India

Tapan Sharma

Department of Biochemistry and Molecular Pharmacology, University of Massachusetts Medical School, Worcester, MA, United States

Academic Press is an imprint of Elsevier
125 London Wall, London EC2Y 5AS, United Kingdom
525 B Street, Suite 1650, San Diego, CA 92101, United States
50 Hampshire Street, 5th Floor, Cambridge, MA 02139, United States
The Boulevard, Langford Lane, Kidlington, Oxford OX5 1GB, United Kingdom

Notices
Knowledge and best practice in this field are constantly changing. As new research and experience broaden our understanding, changes in research methods, professional practices, or medical treatment may become necessary.

Practitioners and researchers must always rely on their own experience and knowledge in evaluating and using any information, methods, compounds, or experiments described herein. In using such information or methods they should be mindful of their own safety and the safety of others, including parties for whom they have a professional responsibility.

To the fullest extent of the law, neither the Publisher nor the authors, contributors, or editors, assume any liability for any injury and/or damage to persons or property as a matter of products liability, negligence or otherwise, or from any use or operation of any methods, products, instructions, or ideas contained in the material herein.

ISBN: 978-0-323-91305-8

For information on all Academic Press publications visit our website at
https://www.elsevier.com/books-and-journals

Publisher: Stacy Masucci
Acquisitions Editor: Peter B. Linsley
Editorial Project Manager: Barbara L. Makinster
Production Project Manager: Omer Mukthar
Cover Designer: Mark Rogers

Typeset by TNQ Technologies

Contents

List of contributors

Ayushi Aggarwal
Special Centre for Molecular Medicine, Jawaharlal Nehru University, New Delhi, Delhi, India

Hafiz M. Ahmad
Department of Molecular Cell and Cancer Biology, University of Massachusetts Chan Medical School, Worcester, MA, United States

Ayeman Amanullah
University of Twente, Enschede, Netherlands

Zahid Ashraf
Department of Biotechnology, Jamia Millia Islamia, New Delhi, India

Gargi Bagchi
Amity Institute of Biotechnology, Amity University Haryana, Amity Education Valley, Gurgaon, Haryana, India

Ajay Bahl
Department of Cardiology, PGIMER, Chandigarh, India

Sudha Bhattacharya
Department of Biology, Ashoka University, Sonipat, Haryana, India

Shankar Chanchal
Department of Biotechnology, Jamia Millia Islamia, New Delhi, India

Ishani Dasgupta
Horae Gene Therapy Center, Department of Pediatrics, University of Massachusetts Chan Medical School, Worcester, MA, United States

Anindita Dasgupta
National Centre for Cell Science, S P Pune University Campus, Pune, Maharashtra, India

Savita Devi
Department of Pathology and Laboratory Medicine, Cedars Sinai Medical Center, Los Angeles, CA, United States

Anindita Dutta
Chromatin Remodeling Laboratory, School of Life Sciences, JNU, New Delhi, India

Shyamal K. Goswami
School of Life Sciences, Jawaharlal Nehru University, New Delhi, India

Manish Goyal
Department of Microbiology & Molecular Genetics, The Kuvin Center for the Study of Infectious and Tropical Diseases, IMRIC, The Hebrew University-Hadassah Medical School, Jerusalem, Israel

Saddam Hussain
Chromatin Remodeling Laboratory, School of Life Sciences, JNU, New Delhi, India

Buddhi Prakash Jain
Department of Zoology, Mahatma Gandhi Central University, Motihari, Bihar, India

Yashika Jawa
Special Centre for Molecular Medicine, Jawaharlal Nehru University, New Delhi, India

Bibekananda Kar
Department of Biochemistry and Molecular Biology, Mayo Clinic, Rochester, MN, United States

Yasir Khan
Department of Biotechnology, Jamia Millia Islamia, New Delhi, India

Madhu Khullar
Department of Experimental Medicine and Biotechnology, PGIMER, Chandigarh, India

Avinash Kumar
Department of Biological Sciences, Louisiana State University, Baton Rouge, LA, United States; Pennington Biomedical Research Center, Baton Rouge, LA, United States

Sangeeta Kumari
Special Centre for Molecular Medicine, Jawaharlal Nehru University, New Delhi, India

Pragya Mishra
Special Centre for Molecular Medicine, Jawaharlal Nehru University, New Delhi, Delhi, India

Alapani Mitra
National Centre for Cell Science, S P Pune University Campus, Pune, Maharashtra, India

Debashis Mitra
National Centre for Cell Science, S P Pune University Campus, Pune, Maharashtra, India

Anupam Mittal
Department of Translational and Regenerative Medicine, PGIMER, Chandigarh, India

Chinmay K. Mukhopadhyay
Special Centre for Molecular Medicine, Jawaharlal Nehru University, New Delhi, Delhi, India

Rohini Muthuswami
Chromatin Remodeling Laboratory, School of Life Sciences, JNU, New Delhi, India

Sandeep Ojha
Institute of Physical Chemistry, Westfälische Wilhelms-Universität, Münster, Germany

Shweta Pandey
Govt VYT PG Autonomous College, Durg, Chhattisgarh, India

Rashmi Pathak
Comparative Biomedical Sciences, School of Veterinary Medicine, Louisiana State University, Baton Rouge, LA, United States

Vibha Rani
Department of Biotechnology, Jaypee Institute of Information Technology, Noida, Uttar Pradesh, India

Ankit Sabharwal
Department of Biochemistry and Molecular Biology, Mayo Clinic, Rochester, MN, United States

Tapan Sharma
Department of Biochemistry and Molecular Pharmacology, University of Massachusetts Medical School, Worcester, MA, United States

Swati Sharma
Department of Biotechnology, Jamia Millia Islamia, New Delhi, India

Smriti Shreya
Department of Zoology, Mahatma Gandhi Central University, Motihari, Bihar, India

Karina Simantov
Department of Microbiology & Molecular Genetics, The Kuvin Center for the Study of Infectious and Tropical Diseases, IMRIC, The Hebrew University-Hadassah Medical School, Jerusalem, Israel

Ramandeep Singh
Department of Cardiology, PGIMER, Chandigarh, India

Saba Tabasum
Department of Medical Oncology, Dana-Farber Cancer Institute, Harvard Medical School, Boston, MA, United States; Melanoma Disease Center, Dana-Farber Cancer Institute, Harvard Medical School, Boston, MA, United States

Anjali Tripathi
National Centre for Cell Science, S P Pune University Campus, Pune, Maharashtra, India

Rakesh K. Tyagi
Special Centre for Molecular Medicine, Jawaharlal Nehru University, New Delhi, India

Monika Yadav
Cancer Biology Laboratory, School of Life Sciences, Jawaharlal Nehru University, New Delhi, India

Sameeksha Yadav
Special Centre for Molecular Medicine, Jawaharlal Nehru University, New Delhi, Delhi, India

A note from the editor

The quest for understanding diseases and finding their cure is as old as human civilization. However, before the emergence of the era of modern medicine, therapeutics was primarily natural compounds in impure or crude forms. The past century has seen a significant boost in drug discovery with advancements in synthetic chemistry, pharmacology, diagnostics, cell biology, etc. Further, what has expedited our knowledge of human pathobiology is the deciphering of the human genome and its regulation. Understanding of cellular and molecular biology has expanded the horizon of our understanding of human diseases and their potential targets to an unprecedented level. The human genome is a large and complex storehouse of information that is meticulously propagated and deciphered by the macromolecular assemblies of gene regulatory proteins and RNAs. An intricate network of feed-forward and feedback interactions among these entities ensures the precise levels of gene expression in any cell type in a given context; while its aberration leads to diseases. Molecular biology, the sub-discipline that studies the process of gene expression, has grown phenomenally since the discovery of the DNA structure in 1953. The other disciplines of biology viz., biochemistry, structural biology, genetics, epigenetics, developmental biology, proteomics, transcriptomics, and bioinformatics have also significantly contributed to the accumulation of a large volume of information on gene regulation. It would thus be a gargantuan effort to correlate various aspects of gene regulation to the repertoire of diseases. Therefore, in this book, we have primarily focused on one aspect, "*the role of post-transcriptional gene regulation*" in model diseases. Each of the chapters in the book is written by researchers associated with the respective disorders for a long time. As it is not possible to cover all major human diseases, we have made efforts to include only those that are extensively researched and sufficient information is already available for compilation. Also, to help readers comprehend the thematic chapters, two chapters are included in the beginning that describe the basic principles of gene regulation. I hope this review will give the readers a platform to realize the essence of contemporary developments in human biology in general and diseases in particular.

CHAPTER

Regulation of gene expression in mammals: an overview

1

Shyamal K. Goswami

School of Life Sciences, Jawaharlal Nehru University, New Delhi, India

Introduction

In the year 1958, Francis Crick proposed the "Central Dogma of Molecular Biology." He was awarded the Nobel Prize in 1962 for his contributions toward the understanding of the DNA structure and its functions. Since then, phenomenal progress has been made in analyzing the structure and the organization of the genomes and the mechanisms of their expression across species. Although the basic tenets of the molecular mechanisms by which the DNA sequence is transcribed into the mRNA, which in turn is translated into the proteins remain unchallenged, the past 60 years have witnessed the discovery of newer and newer regulatory modules and their complex interactions that precisely control the entire process of gene expression, especially in the mammals. The progress of our understanding of the mechanisms of gene expression can be divided into two distinct phases, viz., the pre- and the postgenomics era that also happens to be at the juncture of the past and the present millennium. In the 1970s, with the advent of breakthrough tools like recombinant DNA technology, DNA sequencing, Southern and Northern hybridizations, construction and screening of cDNA and genomic libraries, etc., numerous genes were isolated and studied. In the following decades, using biochemical, genetic, and molecular biological approaches, different RNA polymerases and a large number of general and tissue-specific transcription factors across species were also isolated and characterized. Among the breakthrough discoveries were the post-transcriptional capping, splicing, and polyadenylation of the pre-mRNAs; and the role of enhancers and long noncoding RNAs in the metazoan gene expression. These created a conceptual framework that is, the interactions of the RNA polymerase, general and gene-specific transcription factors, chromatin modifiers, and RNA processing enzymes with the segments of DNA called promoters, and the enhancers govern the expression of the cognate gene. Over the years, it became evident that the transcription of the mRNA coding genes and the post-transcriptional processing of the pre-mRNA does not occur sequentially as envisaged earlier rather it occurs simultaneously for most of the genes. Also, the covalent modifications of the histones play a major role in the selective expression of various genes or otherwise. These observations heightened the complexities of gene regulation, especially in mammals. Completion of genome sequences in the late 1990s followed the emergence of various tools of bioinformatics and the expressed sequence tag (EST) database: methods for high throughput RNA sequencing, chromatin immunoprecipitation with DNA microarray (ChIP-on-chip), etc. Based on these new tools of genome

Post-Transcriptional Gene Regulation in Human Disease, Volume 32. https://doi.org/10.1016/B978-0-323-91305-8.00019-3

analyses, there was a paradigm shift in our understanding of gene regulation. It astonishingly revealed that almost 90% of the genomes are transcribed from both the strands of the DNA (hence named "Pervasive transcription), transcription can also start from many nucleotides spreading over several hundred base pairs around the promoter (thereby challenging the concept of the "transcription start site"); and long noncoding-, enhancer derived- and micro-RNAs play a major role in gene regulation. These studies established that the RNA, originally envisaged as being a product of transcription per se, is indeed the major regulator of gene expression.

Another fascinating development in the past 50 years is the understanding of the 3D structure of the genome and the epigenetic regulation that ensures the selective expression of a set of genes in a given cell type. Although many fundamental discoveries in gene regulation in the early days had considered DNA as a string of nucleotides interacting with the transcription factors and generating RNAs, over the years, it emerged that the histone proteins that wrap the DNA molecules around and regulate the access of transcription, splicing, and other factors are the key determinants of the profile of gene expression in a given condition. Let us consider the simple fact that the human genome is 2 m long and it is packaged into a 10-μm nucleus in such a way that a fraction of it is dynamically as well as selectively exposed for its expression in a given context. Therefore, the mechanism by which the nucleosomes are selectively modified and regulate the expression of various genes became more paramount than the genome itself. In contrast to the belief that the nucleotide sequence is the final determinant of a cell's fate, it has now been found that certain diseases are caused by changes in the local chromatin conformations and the 3D structure of the genome.

Although this book primarily deals with the role of post-transcriptional regulation of gene expression in diseases, this chapter is aimed toward giving a general background on gene regulation in human so that the comprehension of the specific aspects of human diseases as described in other chapters become easier for the readers with different backgrounds. However, considering the hugeness of literature on the mechanisms of gene expression, the information to be provided in the following sections would be a brief overview than a comprehensive compilation on this immensely vast subject. The readers are also requested to follow an excellent review by Klaus Scherrer wherein several fundamental concepts of gene expression have been critically assessed [1].

The genome, genes, and the regulatory elements: how many genes do we have?

The human genome comprises ~3,200,000,000 nucleotides assorted into 23 independent chromosomes. Among those, the smallest chromosome is ~50,000,000 and the longest one is of ~260,000,000 nucleotides [2]. Each somatic cell contains pairs of those chromosomes of which 22 pairs are called autosomes while the 23rd pair called the sex chromosomes differ between the male and the female. Despite such enormity of the genome and the complexity of the human species, the number of genes it harbors is surprisingly low. As per the most recent analyses, out of a total of 42,611 genes, 20,352 encode for the proteins, 18,887 for the long noncoding RNAs (lncRNAs), and the remaining code for the transfer, ribosomal, antisense, and small regulatory RNAs (http://ccb.jhu.edu/chess). These genes are transcribed into 323,258 transcripts of which 266,331 represent isoforms of protein-coding transcripts and the rest are the noncoding regulatory RNAs. Surprisingly, there are also over 30 million transcripts spread over ~650,000 genomic loci that are either nonfunctional (noise), or their functions are yet to be defined [3]. The question that is quite obvious: how many proteins are produced by this small number of genes, and does it explain the complexity of the human species? Proteomic analyses, especially detecting all the constituent proteins, both abundant and scarce; in a cell or tissue are a major challenge. Detecting all the constituent proteins, both abundant and scarce, in a cell or tissue, with the available tools of proteomics is still a major challenge. As per the UniProt Knowledgebase (UniProtKB), the central repository for the information on all proteins, currently there are 20,386 human proteins corresponding to the protein-coding genes (https://www.uniprot.org/help/uniprotkb; https://www.uniprot.org/). According to the Human Proteome Map, another resource portal that records protein products from multiple organs, tissues, and cell types, there are 30,057 proteins corresponding to 17,294 genes (https://www.humanproteomemap.org/). However, these estimates are highly conservative and primarily aimed toward validating the identities of the protein-coding genes rather than making a comprehensive catalog of the human proteome. Considering the alternative splicing and editing of the pre-mRNAs, the existence of alternate open reading frames, variability of translational start sites, and post-translational modifications, each of which would yield functionally different proteins from the same transcript; the total number of unique proteins in humans is likely to be very high. To date, the determination of the number and quantity of unique proteins in any human tissue is highly challenging; and the estimate is speculative, varying from researcher to researcher. While some experts in the field estimate it to be 80,000−400,000, some others even anticipate about 10,00,000 different proteins in the human proteome (https://www.mpg.de/11447687/W003_Biology_medicine_0594-05.pdf [4]). Therefore, the transmission of the genetic information from the DNA to RNA to proteins seems highly complex and not in conformity with the "one gene-one enzyme" hypothesis proposed by George Beadle in 1941 (awarded Nobel Prize in 1958) and that was later revised as "one gene-one peptide" hypothesis. Our present understanding of the complexities of the mechanisms of gene expression largely came from studying the expression of the protein-coding genes.

Promoters, enhancers, and their regulation

The core promoter and the preinitiation complex

The first step of gene expression is transcription. RNA Polymerase II transcribes all protein-coding genes, and in addition, many noncoding genes are also transcribed by it. Early methodologies like S1 nuclease mapping and primer extension studies suggested that RNA Polymerase II-mediated transcription starts at a particular nucleotide named "transcription start site (TSS)." The TSS is central to the "Core promoter" that encompasses ~50 nucleotides upstream and ~50 nucleotides downstream. The RNA Polymerase II along with the general transcription factors (GTFs) uses the core promoter as the platform for its binding, followed by the initiation of transcription. The core promoters harbor combinations of several conserved motifs (small nucleotide sequences) that are targeted by the GTFs and the RNA Polymerase II. Common among these motifs are the TATA-box, Initiator elements (Inr), Downstream Promoter Element (DPE), Motif 10 Element (MTE), TFIIB Recognition Elements (BREs), and downstream core elements (DCE). These sequences assist the recruitment of the RNA Polymerase II and the GTFs on the core promoter to form the preinitiation complex (PIC), a large multimeric assembly involving about a hundred proteins [5]. Also, these motifs are located at specific distances on the core promoters so that the PIC can align properly and initiate transcription at the TSS.

All these motifs are not necessarily found together in a promoter, as they occur in different combinations in different classes of genes, providing the first layer of the regulation of transcription. The characteristics of these motifs and the GTFs are summarized in Tables 1.1A and 1.1B and Fig. 1.1.

The PIC consists of six GTFs viz., TFIIA, TFIIB, TFIID, TFIIE, TFIIF, TFIIH, plus the RNA Polymerase II and the Mediators [6]. In addition, long noncoding RNAs also play roles in their assembly and functions [6]. To be noted that the subunit compositions of the PICs are not absolute and it

Table 1.1A Core promoter elements found in the mammalian genes transcribed by the RNA Polymerase II.

Motif	Relative position	Consensus sequence	Targeted by
TATA box	−30/−31 to −23/−24	TATAWAAR	TBP
BRE^{u}	Immediately upstream of the TATA box	SSRCGCC	TFIIB
BRE^{d}	Immediately downstream of the TATA box	RTDKKKK	TFIIB
Inr	−2 to +5	YYA^{+1}, $NWYY/CCA^{+1}$, TYTT	TAF1 and TAF2/TAF1
MTE	+18 to +29	CSARCSSAACGS	TAF6 and TAF9
DPE	+28 to +33	DSWYVY	TAF6 and TAF9
DCE	Subelements Part I: CTTC Pat II: CTGT Part III: AGC	Part I: + 6 to +11 Part II:+ 16 to +21 Part III:+ 30 to +34	TAF1

D: A, G, or T; K: G or T; N: A, C, G, or T; R: A or G; S: G or C; V: A, C, or G; W: A or T; Y: C or T.

Table 1.1B General transcription factors (HeLa cells) involved in the formation of the preinitiation complex by the RNA Polymerase II in the core promoters.

Name	No of subunits	Function
TFIIA	3	Stabilization of binding of TFIID to the core promoter
TFIIB	1	Recruitment of RNA Polymerase II-TFIIF
TFIID (TBP and TAFs)	1, 12	Binding to TATA, recruitment of TFIIB Recognition of the core promoter
TFIIE	2	Recruitment of TFIIH and regulation of its activities, formation of the open complex and promoter clearance, RNA Polymerase II activity.
TFIIF	2	Modulation of the binding of RNA Polymerase II and elongation
TFIIH	9	Formation of the open complex and promoter clearance, RNA Polymerase II activity.
RNA Polymerase II	12	DNA-dependent RNA synthesis

Reconstituted from Yoshiaki Ohkuma, J Biochem 1997;122, 481−489

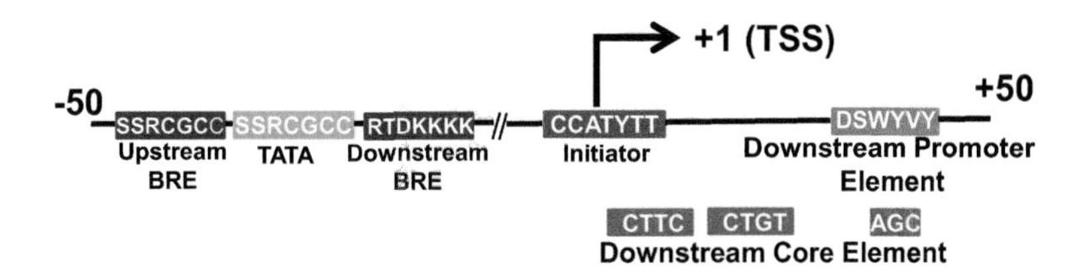

FIGURE 1.1

Schematic representation of a typical core promoter of a mammalian protein coding gene. The solid line represents the segment of the DNA spanning about 50 base pair upstream (-) to 50 base pair downstream (+) of the transcription start site (TSS). Various sequence motifs described in Table 1 are represented by boxes and the consensus sequences are shown. None of the core promoters carry all these motifs together. The relative positions of the elements in the promoter are maintained but are not in absolute scale.

Reproduced from Vo Ngoc L, Wang Y.-L, Kassavetis GA, Kadonaga JT, The punctilious RNA polymerase II core promoter, Genes Dev, 2017;31(13):1289–1301. https://doi.org/10.1101/gad.303149.117.

rather varies for different types of genes (constitutive, housekeeping, inducible, etc.). The assembly of the PIC is a highly complex process that has been studied extensively during the past 50 years [7,8]. Earlier studies on the formation of PIC and the initiation of transcription were done in vitro using biochemically purified GTFs and RNA Polymerase II, using prototype promoters as the templates. It showed that the formation of the PIC is hierarchical and involves cooperative interactions between various GTFs. Several salient features of this process are as follows: it is initiated by the engagement of the TATA-box binding protein (TBP), a component of the TFIID complex, to the core promoter. It is a crescent-shaped molecule that binds through its concave surface to the TATA or TATA-like sequences located ~30 base pairs upstream of the TSS and bends the DNA by ~90 degrees. The bent DNA structure is then stabilized by the engagement of TFIIB and TFIIA and thereafter, the TFIIF and the RNA Polymerase II are brought in. The recruitment of TFIIE and TFIIH then completes the formation of the PIC ([8], Fig. 1.2). In promoters devoid of the TATA box, the PIC assembles on the Inr site. Since the promoter DNA is enwrapped by the nucleosomes, RNA Polymerase II-mediated initiation of transcription also involves ATP-dependent chromatin remodeling complexes and histone-modifying enzymes (will be discussed later in detail). ATP-dependent translocase xeroderma pigmentosum type B (XPB), a component of TFIIH, separates the DNA strands at the promoter, enabling the single-stranded template to be transcribed. Once the transcription is initiated, the RNA Polymerase II leaves behind the GTFs and the Mediator complex and moves forward for the elongation, allowing the entry of another molecule of RNA Polymerase II for the reinitiation of transcription of the same gene. To ensure that the RNA Polymerase II does not pause or stall due to the nucleosomes or the DNA structures, elongation is facilitated by the elongation factors. During the elongation process, the transcripts are simultaneously processed by capping, splicing, and polyadenylation, preceding termination [6–8].

One key aspect of the transcription process is the role of the carboxy-terminal domain (CTD) of the largest subunit of the RNA polymerase II. In humans, the CTD domain comprises 52 repeats of the heptad Tyr-Ser-Pro-Thr-Ser-Pro-Ser (YSPTSPS), in which the phosphorylation of the serine residues

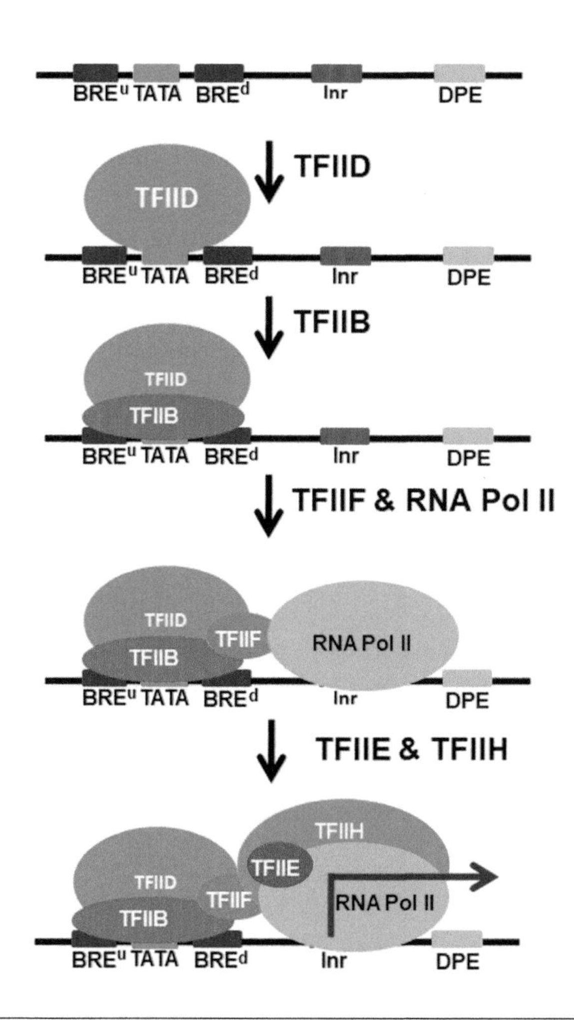

FIGURE 1.2

Sequential assembly of the preinitiation complex (PIC) on the core promoter of a mammalian protein-coding gene. The assembly starts with the engagement of TFIID with the core promoter at the TATA sequence ~23–30 base pair upstream of the transcription ion start site at the Inr. Thereafter, TFIIB, TFIIF, and RNA Polymerase II enter the complex. The assembly is completed by the engagement of TFIIE and TFIIH. The involvement of TFIIA is not shown in this scheme as it is nonessential. It engages along with TFIIB to stabilize the binding of TFIID to the promoter. The CTD domain of RNA Polymerase II brings in the mediator complexes that coordinate with the transcription factors bound further upstream initiating transcription (not shown).

at second and fifth positions plays a critical role in transcription initiation, elongation, and pre-mRNA processing. Unphosphorylated/hypophosphorylated CTD is necessary for the formation of the PIC, promoter melting, and the initiation of transcription. The phosphatase activity associated with the TFIIF ensures that the CTD domain remains dephosphorylated during the initiation of transcription [9]. Phosphorylated CTD is required for the elongation and concurrent capping, splicing, and polyadenylation of the transcript. Phosphorylated CTD is also required for the interaction of RNA Polymerase II with the Mediator complexes through which it communicates with the transcription factors bound to the upstream promoters and the enhancers. Following transcription initiation, the CTD domain becomes phosphorylated by CDK7, a TFIIH associated kinase, and by CDK9, a constituent of P-TEFb (Positive Transcription Elongation Factor), facilitating the elongation process [10].

Although during the pregenomic era, the core promoters were viewed as of generic type having similar mechanisms of activation for all genes, following the extensive sequencing of total cellular RNA vis-à-vis analyses of the genome, it became clear that there is substantial diversity in the core promoters and the mechanisms of transcription initiation. Apart from the presence of different combinations of the motifs in the core promoters, it is also now established that there are two distinct modes of the initiation transcription in vertebrates, viz., "focused" and "dispersed". In focused transcription, RNA synthesis starts at a single nucleotide as discussed earlier, but in dispersed transcription, there are several start sites over a span of about 50–100 nucleotides. Interestingly, focused transcription is more prevalent in the lower metazoans like *Drosophila,* while in vertebrates, ~70% of the promoters are dispersed [5,6]. Dispersed promoters are generally associated with the constitutive genes, are devoid of the core promoter motifs like TATA, BRE, DPE, and MTE, and generally have the CpG islands (~200-bp region in the proximal promoter with a higher GC content than that are commonly found in the genome). Thus, the mechanisms of transcription from focused versus dispersed promoters are likely to be fundamentally different and a better understanding of this process needs further investigations. In the past 2 decades, with the advent of powerful tools of structural biology, especially cryo-electron microscopy, it is evident that the initiation of transcription is an enormously complex but precise process that ensures the appropriate level of expression of every gene in a spatiotemporal manner [11].

The proximal- and the distal promoters

In vitro and ex vivo assays had shown that although the PIC assembled on a core promoter can initiate transcription, the output is quite low. To understand the overall process of gene expression, it is thus necessary to have an understanding of the anatomy of the protein-coding genes. The segment of DNA that is several hundred base pairs upstream of the core promoter is called the "Proximal Promoter" and that up to about 2000 base pairs further upstream is the "Distal Promoter." The optimum expression of any gene depends on the regulatory sequences present on both the proximal- and the distal promoters [11,12]. In addition, segments of a few hundred base-pair lengths located as far as hundreds of kilobases or even more from the promoter, often significantly boost the expression of many genes. These sequences are called "Enhancers" Enhancers have two distinct characteristics: firstly, their functions are position-independent as they can increase transcription even if they are shifted to a different location closer to or further away from the coding sequence. Many enhancers also work when they are placed even downstream of the gene sequences. Enhancers have also been found in the intron segments of the genes they regulate. Secondly, they remain functional even when the 5′-3′ orientation of the two strands is reversed. The reason for such unique characteristics of the enhancers is the 3D structure of the genome that brings them to the proximity of the core promoter even if their positions or orientations are altered [13,14]. A schematic representation of the proximal-, distal-promoters, and the enhancers in the context of gene expression is shown in Fig. 1.3.

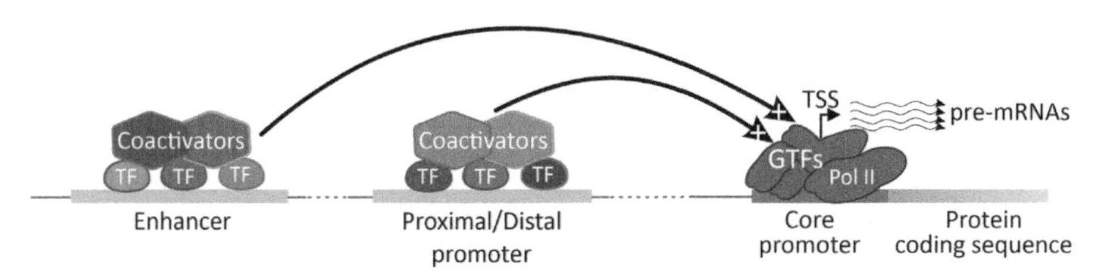

FIGURE 1.3

A schematic representation of the role of proximal-, distal-promoters and the enhancers in gene expression. The enhancer and the promoter (both proximal and distal) are occupied by various transcription factors (TF). These factors along with the coactivators make a local assembly that then communicates with the RNA Polymerase II complex through the mediators.

Reproduced from Haberle V, Stark A., Eukaryotic core promoters and the functional basis of transcription initiation. Nat Rev Mol Cell Biol. 2018; 19(10):621–637. https://doi.org/10.1038/s41580-018-0028-8.

Transcription factors are the drivers of gene expression

The promoters (both distal and proximal) and the enhancers exert their effects through a series of small (4–10 base pairs) *cis*-acting sequences targeted by a family of proteins called transcription factors. These transcription factors are not generic as the GTFs described above. Rather, they are quite specific for the set of genes they regulate in various tissues. In humans, ~1600 such transcription factors have been predicted, of which a few hundred have been extensively studied to date [15]. While many of these transcription factors are ubiquitous (e.g., SP-1) and are involved in the regulation of a diverse set of genes in many tissues, some others (e.g., NFκB), are quite exclusive for a group of genes.

Depending on their secondary structures, transcription factors are often classified as the Helix-loop-Helix (HLH), Leucine zipper (LZ), Zn-finger, homeodomain, etc. The leucine zipper family of transcription factors are characterized by the presence of about 30 amino acid segment with a repetition of leucine residues at every seventh position so that they are aligned with each other on the outer surface of the α-helix it forms. These factors form homo- or heterodimers through the leucine residues (hence named leucine zipper) and they bind the DNA as the dimers only. That way, about 50 such potential leucine zipper transcription factors in humans can create a larger repertoire of functional dimers [16]. Nuclear hormone receptors are also a large group of transcription factors that are activated by the steroid hormones, lipophilic vitamins, sterols, and many genotoxic compounds. Many of the nuclear hormone receptors are zinc finger proteins. The Homeobox (Hox) transcription factors are involved in the development of the body plan in the early embryo that eventually leads to the complex structures of the limb and the organs. This family of factors harbors a conserved Homeodomain of 60 amino acids (the corresponding DNA sequence of 180 bases is called the homeobox) with a helix-turn-helix secondary structure that binds the target DNA. In mammals, 39 Hox genes have been predicted and their functional specificity lies with the variable amino-acid residues within the homeodomains [17]. *Drosophila* genetics has enormously contributed to the understanding of the role of Hox genes in embryonic development.

Transcription factors have certain common characteristics. Each factor has one well-defined DNA binding domain that is rich in basic amino acids and it is called the basic domain. Due to the presence of this basic domain, helix–loop–helix and leucine zipper transcription factors are called bHLH and bLZ (or bzip) factors. Most transcription factors have structured dimerization domains and function as hetero- or homodimers (e.g., the leucine zipper family). They also have compatible protein–protein interaction domains that are involved in the interaction with the neighboring transcription factors being bound to the respective DNA sequences. Finally, each transcription factor is characterized by an activation domain through which they communicate with the RNA Polymerase II to drive transcription (Fig. 1.4).

Many transcription factors are expressed in the specific locations of the developing embryos and determine the cell lineage. The bHLH transcription factors Neurog1/2/4 are involved in the initiation of neuronal specification and differentiation [18]. Myogenic Regulatory Factors viz., MyoD, myogenin, Myf5, and MRF4 are also members of the bHLH family and they orchestrate the determination and differentiation of skeletal muscle cells during embryogenesis. They also coordinate the expression of muscle-specific genes during postnatal development [19]. Ectopic expression of the myogenic factors in certain nonmuscle cells converts those into myoblasts, a process called transdifferentiation.

Another class of transcription factors named "Pioneer Transcription Factors" can bind specific DNA sequences in the heterochromatin stage (heterochromatin is the part of the genome where gene expression is silenced by the epigenetic modifications of the histones). Upon binding, they initiate the opening of the chromatin structure by erasing those epigenetic marks, paving the way for other factors to bind their target sequences and initiating transcription. NeuroD1 is one such pioneer factor that

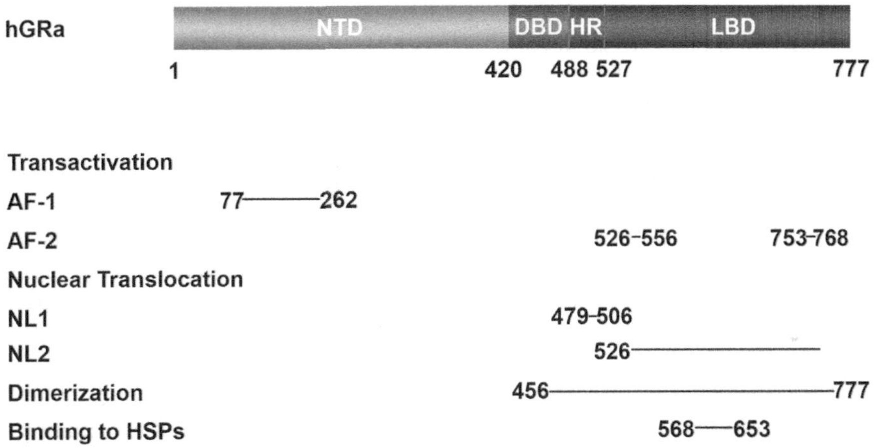

hGRa NTD DBD HR LBD
1 420 488 527 777

Transactivation
AF-1 77—262
AF-2 526–556 753–768
Nuclear Translocation
NL1 479–506
NL2 526—
Dimerization 456—777
Binding to HSPs 568—653

FIGURE 1.4

A schematic (linearised protein structure) representation of the functional domains of human glucocorticoid receptor alpha (hGRα). NTD: N-terminal domain; DBD: DNA binding domain; HR: Hinge region; LBD: ligand (glucocorticoid) binding domain. These domains are further divided into several subdomains as shown below. AF: activation function. The AF-1 subdomain plays an important role in the interaction of the receptor with the coactivators, chromatin modulators and basal transcription factors. The DNA-binding domain (DBD) contains two zinc finger motifs through which the receptor binds to specific DNA sequences called the glucocorticoid-response elements (GREs) in the promoter region (s) of the target genes. The DBD also contains sequences important for receptor dimerization and nuclear translocation. The hinge region (HR) confers structural flexibility in the receptor dimmers. The ligand-binding domain (LBD) binds to glucocorticoids and plays a critical role in the ligand-induced activation of the receptor. The LBD also contains a second transactivation domain termed AF-2, which is ligand-dependent. LBD also harbor sequences important for the dimerization of the receptor, its translocation to the nucleus, binding to the heat shock proteins and interaction with coactivators.

Adapted with permission from Nicolas C. Nicolaides, Zoi Galata, Tomoshige Kino, George P. Chrousos, and Evangelia Charmandar, The Human Glucocorticoid Receptor: Molecular Basis of Biologic Function, Steroids. 2010 January; 75(1): 1. https://doi.org/10.1016/j. steroids. 2009.09.002 alters the epigenetic and transcriptional program during neurogenesis, converting microglia to neurons. These factors play a major role in tissue differentiation during embryonic development [20,21].

Although the number of genes for the transcription factors in mammals lies around 1600, the total number of functional factors is much higher as many of those factors have multiple isoforms arising out of alternative splicing. Various domains of the transcription factors are post-translationally modified by ubiquitination, acetylation, phosphorylation, methylation, etc., by a range of extra- and intracellular stimuli [22]. Such modulations regulate their nuclear-cytoplasmic localization, stability, DNA binding activity, interaction with other transcription factors, and transcriptional activity ([23], Fig. 1.4). While many transcription factors are constitutively expressed, many are induced at the level of transcription in response to specific stimuli.

Interestingly, the target DNA sequences of the mammalian transcription factors are quite short that is, 4–10 nucleotides (the GATA factors involved in hematopoiesis are named after the four nucleotide long target sequence). Considering the total length of the human genome, it is likely that these small sequences would randomly occur in millions. So, how do these factors maintain their target specificities? Generally, many of these transcription factors act as dimers and therefore their target sequences need to be in tandem with proper orientation and spacing (to accommodate each of the binding factors that are often bulky). As an example, the AP-1 transcription factor is a dimer of the two bzip proteins Fos and Jun. It binds to the target sequence 5′-TGACTCA-3′ (TRE element) wherein each monomer binds to the terminal three nucleotides, called the half-sites (underlined). In vitro studies have shown that since the TRE is a palindrome (i.e., the reverse strand read from its 5′ end is also TGACTCA), the Fos: Jun heterodimer can bind it in both orientations (Fig. 1.5). However, when it binds to different AP-1 sites in various promoters, it shows a preference for one orientation over the other and that depends upon the sequences flanking the core TRE sequence as well as certain amino acids beyond their DNA binding (basic) domains [24]. Such preference also plays a role in their interactions with the transcription factors bound to the adjoining sequences in the same promoter. The cAMP response element-binding protein (CREB) family of bzip factors binds to the CRE sequences and mediates the cAMP response. It is an eight nucleotide sequence (5′-TGA**CG**TCA-3′) that is identical to the TRE, except having one extra **G** in between the two terminal half-sites. CRE sites are occupied by either homodimer of CREB or CREB-ATF-1 heterodimer, but not by AP-1 (Jun-Fos heterodimer) (Fig. 1.5). Such selectivity in the binding sequence is conferred by certain amino acid residues in the DNA binding, leucine zipper, and the intervening domains of the CREB/ATF-1 proteins [25]. Another example of small nucleotide sequences with high functional specificity is the binding sites of the nuclear receptors. Nuclear receptors are a large family of transcription factors that are activated by small lipophilic molecules such as steroid hormones, retinoic acids, thyroid hormone, vitamins D3, and genotoxic compounds. They regulate numerous physiological and developmental processes through the modulation of their respective target genes. Their target sites consist of two hexanucleotide half-sites that are arranged in two different orientations. One is the inverted repeats where the two half-sites are arranged in the opposite orientation (head-to-head) forming a palindrome with an intervening sequence of three nucleotides. The other orientation is the direct repeats where the half-sites are oriented toward the same direction (head-to-tail) with an intervening sequence of 1–5 nucleotides [26]. There are different types of binding of nuclear receptors to their respective target sequences. The steroid receptors subfamily, that is, the glucocorticoid, progesterone, mineralocorticoid, androgen, and estrogen receptors, bind to the inverted repeats as homodimers (Fig. 1.5). Another subgroup comprising the retinoic acid, the vitamin D, and the peroxisome proliferator-activated receptors heterodimerize with the retinoid X receptors and the heterodimers bind the direct repeats

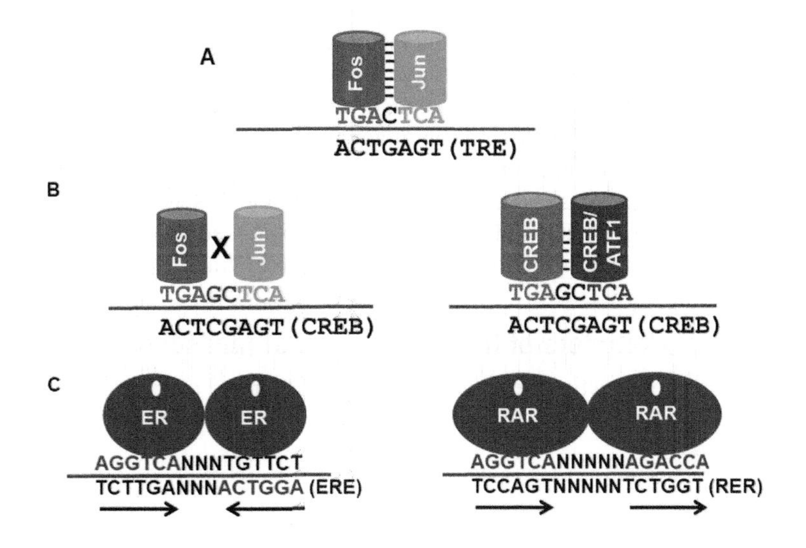

FIGURE 1.5

Dimeric binding of various transcription factors to their cognate DNA sequences. **(A)** The Fos: Jun heterodimer binds to the TRE (TGACTCA) where each half-site contacts the respective DNA binding domains. **(B)** The CRE element has the same half-site as the TRE but it has one extra G in the intervening sequence. While Fos: Jun dimer does not bind to it (left), the CREB: ATF-1 binds to the CRE (right) that are found in the promoter region of many genes regulated by cAMP. **(C)** Two estrogen receptor monomers bind to the estrogen receptor response element (ERE) found in estrogen-responsive genes (left). The ERE comprises two canonical half-sites (AGGTCA) arranged in a head-to-head orientation (marked by the *arrows*) with a spacer of three nucleotides. Two retinoic acid receptors bind to the retinoic acid response element (RRE, right) that are found in genes that are regulated by the retinoic acid. This RRE comprises two half-sites arranged in head-to-tail (direct repeat) orientation (marked by the *arrows*) with a spacer of five nucleotides. In this RRE, one half-site is consensus, that is, AGGTCA while the other is a variant, that is, AGACCA. Such variation in certain nucleotides also contributes toward the target specificity of the transcription factors (Fig. 1.4). The respective ligands (estrogen and retinoic acid) are shown as *white circles.*

(Fig. 1.5). The binding sites of the steroid hormone receptors are quite similar, and these receptors recognize each other's target sequences in vitro. In the case of the direct repeats, where the hetero-dimers bind, the target sequences are also quite similar for all and the length of the intervening spacer sequences (DR1-DR5) determines their target specificities (Fig. 1.5). The sequences that are targeted by the homo- and hetero-dimers of various receptors are called the hormone response elements (HRE). It is a generic term as all the nuclear receptors are not regulated by hormones only. Interestingly, while the sequences of individual HREs can vary considerably, all nuclear receptors can strongly bind to two of the consensus half-sites, that is, 5′-AGAACA-3′ and 5′-AGGTCA-3’.

These are examples of how various transcription factors despite having similar or even the same binding sequences selectively regulate their target genes. Taken together, the constraints on the binding sequences, the adjoining sequences, and the amino acids in the DNA binding domains of each transcription factor contribute toward the nonrandom selective binding by the members of the same family having similar binding sites [27–29].

Although the binding sites for the transcription factors are largely conserved, extensive analyses of the target sequences for a particular factor in a large number of promoters have shown that while the nucleotides in certain positions are highly conserved, in certain others it might vary to various extents. Such degeneracy in binding sites plays a role in attaining subtle differences in their conformations upon binding to those sequences. Such conformational changes are important for their interaction with the factors binding to the adjacent sites, thereby conferring promoter-specific functions. The relative frequency of occurrence of each nucleotide is quantitatively expressed by a "position weight matrix" that can be used for judging the authenticity of the binding site for a transcription factor (Fig. 1.6 [30]).

Enhancers, the true arbitrators of the transcriptional landscape

As mentioned in the previous sections, enhancers are several hundred base pair long DNA fragments that strongly augment the transcription of a target gene. They function as a self-sufficient unit

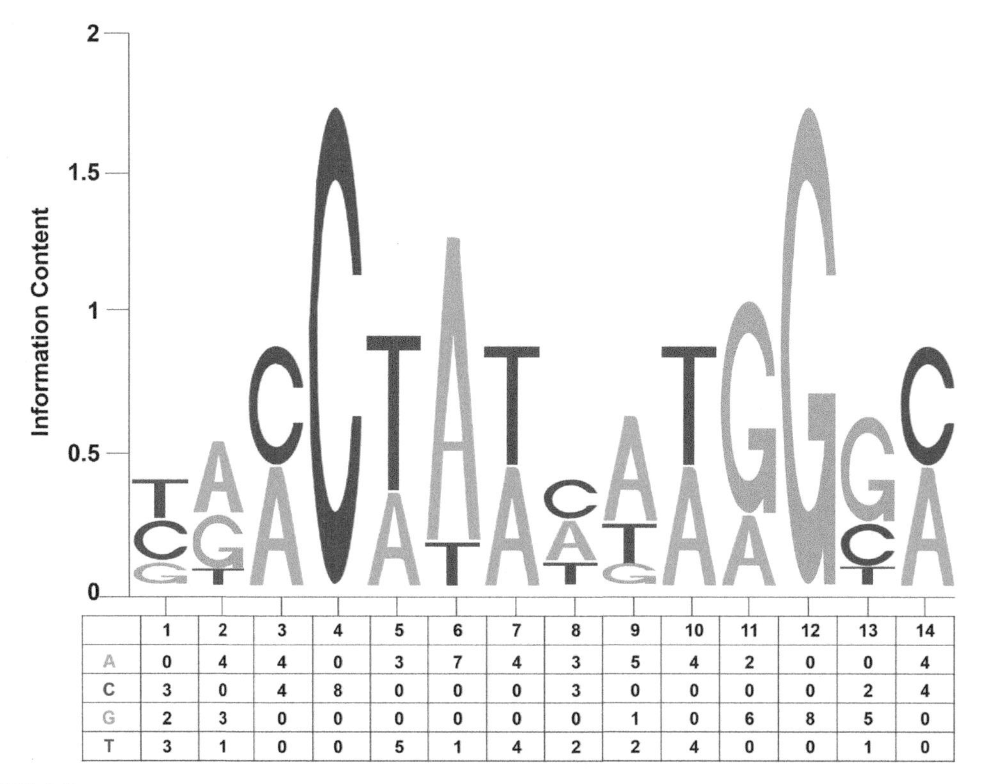

	1	2	3	4	5	6	7	8	9	10	11	12	13	14
A	0	4	4	0	3	7	4	3	5	4	2	0	0	4
C	3	0	4	8	0	0	0	3	0	0	0	0	2	4
G	2	3	0	0	0	0	0	0	1	0	6	8	5	0
T	3	1	0	0	5	1	4	2	2	4	0	0	1	0

FIGURE 1.6

The position weight matrix (PWM) for the binding site of Myocyte enhancer factor-2 (MEF2), a family of transcription factors which regulate of cell differentiation and embryonic development. Lower panel: the Position Frequency Matrix (PFM) as done by collecting experimentally validated binding sites from 8 published studies. The PFM was then converted to the corresponding PWM (upper panel) by normalizing the frequency values on a log-scale. The relative frequency of each base at each nucleotide position is represented by the height of its symbol.

independent of their locations and orientations in the genome. Enhancer activity was first discovered in 1981 in the Simian virus 40 (SV40, a DNA virus) genome that was being exploited as a vector for the expression of the β-globin gene in cultured cells [31]. In subsequent years, many enhancers were identified and their paramount role in tissue-specific gene expression was established [13]. Conventionally, enhancers can exert their effects over a distance of several thousand to several megabase pairs away, either upstream or downstream of the genes they activate. There are instances: enhancers can even activate a gene located on a different chromosome [32]. *Drosophila* genetics had significantly contributed toward the understanding of enhancer functions.

In the early years, detection of enhancers was gene-centric, quite tedious, and of low output. With the incremental understanding, the consensus on the signatures of enhancers are as follows: (1) they do not harbor any coding sequence; (2) active enhancers (a) lack nucleosomes, (b) are occupied by multiple transcription factors, and (c) are characterized by certain histone modifications in the adjoining nucleosomes (to be discussed later in details). In the postgenomic era, novel enhancers are being explored by (1) searching for the noncoding sequences harboring binding sites for transcription factors that are conserved across species, (2) chromatin immunoprecipitation for the transcription factors and coactivators followed by sequencing (ChIP-seq), (3) sequencing of the regions of the genomic DNA that are sensitive to cleavage by DNase I, and (4) Formaldehyde-Assisted Isolation of Regulatory Elements (FAIRE-Seq), a method based on the principle that in the genome, nucleosome-bound DNA is more amenable to formaldehyde cross-linking than the nucleosome-depleted enhancer regions [33]. In a seminal discovery about a decade ago, it was shown that a large number of enhancers undergo transcription by the RNA polymerase II, generating the enhancer RNAs (eRNAs [34], Fig. 1.7). Simultaneously, it was also reported that many of the enhancer sites are indeed occupied by the RNA Polymerase II and the GTFs [35]. Extensive genome-wide analyses had shown that there are two types of eRNAs. Some are shorter (0.5–2 kb), nonpolyadenylated and are generated by the transcription of the enhancer segments from both ends (bidirectional transcription). Also, there are eRNAs that are longer (>4 kb), polyadenylated, and are generated by the unidirectional transcription of the enhancer regions [36]. As enhancers are also transcribed by the RNA Polymerase II, the differences between the enhancers and the promoters are now becoming blurred and some promoters

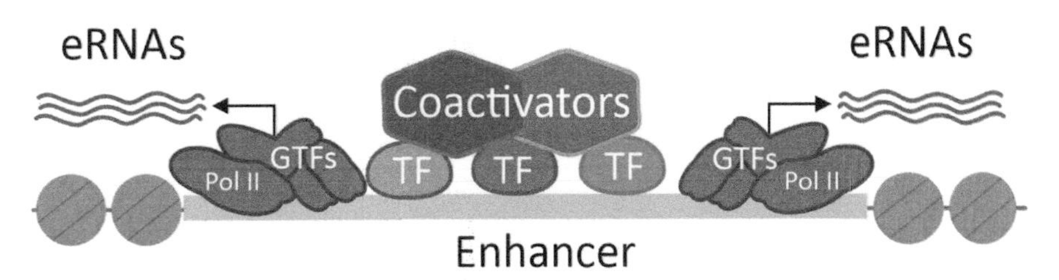

FIGURE 1.7

Schematic representation of transcription of a mammalian enhancer. The segment of enhancer DNA is shown in the pear green bar. The enhancer fragment is free from nucleosomes while those present in the adjoining regions are shown in gray spheres. The RNA Polymerase II and GTF complex at both the ends of the enhancer represents the transcribing complexes for the sense and the antisense strands (with reference to the coding gene) of the enhancer DNA. While most eRNAs are not modified after transcription, some eRNAs are capped but rarely polyadenylated.

Reproduced from Haberle V, Stark A., Eukaryotic core promoters and the functional basis of transcription initiation. Nat Rev Mol Cell Biol. October 2018; 19(10):621–637. https://doi.org/10.1038/s41580-018-0028-8.

have even been found to have enhancer-like activities [37]. Recent data suggest that the number of enhancers in the human genome is hundreds of thousands, and it could be even up to a million [38].

In view of such a large number of enhancers in the human genome, the focus on gene expression has now been shifted from the genes per se to the vast number of enhancers that regulate their expression. There is evidence that one gene can be regulated by multiple enhancers in a hierarchical, additive, or synergistic manner. There are also instances where a single enhancer can regulate multiple genes. Taken together, enhancers play a critical role in determining the tissue and cell-specific gene expression during embryonic development and the intensity and the duration of expression of various genes in a given pathophysiological context [39–41]. The genome-wide study has shown that the engagement of enhancers and the promoters change dynamically during the progression of certain diseases like cancer [42].

As functions of the enhancers do not depend on their chromosomal locations relative to the genes they regulate, the way they communicate to their target promoters has been a subject of intense interest. Early studies on the function of the Locus Control Region of the globin genes during the development of erythroid cells suggested that the enhancers interact with the target promoter by DNA looping [43]. Subsequent studies over the past 3 decades have shown that a highly dynamic 3D organization of the genome ensures that certain enhancers from among a pool are selectively engaged to their cognate promoters in a spatiotemporal manner. In the genome, transcriptionally active regions

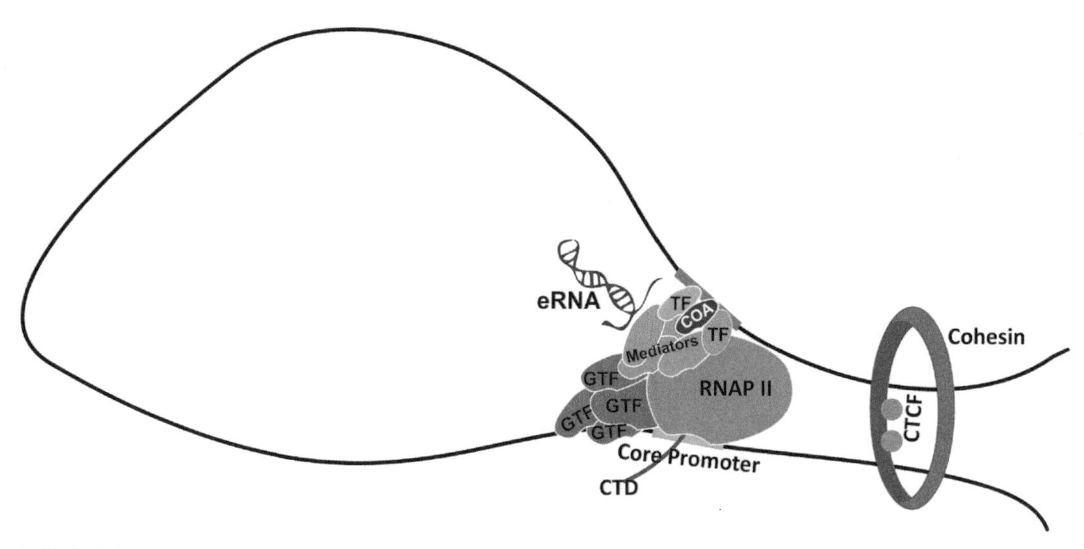

FIGURE 1.8

A representative diagram of the 3D organization of the transcription complex on the core promoter and the enhancer. Transcriptionally active regions in the genome form the topologically associating domains (TADs) in which a chromatin loop brings together a particular enhancer and the cognate promoter for direct physical interaction. The cohesion complex along with the CCCTC-binding factor (CTCF) facilitates the stability and the formation of the chromatin loops. The mega-assembly involves the general transcription factors (GTF) and the RNA Polymerase II forming the PIC on the core promoter; and the specific transcription factors, coactivators, and chromatin modifiers assembled on the enhancer. The mediator complex cements the PIC and the transcription factors and the coactivators bound to the enhancer. Often an eRNA act as a scaffold to complete the assembly. While the enhancer and the core promoter region are free from the nucleosomes, the intervening loop is not (not shown).

harboring a pool of genes interact with each other and form the topologically associating domains (TADs) in which specific regions of the genome are brought together forming chromatin loops (Fig. 1.8). The cohesion complex along with the CCCTC-binding factor (CTCF) facilitates the formation of the chromatin loops [44]. The formation of the chromatin loops with the TADs is a dynamic process playing a major role in gene regulation during cell differentiation, embryonic development, and the onset of diseases [45]. Besides cohesion and CTCF, several other factors viz., YY1, EZH2, IKAROS, condensin, etc. are also involved in chromatin looping, facilitating the enhancer–promoter interactions [46]. Also, at the edge of the loops, there are boundary elements or insulators that prevent indiscriminate interactions of enhancers in one loop to promoters in the adjoining ones [47]. Taken together, the highly complex chromosomal organization ensures the appropriate engagement enhancers to their target genes in various pathophysiological contexts [48].

Several models have been proposed on the mechanisms by which enhancers get activated. In one model, a pioneer transcription factor contacts the nucleosomes of a dormant enhancer initiating their clearance by the chromatin remodeling factors and histone acetyltransferases. Once freed from the nucleosomes, other transcription factors then bind to their target sites followed by the coactivators, chromatin remodelers, mediators, and RNA Polymerase II, creating a mega assembly of about 200 proteins, far more than the number of proteins assembling on the PIC. The enhancer then communicates with the promoter and transfers the mega-complex to the target promoter, to initiate transcription [49]. In recent years, a subclass of enhancers called "super-enhancer" has emerged. These are a cluster of enhancers that are heavily occupied by the Mediators, pioneer- (Oct4, Sox2, Nanog, etc.), and other transcription factors and play a key role in determining the cell fate and identity [50,51]. Super-enhancers also play a role in cancer progression [42].

The role of long noncoding and enhancer RNA in gene expression

In the past decade, extensive analyses of the transcriptome of various eukaryotic organisms have revealed that the genomes are pervasively transcribed, producing a vast array of noncoding RNAs ([52], Fig. 1.9). The functions of these diverse RNAs are highly debated as most are degraded by the RNA exosome [53]. A subset of the noncoding transcripts that are generated by the pervasive transcription of ~98% of the genome is long noncoding RNA (lncRNA). LncRNAs are broadly divided into two groups: those that are transcribed from regions in between genes are called intergenic lncRNAs; whereas, those transcribed from the introns of protein-coding genes (both sense and antisense) are intronic lncRNAs. They are generally more than 200 nucleotides long, capped, spliced; and many are polyadenylated at the 3′-end [54]. Although the role of various RNA in the regulation of gene expression was proposed in the 19 sixties, the gene-specific regulatory role of lncRNAs first emerged in the early 1990s [55]. LncRNAs bind to the DNA, RNA, or proteins and can modulate almost all aspects of gene expression by (1) regulating the assembly and the functions of the transcriptional complexes; (2) modulating chromatin organizations; (3) interacting with the nuclear matrix proteins, influencing the organization of the genome; (4) forming ribonucleoprotein complexes; (5) acting as decoy molecules by binding and sequestering proteins and miRNAs [56,57]. LncRNAs have been associated with a plethora of cellular processes such as cell proliferation, differentiation, apoptosis, embryonic development, and genomic imprinting [56–58]. They have been associated with diseases such as Alzheimer's, cancer, and diabetes [59]. The biological functions of lncRNAs are of extensive interest.

Enhancer RNAs (eRNAs) were discovered almost 20 years after the discovery of the lncRNAs and their cellular functions are still emerging [34]. Due to their high degree of heterogeneity and sparse abundance, they were even considered as the background noise of transcription. Numerous studies

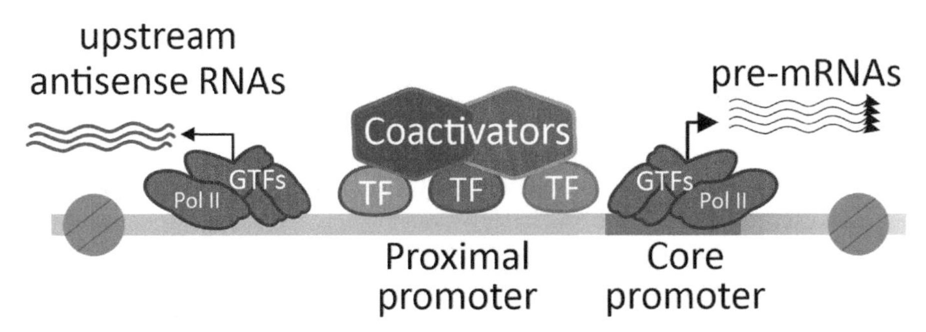

FIGURE 1.9

Pervasive transcription of the upstream region of a protein-coding gene. Recent transcriptomic data suggests that while the three RNA polymerases transcribe various protein-coding and noncoding genes, almost 98% of the remaining part of the genome is also transcribed producing various noncoding RNAs. The mechanisms of such transcription are not understood yet. An example of such transcription, that occurs from the upstream region of a protein-coding gene is shown here. Although only the antisense transcripts are shown, such transcription generally occurs from both the strands producing sense and antisense RNAs. Similarly, pervasive transcription occurs throughout the genome, including the repeat sequences, the intronic, and the intergenic regions.

Reproduced from Haberle V, Stark A., Eukaryotic core promoters and the functional basis of transcription initiation. Nat Rev Mol Cell Biol. October 2018; 19(10):621–637. https://doi.org/10.1038/s41580-018-0028-8.

over the past decade have shown physiological roles of the transcription of enhancers and that of eRNAs [60]. Evidence supporting the role of eRNAs in cellular functions are as follows: (1) their transcription is triggered by the external stimuli and precedes the transcription of the adjoining genes, (2) certain histone marks like H3K27ac and H3K4me1/2 are closely associated with the enhancer transcription, (3) it is associated with a higher level of phosphorylation of Ser5 and the tyrosine residues in the CTD domain of RNA Polymerase II that ensures shorter transcripts, (4) although many enhancers are transcribed bidirectionally, transcription from one of the two strands normally prevails, (5) certain long intergenic eRNAs undergo post-transcriptional modifications like the methylation of adenosine (6-methyladenosine [m6A]) that alter their structures and stability, and (6) modified eRNAs colocalize with the proteins that recognize those modifications [61–63]. The involvement of these regulatory processes in the synthesis of eRNAs suggests that eRNAs have certain designated functions. The possible advantage of enhancer transcription is; it makes the dormant enhancers active by freeing it from the nucleosomes and accessible to transcription factors and other regulatory proteins. There are instances where the antisense eRNAs from the intragenic enhancers act as the moderators of transcriptional elongation of the target genes [64]. It has been demonstrated that when the abundance of an eRNA is modulated by the CRISPR/Cas9-mediated deletions, it affects the level of transcription of the cognate gene [65]. Mass spectrometry with certain eRNAs has shown their direct association with numerous proteins including chromatin remodelers and sequence-specific DNA binding transcription factors. It is hypothesized that eRNAs might interact with those proteins as a scaffold to facilitate enhancer-promoter looping, chromatin modification, and the function of transcriptional machinery by a process called phase separation, an event wherein a three-dimensional mega assembly of proteins and nucleic acids are made [49,62–66]. However, due to the extreme heterogeneity, a single canonical function of eRNAs in gene regulation is unlikely.

Epigenetic regulation of gene expression

Although the Central Dogma of Molecular Biology had envisaged that the gene segments in association with various protein factors would determine its expression in a given cellular context, another mode of gene regulation, that is, epigenetic, was conceptualized much earlier. In the early 20th century, it was known that in *Drosophila,* the expression of a gene would depend on its location in the heterochromatin, that is, a part of the chromosome that is darkly stained for the DNA, more condensed, and transcriptionally inactive; or in the euchromatin, the remaining part of the chromosome that is rich in genes and is actively transcribed. While explaining the relationship between the genes and the cell phenotypes governing embryonic development, British developmental biologist Conrad Waddington coined the term epigenetics in 1942. It postulates that following cell division, although the daughter cells inherit the entire genome, specific gene expression patterns are inherited and maintained through the "epigenetic memory" encoded by the epigenetic marks [67]. Epigenetic regulations thus play a major role in the profile of gene expression, including turning "off" and "on" of a set of genes in any cell type.

Epigenetic regulations are of diverse types including the methylation of DNA, covalent modifications of histones, remodeling of chromatin, and regulation through the noncoding RNAs (ncRNAs), etc. A unique combination of these modifications and the modifiers define the epigenetic code for each gene dictating the state of its expression or otherwise [68]. The combination of these codes then defines the epigenome due to which the same set of DNA (i.e., the genome) is differentially expressed in different cell types and also in cells affected by diseases [69].

The most common modification of DNA that significantly affects the function of the genome is the enzymatic methylation of the fifth carbon of the cytosine residues (5-methylcytosine, 5 mC). It primarily occurs at the C residue preceding a G, that is, CpG. The human genome has about 28 million CpGs of which $<10\%$ occur in the small stretches (0.5–2 kb) that have a higher abundance of the CpGs and are called the CpG islands. The number of CpG islands in the human genome is around 30,000; and they are found in the proximal promoter regions of the housekeeping and developmentally regulated genes, within the gene bodies (intragenic), and in between the coding genes (intergenic). While 70%–80% of the genome-wide CpGs are methylated, those in the CpG islands in the promoters and the intergenic regions are constitutively non- or hypo-methylated. Although these promoters are not generally repressed upon methylation of the CpGs, there are also instances where some of the promoters that are repressed during embryonic development remain methylated [70]. There are proteins containing methylated CpG binding domains, which upon binding to ^{5m}CpG, recruit histone deacetylases, causing the compaction of chromatin and gene silencing. DNA methylation also prevents the binding of the transcription factors to the promoter, preventing its expression [71,72]. The CpG islands associated with the promoters and the intragenic regions are conserved in evolution from mice to humans, suggesting their important roles in gene expression the details of which are still emerging. Like the promoter-associated CpG islands, the intragenic CpG islands can also be independently transcribed and they play an important role in the expression of their host genes. Methylation of intragenic CpG islands has been associated with the alternative promoter usage and the processing of the pre-mRNA for the genes harboring them [73]. Many segments in the mammalian genome like the pericentromeric repeats, the long and short interspersed nuclear elements (LINEs and SINEs), and long terminal repeats are transcribed to noncoding RNAs; and genome-wide methylation has important roles in regulating their transcription [74]. Unlike the promoters, most of the enhancers have less CpG content, except a few that are rich in the CpGs [75]. Demethylation of enhancers DNA has been seen during vertebrate embryogenesis that requires rapid transcriptional reprogramming. Taken together, CpG methylation has major roles in the spatiotemporal expression of the genome during the development and during the onset of diseases like cancer [76].

The nucleosomes are the structural units of chromatin that wrap around ~146 base pairs of DNA. They consist of an octamer of histones (a tetramer of H3 and H4 histones and two peripheral heterodimers of H2A and H2B histones or their variants). Histones undergo several post-translational modifications viz., phosphorylation of serine and threonine residues, acetylation and methylation of the lysine and arginine residues, ubiquitinylation and sumoylation of lysines, and poly-ADP-ribosylation of glutamic acid. These modifications alter the chromatin structure and provide the platforms for the engagement of diverse gene regulatory proteins viz., the transcription factors, chromatin remodelers, enzymes for the modifications of DNA and RNA (e.g., DNA methyltransferase and splicing factors), etc., to modulate the transcription process. Although it was known since the 1960s that histones are acetylated, only in the 1980s it became apparent that the acetylation of histones at specific lysine residues decreases the degree of condensation of chromatin facilitating gene expression [77]. Acetylation of lysine residues in histones by the histone acetyltransferases neutralizes the positive charges of lysine and decreases histone–DNA interactions, favoring transcription. Several histone acetyltransferases viz., E1a-binding protein p300, CREB-binding protein (CBP), TAF (II) 250 subunit of TFIID have been extensively studied for their roles in transcription [78]. Histone acetylation is reversed by histone deacetylases. Acetylated Lys residues are specifically identified by the bromodomains that are present in the 'writers' (Histone acetyltransferases), 'readers' (that recognizes the acetylated histones), and 'erasers' (Histone deacetylases) governing transcription. Large-scale proteomic studies have shown that acetylated histones are involved in macromolecular complexes that regulate chromatin remodeling, splicing, capping, and polyadenylation [79]. A number of nonhistone proteins, like transcription factor p53, are also acetylated, thereby suggesting a wider role of acetylation in gene expression.

Histone methylations play an even bigger role in gene regulation. It occurs on specific lysine and arginine residues and the same residue may be mono-, di-, or tri-methylated with distinct functions. Based on the residue and the extent of methylation, it could both facilitate or repress transcription. The genome-wide study has suggested that while methylation at H3K4, H3K36, and H3K79 facilitates transcription; that at H3K9, H4K20, and H3K27 causes repression [80]. Like histone acetylation, histone methylation is also a dynamic process mediated by various methylases ("writers") while demethylases remove it ("erasers" [81]). Extensive study has linked H3K4 methylation to the functions of both the enhancers and promoters. Among the three states of methylation of H3K, H3K4me1 is more abundant in the enhancers, H3K4me2 is the highest toward the 5′ end of the transcribing gene and H3K4me3 is prevalent in the promoters of the poised or the actively transcribing genes. In mammals, the MLL family of histone methyltransferases (HMT) binds to the coactivators p300/CBP and the transcription factors CREB or MYB, thereby recruiting them to the target genes [82]. Mutations in H4K3 methyltransferases have been associated with several cancers [83]. Like the methylation of lysine 4 of H3, that of other lysine residues, that is, lysine 9, 20, 27, and 36 also play a role in gene expression with distinct functions. To avoid redundancies, rather than separately describing their roles, it is summarized in Table 1.2.

Table 1.2 Role of methylation of lysine residues of histones in gene expression.

Histone residue	Modification	Function
1. H3K9	Trimethylation	Heterochromatin organization and genome integrity.
2. H3K27	Methylation	Transcriptional repression
3. H3K36	Methylation	Transcriptional activation
4. H4K20	Monomethylation	Transcriptional activation

MicroRNA and post-transcriptional gene silencing

Micro RNAs (miRNAs) are small (~22 nucleotides) single-stranded non-coding RNAs that play a major role in gene regulation. The first miRNA, lin-4, was reported in *Caenorhabditis elegans* in the year 1993, and the first human miRNA Let-7 was reported in 2001 [84]. To date, ~2675 mature human miRNAs have been reported in the miRNA database (www.mirbase.org). MicroRNAs are conserved in evolution and have been reported in animals, plants, and some viruses, having regulatory roles in gene expression. Bioinformatics analyses suggest that the expression of most human protein-coding genes is regulated by one or more miRNAs, primarily by inducing the degradation of the target mRNAs or by inhibiting their translation (Fig. 1.10 [85]). When transcribed, the precursors of miRNAs are much longer and are called pri-miRNA. In the canonical pathway of biogenesis, the pri-miRNAs are processed stepwise, first in the nucleus and in then in the cytoplasm. Briefly, the pri-miRNAs are cleaved into pre-miRNAs by DROSHA, an RNase III enzyme in the nucleus, and the pre-miRNA with the typical hairpin loop structure is then exported to the cytoplasm. The pre-miRNA is then processed by another RNase III enzyme DICER, generating ~22 nucleotides long duplex miRNA that interacts with the RNA-induced silencing complex (RISC) loaded complex (RLC). There the passenger strand

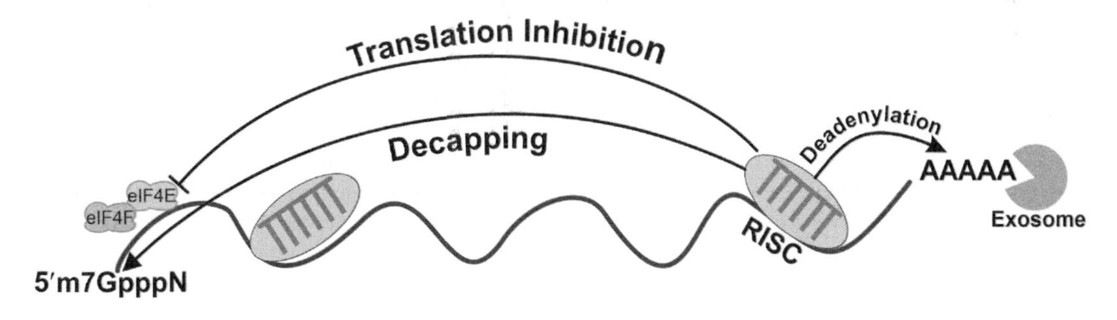

FIGURE 1.10

An illustration of miRNA-induced silencing of gene expression. MicroRNAs bind to their target sequences as the miRNA-induced silencing complex (miRISC) that contains Argonaute and GW182 proteins. MicroRISC can induce the decapping and deadenylation of the mRNA followed by its degradation by the exosome. It also can inhibit translation by interfering with the initiation factors eIF4E and eIF4E. Normally, miRNAs target the 3′-UTR of the mRNA, but they also can exert their effects by binding to the 5′-UTR and the coding regions.

of the duplex is degraded, and the RISC-loaded mature miRNA (miRISC) binds to the reverse complementary sequences in their target mRNAs [86]. Almost half of all known mammalian miRNAs are of intragenic origin as their sequences are found in the introns of the protein-coding genes. They are transcribed by the RNA Polymerase II as a part of the pre-mRNAs from the host genes and are initially processed by the nuclear splicing machinery rather than the DROSHA [87]. Once transported to the cytoplasm, they are processed by the DICER as in the canonical pathway. These miRNAs are also called Mirtrons and they are often linked to the expression of their host genes [88]. Many miRNAs are also intergenic (in between annotated genes) and are transcribed independently as the pri-mRNAs from their promoters [89].

MiRNAs generally bind to the 3′-UTR of their target mRNAs but they also can bind to the 5′-UTR and the coding regions. Upon binding their targets in association with the RNA-induced silencing complex (RISC), they recruit other proteins either to induce the degradation or inhibit the translation of the mRNAs that occur in the cytoplasm [90]. There are also instances: miRNAs in association with other proteins (microribonucleoproteins) post-transcriptionally stimulate gene expression by multiple mechanisms [91]. Perfect sequence complementarity is not essential for the binding of mammalian miRNAs to their targets, and that gives versatility to their functions. The nucleotides at positions 2–7 at the 5′-end, called the seed region, play a critical role in their target specificities. Only the seed sequence of the RISC-bound miRNAs remains available for the initial pairing to a target site. Once the miRNA binds the target RNA through the seed sequence, the Argonaut protein, as an essential component of the RISC, undergo a conformational change that exposes the remaining part of the miRNA (the 3′- 13–16 nucleotides) for additional interactions with the target [92]. Being short oligonucleotides with lesser stringency of target selection, one miRNA can regulate several genes, while one gene can be targeted by several miRNAs, making their regulatory networks quite complex.

Besides regulating the stability and translation of their target mRNAs that occur in the cytoplasm, nuclear-localized miRNA can directly inhibit or activate transcription. There are evidences that certain gene promoters have complete or partial complementarity with the miRNAs, enabling those to inhibit transcription either by blocking the RNA Polymerase-II activity or by recruiting corepressors [93]. Nuclear miRNAs also stimulate transcription by activating the enhancers through chromatin remodeling [94]. Taken together, miRNAs regulate gene expression both at transcriptional and post-transcriptional levels through multiple mechanisms that are still emerging.

Deep sequencing techniques for the identification of novel RNA species have revealed the existence of sequence variants of miRNAs called isomiRs that are generated by alternative post-transcriptional processing of pri-miRNAs. These isomiRs have been associated with tissue specificity, diseases, aging, etc [95]. Certain miRNAs are not generated by the canonical pathway described above, and those are called non-canonical miRNAs. Small nucleolar RNAs (snoRNAs) are a class of ~60–300 nucleotide long non-coding RNAs that are among the sources of non-canonical miRNAs. The miRNAs derived from the snoRNAs are ~21 nucleotides long, and they repress their target mRNAs in a manner similar to that of the miRNAs [96]. Another type of non-canonical miRNAs is those derived from the tRNAs which are also the substrates for the DICER [97]. Therefore, miRNAs are generated from a highly diverse group of non-coding RNAs by multiple mechanisms in a cell type and context-specific manner. Their synthesis is linked to various pathophysiological conditions such as oncogenic transformation, DNA damage response, oxidative stress, and metabolic shifts, allowing the cells to modulate their gene expression profile appropriate to the altered milieu [98]. They are also involved in cell−cell communication as the naked miRNAs can directly transfer to the adjacent cells by the Gap junctions. Individual miRNAs are also secreted from one cell via the exosomes, microvesicles, lipid droplets, etc. and move through the circulation to reach the target recipient cells [99].

Thus, miRNAs regulate each and every aspect of gene expression in an all-encompassing manner, creating a sophisticated and elaborate regulatory system. They have been associated with diverse biological functions such as embryonic development, cell differentiation, organogenesis, and metabolism [100]. The role of miRNAs in cardiovascular diseases, diabetes, cancer, and metabolic disorders has been established, and therapeutic applications of their mimics or their quenchers (anti-miRs) are being explored [101].

Post-transcriptional processing of the pre-mRNA transcripts

In eukaryotes, the nascent pre-mRNA undergoes three major covalent modifications, that is, capping of the 5′-terminus, removal of the intron sequences by splicing, and cleavage/polyadenylation at the 3′-tail, all occurring cotranscriptionally, that is, while transcription is still occurring. Capping is a three-step enzymatic reaction through which the 7-methylguanosine cap is added to the 5′-hydroxyl group of the first nucleotide of the nascent pre-mRNA. In the first step, the RNA 5′-triphosphatase hydrolyzes the triphosphate group of the 5′-nucleotide of the newly synthesized pre-mRNA, making it a diphosphate. Thereafter, the guanylyltransferase adds one GMP to it, producing the guanosine cap. Finally, the RNA methyltransferase adds a methyl group to the guanosine cap to yield a 7-methylguanosine cap covalently attached to the 5′-end of the mRNA. These three enzymes together are called the capping enzymes and they function only when in association with the CTD of the transcribing RNA Polymerase II. During splicing, segments of the primary transcripts, called introns, are excised and the remaining fragments called exons, are covalently joined, creating the mature mRNA. Among these three processes, splicing is the most complex one. Splicing was first reported in 1977, and it was as fascinating as the discovery of enhancers a few years later [102]. Polyadenylation is the addition of a series of adenosine residues (poly-A tail) at the 3′-end of the mRNA. Polyadenylation occurs when the transcription is about to be completed. A multiprotein complex called the cleavage and polyadenylation specificity factor (CPSF), in association with several other proteins, first cleaves at the 3′-end region of the transcribing pre-mRNA. CPSF binds to the polyadenylation signal sequence AAUAAA (or its variants) and cleaves the RNA 10−30 nucleotides further downstream. Once the RNA is cleaved, another enzyme called polyadenylate polymerase adds about 250 adenosine monophosphates creating the poly(A) tail. CPSF coordinates with the RNA Polymerase II and signals it to terminate the transcription when it reaches 5′-AAUAAA-3′ (called "termination sequence"). The polyadenylation machinery is recruited to the transcribing RNA Polymerase II at its CTD. The poly(A) tail is required for nuclear export, efficient translation, and the stability of the mRNA [103].

Splicing occurs at precise nucleotide positions (located within the 3′- and 5′-splice sites of the intron) as any mistaken addition or deletion of an extra nucleotide would alter or destroy the coding sequence. There are two types of splicing: constitutive and alternative. In constitutive splicing, the introns are removed from the pre-mRNA and the exons are joined, generating one mature mRNA from each primary transcript. Alternative splicing is far more complex as it involves the inclusion or exclusion of the introns and exons in multiple ways. The most prevalent mode of alternative splicing is "exon skipping" wherein one or more exons are removed along with the adjoining introns. In another type of alternative splicing, either one or the other exon is included in the mature mRNA, and it is called "mutually exclusive exon usage" The third type is called "alternative splice site selection" in which different splice sites at the 5′- or the 3′-end of an exon are processed, resulting in the generation of several coding sequences from the same segment of an exon. The fourth type is called "intron retention" when an intron or a part of it is retained in the mature mRNA. Alternative splicing thus creates an assortment of mRNAs from one primary transcript, thereby increasing the coding potential of the fewer number of genes (Fig. 1.11 [104,105]). The most fascinating example of alternative splicing is that of the Down's syndrome cell adhesion molecule (Dscam) transcript in *Drosophila melanogaster.* It generates 38,016 different isoforms through mutually exclusive splicing [106]. Alternative splicing thus significantly increases the diversity of the transcriptome (and the proteomes) in the metazoan tissues. Large-scale RNA-sequence analyses found that ~95% of the transcripts from

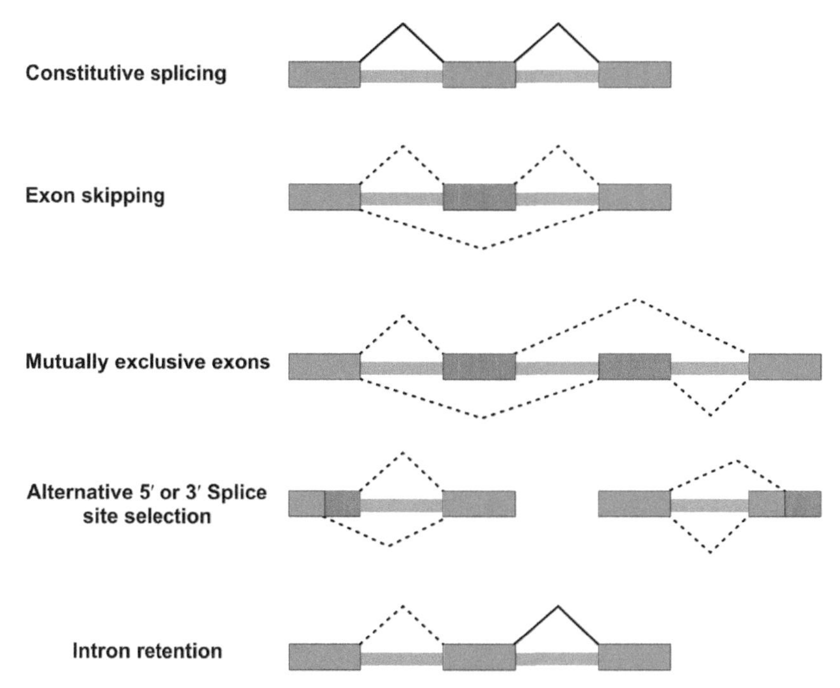

FIGURE 1.11

Different modes of splicing of mammalian pre-mRNAs. In constitutive splicing, introns are removed, and the exons are covalently joined. In exon skipping, one exon is spliced out along with two adjoining introns. In mutually exclusive exons, among the two consecutive exons, one or the other is included. During alternative 5′- or 3′-splice site selection, a segment of the sequence has two splice sites, of which either one is used, resulting in shorter or longer exons. In intron retention, an intron is included in the mature mRNA. A solid line means where the splicing invariably occurs. In the case of dotted lines, depending on the alternative possibilities, splicing may or may not occur. Intron: thinner line in coral color; exon: thicker line in stone color, alternative exon: thicker line in orange color.

about 20,000 protein-coding genes in humans undergo alternative splicing, producing as many as ~150,000 transcript variants, or even more. It means that in humans, on average, there are more than seven splice variants per annotated gene [107]. Interestingly, proteomics data from 6 primary hematopoietic cells, 17 adult and 7 fetal tissues has recently been compiled and a substantial number of proteins predicted from various alternatively spliced mRNAs were found missing. It was thus suggested that the mass spectrometric analyses of a complex mixture of human proteins might not be sensitive as yet to identify the low abundance proteins that are produced from most of the alternatively spliced transcripts [108]. Highly efficient and quantitative proteomic studies have also shown that although many primary transcripts undergo alternative splicing, most have a single dominant variant that is translated in a given physiological condition [109].

The splicing process is mediated by large (megadalton) ribonucleoprotein complexes called the spliceosome. Small nuclear ribonucleoproteins (snRNPs) are the RNA-protein complexes that in association with other proteins bind to the pre-mRNA to form the spliceosome. The RNA components of the snRNP particles are called small nuclear RNA (snRNA) that are about 150 nucleotides long. The snRNAs recognize the signal sequences at the splice sites located at the 5′- and 3′-ends of the introns. There are five major snRNAs viz., U1, U2, U4, U5, and U6, and in addition, there are several minor snRNAs viz., U11, U12, U4atac, U6atac, and U5. These snRNAs in association with more than 150 proteins form various types of snRNPs. Sequential assembly of several snRNP at the splice sites leads to the formation of the spliceosomes. It has been demonstrated that the small nuclear RNAs devoid of any proteins can also perform splicing reactions in vitro with synthetic oligo-ribonucleotides as the substrates. These studies thus showed that the snRNAs not only recognize the splicing signals on the pre-mRNAs, they also can function as enzymes (ribozymes, the Nobel Prize in Chemistry 1989 was awarded to Sidney Altman and Thomas R. Cech for their discovery of the catalytic properties of RNA). During the elongation of transcription, splicing factors are recruited by the RNA Polymerase II which binds to the nucleotide sequences on the newly transcribed RNA and carries out the splicing process [110]. However, there are also instances where splicing occurs after the completion of the transcription, and it generally happens to the splice sites at the 3′-end of the transcripts. Spliceosomes are uniquely capable of precisely identifying short splice sites from among large nucleotide sequences in the exons and the introns. Numerous studies have shown that transcription and splicing are highly integrated through spatiotemporal mechanisms. The macromolecular RNA Polymerase II complex that performs the initiation, elongation, and termination, also harbors proteins that are necessary for the simultaneous processing of the nascent transcript [111]. Alternative splicing often requires the preferential engagement of the splicing machinery to the suboptimum (weaker) sites against the stronger ones. To ensure that, splicing is regulated through the rate of transcriptional elongation. Following the initiation, the RNA Polymerase II in association with the elongation and other factors traverse (0.5−4 kb/min) through the entire length of the gene transcribing it. Elongation is modulated by histone modifications and the nucleosomal structures in and around the transcribed section of the gene. During elongation, the polymerase often accelerates, decelerates, and even pauses, which in turn regulates the rate of emergence of the nascent RNA. Since splicing factors assemble on the transcript as and when it emerges, a fast-moving RNA Polymerase II would synthesize and rapidly release segments of RNA harboring two consecutive introns, exposing them to the splicing machinery. In case the upstream intron contains a suboptimal splice site and the downstream intron harbors a strong one, splicing would occur in the downstream intron, resulting in removing both the introns along with them in between exon (exon skipping). On the contrary, when the elongation is slow, as soon as the first intron emerges, the spliceosome would assemble there, remove it, and the downstream exon would be retained. The slow-moving RNA Polymerase would then move forward, exposing the next intron with stronger splice sites which would then be spliced out [112]. Splicing is also facilitated by the exonic and intronic splicing enhancers (ESEs and ISEs) that are targeted by the positively splicing acting factors like the SR proteins. There are also exonic and intronic splicing silencers (ESSs and ISSs) where the negative acting factors (e.g., hnRNPs) bind and prevent splicing therein [113]. The recruitment of splicing factors to the transcription elongation complex occurs through the CTD of RNA Polymerase II that acts as the "landing pad" for the splicing and other factors for the post-transcriptional modifications of the pre-mRNA. The dynamic phosphorylation of heptad repeats in the CTD regulates the recruitment of the splicing and other factors involved in the modifications of the pre-mRNA [111]. Besides the strength of the splice sites, alternative splicing also depends on the promoter elements and the secondary structure of the transcripts [114]. Numerous studies have shown that the engagement of certain transcription factors to the splice sites also affects alternative splicing [115]. Histone modifications are also linked to splicing regulation. Histone marks like H3K36me3 and H3K4me3 are recognized by the chromatin-binding proteins that recruit the splicing regulators and factors, modulating the splicing output [116].

While a majority of splicing events result in partial changes in the protein-coding sequences, frame-preserving splicing events are more conserved in evolution than frame-shifting splicing. Also, while preserving the reading frame, there are evolutionary preferences for smaller exons that cause mild changes in the protein sequences (and structure) due to alternative splicing [117]. Taken together, it appears that alternative splicing is a nuanced way to diversify protein functions but not make it an entirely different entity. About 15% of hereditary diseases and cancer have now been associated with defects in alternative splicing [118]. Due to the tremendous importance of splicing in defining cellular proteomes, especially in the context of diseases, it is an active area of research.

Concluding remarks

For any cell type, be it unicellular or metazoan, the profile of gene expression defines its identity. Despite phenomenal growth in our knowledge of gene expression during the past three-quarters of a century, it is still quite inadequate to explain the biology of complex species like humans. Besides the well-studied modules of gene expression as discussed above, newer and newer aspects are emerging fast. With the advanced tools of genomics and proteomics, interspecies and intraspecies evolution of gene expression are now being studied. Highly sophisticated tools of cell biology can now help us visualize the chromatin architecture and the momentary burst of transcription in a single cell. While the effects of reactive oxygen species and novel RNA modifications on gene expression are being investigated, small molecular regulators are being explored for therapeutic purposes. Gene expression analyses are also moving out from the domains of basic biology and biotechnology to cancer diagnostics and forensic research. Highly sophisticated analytical tools now show that almost any perturbation of a cell by the physiological modulators, toxicants, physical insult, light, etc., have profound effects on the gene expression profile. It is thus likely that, in the coming years, more and more captivating aspects of gene regulation across species will enlighten our understanding of this fascinating aspect of biology.

References

[1] Scherrer K. Primary transcripts: from the discovery of RNA processing to current concepts of gene expression. Exp Cell Res December 15, 2018;373(1–2):1–33. https://doi.org/10.1016/j.yexcr.2018.09.011.
[2] Brown TA. Genomes. 2nd ed. Oxford: Wiley-Liss; 2002.
[3] Pertea M, Shumate A, Pertea G, Varabyou A, Breitwieser FP, Chang Y-C, Madugundu AK, Pandey A, Salzberg SL. CHESS: a new human gene catalog curated from thousands of large-scale RNA sequencing experiments reveals extensive transcriptional noise. Genome Biol 2018;19:208. Article number.
[4] Kroehling L, Khitun A, Bailis W, Jarret A, York AG, Khan OM, Brewer JR, Skadow MH, Duizer C, Harman CCD, Chang L, Bielecki P, Solis AG, Steach HR, Slavoff S, Flavell RA, Jackson R. The translation of non-canonical open reading frames controls mucosal immunity. Nature December 2018;564(7736): 434–8. https://doi.org/10.1038/s41586-018-0794-7.
[5] Vo Ngoc L, Wang Y-L, Kassavetis GA, Kadonaga JT. The punctilious RNA polymerase II core promoter. Genes Dev July 1, 2017;31(13):1289–301. https://doi.org/10.1101/gad.303149.117.
[6] Danino YM, Even D, Ideses D, Juven-Gershon T. The core promoter: at the heart of gene expression. Biochim Biophys Acta August 2015;1849(8):1116–31. https://doi.org/10.1016/j.bbagrm.2015.04.003.
[7] Haberle V, Stark A. Eukaryotic core promoters and the functional basis of transcription initiation. Nat Rev Mol Cell Biol October 2018;19(10):621–37. https://doi.org/10.1038/s41580-018-0028-8.
[8] Gupta K, Duygu Sari -A, Haffke M, Simon T, Berger I. Zooming in on transcription preinitiation. J Mol Biol June 19, 2016;428(12):2581–91. https://doi.org/10.1016/j.jmb.2016.04.003.
[9] Harlen KM, Churchman LS. The code and beyond: transcription regulation by the RNA polymerase II carboxy-terminal domain. Nat Rev Mol Cell Biol April 2017;18(4):263–73. https://doi.org/10.1038/nrm.2017.10.
[10] Shilatifard A. Factors regulating the transcriptional elongation activity of RNA polymerase II. Faseb J November 1998;12(14):1437–46. https://doi.org/10.1096/fasebj.12.14.1437.
[11] Maston GA, Evans SK, Green MR. Transcriptional regulatory elements in the human genome. Annu Rev Genom Hum Genet 2006;7:29–59. https://doi.org/10.1146/annurev.genom.7.080505.115623.
[12] Haberle V, Lenhard B. Promoter architectures and developmental gene regulation. Semin Cell Dev Biol September 2016;57:11–23. https://doi.org/10.1016/j.semcdb.2016.01.014.
[13] Field A, Adelman K. Evaluating enhancer function and transcription. Annu Rev Biochem June 20, 2020;89: 213–34. https://doi.org/10.1146/annurev-biochem-011420-095916.
[14] Taher L, Smith RP, Kim MJ, Ahituv N, Ovcharenko I. Sequence signatures extracted from proximal promoters can be used to predict distal enhancers. Genome Biol 2013;14. Article number: R117.
[15] The Human Transcription Factors February 08, 2018;Volume 172(ISSUE 4):P650–65.
[16] Vinson C, Myakishev M, Acharya A, Mir AA, Moll JR, Bonovich M. Classification of human B-ZIP proteins based on dimerization properties. Mol Cell Biol September 2002;22(18):6321–35. https://doi.org/10.1128/MCB.22.18.6321-6335.2002.
[17] Rezsohazy R, Saurin AJ, Maurel-Zaffran C, Graba Y. Cellular and molecular insights into Hox protein action. Development April 1, 2015;142(7):1212–27. https://doi.org/10.1242/dev.109785.
[18] Dennisa DJ, Sisu H, Schuurmans C. bHLH transcription factors in neural development, disease, and reprogramming. Brain Res 2019;1705(15 February):48–65.
[19] Hernández-Hernández JM, García-González EG, Brun CE, Rudnicki MA. The myogenic regulatory factors, determinants of muscle development, cell identity and regeneration. Semin Cell Dev Biol 2017;(72): 10–8. https://doi.org/10.1016/j.semcdb.2017.11.010.

[20] Mayran A, Jacques Drouin. Pioneer transcription factors shape the epigenetic landscape. J Biol Chem 2018;293(36):13795–804. https://doi.org/10.1074/jbc.R117.001232.
[21] Zaret KS. Pioneer transcription factors initiating gene network changes. Annu Rev Genet 2020;54:367–85. https://doi.org/10.1146/annurev-genet-030220-015007.
[22] Calkhoven CF, Ab G. Multiple steps in the regulation of transcription-factor level and activity. Biochem J July 15, 1996;317(Pt 2):329–42. https://doi.org/10.1042/bj3170329 (Pt 2).
[23] Chaturvedi MM, Sung B, Yadav VR, Kannappan R, Aggarwal BB. NF-κB addiction and its role in cancer: 'one size does not fit all. Oncogene 2011;30:1615–30.
[24] Ramirez-Carrozzi VR, Kerppola TK. Control of the orientation of Fos-Jun binding and the transcriptional cooperativity of Fos-Jun-NFAT1 complexes. J Biol Chem 2001;276(24):21797–808. https://doi.org/10.1074/jbc.M101494200.
[25] Roesler WJ. What is a cAMP response unit? Mol Cell Endocrinol April 25, 2000;162(1–2):1–7. https://doi.org/10.1016/s0303-7207(00)00198-2.
[26] Helsen 1 C, Kerkhofs S, Clinckemalie L, Spans L, Laurent M, Boonen S, Vanderschueren D, Frank C. Structural basis for nuclear hormone receptor DNA binding. Mol Cell Endocrinol 2012;348(2):411–7. https://doi.org/10.1016/j.mce.2011.07.025.
[27] Todeschini AL, Georges A, Veitia RA. Transcription factors: specific DNA binding and specific gene regulation. Trends Genet 2014;30(6):211–9. https://doi.org/10.1016/j.tig.2014.04.002.
[28] Georges AB, Benayoun BA, Caburet S, Veitia RA. Generic binding sites, generic DNA-binding domains: where does specific promoter recognition come from? Faseb J February 2010;24(2):346–56. https://doi.org/10.1096/fj.09-142117.
[29] Dubois-Chevalier J, Mazrooei P, Lupien M, Staels B, Lefebvre P, Eeckhoute J. Organizing combinatorial transcription factor recruitment at cis-regulatory modules, 2018. Transcription 2018;9(4):233–9. https://doi.org/10.1080/21541264.2017.1394424.
[30] Eggeling R. Disentangling transcription factor binding site complexity. Nucleic Acids Res November 16, 2018;46(20):e121. https://doi.org/10.1093/nar/gky683.
[31] Schaffner W. Enhancers, enhancers - from their discovery to today's universe of transcription enhancers. Biol Chem April 2015;396(4):311–27. https://doi.org/10.1515/hsz-2014-0303.
[32] Lomvardas S, Barnea G, Pisapia DJ, Mendelsohn M, Kirkland J. Richard Axel Interchromosomal interactions and olfactory receptor choice. Cell 2006;126:403–13.
[33] May D, Blow MJ, Kaplan T, McCulley DJ, Jensen BC, Akiyama JA, Holt A, Plajzer-Frick I, Shoukry M, Wright C, Afzal V, Simpson PC, Rubin EM, Black BL, Bristow J, Pennacchio LA, Visel A. Large-scale discovery of enhancers from human heart tissue. Nat Genet December 4, 2011;44(1):89–93. https://doi.org/10.1038/ng.1006.
[34] Kim T-K, Martin H, Gray JM, Costa AM, Bear DM, Wu J, Harmin DA, Laptewicz M, Barbara-Haley K, Scott K, Markenscoff-Papadimitriou E, Kuhl D, Bito H, Worley PF, Gabriel K, Greenberg ME. Widespread transcription at neuronal activity-regulated enhancers. Nature May 13, 2010;465(7295):182–7. https://doi.org/10.1038/nature09033.
[35] De Santa F, Barozzi I, Mietton F, et al. A large fraction of extragenic RNA pol II transcription sites overlap enhancers. PLoS Biol 2010;8:e1000384.
[36] Ye R, Cao C, Xue Y. Enhancer RNA: biogenesis, function, and regulation. Essays Biochem December 7, 2020;64(6):883–94. https://doi.org/10.1042/EBC20200014.
[37] Medina-Rivera A, Santiago-Algarra D, Puthier D, Spicuglia S. Widespread enhancer activity from core promoters. Trends Biochem Sci June 2018;43(6):452–68. https://doi.org/10.1016/j.tibs.2018.03.004.
[38] Gasperini M, Tome JM, Shendure J. Towards a comprehensive catalogue of validated and target-linked human enhancers. Nat Rev Genet May 2020;21(5):292–310. https://doi.org/10.1038/s41576-019-0209-0.

[39] Ong C-T, Corces VG. Enhancer function: new insights into the regulation of tissue-specific gene expression. Nat Rev Genet April 2011;12(4):283–93. https://doi.org/10.1038/nrg2957.

[40] Jindal 1 GA, Farley EK. Enhancer grammar in development, evolution, and disease: dependencies and interplay. Dev Cell March 8, 2021;56(5):575–87. https://doi.org/10.1016/j.devcel.2021.02.016.

[41] Rickels R, Shilatifard A. Enhancer logic and mechanics in development and disease. Trends Cell Biol August 2018;28(8):608–30. https://doi.org/10.1016/j.tcb.2018.04.003.

[42] Sengupta S, George RE. Super-enhancer-driven transcriptional dependencies in cancer. Trends Cancer April 2017;3(4):269–81. https://doi.org/10.1016/j.trecan.2017.03.006.

[43] Deng W, Rupon JW, Krivega I, Breda L, Motta I, Kristen SJ, Reik A, Gregory PD, Rivella S, Dean A, Blobel GA. Reactivation of developmentally silenced globin genes by forced chromatin looping. Cell August 14, 2014;158(4):849–60. https://doi.org/10.1016/j.cell.2014.05.050.

[44] Racko D, Benedetti F, Dorier J, Stasiak A. Transcription-induced supercoiling as the driving force of chromatin loop extrusion during formation of TADs in interphase chromosomes. Nucleic Acids Res February 28, 2018;46(4):1648–60. https://doi.org/10.1093/nar/gkx1123.

[45] Tena JJ, Santos-Pereira JM. Topologically associating domains and regulatory landscapes in development, evolution and disease. Front Cell Dev Biol July 06, 2021;9:702787. https://doi.org/10.3389/fcell.2021.702787.

[46] Michael L. Atchison, function of YY1 in long-distance DNA interactions. Front Immunol 2014;5:45. https://doi.org/10.3389/fimmu.2014.00045. Published online 2014 Feb 10.

[47] Chen D, Lei EP. Function and regulation of chromatin insulators in dynamic genome organization. Curr Opin Cell Biol June 2019;58:61–8. https://doi.org/10.1016/j.ceb.2019.02.001.

[48] Snetkova V, Jane AS. Enhancer talk. Epigenomics April 1, 2018;10(4):483–98. https://doi.org/10.2217/epi-2017-0157.

[49] Panigrahi A, Bert W. O'Malley Mechanisms of enhancer action: the known and the unknown. Genome Biol 2021;22:108.

[50] Gurumurthy A, Shen Y, Gunn EM, Bungert J. Phase separation and transcription regulation: are super-enhancers and Locus control regions primary sites of transcription complex assembly? Bioessays January 2019;41(1):e1800164. https://doi.org/10.1002/bies.201800164.

[51] Pott S, Lieb JD. What are super-enhancers? Nat Genet 2015;47:8–12.

[52] Palazzo AF, Koonin EV. Functional long non-coding RNAs evolve from junk transcripts. Cell November 25, 2020;183(5):1151–61. https://doi.org/10.1016/j.cell.2020.09.047.

[53] Ogam K, Suzuki HI. Nuclear RNA exosome and pervasive transcription: dual sculptors of genome function. Int J Mol Sci 2021;22(24):13401. https://doi.org/10.3390/ijms222413401.

[54] St Laurent G, Wahlestedt C, Kapranov P. The Landscape of long noncoding RNA classification. Trends Genet May 2015;31(5):239–51. https://doi.org/10.1016/j.tig.2015.03.007.

[55] Brannan CI, Dees EC, Ingram RS, Tilghman SM. The product of the H19 gene may function as an RNA. Mol Cell Biol January 1990;10(1):28–36.

[56] Chen LL. Linking long noncoding RNA localization and function. Trends Biochem Sci September 2016; 41(9):761–72. https://doi.org/10.1016/j.tibs.2016.07.003.

[57] Schmitz SU, Grote P, Herrmann BG. Mechanisms of long noncoding RNA function in development and disease. Cell Mol Life Sci July 2016;73(13):2491–509. https://doi.org/10.1007/s00018-016-2174-5.

[58] Bridges MC, Daulagala AC, Kourtidis A, LNCcation. lncRNA localization and function. J Cell Biol February 1, 2021;220(2):e202009045. https://doi.org/10.1083/jcb.202009045.

[59] Beermann J, Piccoli MT, Viereck J, Thum T. Non-coding RNAs in development and disease: background, mechanisms, and therapeutic approaches. Physiol Rev October 2016;96(4):1297–325. https://doi.org/10.1152/physrev.00041.2015.

[60] Arnold PR, Wells AD, Li XC. Diversity and Emerging Roles of Enhancer RNA in Regulation of Gene Expression and Cell Fate Front. Cell Dev Biol 2020;7:377. https://doi.org/10.3389/fcell.2019.00377.
[61] Tippens ND, Vihervaara A, Lis JT. Enhancer transcription: what, where, when, and why? Genes Dev January 1, 2018;32(1):1–3. https://doi.org/10.1101/gad.311605.118.
[62] Sartorelli V, Lauberth SM. Enhancer RNAs are an important regulatory layer of the epigenome. Nat Struct Mol Biol 2020;27:521–8.
[63] Lee J-H, Wang R, Xiong F, Krakowiak J, Liao Z, Nguyen PT, Moroz-Omori EV, Shao J, Zhu X, Bolt MJ, Wu H, Singh PK, Bi M, Shi CJ, Jamal N, Li G, Mistry R, Jung SY, Tsai K-L, Ferreon JC, Stossi F, Caflisch A, Liu Z, Mancini MA, Li W. Enhancer RNA m6A methylation facilitates transcriptional condensate formation and gene activation. Mol Cell August 19, 2021;81(16):3368–85. https://doi.org/10.1016/j.molcel.2021.07.024. e9.
[64] Cinghu S, Yang P, Kosak JP, Conway AE, Kumar D, Oldfield AJ, Adelman K, Jothi R. Intragenic enhancers attenuate host gene expression. Mol Cell October 5, 2017;68(1):104–17. https://doi.org/10.1016/j.molcel.2017.09.010. e6.
[65] Rahnamoun H, Lee J, Sun Z, Lu H, Ramsey KM, Komives EA, Lauberth SM. RNAs interact with BRD4 to promote enhanced chromatin engagement and transcription activation. Nat Struct Mol Biol August 2018; 5(8):687–97. https://doi.org/10.1038/s41594-018-0102-0.
[66] Bose 1 DA, Donahue 1 G, Reinberg 2 D, Shiekhattar 3 R, Bonasio 1 R, Berger SL. RNA binding to CBP stimulates histone acetylation and transcription. Cell January 12, 2017;168(1–2):135–49. https://doi.org/10.1016/j.cell.2016.12.020. e22.
[67] Van Speybroeck L. From epigenesis to epigenetics: the case of C. H. Waddington. Ann N Y Acad Sci December 2002;981:61–81.
[68] Turner B. Defining an epigenetic code. Nat Cell Biol 2007;9(1):2–6. https://doi.org/10.1038/ncb0107-2. PMID 17199124.
[69] Yan J, Huangfu D. Epigenome rewiring in human pluripotent stem cells. Trends Cell December 23, 2021; 32(3):259–71. https://doi.org/10.1016/j.tcb.2021.12.001.
[70] Smith ZD, Meissner A. DNA methylation: roles in mammalian development. Nat Rev Genet March 2013; 14(3):204–20. https://doi.org/10.1038/nrg3354.
[71] Cusack M, King HW, Spingardi P, Kessler BM, Klose RJ, Kriaucionis S. Distinct contributions of DNA methylation and histone acetylation to the genomic occupancy of transcription factors. Genome Res October 2020;30(10):1393–406. https://doi.org/10.1101/gr.257576.119.
[72] Zhu H, Wang G, Jiang Q. Transcription factors as readers and effectors of DNA methylation. Nat Rev Genet August 1, 2016;17(9):551–65. https://doi.org/10.1038/nrg.2016.83.
[73] Kulis M, Queirós AC, Beekman R, Martín-Subero JI. Intragenic DNA methylation in transcriptional regulation, normal differentiation and cancer. Biochim Biophys Acta November 2013;1829(11):1161–74. https://doi.org/10.1016/j.bbagrm.2013.08.001.
[74] Sundaram V, Cheng Y, Ma Z, Li D, Xing X, Edge P, Snyder MP, Wang T. Widespread contribution of transposable elements to the innovation of gene regulatory networks. Genome Res December 2014;24(12): 1963–76. https://doi.org/10.1101/gr.168872.113.
[75] Steinhaus R, Gonzalez T, Seelow D, Robinson PN. Pervasive and CpG-dependent promoter-like characteristics of transcribed enhancers. Nucleic Acids Res June 4, 2020;48(10):5306–17. https://doi.org/10.1093/nar/gkaa223.
[76] Nishiyama A, Nakanishi M. Navigating the DNA methylation landscape of cancer. Trends Genet November 2021;37(11):1012–27. https://doi.org/10.1016/j.tig.2021.05.002.
[77] Gottesfeld JM. Milestones in transcription and chromatin published in the journal of biological Chemistry. J Biol Chem February 1, 2019;294(5):1652–60. https://doi.org/10.1074/jbc.TM118.004162.

[78] Eric Verdin 1, Melanie Ott 150 years of protein acetylation: from gene regulation to epigenetics, metabolism and beyond. Rev Mol Cell Biol April 2015;16(4):258–64. https://doi.org/10.1038/nrm3931.

[79] Tanny JC. Chromatin modification by the RNA Polymerase II elongation complex. Transcription 2014; 5(5):e988093. https://doi.org/10.4161/21541264.2014.988093.

[80] Kimura H. Histone modifications for human epigenome analysis. J Hum Genet July 2013;58(7):439–45. https://doi.org/10.1038/jhg.2013.66.

[81] Hyun K, Jeon J, Park K, Kim J. Writing, erasing and reading histone lysine methylations. Exp Mol Med April 28, 2017;49(4):e324. https://doi.org/10.1038/emm.2017.11.

[82] Crump NT, Milne TA. Why are so many MLL lysine methyltransferases required for normal mammalian development? Cell Mol Life Sci 16 May 2019;76(2019):2885–98.

[83] Xhabija B, Kidder BL. KDM5B is a master regulator of the H3K4-methylome in stem cells, development and cancer. Semin Cancer Biol August 2019;57:79–85. https://doi.org/10.1016/j.semcancer.2018.11.001.

[84] Lagos-Quintana M, Rauhut R, Lendeckel W, Tuschl T. Identification of novel genes coding for small expressed RNAs. Science October 26, 2001;294(5543):853–8. https://doi.org/10.1126/science.1064921.

[85] Dedeoğlu BG, Noyan S. Experimental MicroRNA targeting validation. Methods Mol Biol 2022;2257: 79–90. https://doi.org/10.1007/978-1-0716-1170-8_4.

[86] O'Brien J, Hayder H, Zayed Y, Peng C. Overview of MicroRNA biogenesis, mechanisms of actions, and circulation. Front Endocrinol 2018;9:402. https://doi.org/10.3389/fendo.2018.00402.

[87] Morlando M, Ballarino M, Gromak N, Pagano F, Bozzoni I, J Proudfoot N. Primary microRNA transcripts are processed co-transcriptionally. Nat Struct Mol Biol September 1, 2008;15(9):902–9. https://doi.org/10.1038/nsmb.1475.

[88] Salim U, Kumar A, Kulshreshtha R, Vivekanandan P. Biogenesis, characterization, and functions of mirtrons. Wiley Interdiscip Rev RNA January 2022;13(1):e1680. https://doi.org/10.1002/wrna.1680.

[89] Schanen Xiaoman Li BC. Transcriptional regulation of mammalian miRNA genes. Genomics Volume January 2011;97(Issue 1):1–6.

[90] Jonas S, Izaurralde E. Towards a molecular understanding of microRNA-mediated gene silencing. Nat Rev Genet July 2015;16(7):421–33.

[91] Vasudevan S. Posttranscriptional upregulation by microRNAs. Wiley Interdiscip Rev RNA May-Jun;3(3): 311–30. https://doi.org/10.1002/wrna.121.

[92] Chandradoss SD, Schirle NT, Szczepaniak M, MacRae IJ, Joo C. A dynamic search process underlies MicroRNA targeting. Cell July 2, 2015;162(1):96–107. https://doi.org/10.1016/j.cell.2015.06.032.

[93] Miao L, Yao H, Li C, Pu M, Yao X, Yang H, Qi X, Ren J, Wang Y. A dual inhibition: microRNA-552 suppresses both transcription and translation of cytochrome P450 2E1. Biochim Biophys Acta April 2016;1859(4):650–62.

[94] Xiao M, Li J, Li W, Wang Y, Wu F, Xi Y, Zhang L, Ding C, Luo H, Li Y, Peng L, Zhao L, Peng S, Xiao Y, Dong S, Cao J, Yu W. MicroRNAs activate gene transcription epigenetically as an enhancer trigger. RNA Biol October 3, 2017;14(10):1326–34.

[95] Tomasello L, Distefano R, Nigita G, Croce CM. The MicroRNA family gets wider: the IsomiRs classification and role. Front Cell Dev Biol June 9, 2021;9:668648. https://doi.org/10.3389/fcell.2021.668648.

[96] Li W, Saraiya AA, Wang CC. Gene regulation in giardia lambia involves a putative MicroRNA derived from a small nucleolar RNA. October 18, 2011. https://doi.org/10.1371/journal.pntd.0001338. Published.

[97] Maute RL, Schneider C, Sumazin P, Holmes A, Califano A, Basso K, Dalla-Faverat R. RNA-derived microRNA modulates proliferation and the DNA damage response and is down-regulated in B cell lymphoma. Proc Natl Acad Sci USA January 22, 2013;110(4):1404–9. https://doi.org/10.1073/pnas.1206761110.

[98] Rupaimoole R, Slack FJ. MicroRNA therapeutics: towards a new era for the management of cancer and other diseases. Nat Rev Drug Discov 2017;16:203–22.

[99] Syazana NN, Kamal BNM, Shahidan WNS. Non-exosomal and exosomal circulatory MicroRNAs: which are more valid as biomarkers? Front Pharmacol January 20, 2020;10:1500. https://doi.org/10.3389/fphar.2019.01500.

[100] Saliminejad 1 K, Khorshid 2 HRK, Fard 1 SS, Ghaffari SH. An overview of microRNAs: biology, functions, therapeutics, and analysis methods. J Cell Physiol May 2019;234(5):5451−65. https://doi.org/10.1002/jcp.27486.

[101] Rupaimoole R, Slack FJ. MicroRNA therapeutics: towards a new era for the management of cancer and other diseases. Nat Rev Drug Discov March 2017;16(3):203−22. https://doi.org/10.1038/nrd.2016.246.

[102] Suran M. Finding the tail end: the discovery of RNA splicing. Proc Natl Acad Sci USA January 28, 2020;117(4):1829−32. https://doi.org/10.1073/pnas.1919416116.

[103] Fusby B, Kim S, Erickson B, Kim H, Peterson ML, Bentley DL. Coordination of RNA polymerase II pausing and 3' end processing factor recruitment with alternative polyadenylation. Mol Cell Biol November 2, 2015;36(2):295−303. https://doi.org/10.1128/MCB.00898-15.

[104] Ule J, Blencowe BJ. Alternative splicing regulatory networks: functions, mechanisms, and evolution. Mol Cell October 17, 2019;76(2):329−45. https://doi.org/10.1016/j.molcel.2019.09.017.

[105] Baralle FE, Giudice J. Alternative splicing as a regulator of development and tissue identity. Nat Rev Mol Cell Biol July 2017;18(7):437−51. https://doi.org/10.1038/nrm.2017.27.

[106] Jin Y, Dong H, Shi Y, Bian L. Mutually exclusive alternative splicing of pre-mRNAs. Wiley Interdiscip Rev RNA May 2018;9(3):e1468. https://doi.org/10.1002/wrna.1468.

[107] Jiang W, Chen L. Alternative splicing: human disease and quantitative analysis from high-throughput sequencing. Comput Struct Biotechnol J 2021;19:183−95. https://doi.org/10.1016/j.csbj.2020.12.009.

[108] Hu Z, et al. Revealing missing human protein isoform based on ab initio prediction, RNA-seq and proteomics. Sci Rep 2015;5:10940.

[109] Michael L. Tress, federico abascal and alfonso valencia, alternative splicing may not Be the key to proteome complexity. Trends Biochem Sci February 2017;42(2). https://doi.org/10.1016/j.tibs.2016.08.008.

[110] Herzel L, Ottoz DSM, Alpert T, Neugebauer KM. Splicing and transcription touch base: co-transcriptional spliceosome assembly and function. Nat Rev Mol Cell Biol October 2017;18(10):637−50. https://doi.org/10.1038/nrm.2017.63.

[111] Maita H, Nakagawa S. What is the switch for coupling transcription and splicing? RNA Polymerase II C-terminal domain phosphorylation, phase separation and beyond. Wiley Interdiscip Rev RNA January 2020;11(1):e1574. https://doi.org/10.1002/wrna.1574.

[112] Muniz L, Nicolas E, Trouche D. RNA polymerase II speed: a key player in controlling and adapting transcriptome composition. EMBO J August 2, 2021;40(15):e105740. https://doi.org/10.15252/embj.2020105740.

[113] Schaal TD, Maniatis T. Multiple distinct splicing enhancers in the protein-coding sequences of a constitutively spliced pre-mRNA. Mol Cell Biol January 1999;19(1):261−73. https://doi.org/10.1128/MCB.19.1.261.

[114] Kolathur KK. Role of promoters in regulating alternative splicing. Gene May 25, 2021;782:145523.

[115] Rambout X, Dequiedt F, Maquat LE. Beyond transcription: roles of transcription factors in pre-mRNA splicing. Chem Rev April 25, 2018;118(8):4339−64. https://doi.org/10.1021/acs.chemrev.7b00470.

[116] Zhang J, Zhang YZ, Jiang J, Duan CG. The crosstalk between epigenetic mechanisms and alternative RNA processing regulation. Front Genet August 20, 2020;11:998. https://doi.org/10.3389/fgene.2020.00998.

[117] Zhang C, Krainer AR, Zhang MQ. Evolutionary impact of limited splicing fidelity in mammalian genes. Trends Genet 2007;23:484−8.

[118] Titus MB, Chang AW, Olesnicky EC. Exploring the diverse functional and regulatory consequences of alternative splicing in development and disease. Front Genet November 24, 2021;12:775395. https://doi.org/10.3389/fgene.2021.775395.
[119] Ohkuma Y. Multiple functions of general transcription factors TFIIE and TFIIH in transcription: possible points of regulation by trans-acting. J. Biochem 1997;122:481–9. 1997.
[120] Scharer CD, McCabe CD, Ali-Seyed M, Berger MF, Bulyk ML, Moreno CS. Genome-wide promoter analysis of the SOX4 transcriptional network in prostate cancer cells. Cancer Res January 15, 2009;69(2): 709–17. https://doi.org/10.1158/0008-5472.CAN-08-3415.

CHAPTER

Post-transcriptional gene regulation: an overview

2

Shweta Pandey[1], Smriti Shreya[2] and Buddhi Prakash Jain[2]

[1]*Govt VYT PG Autonomous College, Durg, Chhattisgarh, India;* [2]*Department of Zoology, Mahatma Gandhi Central University, Motihari, Bihar, India*

Introduction

We share the planet Earth with a myriad of living organisms, which can grow, reproduce, respire, respond to stimuli, inherit genetic information, and can carry out numerous chemical reactions. These organisms comprise cells as the fundamental unit of life, which can vary in number, ranging from a single cell in a unicellular organism to billions and trillions of cells in any multicellular organism. These cells carry biological information in DNA, in the form of genes, the heredity unit. The information stored is expressed in cells via the formation of two more polymers, naming, RNA and Proteins. The process of formation of RNA from DNA is known as transcription, while that of proteins from RNA is termed translation. Each organism enjoys its uniqueness and all credit goes to its DNA, which is unique to each organism and characterizes them. Therefore, it should be expressed accurately and hence must be coordinated and tightly regulated at various stages. Earlier, the regulation of genes was thought to be controlled by various proteins that interact with the regulatory elements present near the promoter of a gene. Now, it has been established that it involves various processes such as chromatin remodeling, nucleosome positioning, modification of histone proteins, binding of transcription factors and other proteins to DNA, RNA modification and localization, the interplay of various proteins that bind RNA and noncoding RNA [1].

The post-transcriptional gene regulation operates after the beginning of RNA synthesis when polymerase has bound the promoter. Although the gene is mainly controlled at the stage of transcription initiation, these modifications also play a crucial role in regulating the expression of some genes. In this chapter, we will have a look at the events that occur after the initiation of transcription and are known to regulate the expression of various genes. It should be noted that all the mechanisms do not necessarily occur in all the RNA molecules.

Attenuation of transcription

Transcription attenuation causes premature termination of premature RNA molecules, mainly in certain bacteria and archaea, although some examples have also been seen in eucaryotes. In some cases, a short hairpin-like structure is formed by nascent RNA that interacts with the RNA polymerase

Post-Transcriptional Gene Regulation in Human Disease, Volume 32. https://doi.org/10.1016/B978-0-323-91305-8.00018-1

so that the transcription of the gene aborts. Whenever the product of a gene is required, then regulatory proteins cease the process of attenuation by binding the nascent RNA, this results in the formation of a complete RNA molecule. Yanofsky [2] discovered the mechanism of attenuation in *E. coli* tryptophan biosynthetic operon. In the trp operon, there exists a short region between the operator and the start of the ORF region of the first structural gene. This region is known as a leader, which is responsible for attenuation. It is approx. 162 bp long and has several inverted repeats that can be numbered as 1–4. These repeats can form three different base-paired secondary structures of RNA. The structures include a pause structure (1:2), an overlapping antiterminator (2:3), and an intrinsic transcription terminator (3:4). The leader region also contains a small ORF of 14 amino acids comprising two UGG Trp codons. Translation of the two Trp codons is the key determining factor for the type of secondary structure of RNA formed in the leader region. As soon as the transcription starts from the trp promoter, the 1:2 structure forms that pause the transcription after nucleotide 92. This pausing of transcription allows time to ribosome so that it initiates translation of the leader peptide. The initiation of translation releases the RNA polymerase from pause to resume transcription. Thus, the process of transcription and translation are now coupled where the ribosome is closely following the RNA polymerase. Now, there are two possible pathways for attenuation that depend on the level of tryptophan in the cell. The efficiency of translation of Trp codon in leader region reflects the availability of aminoacylated tRNATrp in the cell. When tryptophan is low, the amount of charged tRNATrp is low and the translation of Trp codon is inefficient, so ribosome stall in one of the two codons. The associated RNA polymerase continues the transcription; translation and transcription become uncoupled. As RNA polymerase passes the 2 and 3 regions of leader, antiterminator structure (2:3) forms thereby preventing the formation of intrinsic terminator structure (3:4). On the other hand, when there is abundant tryptophan, the amount of charged tRNATrp is high, which allows the efficient translation of Trp codons in the leader region and it covers part of region 2, thereby preventing the formation of antiterminator structure (2:3), while the terminator structure (3:4) is formed. Formation of terminator structure leads to termination of transcription in the leader region, therefore, structural genes are not transcribed (Fig. 2.1) [3].

Similarly, in eucaryotes, the process of attenuation has been observed during the life cycle of the human immunodeficiency virus (HIV) when its genome gets integrated into that of the host. After integration, the RNA polymerase II of the host transcribes the viral DNA. The transcription of the viral genome is attenuated after a few hundred base pairs due to the formation of secondary structures in the RNA. However, when the conditions are optimal for viral growth, a viral protein, Tat, binds the secondary structure of RNA, known as Tar, and recruits various cellular proteins that allow the transcription to continue thereby preventing its premature termination.

Riboswitches, short RNA sequences that can change their conformation when bound to small molecules like metabolites, too can regulate the gene expression. Riboswitches that are present at the 5′-UTR of mRNA recognize their specific molecule and bind it with high specificity and affinity. The binding causes conformational changes in the structure of RNA and can block or permit the process of transcription. It is commonly found in bacteria, for example, a riboswitch controls the expression of purine biosynthetic genes in bacteria. When guanine level is low in cells, then RNA polymerase continues the transcription of guanine. But when the guanine is abundant, it binds the riboswitch present in the 5′-UTR of the RNA and terminates its transcription. The riboswitches represent one the most economical example of gene control features as they do not require any regulatory protein.

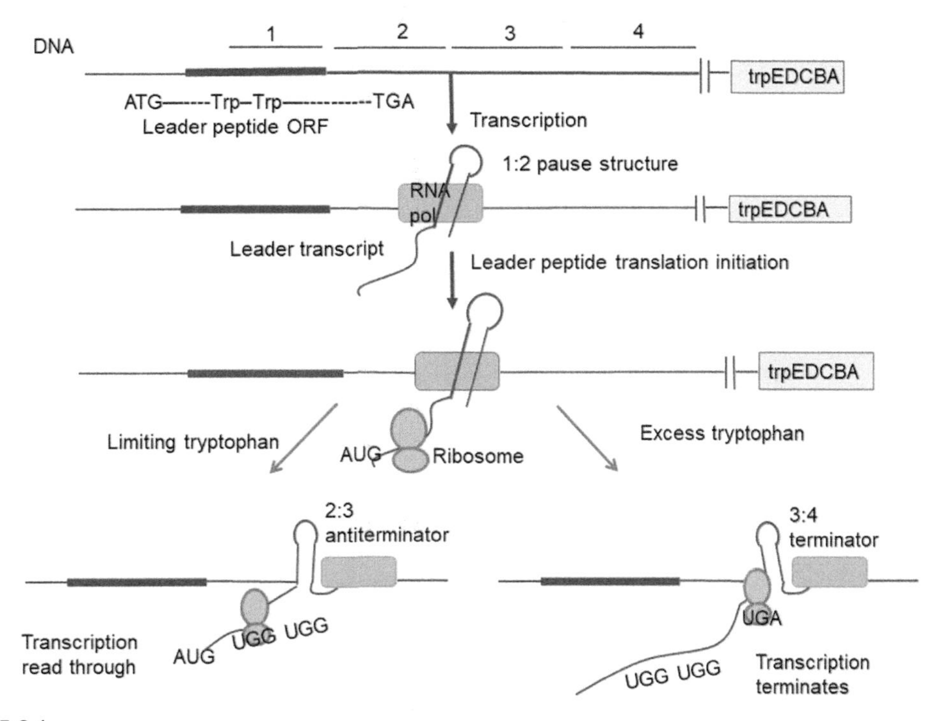

FIGURE 2.1

A model of transcription attenuation in trp operon of *E. coli*: In trp operon, the RNA polymerase pauses as the 1:2 pause structure is formed. The pausing of RNA polymerase provides time for the ribosome to start the translation of the leader peptide. When tryptophan is limiting, the ribosome stalls at tandem codons for tryptophan (UGG), which causes the formation of 2:3 antiterminator structure and the transcription read through occurs. On thre other hand, when the tryptophan is present in excess, the amount tRNATrp is high and translation of Trp codons in the leader region occurs. Due to this the position of the ribosome hinders the formation of 2:3 antiterminator structure resulting in the formation of 3:4 terminator structure and termination of transcription occur.

Untranslated region

In eukaryotes, the transcription of protein-coding genes starts many nucleotides upstream to the first amino acid, this region is known as 5′-UTR. Similarly, the transcription continues beyond the codon that encodes the last amino acid and is termed as 3′-UTR. The typical length of 5′-UTR ranges between 100 and 200 nucleotides while that of 3′-UTR varies a lot and ranges from 200 to 800 in plants and animals, respectively. During the processing of mRNA, a 5′-cap is added (7-methylguanosine) at the 5′-end, through a 5′-5′ phosphate bond, and a poly-A tail is added at the 3′-end [4]. The processing also involves splicing and may include RNA editing. Earlier the UTR and noncoding regions were considered as "junk DNA," but now it has been proven that they regulate the expression of various genes in a spatiotemporal manner. The mechanism of "pervasive transcription" refers to the wide-spread transcription that is responsible for transcribing approx. 80%–90% of genome results in the

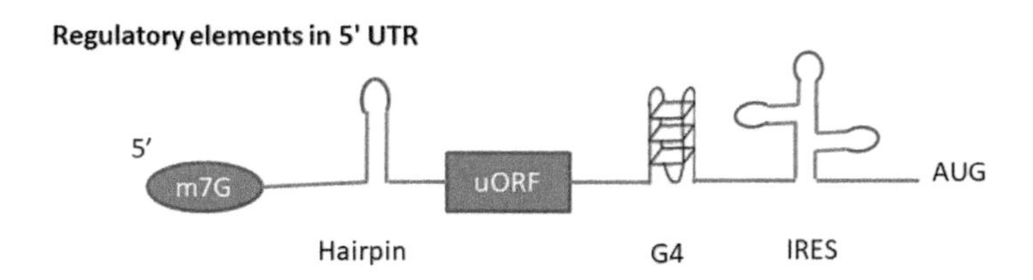

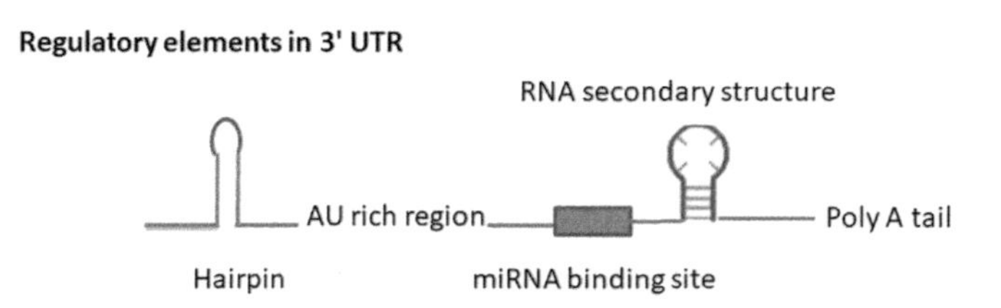

FIGURE 2.2

Regulatory elements in the noncoding regions of the gene: There are various regulatory elements in the 5′- and 3′-UTR regions, in 5′-UTR it involves hairpin structure, G4 structure, and internal ribosome entry site (IRES) elements. Lower panel: In the 3′-UTR hairpin structure, the binding site for miRNAs and poly-A tail are present, which play a prime role in the regulation of gene expression.

formation of RNA that neither code for proteins nor have any established function like tRNA, rRNA, etc. In complex genomes, pervasive transcription results in the production of various noncoding transcripts that interact with UTRs and introns, and together these are responsible for the fine-tuning of gene expression. Various features and interactions of UTRs with different factors can regulate the genetic expression, these can be listed as follows (Fig. 2.2):

(i) **Length and structure of 5′-UTR:** Efficient translation of mRNA is affected by various properties of 5′-UTR. In general, the 5′-UTRs that are short in length, do not form any structure, have a low GC content, and do not comprise any upstream AUG codon (uAUG) enable efficient translation of its mRNA [5]. The complex structure of 5′-UTR is associated with many genes that are expressed in specific tissues only. The complex structure is likely to interact with various RNA-binding proteins. For example, the gene of peroxisome proliferator-activated receptor c (PPAR-c) expresses various splice variants that vary in their 5′-UTR. PPARs are the transcription factor of the nuclear hormone receptor superfamily. In three splice variants of PPARs, the expression level is increased while in two the same is decreased. This difference is due to the nucleotide content and structure of the 5′-UTR, which promote the binding of different proteins that have an enhancive or repressive effect on the expression of proteins [6]. Similarly, G-quadruplex (G4), a well-characterized guanine-rich secondary structure, can fold in a tetra helical structure and is stabilized by Hoogsteen hydrogen bonds. It can strongly repress translation, for example, the telomeric repeat binding protein 2 (TRF2) gene, which plays a major role in the control of the function of telomere has G4 structure in its 5′-UTR. This structure represses the translation of TRF2 by translational suppression mechanism [7,8]. This structure is not confined to 5′-UTR only but is also present in promoters, telomeres, and 3′-UTR.

(ii) Alternative 5′-UTRs: Alternative 5′-UTRs can be generated by using alternative promoters as well as by alternative splicing. Usage of alternative 5′-UTR can cause variation in gene expression depending on the nature of regulatory elements confined in the 5′-UTR. For instance, the gene for estrogen receptor β (Erβ) is responsible for the proper functioning of estrogen and is expressed as multiple isoforms possessing alternative 5′-UTRs. The isoforms of the Erβ gene is misregulated in various cancers as the expression of different isoforms depends on the usage of alternative [9]. Also, axin2 protein plays a prominent role in early postnatal development as well as tumor suppression. It comprises three promoters that enable the expression of three isoforms in UTRs that differ their 5′-UTR. Each isoform has a different secondary structure and upstream open reading frames (uORFs).

(iii) IRES and cap-independent translation initiation: Under normal conditions, initiation of translation requires the recognition of 5′-cap by initiation factors. Under conditions like stress, hypoxia, or apoptosis, internal ribosome entry sites (IRESs) ensure the recruitment of ribosomes to the mRNA, thereby facilitating the cap-independent initiation of translation [10]. IRESs are the regulatory motifs that are present on mRNA and ensure constitutive expression of essential proteins as c-Myc, Apaf-1 (Fig. 2.2). These proteins are expressed at a low level in normal conditions while their expression level is increased via IRES pathway under stressed conditions. Based on the mechanism of the recruitment of ribosome, cellular IRES can be classified into two types: in IRES type I, the cis-elements in cellular IRESs, for example, RNA binding motifs, are bound by IRES transacting factors (ITAFs), the ITAFs interact directly with the 40S ribosomal subunit or recruit the 40S subunit with the help of initiation factors (IFs) and finally, 60S ribosome is recruited. In IRES type II, a short cis-element is present in IRESs that can base pair with 18S rRNA, leading to recruitment of 40S subunit followed by 60S ribosome.

(iv) Upstream open reading frame: The uORF is an ORF that is situated in 5′-UTR. It is composed of a start codon in the 5′-UTR region and an in-frame stop codon. Its typical length is $\geq$9 nucleotides. This is followed by the main coding sequence, while some uORFs may be fully upstream some may be overlapping with the coding sequence (CDS) [11] (Fig. 2.3). These are present in approximately 50% of 5′-UTR in humans and can reduce the mRNA levels by 30% and the expression of proteins by 30%−80% [12]. If the uORF is bound by ribosome then it may get translated by the ribosome, however, this can alter the efficiency of translation of the main ORF. If the uORF is not bound by the ribosome then the protein expression from the gene will be reduced. For example, the gene poly(A)polymerase-α, which encodes a protein required for the addition of poly-A tail to the mRNA, comprises two conserved uORFs in 5′-UTR. Mutation in the uAUG codon of the gene results in the increased translation efficiency of the main CDS. This depicts that there is an inhibitory effect of uORF on the expression of the gene and any mutation in the uORF may be detrimental as it will result in aberrant gene expression [13].

(v) *Introns in 5′-UTRs*: In humans, about 35% of genes comprise an intron in their 5′-UTR [14]. They are twice the length of the introns that are found in the coding region. The genes that are

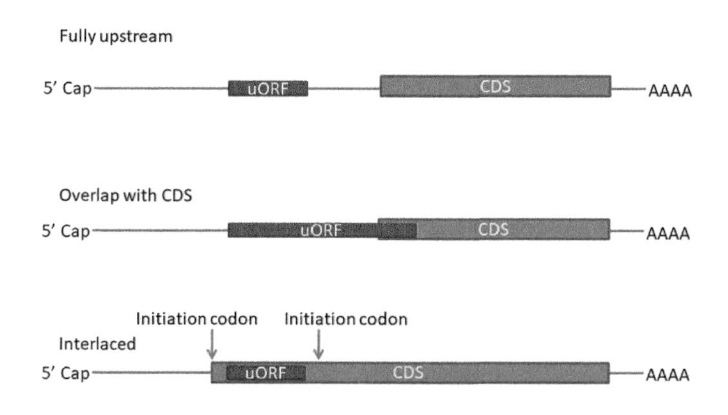

FIGURE 2.3

Three main classes of upstream open reading frame: In fully upstream uORFs, one or more ATG is present in frame with the main coding sequence (CDS) and the stop codon lies in the UTR only. In CDS overlap uORFs, the start codon of ORF is in 5′-UTR while the stop codon lies beyond the initiation codon of the main CDS. In interlaced uORFs, the uORFs may be situated between the alternative initiation codons within the CDS. Apart from these types, uORFs may overlap each other in or out of frame too (not shown in the figure).

highly expressed usually comprise short introns in 5′-UTR. The presence of a short intron in the 5′-UTR enhances the gene expression by enhancing the transcription of mRNA or by stabilizing the mRNAs. Deletion analyses of the ubiquitin gene showed that upon deletion of the intron in 5′-UTR the activity of the promoter is reduced to half [15].

(vi) ***Poly (A) tail*****:** During the cotranscriptional modification of mRNA, the addition of a series of adenosine bases at its 3′-end of an mRNA, results in the formation of its poly(A) tail. In eukaryotes, it starts in the nucleus and is a process of two steps: (a) cleavage at poly (A) site of pre-mRNA, (b) addition of poly-A tail upstream to the cleaved product. All mRNA comprises a hexanucleotide signal sequence, AAUAA, located 10–30 nucleotides upstream to the cleavage site and a guanine-uridine (GU)-rich motif present at 20–40 bases downstream to the cleavage site. The cleavage site lies immediately after CA dinucleotide. The hexanucleotide signal sequence binds to the cleavage and polyadenylation specificity factor (CPSF), while the GU-rich region is the binding site of the cleavage stimulating factor (CstF). Their binding causes the bending of DNA. This is followed by the binding of cleavage factors I and II (CF I and CF II) and poly (A) polymerase (PAP) that stabilize the whole complex as well as stimulate the cleavage at the cleavage site. The downstream cleavage product is released and degraded, while the CF I and CFII are released. The bound PAP then adds approx. 12 A residues to the 3′OH of the cleavage product. The addition of poly (A) is accelerated by the binding of poly (A) binding protein II (PABP II) (Fig. 2.4). The addition of poly-A stops after the addition of 200–250 A residues. The poly (A) tail is a binding site of poly-A binding proteins (PABP), which play a crucial role in the regulation of gene expression, stability, and decay of mRNA, its export, and translation [16]. PABPs act as scaffolds that provide a platform for the interaction of various factors that can modulate the expression of genes. It has been shown that the addition of poly (A) tail to the luciferase gene increases the protein expression to 97 folds [17].

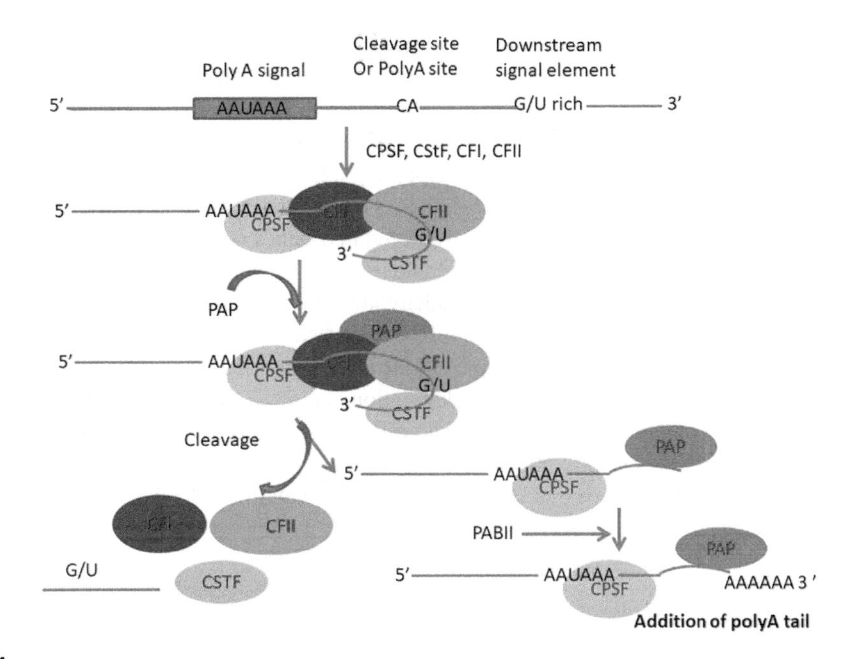

FIGURE 2.4

Model for cleavage and polyadenylation of pre-mRNA during their processing: mRNA contains a hexanucleotide AAUAAA poly (A) signal, which is bound by CPSF. Cleavage stimulating factor (CstF) interacts with GU or U rich sequence that is present downstream to the hexanucleotide. CstF interacts with CPSF, thereby formation of an RNA loop occurs. This is stabilized by binding of CFI and CFII. This is followed by the binding of poly (A) polymerase (PAP) and cleavage of mRNA 10–35 nucleotide downstream to poly (A) signal. The cleavage factors are released and PAP adds approx. 12 A residues to the 3′-hydroxyl end of the cleavage site. The rate of addition of poly-A is increased after binding of poly (A) binding protein II (PABPII). PABPII signals PAP to stop adding A after the addition of 200–250 A residues.

(vii) ***Length and structure of 3′-UTR*****:** 5′-UTR and poly (A) tail function in a synergistic manner to control the expression of a gene. Interaction of PABPs with poly (A) tail facilitates the binding of eIF4F to the 5′-cap structure. The length of the 3′-UTR also affects the expression of the gene, for example, the increase in the length of 3′-UTR of luciferase mRNA from 19 to 156 nucleotides decreases its expression by 45 folds [18]. This shows that 3′-UTR is a prime determinant of mRNA expression because of its interaction with 5′-UTR and with miRNA. In addition to the length, the structure of 3′-UTR also determines the efficiency of translation. The most common secondary structure is the stem-loop structure. Many proteins can bind the 3′-UTR and can change its spatial configuration by disruption of folding of mRNA or by interacting with other proteins. For example, in response to calcium ions (Ca^{2+}), the stem-loop structure of brain-derived neurotrophic factor (BDNF) transcript (BDNF supports the survival of existing neurons, growth, and differentiation of new neurons) acts as a scaffold and binds to the RNA-binding proteins, noncoding RNAs, and various polyadenylation factors and thus stabilizes the transcript [19].

(viii) Alternative 3′-UTRs: Various isoforms of mRNA that vary in their 3′-UTR can be produced by the process of alternative polyadenylation (APA). It occurs due to the presence of multiple polyadenylation sites, that may lie in the 3′-UTR region, intron, or exon. About 50% of human genes exhibit APA sites [20]. Polyadenylation can be categorized into 2 classes, namely constitutive polyadenylation and alternative polyadenylation (Fig. 2.5). Constitutively polyadenylated genes contain a unique polyadenylation site, for example, the INS gene that encodes for insulin [21]. Alternatively, polyadenylated genes comprise more than one polyadenylation site, which is again classified as

(a) UTR-APA—when the alternative polyadenylation occurs at the poly (A) sites located in 3′-UTR of the last exon. It results in mRNAs that have the same coding region but the varied length of 3′-UTR, for example, INSR gene that codes for the insulin receptor.

(b) CR-APA—when the alternative polyadenylation occurs at the poly (A) sites located in the coding region of exons or in introns. The mRNAs resulting from it differ in both C terminal coding region as well as 3′-UTR, for example, CALCA (calcitonin related polypeptide-α) gene that encodes peptide hormone calcitonin/CGRP (calcitonin gene-related peptide).

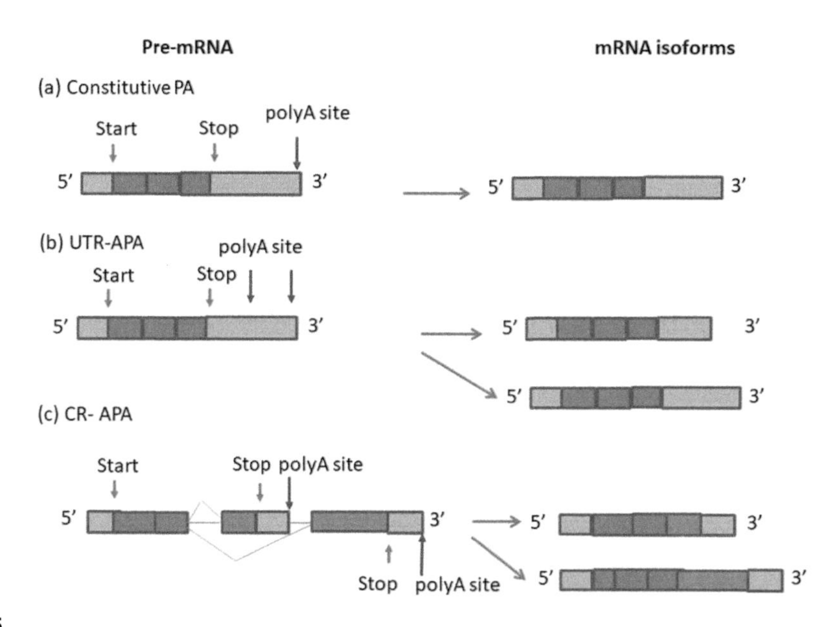

FIGURE 2.5

Types of polyadenylation: constitutive polyadenylation: genes contain only poly (A) site, so alternative polyadenylation does not take place. Untranslated region (UTR) APA—Genes contain multiple poly (A) sites in their 3′-UTR, this results in the formation of various mRNAs producing the same proteins but with different lengths of their 3′-UTRs. If the poly(A) tail is added at the proximal polyadenylation site, then a shorter mRNA is formed with less post-transcriptional modification and increased production of protein. Coding region (CR) APA: genes contain multiple poly (A) sites that are located in either exons or introns. This results in the formation of mRNAs that differ in their C-terminal CRs as well as 3′-UTRs that code for different isoforms of proteins. In the figure, UTRs are shown as *green boxes*, exons are *blue boxes*, and the poly A sites are in *red arrows*.

Alternative polyadenylation enables a transcript to be expressed with varying levels of expression and/or also in a tissue or developmental specific manner. Changes in cellular conditions can also change the usage of the poly (A) site, for example, increased proliferation and dedifferentiation are linked with the usage of the proximal poly (A) site [22,23].

Alternative polyadenylation also plays a role in isoform localization. For example, the ELAVL1 (ELAV like RNA binding protein 1) gene encodes a Human antigen R (HuR) protein containing three RNA binding domains and binds the AU-rich elements (ARE) in various mRNA. Its main function is to stabilize the ARE containing mRNAs. Alternative polyadenylation of HuR mRNA results in the formation of many variants that differ from each other in expression level. The prime transcript lacks AREs while a variant contains functional AREs in their 3′-UTR. This results in the stabilization of the variant with ARE and the process results in a self-up-regulation loop [24].

Alternative polyadenylation also results in tissue-specific expression of various proteins by expression of mRNAs that differ in their 3′-UTR. Alternative polyadenylation results in the formation of three isoforms of the mRNA of parathyroid hormone-related protein (PTHrP) encoded from the gene PTHLH. The three isoforms differ in their 3′-UTR and are concerned with the development of humoral hypercalcemia of malignancy, cancer-related syndrome. The expression of the isoforms depends on the cell type. When cells are subjected to cytokine transforming growth factor b1 (TGF-β1), one isoform is stabilized owing to its 3′-UTR (while others were not affected) resulting in TGF-β1 induced increased expression of PTHrP only in the cells that harbor the responsive isoform [25]. Similarly, the calcitonin-related polypeptide-α gene (CALCA) has two transcript isoforms. The isoform that uses proximal poly(A) signal forms an mRNA that encodes for calcitonin protein. It is expressed in the thyroid. The isoform that uses distal poly (A) signal, results in the formation of an mRNA encoding calcitonin gene-related peptide 1 (CGRP) and is expressed in the hypothalamus [26]. A change in concentration of mRNA processing protein and alternative polyadenylation may also lead to diversification of proteins, as seen in B cells. In eukaryotic mRNA molecule, the 3′-end results from an RNA cleavage reaction that is catalyzed by additional factors. The cleavage site can be controlled by the cell to change the C terminus of the resultant protein. For example, during the development of B cells, the switch from membrane-bound antibody synthesis to secreted antibody synthesis. In naïve B cells, the membrane-bound antibody is produced, which is anchored in the plasma membrane and serves as a receptor for antigen. Upon antigenic stimulation, the B lymphocytes multiply and start secreting the antibodies. The membrane-bound and secreted antibodies differ only in their C-terminus. The membrane-bound antibody has a long stretch of hydrophobic amino acid that traverses the lipid membrane while the secreted form has a relatively short hydrophobic amino acid stretch. A change in the site of cleavage of mRNA results in the formation of two different nucleotide sequences with different 3′-ends of the mRNA. The varied length of the nucleotide sequence is formed by the change in the concentration of CstF protein that binds the GU rich sequence of mRNA cleavage and poly-A addition site. In an unstimulated B cell, the concentration of CstF is low and the first cleavage-poly-A addition site is suboptimal and is skipped to form a longer transcript. When the B cells get activated the concentration of CstF is increased and cleavage occurs at the suboptimal site and results in a shorter transcript [27].

RNA editing

Some cells can alter the genetic information by making distinct changes to specific nucleotide sequences in an mRNA, after their synthesis, by the process of transcription. This results in the change of nucleotides in RNA that is different from what was transcribed from DNA. In the course of evolution, RNA editing is one of the most conserved phenomenons that occur in all living organisms. The most drastic form of editing is seen in mitochondrial mRNAs of trypanosomes, where one or more U are inserted in the mRNA that results in the change in the coding frame, the sequence, and overall message in the mRNAs. RNA editing can result from insertions, deletions, and substitution of nucleotides [28,29]. In eucaryotic mRNAs, the most common and widespread form of editing is the base modification of adenosine (A) to inosine (I). In addition to changing the codes of mRNA, tRNA, and rRNA, it is also known to change the bases in UTRs and other noncoding RNAs.

The editing of RNA can be broadly classified into two types: (a) editing through addition or deletion—this type of editing is mainly found in kinetoplast (mitochondrial DNA) of *Trypanosoma brucei* [30]. The unedited primary transcript is base-paired with guide RNA (gRNA). gRNA is complementary to the region of insertion or deletion and thus hybridizes with it. This is surrounded by editosome (multiprotein complex that catalyzes editing) [31,32]. Editosome comprises approx. 20 proteins, an endonuclease, a terminal uridyl transferase (TUTase), 3′-5′ exonuclease specific to U (exoUase), and RNA ligase. The editosome causes the opening of the transcript, cuts it endonucleolytically, and starts incorporating uridines, till there is A on the guide RNA [33]. The insertion will stop as soon as other bases are encountered. The incorporation of uridines is catalyzed by terminal uridyltransferase. Deletion of uridine is mediated by exoUase that works along with 3′-phosphatase, so that ligation can occur by ligase enzyme, in the newly edited RNA [34]. The added U is retained only if it is complementary to the guide RNA otherwise it is removed (Fig. 2.6). (b) Deamination that involves C to U editing and A to I editing. In the C to U editing process deamination of cytidine base to uracil base occurs with the help of enzyme cytidine deaminase. In humans, the apolipoprotein B gene is expressed as two isoforms. In the liver apoB100 and in the intestine apo48 is expressed. In the intestine, editing results in changing of CAA codon to UAA, which is a stop codon, thus resulting in early termination of the mRNA and formation of apoB48 [35]. Approx. 90% of the editing that occurs in the cell is A to I editing. The deamination is catalyzed by adenosine deaminase (ADAR), which acts specifically on double-stranded RNAs. ADARs recognize the double-stranded structure formed between the site of mRNA to be edited and a complementary sequence present somewhere else (typically in 3′-intron) in the same RNA molecule. An example of A to I editing is the editing that occurs in the mRNA that codes for a transmitter gated ion channel in the brain. The editing results in the change of glutamic acid to arginine that results in the altered Ca^{2+} ion permeability. The absence of editing results in the mutant mice that are more prone to epileptic seizures [36]. Many theories are explaining the existence of editing. One theory is that it arose in the genome to correct the mistakes. The second is that the editing enables the cell to form different proteins from the same gene. Third is that it evolved originally as a defense system against retroviruses and retrotransposons, for example, after infection, HIV and other retrovirus get edited extensively, which creates many deleterious effects in their genome.

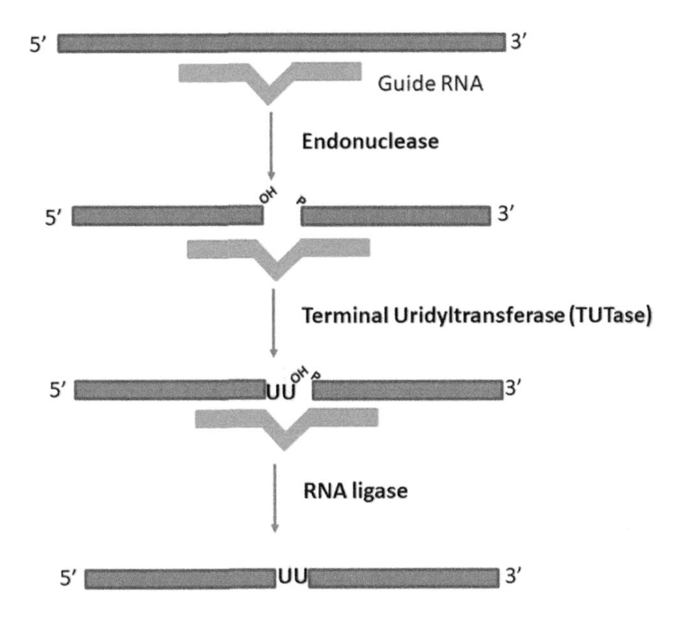

FIGURE 2.6

RNA editing: The unedited primary transcript is base paired with guide RNA (gRNA), which is complementary to the region of insertion or deletion. This is surrounded by editosome comprising approx. 20 proteins, an endonuclease, a terminal uridyl transferase (TUTase), 3′-5′-exonuclease specific to U (exoUase) and RNA ligase. The editosome causes the opening of the transcript, cuts it endonucleolytically, and starts incorporating uridines, till there is A on the guide RNA. The ligase enzyme catalyzes the formation of a phosphodiester bond between the last U and 5′ P of the downstream fragment.

Alternative splicing

Alternative splicing is another mechanism that gives rise to various isoforms of proteins from a single mRNA, that differ from each other only in specific domains. In humans, approximately, 95% of genes undergo splicing [37]. This change can affect the function of a protein, its localization, and enzymatic activity. If such changes occur in the protein that are transcription factors, ion channels, or any regulatory protein, then the biological significance of such changes is very crucial. Alternative splicing can quantitatively regulate the gene expression by producing prematurely truncated protein, by producing mRNA that differs in their UTRs thus regulating the stability of mRNA, or affecting the translational efficiency. In constitutive splicing removal of the intron is followed by ligation of exons in the same order as they appear in the gene. In alternative splicing, there is a deviation from the preferred sequence. This occurs due to weaker splicing signals, shorter length of exons, or high conservation of sequence that surrounds orthologous alternative exons [38].

Seven main types of alternative splicing have been discovered so far (Fig. 2.7). In vertebrates and invertebrates, cassette-type alternative exons, while, in lower metazoans, intron retention is most prevalent. During the process of splicing, assembly of spliceosome occur at mRNA. A typical intron in eukaryotes has GU at its 5′-end and AG at the 3′-end. In higher eukaryotes, a polypyrimidine tract is

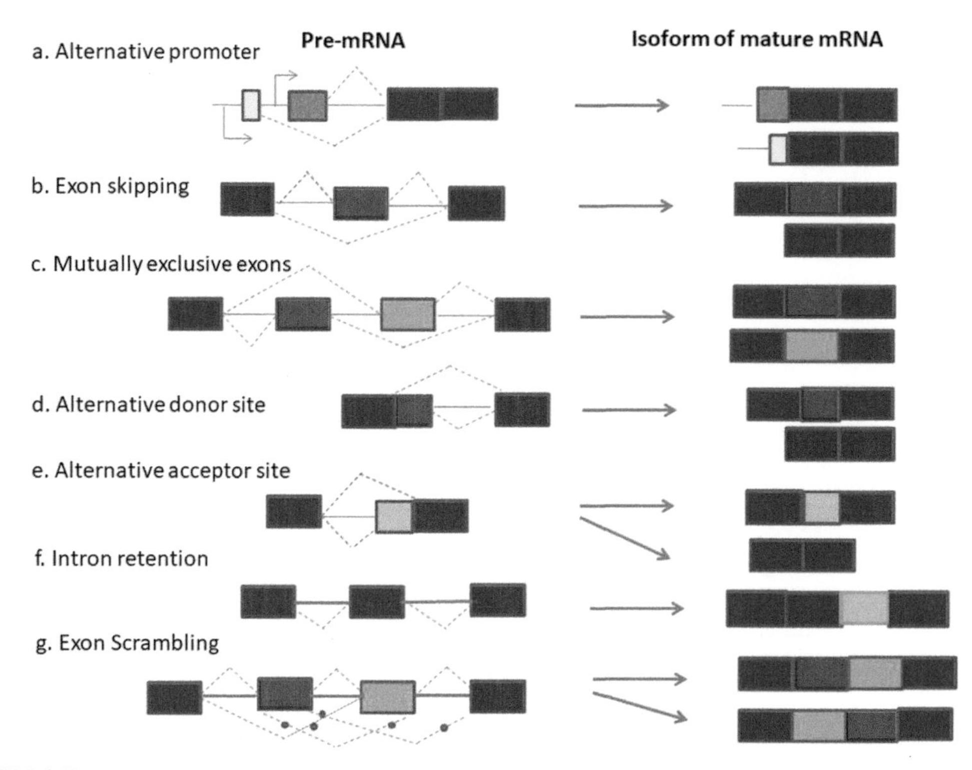

FIGURE 2.7

Common mechanisms and types of alternative splicing: Alternative splicing results in different transcripts by different processes and results in the formation of various mRNAs with different coding regions and ultimately different isoforms of proteins. Various factors like splicing enhancers and silencers bind to specific sites of mRNAs and result in alternative splicing. Purple color shows constitutive exon, blue or red color depicts alternatively spliced exons, retained introns have been shown in light green color.

present just upstream to the 3′-end of the intron sequence. The splicing pathway can be divided into two steps (Fig. 2.8): (A) The hydroxyl group at 2′ C of adenosine in intron sequence attacks the 5′ splice site that results in the cleavage of a phosphodiester bond at the 5′ splice site by a trans-esterification reaction. This results in the formation of a new 5′-2′ phosphodiester bond linking the first nucleotide of the intron with internal adenosine. (B) The 3′-OH group attached to the end of the upstream exon attacks the phosphodiester bond at the 3′-splice site resulting in its cleavage and release of intron as a lariat structure. Simultaneously, the 3′-end of the upstream exon is joined to the 5′-end of the downstream exon. sn RNAs namely, U1, U2, U3, U4, U5, and U6 form the central component of the splicing apparatus. snRNAs are short nucleotide molecules (106–185 nucleotides) in association with proteins. Splice site selection is done by various interactions between cis-acting elements and trans-acting factors, where the cis-acting elements include exonic splicing enhancers (ESEs) and intronic splicing enhancers. Both are bound by SR proteins (positive trans-acting factors). SR proteins are nucleophosphoprotein rich in serine and arginine. Whereas, the exonic splicing silencers (ESSs)

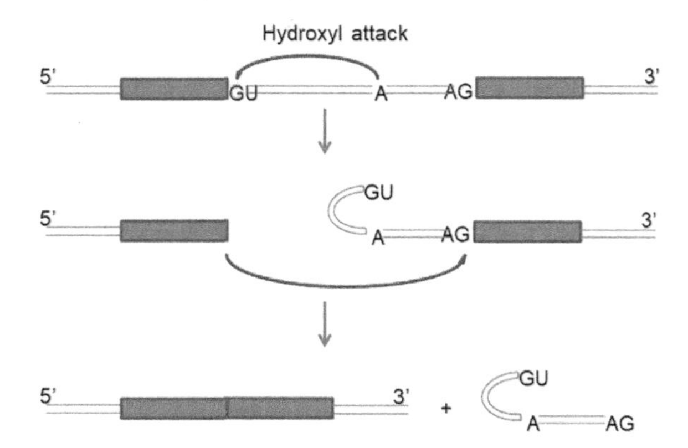

FIGURE 2.8

Mechanism of splicing (outline): The hydroxyl group attached to the 2′ C of the adenosine nucleotide in intron sequence promotes the cleavage of 5′ splice site, this results in the formation of a lariat structure. Next, the 3′-hydroxyl group of upstream exon induces the cleavage of 3′-splice site, enabling the ligation of two exons; the intron is released and then degraded.

and intron splicing silencers are bound by negative trans-acting factors as heterogeneous nuclear ribonucleoproteins (hnRNPs). While the enhancers play a prominent role in constitutive splicing, the silencers are more crucial in the control of alternative splicing [39].

Alternative splicing has been shown to affect various processes that can be grouped as follows:

(a) localization of protein-in immunoglobulin M (Ig μ) in early B lymphocytes, the heavy chain is present as a membrane-bound (Igμm) form, while on matured B cells, on antigenic stimulation the expression of membrane-bound form get decreased while that of the secreted form (Igμs) is augmented. This switch from membrane-bound to secreted form of antibody is achieved by alternative usage of 3′-end exons, which encodes the hydrophobic membrane binding segment [40]. In the secreted form, the termination of transcription before poly(A) signal.

(b) deletion of function: it can result in loss of one or more functions of an alternatively spliced protein product. For example, sxl pre-mRNA of Drosophila undergoes sex-specific alternative splicing. Exon 3 of the sxl gene has a termination codon, in males all the exons including exon 3 are present, this results in the formation of a truncated protein, while in female's exon 3 is skipped resulting in the formation of a fully functional SXL protein. In females, in tra pre mRNA, the binding of SXL protein in the first exon blocks the 3′-splice site. Therefore, the site is not available for binding of U2AF65 and it binds to the exon 2. This results in splicing to a cryptic site in exon 2 and generation of fully functional TRA protein occurs. In males, there is no formation of SXL, so, no binding of SXL to the 3′-splice site of intron 1 occurs and the formation of a dysfunctional mRNA occurs. In females, in dsx mRNA, there exists an exonic splicing enhancer in exon 4 to which binding of TRA protein occurs which in turn stabilizes the attachment of SR proteins. So, exon 4 is not skipped, and mRNA encodes female-specific DSX protein. While, in males, the dysfunctional TRA protein does not bind the exon 4 of the dsx mRNA, and the exon 4 is skipped, resulting in the formation of an mRNA that encodes for male-specific DSX protein (Fig. 2.9).

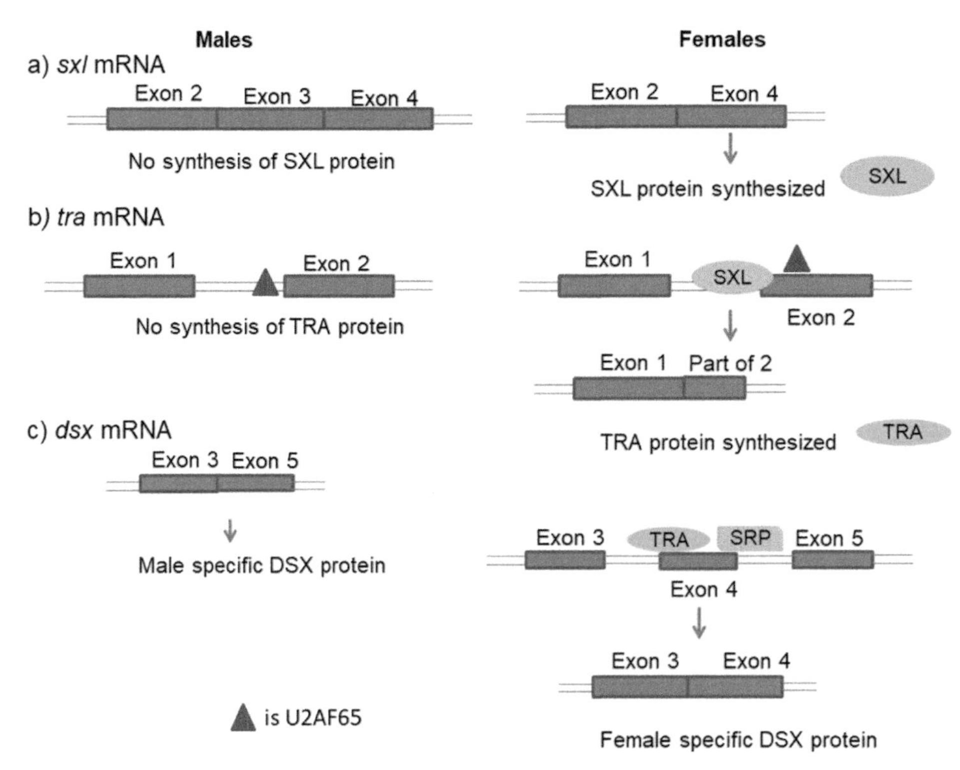

FIGURE 2.9

Alternative splicing in sex determination of Drosophila: Exon 3 of the sxl pre-mRNA contains a termination codon. In males, all the exons are present in the mRNA resulting in the formation of a truncated protein. In females, alternative splicing results in the skipping of exon 3 thereby formation of a fully functional protein occurs in females. In females, SXL protein binds to the 3′-splice site of the first intron of *tra* pre-mRNA and blocks the recruitment of U2AF65 to this site. So U2AF65 bind to exon 2 and directs the splicing to a cryptic site of exon 2 of *tra* pre-mRNA resulting in the formation of functional TRA protein. In males, in the absence of SXL protein 3′-splice site of first intron is not occupied and formation of a dysfunctional TRA protein occurs. In males, exon 4 of *dsx* pre-mRNA is skipped and mRNA that codes for male-specific DSX protein, while in females, TRA protein stabilizes the attachment of SR proteins at exon splicing enhancers in exon 4, so the exon is not skipped and female-specific DSX protein is formed.

Noncoding RNAs

RNA molecules that are not translated into protein are known as noncoding RNA (ncRNA). They include tRNAs, rRNAs, miRNAs, siRNAs, snoRNAs, snRNAs, piRNAs, and long ncRNAs. They can be classified into short noncoding RNAs, which are less than 200 nucleotides, and long noncoding RNAs (lncRNAs) with lengths of more than 200 nucleotides. These RNAs have been known to regulate gene expression, especially siRNAs, miRNAs, and lncRNAs. RNA interference (RNAi) is the process in which sequence-specific suppression of gene expression mediated by RNA molecules occurs. It is controlled by RNA induced silencing complex (RISC). miRNA are approx. 22 nucleotides long small ncRNAs present in metazoan and plants. They are originated from endogenous precursor molecules and then get processed by protein complexes. miRNA is processed from stem-loop structures and its maturation occurs in two steps that are catalyzed by Drosha and Dicer (RNase III enzymes). Drosha has a mol weight of 130–160 kDa and recently it has been shown that Drosha cannot work in isolation, it is present in a complex with double-stranded RNA binding domain protein Pasha (in Drosophila) or DGCR8 (in mammals). In the first step, the primary mRNA transcript (pri-mRNA) is converted to pre-mRNA (approx. 70 nucleotide long) by the action of Pasha/DGCR8 [41]. This step occurs in the nucleus. Then the pre-mRNA is transported to the cytoplasm where it is acted upon by the Dicer enzyme that results in the formation of approx. 22 nucleotides mi-RNA molecules. In mammals, siRNAs are originated from the cleavage of long double-stranded RNA molecules, which too are acted upon by RNase III endonucleases Dicer. They can also be introduced chemically. Both siRNAs and miRNAs function with RISC and miRNPs respectively, both are ribonucleoprotein complexes. Argonaute proteins are the common component of both complexes and are also found to interact with Dicer [42]. The number of argonaute paralogs differs in various organisms (1 in *S. pombe* and 27 in *C. elegans*). They can be divided into two subfamilies: Ago and Piwi. Humans have four Ago (Ago1-4) and four piwi proteins.

Exogenous double-stranded RNA activates Dicer that binds and cleaves dsRNA to produce approx. 20 nucleotide base pair fragments comprising a 2 nucleotide overhang at 3′end. They are separated into single strands and with the help of the RISC-loading complex (RLC) get integrated into RISC. RLC comprises Dicer and a protein R2D2. R2D2 carries domains that bind to the double-stranded RNA and recognize the terminus of siRNA that is thermodynamically more stable, while, Dicer-2 binds to the less stable strand. Loading of more stable strand occurs at RISC. Guide strand (antisense) base pairs to targeted mRNA and cleaves it. On the other hand, the miRNA-loaded RISC complex scans the mRNAs that are present in the cytoplasm for the potential complementarity, binds their UTR region, and inhibits its translation (Fig. 2.10) [43]. Thus, the binding of miRNAs or siRNAs can modulate the expression of the mRNA in a post-transcriptional manner.

The other class of noncoding RNA includes long noncoding RNAs (lncRNA), these are not well characterized. To date, more than 96,000 human lncRNAs have been listed in the noncoding database. They play a role in the regulation of gene expression at the transcriptional and post-transcriptional levels [44]. LncRNAs have various functions, at the level of transcription, they may serve as transcription signals or serve as a scaffold to recruit various components of the transcription complex, including RNA polymerase II or it may take part in the modification of chromatin structure (Fig. 2.11). For example, lncRNA HOTAIR recruits the polycomb chromatin remodeling complex to the HOXD locus (a subset of homeobox genes that are involved in the developmental process) and mediates its silencing [45]. At the post-transcriptional level, it may act as a scaffold to recruit various proteins to form ribonucleoprotein complexes, comprising lncRNAs and RNA binding proteins (RBPs), which may affect the stability of mRNA, its splicing, stability of protein, and its localization. RBPs contain an RNA recognition motif (RRM) that binds to a specific sequence or secondary structure of transcript. The function of lncRNA at the post-transcriptional level can be categorized as the following:

(a) Regulation of splicing: The process of splicing requires a spliceosome, which is comprised of various proteins and factors required to complete the process. lncRNA can interact with enhancing splicing factor and bring it to the nearby exon to include it in the spliced isoform or it

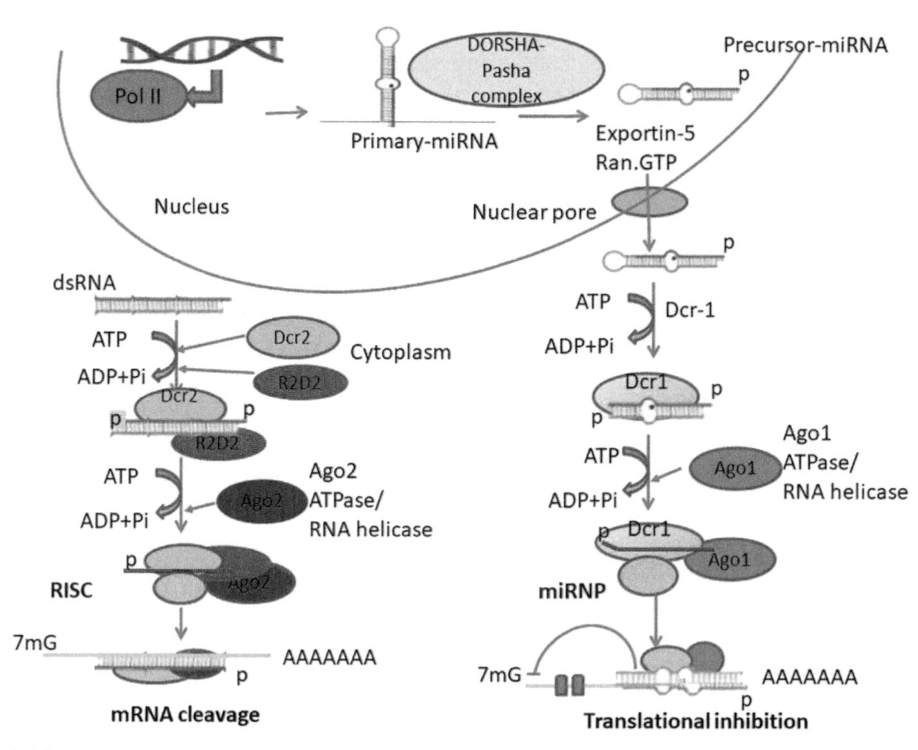

FIGURE 2.10

Pathway of miRNA biogenesis and post-transcriptional regulation by siRNAs and miRNAs: In Drosophila, the stem looped pri-miRNA is acted upon by Drosha-Pasha and results in the formation of pre-miRNA. This is exported out of the nucleus and is acted upon by the enzyme Dicer. Both siRNAs and miRNAs function with RISC and miRNPs, respectively, both are ribonucleoprotein complexes. Argonaute proteins are the common component of both complexes and are also found to interact with Dicer. For details see the text.

can bring silencing splicing factor to the nearby exon to exclude it in the spliced isoform of mRNA. Serine/arginine-rich splicing factor (SRSF1) too plays a crucial role in splicing. lncRNA MALAT1 has been shown to interact with SRSF1, through its RRM domain and is responsible for its localization to nuclear speckles. In nuclear speckles, MALAT1 changes the phosphorylation status of the SRSF1 and modulates the pattern of alternative splicing [46]. Similarly, suppression of MALAT1 results in decreased expression of a splicing factor termed RBFOX2, which promotes the alternative splicing of proapoptotic tumor suppressor gene K1F1B in ovarian cancer. lncRNAs also interact with hnRNP and other proteins that take part in splicing.

(b) Stability of mRNA: Various types of RBPs play a prominent role in determining the stability of mRNAs. lncRNAs have been known to interact with many RBPs to form ribonucleoprotein complexes thereby influencing mRNA stability. For example, lncRNA OCC-1 can interact with HuR and recruit β-TrCP1 (ubiquitin E3 ligase) to it. HuR is an RNA binding protein, it stabilizes

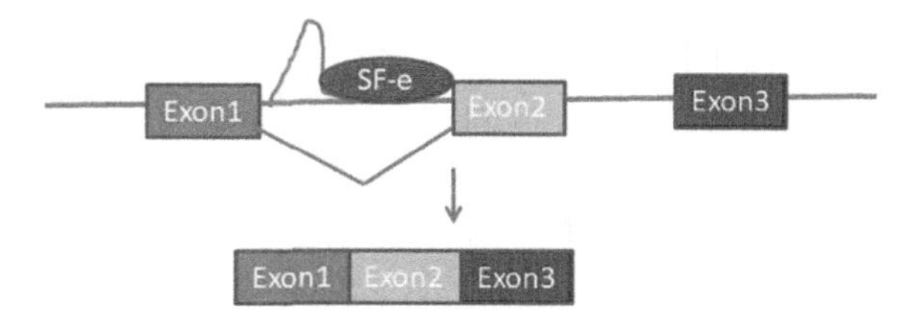

Splicing variant 1

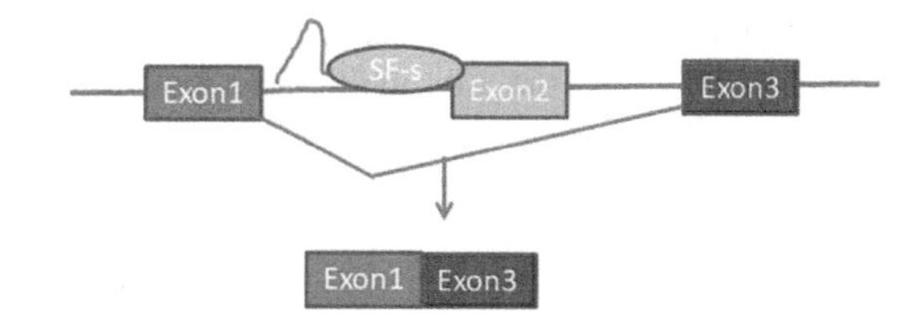

Splicing variant 2

FIGURE 2.11

lnc mediated alternative splicing (simplified model): lncRNA can interact with enhancing splicing factor (SF-e) and promote the inclusion of exon 2 to form a splice variant 1. It can also interact with silencing splicing factor (SF-s) and promote skipping of exon 2 resulting in the formation of splice variant 2. It can also act as a positive or negative regulator: binding of lncRNAs can facilitate or inhibit the interaction/recruitment of RBPs to the 3′-UTR, resulting in stabilization or destabilization of the mRNA. RBPs can be both stabilizing or destabilizing in nature.

RNAs and is also known as RNA stabilizing protein. Recruitment of β-TrCP1 results in proteasomal degradation of HuR and ultimately causes the downregulation of HuR targeted mRNAs [47]. Similarly, Linc-ROR (long intergenic nonprotein coding RNA, regulator of reprogramming) is a lncRNA and has been shown to interact with a heterogeneous nuclear ribonucleoprotein (hnRNPI) and also with AU-rich element RNA-binding protein 1 (AUF1). Both have an opposite effect on c-Myc mRNA. hnRNP1 acts as stabilizing factor while AUF1 acts as a destabilizing factor. In the case of tumorigenesis, hnRNP1 interacts with c-Myc mRNA (a proto-oncogene) with the help of Linc-ROR. On the other hand, the interaction of Linc-ROR with AUF1 renders it from binding to c-Myc mRNA thereby increasing its stability [48].

Transport of mRNA

In mammals, only 5% of mRNA is exported from the nucleus. Localization of mRNA is a common mechanism that is used by the cells to produce a high level of proteins at any specific site. mRNA localization requires specific signals primarily at their 3′-UTRs. The transport of RNA molecules from the nucleus to the cytoplasm is delayed till its processing has been completed. Although, this checkpoint can be overcome by various mechanisms. For example, in HIV infection, the virus directs the formation of a double-stranded DNA copy of the genome that needs to be inserted into the host's genome. Once the insertion of the viral genome in host DNA is complete, the RNA polymerase II of the host cell transcribes the viral DNA. This is followed by extensive splicing of the mRNA to ultimately produce at least 30 types of mRNA molecules that ultimately are translated to various proteins. On the other hand, to make progeny virus, the whole unspliced transcript must be transported from nucleus to cytosol. This unspliced transcript is later packaged into a viral capsid thereby serving as the genome of the virus. The host cell normally blocks the export of the transcript that comprises introns, from the nucleus. To overcome this blockage, the virus encodes a protein known as Rev, which binds to Rev responsive element (RRE), which is located in its intron. The Rev protein interacts with exportin 1, which is a nuclear export receptor, and directs the export of unspliced mRNA from the nucleus to cytoplasm through the nuclear pores. Thus, the viral multiplication continues.

Localization of mRNA to specific regions of cytoplasm before the beginning of translation has been shown to regulate gene expression. There are various advantages of this type of arrangement—it establishes asymmetry in the cytosol that may be crucial in many stages of development; it may also allow the cell to control the expression of genes independently in different parts of the cells [49]. One such example is seen in the Drosophila egg, where, the localization of bicoid gene regulatory protein occurs. It is associated with the cytoskeleton located at the anterior tip of the egg. Translation of this mRNA starts as the process of fertilization occurs and results in the formation of a gradient of bicoid protein. This gradient plays an important role in directing the development of the anterior part of the embryo.

Summary

The phenotype of a cell is determined by its protein and hence the expression of genes is controlled at various stages including at the post-transcriptional level. The regulatory mechanisms include (a) the attenuation process that causes premature termination of the transcripts, (b) the regulatory elements present in the UTRs of the transcripts and their interaction with various factors or proteins, (c) length and structure of the UTRs, (d) RNA editing, (e) alternative splicing and polyadenylation, (f) regulation of mRNA stability and its degradation, and (g) localization and transport of mRNAs. These processes control the expression of genes after the RNA polymerase starts the transcription process. Although the main control of gene expression occurs at the stage of activation or repression of initiation of transcription, the post-transcriptional modification mechanisms control the fine-tuning of gene expression in a spatiotemporal manner and are crucial to all living systems.

References

[1] Ghedira K. Introductory chapter: a brief overview of transcriptional and post-transcriptional regulation. 2018.

[2] Yanofsky C. Attenuation in the control of expression of bacterial operons [Internet, Cited 2022 January 13] Nature February 26, 1981;289(5800). Available from: https://pubmed.ncbi.nlm.nih.gov/7007895/.

[3] Gollnick P. Trp operon and attenuation [internet, Cited 2022 January 13]. In: Lennarz WJ, Lane MD, editors. Encyclopedia of biological chemistry. New York: Elsevier; 2004. p. 267–71. Available from: https://www.sciencedirect.com/science/article/pii/B0124437109002477.

[4] Banerjee AK. 5′-terminal cap structure in eucaryotic messenger ribonucleic acids. Microbiol Rev June 1980; 44(2):175−205.

[5] Kochetov AV, Ischenko IV, Vorobiev DG, Kel AE, Babenko VN, Kisselev LL, et al. Eukaryotic mRNAs encoding abundant and scarce proteins are statistically dissimilar in many structural features. FEBS Lett December 4, 1998;440(3):351−5.

[6] McClelland S, Shrivastava R, Medh JD. Regulation of translational efficiency by disparate 5′ UTRs of PPARgamma splice variants. PPAR Res 2009;2009:193413.

[7] Beaudoin J-D, Perreault J-P. 5′-UTR G-quadruplex structures acting as translational repressors. Nucleic Acids Res November 2010;38(20):7022−36.

[8] Gomez D, Guédin A, Mergny J-L, Salles B, Riou J-F, Teulade-Fichou M-P, et al. A G-quadruplex structure within the 5′-UTR of TRF2 mRNA represses translation in human cells. Nucleic Acids Res November 2010; 38(20):7187−98.

[9] Smith L, Brannan RA, Hanby AM, Shaaban AM, Verghese ET, Peter MB, Pollock S, Satheesha S, Szynkiewicz M, Speirs V, Hughes TA. Differential regulation of oestrogen receptor β isoforms by 5' untranslated regions in cancer. J Cell Mol Med 2010;14(8):2172−84. https://doi.org/10.1111/j.1582-4934.2009.00867.x.

[10] Meijer HA, Thomas AAM. Control of eukaryotic protein synthesis by upstream open reading frames in the 5′-untranslated region of an mRNA. Biochem J October 1, 2002;367(Pt 1):1−11.

[11] Wethmar K. The regulatory potential of upstream open reading frames in eukaryotic gene expression. Wiley Interdiscip Rev RNA December 2014;5(6):765−78.

[12] Calvo SE, Pagliarini DJ, Mootha VK. Upstream open reading frames cause widespread reduction of protein expression and are polymorphic among humans. Proc Natl Acad Sci USA May 5, 2009;106(18):7507−12.

[13] Rapti A, Trangas T, Samiotaki M, Ioannidis P, Dimitriadis E, Meristoudis C, et al. The structure of the 5′-untranslated region of mammalian poly(A) polymerase-alpha mRNA suggests a mechanism of translational regulation. Mol Cell Biochem July 2010;340(1−2):91−6.

[14] Cenik C, Derti A, Mellor JC, Berriz GF, Roth FP. Genome-wide functional analysis of human 5′ untranslated region introns. Genome Biol 2010;11(3):R29.

[15] Bianchi M, Crinelli R, Giacomini E, Carloni E, Magnani M. A potent enhancer element in the 5′-UTR intron is crucial for transcriptional regulation of the human ubiquitin C gene. Gene December 1, 2009;448(1): 88−101.

[16] Gorgoni B, Gray NK. The roles of cytoplasmic poly(A)-binding proteins in regulating gene expression: a developmental perspective. Briefings Funct Genomics Proteomics August 2004;3(2):125−41.

[17] Gallie DR. The cap and poly(A) tail function synergistically to regulate mRNA translational efficiency. Genes Dev November 1991;5(11):2108−16.

[18] Tanguay RL, Gallie DR. Translational efficiency is regulated by the length of the 3′ untranslated region. Mol Cell Biol January 1996;16(1):146−56.

[19] Fukuchi M, Tsuda M. Involvement of the 3′-untranslated region of the brain-derived neurotrophic factor gene in activity-dependent mRNA stabilization. J Neurochem December 2010;115(5):1222−33.

[20] Dickson AM, Wilusz J. Polyadenylation: alternative lifestyles of the A-rich (and famous?). EMBO J May 5, 2010;29(9):1473−4.

[21] Garin I, Edghill EL, Akerman I, Rubio-Cabezas O, Rica I, Locke JM, et al. Recessive mutations in the INS gene result in neonatal diabetes through reduced insulin biosynthesis. Proc Natl Acad Sci USA February 16, 2010;107(7):3105−10.

[22] Ji Z, Tian B. Reprogramming of 3′ untranslated regions of mRNAs by alternative polyadenylation in generation of pluripotent stem cells from different cell types. PLoS One December 23, 2009;4(12):e8419.

[23] Sandberg R, Neilson JR, Sarma A, Sharp PA, Burge CB. Proliferating cells express mRNAs with shortened 3′ untranslated regions and fewer microRNA target sites. Science June 20, 2008;320(5883):1643−7.

[24] Al-Ahmadi W, Al-Ghamdi M, Al-Haj L, Al-Saif M, Khabar KSA. Alternative polyadenylation variants of the RNA binding protein, HuR: abundance, role of AU-rich elements and auto-regulation. Nucleic Acids Res June 2009;37(11):3612–24.
[25] Sellers RS, Luchin AI, Richard V, Brena RM, Lima D, Rosol TJ. Alternative splicing of parathyroid hormone-related protein mRNA: expression and stability. J Mol Endocrinol August 2004;33(1):227–41.
[26] Amara SG, Jonas V, Rosenfeld MG, Ong ES, Evans RM. Alternative RNA processing in calcitonin gene expression generates mRNAs encoding different polypeptide products. Nature July 15, 1982;298(5871): 240–4.
[27] Takagaki Y, Seipelt RL, Peterson ML, Manley JL. The polyadenylation factor CstF-64 regulates alternative processing of IgM heavy chain pre-mRNA during B cell differentiation. Cell November 29, 1996;87(5): 941–52.
[28] Farajollahi S, Maas S. Molecular diversity through RNA editing: a balancing act. Trends Genet TIG May 2010;26(5):221–30.
[29] Nishikura K. Functions and regulation of RNA editing by ADAR deaminases. Annu Rev Biochem 2010;79: 321–49.
[30] Benne R. RNA editing in trypanosomes. Eur J Biochem April 1, 1994;221(1):9–23.
[31] Alfonzo JD, Thiemann O, Simpson L. The mechanism of U insertion/deletion RNA editing in kinetoplastid mitochondria. Nucleic Acids Res October 1, 1997;25(19):3751–9.
[32] Arts GJ, Benne R. Mechanism and evolution of RNA editing in kinetoplastida. Biochim Biophys Acta June 3, 1996;1307(1):39–54.
[33] Blum B, Bakalara N, Simpson L. A model for RNA editing in kinetoplastid mitochondria: "guide" RNA molecules transcribed from maxicircle DNA provide the edited information. Cell January 26, 1990;60(2): 189–98.
[34] Stuart K. RNA editing in mitochondrial mRNA of trypanosomatids. Trends Biochem Sci February 1991; 16(2):68–72.
[35] Johnson DF, Poksay KS, Innerarity TL. The mechanism for apo-B mRNA editing is deamination. Biochem Biophys Res Commun September 30, 1993;195(3):1204–10.
[36] Zaidan H, Ramaswami G, Golumbic YN, Sher N, Malik A, Barak M, et al. A-to-I RNA editing in the rat brain is age-dependent, region-specific and sensitive to environmental stress across generations. BMC Genom January 8, 2018;19(1):28.
[37] Nilsen TW, Graveley BR. Expansion of the eukaryotic proteome by alternative splicing. Nature January 28, 2010;463(7280):457–63.
[38] Zheng CL, Fu X-D, Gribskov M. Characteristics and regulatory elements defining constitutive splicing and different modes of alternative splicing in human and mouse. RNA December 2005;11(12):1777–87.
[39] Wang Z, Burge CB. Splicing regulation: from a parts list of regulatory elements to an integrated splicing code. RNA May 2008;14(5):802–13.
[40] Rogers J, Early P, Carter C, Calame K, Bond M, Hood L, et al. Two mRNAs with different 3′ ends encode membrane-bound and secreted forms of immunoglobulin mu chain. Cell June 1980;20(2):303–12.
[41] Han J, Lee Y, Yeom K-H, Kim Y-K, Jin H, Kim VN. The Drosha-DGCR8 complex in primary microRNA processing. Genes Dev December 15, 2004;18(24):3016–27.
[42] Meister G, Tuschl T. Mechanisms of gene silencing by double-stranded RNA. Nature September 16, 2004; 431(7006):343–9.
[43] Nakanishi K. Anatomy of RISC: how do small RNAs and chaperones activate Argonaute proteins? Wiley Interdiscip Rev RNA September 2016;7(5):637–60.
[44] Kornienko AE, Guenzl PM, Barlow DP, Pauler FM. Gene regulation by the act of long non-coding RNA transcription. BMC Biol May 30, 2013;11:59.

[45] Wang KC, Chang HY. Molecular mechanisms of long noncoding RNAs. Mol Cell September 16, 2011; 43(6):904–14.

[46] Tripathi V, Ellis JD, Shen Z, Song DY, Pan Q, Watt AT, et al. The nuclear-retained noncoding RNA MALAT1 regulates alternative splicing by modulating SR splicing factor phosphorylation. Mol Cell September 24, 2010;39(6):925–38.

[47] Lan Y, Xiao X, He Z, Luo Y, Wu C, Li L, et al. Long noncoding RNA OCC-1 suppresses cell growth through destabilizing HuR protein in colorectal cancer. Nucleic Acids Res June 20, 2018;46(11):5809–21.

[48] Huang J, Zhang A, Ho T-T, Zhang Z, Zhou N, Ding X, et al. Linc-RoR promotes c-Myc expression through hnRNP I and AUF1. Nucleic Acids Res April 20, 2016;44(7):3059–69.

[49] Kindler S, Wang H, Richter D, Tiedge H. RNA transport and local control of translation. Annu Rev Cell Dev Biol 2005;21:223–45.

CHAPTER

3

Methods to study post-transcriptional regulation of gene expression

Tapan Sharma

Department of Biochemistry and Molecular Pharmacology, University of Massachusetts Medical School, Worcester, MA, United States

Introduction

The central dogma of molecular biology was proposed by Sir Francis Crick in 1958 [1,2]. It was an explanation of how genetic information encoded in DNA is passed on to functional components called proteins with an intermediate messenger stage called RNA (Fig. 3.1). With the advancement in the field of molecular biology, this understanding has developed much beyond this simplistic model. It is now known that not all DNA encodes for proteins [3,4]. Only about 2% of the information in the genome of a eukaryotic organism is eventually converted to a translated protein sequence. The remaining 98% was initially considered as "junk DNA." With the advent of high-throughput techniques, it was later discovered that nearly 80% of this junk is either actively transcribed to noncoding RNA (ncRNA) or is associated with histone modifications and chromatin structure [5]. This noncoding but biochemically active component of a eukaryotic genome plays a crucial role in the regulation of gene expression.

Regulation of gene expression at the transcriptional level is a well-studied field and can be explored in detail here [4,6,7]. This book delves into the regulation of gene expression beyond transcriptional control and previous chapters have provided a detailed understanding of the mechanisms. In this chapter, we will review experimental approaches that allow us to investigate such a mode of regulation. As post-transcriptional regulation of gene expression can affect aspects such as mRNA stability, processing, splicing, and storage, we will explore methods that can address such findings. We will also look at methods to study RNA–protein interactions specifically considering RNA-binding proteins (RBP) and their role in mRNA transport and turnover. Lastly, we will explore techniques to study microRNA-mediated regulation of gene expression.

Methods for determining mRNA stability

The first form of ribonucleic acid generated by transcribing the genetic information from DNA is called pre-mRNA (Fig. 3.2). A pre-mRNA undergoes necessary processing to be utilized by ribosomes

Post-Transcriptional Gene Regulation in Human Disease, Volume 32. https://doi.org/10.1016/B978-0-323-91305-8.00014-4

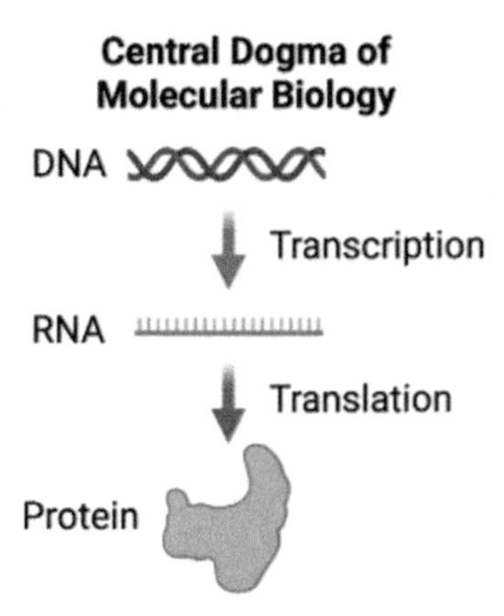

FIGURE 3.1

The flow of genetically encoded information from DNA to RNA to Proteins popularly known as the Central Dogma of Molecular Biology.

to generate proteins. This includes capping the pre-mRNA at the 5′-end, tailing it with poly-A at the 3′-end, and splicing out the introns to produce a continuous reading frame of exons. The mature mRNA resulting from this process is utilized by ribosomal machinery to generate amino acid polymers called peptides.

The regulation of mRNA turnover is defined as a bunch of steps determining the abundance of messenger transcripts that ultimately result in high/low protein expression [8]. The number of copies of a particular mRNA in any given cell at any given time is tightly regulated. The study of mRNA steady-state levels in a cell is determined in terms of its half-life. Half-life, as the name suggests, is defined as the time required to degrade half of the existing mRNA transcripts in the cell [9]. This is an important parameter that is experimentally determined by methods that will be described in the following section. For an understanding of this concept, one can look at the steady-state of transcripts from histone genes. Histone mRNAs are transcribed in all phases of the cell cycle. Histone protein expression, however, occurs strictly during the S-phase [10]. This is enabled by post-transcriptional gene regulation of histone transcripts affecting their steady-state. During the early S-phase, the efficiency of histone pre-mRNA processing goes up which results in a half-life of up to an hour. At the end of the S-phase, histone transcript half-life goes back to basal levels of less than 10 min.

mRNA half-life is calculated using a standard time-course experiment where samples are collected at multiple intervals and the concentration of a bunch of mRNAs is calculated [8,9]. The rate of mRNA decay is computed using the following equation:

$$dC/dt = -k_{decay}C$$

where dC/dt is the rate of decay of any mRNA during a time t, C is the total concentration of any mRNA, and k_{decay} is the decay constant. The negative sign associated with k shows that mRNA is being degraded. To study mRNA half-life, various experimental approaches have been standardized (Table 3.1).

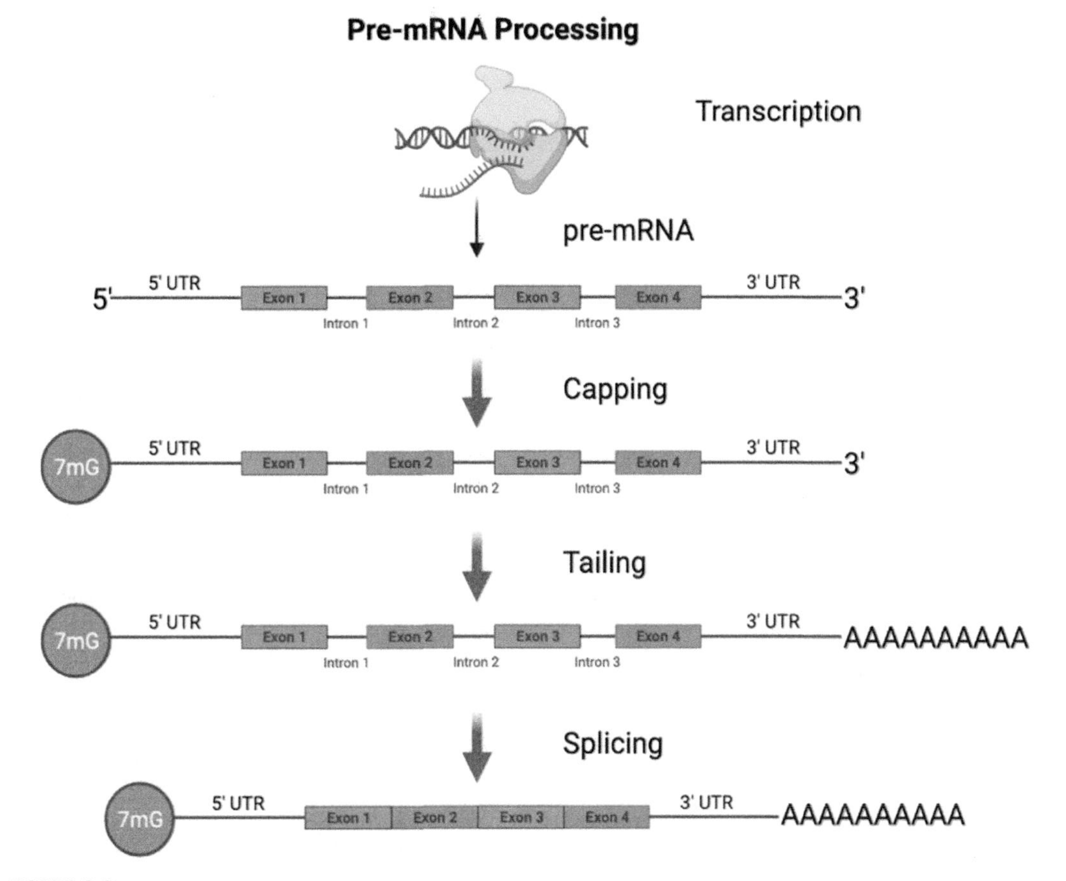

FIGURE 3.2

Pre-mRNA generated after transcription undergoes a few processing steps to be made useable by the translational machinery. These include addition of a 7-methyl Guanosine cap on the 5′ end, polyadenylation at the 3′ end and splicing of noncoding introns to generate a mature uninterrupted transcript.

Inhibition of transcription

One of the classic methods to study mRNA turnover and stability is to temporarily shut down all cellular transcription and monitor the decrease in mRNA copies of a particular kind [8,9]. This can be done using small molecule inhibitors of RNA polymerase. Classically, inhibitors are added to cell culture media for the duration of the experiment. Samples are collected at different time intervals post addition to the inhibitor and mRNA stability is determined. The choice of the inhibitor can vary according to the requirement and specificity of the experiment [11].

Table 3.1 Methods to study mRNA turnover.

mRNA half-life and turnover rate			
Approach	**Method**	**Mode of action**	**Features**
Transcriptional inhibition using small molecules	Actinomycin D (5−10 μg/mL)	Inhibits elongating RNA polymerases	Pros: Fast action; works on all classes of all RNA polymerases; reversible action Cons: Causes DNA breaks, prolonged cytotoxicity after removal
	α-amanitin (2 μg/mL)	Blocks nucleotide incorporation by binding to RNA pol II catalytic site	Pros: Specific to RNA pol II; inhibits RNA pol III at a hundred-fold lower concentration Cons: Irreversible inhibition
	DRB (20 μg/mL)	Inhibits CDK9 subunit of p-TEFb complex	Pros: Specific for RNA pol II; reversible action; fast action Cons: Low solubility allows usage at high concentrations
	Flavopiridol (<1 μg/mL)	Inhibits CDK9 subunit of p-TEFb complex	Pros: Effective at low concentrations Fast action; reversible action Cons: Can inhibit other CDKs
Inhibition using temperature-sensitive alleles	Rpb1 gene	Subunit of RNA pol II	Pros: Avoids use of chemical inhibitors that can be cytotoxic Cons: Genetic manipulation required
Metabolic labeling using labeled nucleotides	Radioactive isotopes	[^{3}H]-adenine or [^{32}P]-phosphate-labelled	Pros: Pulse chasing after prelabeling; approach to equilibrium without prelabeling Cons: 4′-thiouridine can inhibit the production of ribosomal RNAs
	Nonradioactive labels	Bromouracil (BrU) or 4′-thiouridine	
Use of inducible promoters	*c-fos* promoter	Serum-induced	Pros: Rapid and transient system Cons: Noncompatible with some cells
	TET-off/on promoters	Tetracycline-induced	Pros: Universal application in cell lines and animal models
	GAL promoter	Galactose-induced	Pros: Universal application in cell lines and animal models
	HSP promoter	Induced by heat shock	Pros: Universal application in cell lines and animal models

Actinomycin D

Actinomycin D is one of the most used RNA polymerase inhibitors. It is known to intercalate into actively transcribed DNA that causes inhibition of transcriptional elongation [12]. Actinomycin D inhibits all classes of RNA polymerases in the concentration range of 5−10 μg/mL. It can cause double-stranded breaks in DNA and thus can cause prolonged cytotoxicity even after it is removed from the culture medium. The specificity of inhibition follows the order: RNA pol I > RNA pol II > RNA pol III.

α-Amanitin

α-Amanitin is a compound purified from a mushroom of the *Amanita* genus [13]. It is highly specific to RNA polymerase II and can bind close to its active catalytic site to inhibit transcription by blocking nucleotide incorporation. The mode of inhibition is irreversible because the binding of α-amanitin to the polymerase causes it to be ubiquitinated and proteasomally degraded. The optimum concentration for usage is 2 μg/mL. RNA pol II is highly sensitive to α-amanitin while the sensitivity of RNA pol III is about a hundred-fold lower. RNA pol I is not inhibited by α-amanitin.

DRB

DRB (5,6-dichloro-1β-1-ribofuranosyl-benzimidazole) is an inhibitor of CDK9, a component of the positive transcription elongation factor (p-TEFb) complex that is responsible for overcoming transcriptional pausing of RNA polymerase II [11]. This makes DRB a specific candidate to inhibit RNA polymerase II transcription. It is used at a concentration of 20 μg/mL. The advantages of using DRB are that RNA polymerase II transcription is arrested within minutes of adding it to the culture medium. The inhibition is reversible because efficient transcription resumes within minutes once DRB is removed.

Flavopiridol

Flavopiridol is another CDK9 inhibitor that is known to block RNA polymerase II transcription [11]. The advantage of using flavopiridol is that it works at a hundred-fold lower concentration than DRB. The inhibition occurs in a matter of minutes and is reversible. A drawback is that it can also inhibit other CDKs such as CDK1 and CDK4.

Temperature-sensitive alleles

Another method of inhibiting transcription is by using temperature-sensitive alleles of genes encoding one or more subunits of RNA polymerase II. One such example is the use of a mutant version of the Rpb1 gene identified in the Syrian hamster cell line, BHK21 [14]. Mutation in one nucleotide of this gene at position 944 results in a protein that has a modified amino acid residue (A → D). This mutation makes Rpb1 expression temperature sensitive. The use of this system can be done by allowing cells to grow at temperatures nonpermissive for mutant Rpb1 resulting in inhibition of transcription during which samples can be harvested for evaluation of mRNA turnover.

Metabolic labeling

Another approach to study mRNA turnover rate is by in vivo metabolic labeling of cellular transcripts. In earlier days, labeling was done by providing radioactive nucleotides (containing [^{3}H]-adenine or [^{32}P]-phosphate) to cells in culture for a given duration followed by their removal. This is known as **pulse chasing** as the cells are "chased" after radioactive labeling by allowing them to continue transcription in the presence of natural nucleotides [9]. This causes replacement of radioactively labeled transcripts by unlabeled transcripts and such decay could be evaluated over a duration of time by using northern blotting techniques to determine mRNA half-life.

Another version of a similar experiment can be performed by using the **approach to equilibrium** method [8,9]. This method works opposite to the pulse-chase method. Radioactive nucleotides are provided to cells and the increase in radioactivity is monitored over a duration of time until equilibrium is reached.

The advent of nonradioactive nucleoside analogs has allowed for them to replacing.

conventional radioactivity-based labeling techniques. Two such nucleoside derivatives are bromouracil (BrU) and 4′-thiouridine [8,9]. They can be utilized similarly as described earlier. BrU is less cytotoxic than 4′-thiouridine as the latter can inhibit the production of ribosomal RNAs.

Use of inducible promoters

If a study aims to determine mRNA turnover and stability of a particular gene/transcript, the use of regulatable promoters is an approach of choice. This technique has two distinct advantages over the use of small-molecule inhibitors or metabolites in a cell culture medium. Firstly, the experiment can be performed under physiologically normal conditions. The addition of external inhibitors and metabolites can lead to cytotoxicity. Secondly, it makes the study of a gene of interest possible without affecting the turnover of all other transcripts in the cell. Thus, indirect effects of pan-inhibition of cellular transcription can be avoided.

Serum-inducible c-fos *promoter system*

In eukaryotic systems, few inducible promoters have been extensively used to study mRNA turnover. The *c-fos* promoter is induced in response to serum addition rapidly and transiently [15]. To utilize this system, the gene of interest is cloned under the control of a *c-fos* promoter and is transiently or stably transfected in cell lines of interest. To prime the system, target cells are cultured briefly (12–18 h) in the absence/low concentration of serum (0.5%–2%) in culture media. To initiate transcription, culture media containing 10%–20% serum is added to the cells, and samples are harvested at regular intervals. Specific RNA isolated from cells can be quantified using qPCR or hybridization-based probing techniques.

The use of a serum-inducible system can have disadvantages when studying certain cell types that are sensitive to low serum. For example, skeletal muscle differentiation is induced when cells are cultured at low serum concentrations. The *c-fos* promoter system might also render useless for some transformed cell lines that are insensitive to serum deprivation. Due to these disadvantages, other inducible promoter systems have been developed.

TET-off promoter

The tetracycline (TET) promoter is derived from bacterial cells and controls the expression of a tetracycline-resistance gene [8,16,17]. The activity of the promoter is controlled by the expression of another transactivator gene product, tTA, which binds to tetracycline-responsive elements (TRE) in the TET promoter and is positively stimulated in the presence of tetracycline. To make use of this system in eukaryotes, the gene of interest is cloned under the control of a minimal promoter. A stretch of TRE repeats is placed upstream of the minimal promoter. A wild-type version of tetracycline controlled transactivator, tTA, or a mutated reverse tetracycline controlled transactivator, rtTA (works as a repressor) is simultaneously expressed in the system that makes it either TET-off or Tet-on promoter.

To use the TET-off promoter system, tetracycline is added to the cells to turn off transcription for the gene of interest, and a decrease in transcript abundance is monitored. On the other hand, to use the TET-on system, the addition of tetracycline allows an increase in transcription and can be followed in an "approach to equilibrium" manner.

GAL and HSP promoters

Galactose-dependent promoters were identified in yeast and are sensitive to the presence of either galactose (activating) or glucose (repressing). Similarly, promoters of heat shock genes (HSP) are sensitive to temperature and can be induced by temperature. The measurement of mRNA half-life using the GAL or the HSP promoters is done like the TET-promoter systems [8,9].

Understanding the results from mRNA turnover studies

All methods described earlier provide information about the rate of decay of mRNA transcripts specific to the interest of the investigator. Once the mRNA half-life is determined, the next step is to assess what mechanism related to mRNA stability could be causing its differential decay as compared to other cellular transcripts.

There are a few characteristic features of a eukaryotic pre-mRNA that impart its stability by increasing/decreasing its susceptibility to ribonucleases in the cell. The structure of a pre-mRNA is shown in Fig. 3.2. One of the earliest post-transcriptional processing steps performed on a pre-mRNA is capping of its 5′-end [7]. This modification is performed in a series of steps by the action of a few enzymes. Firstly, a phosphatase removes a γ-phosphate from the triphosphate at the 5′-end of a nascent pre-mRNA. To the free 5′-biphosphate group, the guanylyltransferase enzyme catalyzes the binding of GMP in an unusual triphosphate linkage. Lastly, the guanine 7-methyl transferase enzyme transfers a methyl group to the seventh position of the 5′-guanine using S-adenosyl-L-methionine as the donor to form a 7-methyl guanosine capped mRNA.

The 7-methyl guanosine (7 mG) cap at the 5′-end is crucial for several reasons [7,18]. First and foremost, it protects the mRNA from degradation by 5′- to 3′-exonucleases. The 7 mG cap interacts with factors that can transport the mRNA from the nucleus to the cytoplasm to be used by translation machinery. Thirdly, translation initiation factors recognize the 7 mG cap and load it to the ribosomal machinery to enable efficient translation thus making capping a key step of post-transcriptional regulation.

Apart from capping at the 5′-end, a pre-mRNA is also modified at the 3′-end [7]. Once the RNA polymerase reaches a transcription termination signal, an enzyme called poly-A polymerase adds adenosine monophosphates to the 3′-end. Usually, up to 200–300 adenosines are added to the 3′-tail. Studies have shown that the length of the 3′-poly-A tail is directly proportional to its stability. The primary function of the 3′-tailing is to protect the nascent mRNA from 3′- to 5′-exonucleases in addition to contributing to cytoplasmic transport and translational efficiency.

All these modifications contribute to the regulation of gene expression at the post-transcriptional level. Therefore, once mRNA half-life is determined, investigators look at the regulation of these steps to pinpoint the reasons behind it. This can be done by evaluating the gene expression of exonucleases or enzymes involved in capping or polyadenylation. It should however be understood that any such change would affect the stability of all cellular mRNA indifferently. Other more specific modes of post-transcriptional regulation are discussed later.

RNA-binding proteins

In the 1970s, a novel class of proteins called RBP was identified [19–21]. These proteins could interact with single- or double-stranded RNA to form ribonucleoprotein complexes. Due to this ability,

RBPs play a crucial role in the post-transcriptional regulation of gene expression. RBPs may contain sequence-specific, structure-specific domains called RNA-binding domains that allow them to interact with both protein-coding mRNAs and ncRNAs.

Interactions between RNA-binding proteins and coding RNA

Let us begin this section by looking at the example of the 3′-poly-A tailing of mRNA [20,21]. As discussed above, poly-A polymerase adds stretches of adenosines to the nascent mRNA resulting in a 200−300 long poly-A tail. RBPs called poly-A binding proteins (PABP) coat the poly-A tail with an approximate frequency of one protomer every 27 nucleotides. Thus the 3′-end of a nascent mRNA is a ribonucleoprotein complex containing a poly-A tract bound by PABP. The formation of this complex imparts stability, allows transport to the cytoplasm, and contributes to the initiation of translation.

Apart from the 7 mG cap and poly-A tail, mRNA also consists of a 5′- and 3′-untranslated region (UTR), coding sequences called exons, and interrupting noncoding stretches called introns. The generation of mature mRNAs suitable to be used by the translational apparatus requires one more processing step apart from capping and tailing. This step called splicing removes the introns to generate an uninterrupted sequence of exons called coding sequence [22,23]. In certain cases, one or more exons can be skipped, or an intron could be retained during processing. Such an event is called alternative splicing where the same pre-mRNA can result in the generation of different mature mRNAs and is enabled by RBPs called splicing factors. Both splicing and alternative splicing, therefore, contribute to post-transcriptional regulation of gene expression.

The 5′- and 3′-UTRs have also been reported to regulate gene expression [18,24]. They do so due to the presence of secondary structures that could be bound by RBPs in a structure-specific manner. The 5′-UTR can form hairpins or G-quadruplexes that can be bound by specific RBPs, and these interactions can contribute to mRNA stability and translational efficiency. Internal ribosome entry site is one such element that allows for unconventional binding of the translational machinery. Similarly, the 3′-UTR in most mRNAs has an AU-rich element (AURE) that is a variable sequence of 5−9 nucleotides. AUREs exist in the form of repeats and can be bound by RBPs called AU-binding proteins (AUBP). The presence of AUREs bound by AUBPs can have multiple implications on mRNA stability depending on the transcript, cell type, cell cycle stage, presence of other regulatable elements, etc. Similarly, genes related to iron metabolism usually contain iron-responsive elements (IRE) in their 3′-UTR that can be bound by iron-responsive proteins (IRP). IRP is one of the best examples of feedback regulation to understand the role of RBPs in mRNA stability and post-transcriptional regulation of gene expression [25,26]. IRPs are sensitive to iron concentration in the cell. When iron is scarce, IRP binds to IRE with high affinity and stabilizes gene transcripts related to iron metabolism. When iron concentrations are high, iron binds to IRP and blocks its binding to IRE that in turn allows transcript degradation.

Noncoding RNA and RBPs

As mentioned earlier, protein-coding mRNAs form only about 2% of the genome while 80% of the transcriptome consists of ncRNAs [5,23]. Two popularly known ncRNA types are transfer RNA (tRNA) and ribosomal RNA (rRNA) that play key roles in the process of translation. Splicing factors interact with small nuclear RNA (snRNA) and small nucleolar RNA (snoRNA) to regulate the efficient removal of introns and generation of mature transcripts. Other ncRNA includes microRNA (miRNA), short interfering RNA (siRNA), piwi-interacting RNA (piRNA), and long noncoding RNA (lncRNA), all of which have been characterized in direct or indirect post-transcriptional regulation of gene expression. Regulatory ncRNA can interact with specific protein complexes, like miRNA, siRNA and piRNA can interact with dicer, argonaute family of proteins, and the RNA-induced silencing complex. lncRNA like Xist is known to nonspecifically silence one of the X-chromosomes in females leading to the generation of Barr body.

Methods to study post-transcriptional gene regulation due to RNA–protein interactions

RNA–protein interactions had been studied in the last century with conventional approaches such as RNA immunoprecipitation followed by end-point PCR or mass spectroscopy. The advent of high-throughput techniques such as sequencing and bioinformatics has revolutionized the exploration of the RNA–protein interactome. Some of these methods are discussed later (Table 3.2).

Table 3.2 Methods to study RNA–protein interactions.

Alternative splicing and RNA–protein interactions			
Approach	**Method**	**Application**	**Features**
Identification of splice variants	PCR	End-point PCR	Cons: Qualitative
		Quantitative real-time PCR	Pros: Quantitative cons: Targets are predetermined; limited application
	Microarray or DNA chips	Use of probes against known isoforms	Pros: Screens thousands of genes; low cost. Cons: Low reproducibility; noise; targets are predetermined
	High-throughput deep sequencing	cDNA libraries are prepared and sequenced.	Pros: Unlimited usage; identifies novel transcripts Cons: Costly; skilled data interpretation
RNA–protein interactions	RNA immunoprecipitation (RIP)	Pull down known RBPs	Pros: Captures native interactions Cons: Might miss weak/rare interactions
	Cross-linking and immunoprecipitation (CLIP) Variants: e-CLIP, i-CLIP, PAR-CLIP	UV-crosslinking of native interactions followed by IP	SDS-page of IP'ed RNA–protein complexes RNA reverse transcribed to cDNA can be used for RNA-sequencing-based evaluation.
	RNA-affinity capture	Using tagged RNA as bait to capture novel RBPs	Captured RBPs can be identified using MS-based approaches.

Microarray analysis and RNA sequencing

The evaluation of changes in gene expression within a cell or a tissue as a result of genetic manipulation, chemical inhibition, physiological changes, developmental stages, etc. can be done at the transcriptional or translational level. While the latter is performed using techniques like western blotting, immunofluorescence, ELISA, and similar proteomics-based methods, the former is usually done by isolation of RNA followed by reverse transcription to generate complementary DNA (cDNA). The cDNA thus produced can be probed using qualitative methods like end-point PCR in which the presence of a transcript of interest is confirmed using gene-specific primers. An upgrade to end-point PCR is quantitative real-time PCR (qRT-PCR). As the name suggests, qRT-PCR can be used to quantify changes in gene expression, but with a limited application because only predetermined transcripts of interest are probed by using target-specific primers.

To overcome these limitations, cDNA can be used for large-scale screening of gene expression changes by performing an array-based evaluation. Microarrays or DNA chips are microscopic slides with tiny probes against thousands of genes immobilized in predefined spots [27]. In this technique (Fig. 3.3), RNA is isolated from cells/tissues of interest and reverse transcribed into cDNA. RNA from

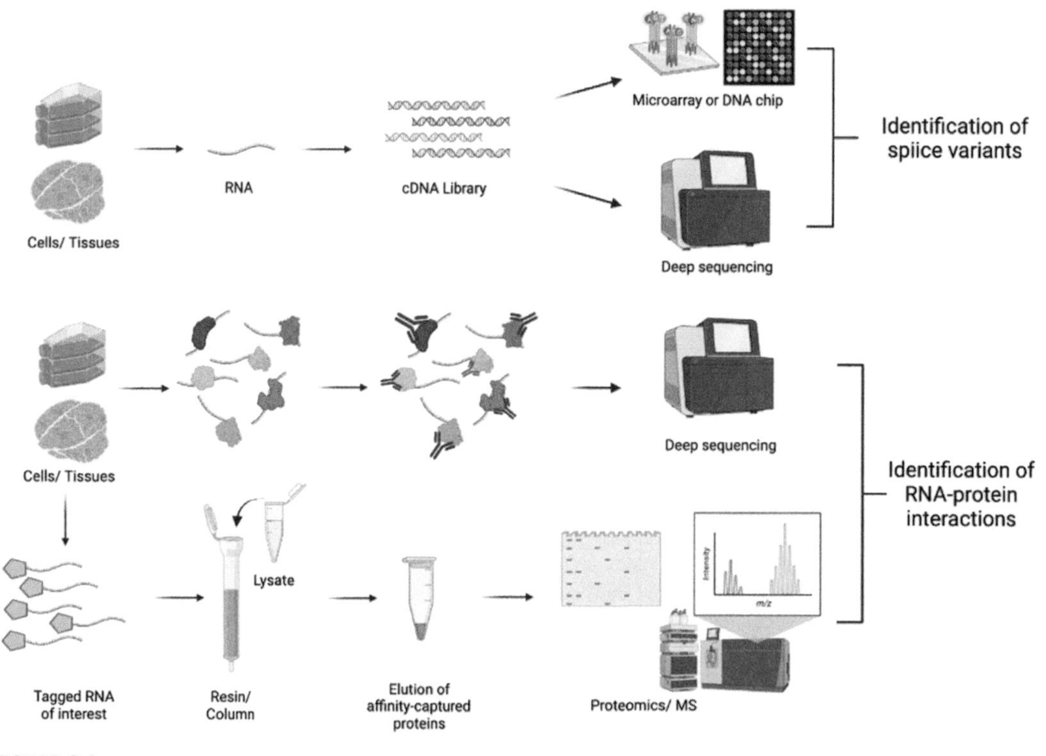

FIGURE 3.3

Schematic illustration of methods used to study alternative splicing and RNA-binding proteins. The methods include microarray and deep sequencing-based approaches to identify the expression of splice variants. Techniques using antibody-based purification/enrichment followed by deep sequencing or proteomics are utilized for identification of RNA—protein interactions.

a control sample is also processed simultaneously. cDNA from control and test samples are differentially labeled with fluorescent or chemiluminescent probes and then allowed to hybridize with the microarray slides. Excess unbound cDNA is then washed off and target gene expression is calculated by computationally evaluating the fluorescent/chemiluminescent signal resulting from hybridization of cDNA with target probes to determine relative gene expression levels.

Microarray-based quantification of gene expression is accompanied by pros and cons. The advantages include low cost and easy data interpretation. Looking at cons, microarrays suffer from background noise and cross-contamination issues. The cDNA quality and transcript abundance can also limit probe hybridization, reproducibility and provide an inaccurate evaluation. Moreover, microarray cannot be used to identify novel transcripts.

High-throughput deep sequencing can also be used to quantify gene expression [7]. For RNA-sequencing (Fig. 3.3), cDNA synthesized from isolated RNA is cloned to generate libraries that can be amplified and sequenced. The sequences generated can be aligned against the genome of interest to identify transcripts. RNA-sequencing provides unlimited choice in terms of the genome of the target organism. Moreover, due to low background noise, RNA-sequencing is highly reproducible, and comparisons can be done between independent experiments separated by time and space. Also, since cDNA is not directly used for sequencing, there is the flexibility of using less starting material that is a drawback in the case of microarray. RNA-sequencing can also identify single nucleotide polymorphism, novel transcripts. Most importantly, previously unknown alternatively spliced transcripts can be identified with ease and precision. The drawbacks of using RNA-seq are high cost, huge data files, and complicated downstream analysis.

Although the use of these two techniques is more applied to look at gene transcription, they can be extrapolated to study post-transcriptional regulation of gene expression. We will see more methods later that can be combined with high throughput deep sequencing techniques for this purpose.

RNA immunoprecipitation

RNA immunoprecipitation is a method used to examine RNA—protein interactions (Fig. 3.3). RNA immunoprecipitation (RIP) uses native purification conditions to identify naturally occurring RNA—protein complexes [28,29]. Any RBP of interest is immunoprecipitated from crude cellular lysate and any bound RNA is eluted and reverse transcribed to generate cDNA. The cDNA can be evaluated using microarray or sequencing methods.

An advantage of using this method is that physiologically existing RNA—protein interactions can be identified. However, this method can overlook RNA that is less abundant due to loss during elution or purification steps. Moreover, there is always contamination from naturally abundant RNA like ribosomal RNA.

Crosslinking and immunoprecipitation

To overcome the drawbacks of RIP, improvisation of the technique was performed to better capture RNA—protein interactions and avoid loss of rare or weak interactions during analysis. As a result, crosslinking of RNA and protein is performed immediately after harvesting samples of interest. The crosslinking agents can vary according to the method used [30]. In most cases, UV is used to covalently link RNA with proteins that are close under physiological conditions (Fig. 3.3). Once crosslinked, the samples can either be resolved on an SDS page gel to cut the band of the protein of interest. Alternatively, the protein of interest can also be immunoprecipitated as done in RIP or can be affinity purified if a tagged version of the RBP is available. This step can overcome the limitation of missing a result due to the rare occurrence of any RBP-RNA interactions.

Variants of the crosslinking and immunoprecipitation (CLIP) method have been developed with minor variations depending on the application [31–33]. For example, in photoactivatable ribonucleoside (PAR)-CLIP, stable cross-linking could be attained by providing 4′-thiouracil (4SU) to cells in a culture medium. 4SU is a PAR analog that can be crosslinked to nearby proteins by using UV. There are other versions where downstream library preparation steps are modified (eCLIP and iCLIP).

RNA affinity capture

RIP and CLIP methods described above depend on immunoprecipitation of a previously known RBP and are used to identify interacting RNAs. If the aim of an investigator is to identify novel RBPs that interact with newly identified transcripts, RNA affinity capture methods can come in handy (Fig. 3.3).

Affinity capture can be used to identify in vitro and in vivo interactions [28]. For in vitro method, RNA of interest is tagged and immobilized onto a resin/column. This is followed by pouring cell lysate onto the column and allowing protein binding to the tagged RNA. Nonspecifically bound proteins are then washed off and captured proteins are eluted. Proteins can then be identified using mass spectroscopy (MS) based methods to identify novel RBPs. These methods can be further modified by using improved MS techniques based on the aim of the study. For example, to perform quantitative mass spectrometry, Stable Isotope Labeling by Amino-acids in cell culture (SILAC)-MS can be used, which requires differential metabolic labeling of samples with isotopes.

Summary

All methods described earlier have vast potential in studying the regulation of gene expression at the post-transcriptional level. Methods utilizing high throughput approaches can be assisted by bioinformatic prediction. Computational tools can predict properties like RNA secondary structures, RNA binding motifs, etc. based on RNA sequence. Moreover, extensive literature can be used to generate a database to predict mRNA stability, half-life, and interacting partners. These approaches can thus be instrumental in furthering our understanding of post-transcriptional regulation of gene expression.

References

[1] Crick FHC. Nucleic Acids Sci Am 1957;197(3):188–203.
[2] Cobb M. 60 years ago, Francis Crick changed the logic of biology. PLoS Biol 2017;15(9).
[3] Kadonaga JT. Eukaryotic transcription: an interlaced network of transcription factors and chromatin-modifying machines. Cell 1998;92(3):307–13.
[4] Li B, Carey M, Workman JL. The role of chromatin during transcription. Cell February 23, 2007;128(4): 707–19.
[5] Ha M, Kim VN. Regulation of microRNA biogenesis. Nat Rev Mol Cell Biol 2014;15(8):509–24.

[6] Petty E, Pillus L. Balancing chromatin remodeling and histone modifications in transcription. Trends Genet 2013;29(11):621–9.
[7] Cramer P. Organization and regulation of gene transcription. Nature September 5, 2019;573(7772): 45–54.
[8] Chen CYA, Ezzeddine N, Shyu AB. Messenger RNA half-life measurements in mammalian cells. Methods Enzymol 2008;448:335–57.
[9] Wada T, Becskei A. Impact of methods on the measurement of mRNA turnover. Int J Mol Sci December 15, 2017;18(2723):1–14.
[10] Marzluff WF, Duronio RJ. Histone mRNA expression: multiple levels of cell cycle regulation and important developmental consequences. Curr Opin Cell Biol December 1, 2002;14(6):692–9.
[11] Bensaude O. Inhibiting eukaryotic transcription: which compound to choose? How to evaluate its activity? Transcription 2011;2(3):103–8.
[12] Ratnadiwakara M, Änkö M-L. mRNA stability assay using transcription inhibition by actinomycin D in mouse pluripotent stem cells. Bio-protocol 2018;8(21):1–5.
[13] Wieland T, Faulstich H. Fifty years of amanitin. Experientia December 1991;47(11):1186–93.
[14] Sugaya K, Sasanuma S, Cook P, Mita K. A mutation in the largest (catalytic) subunit of RNA polymerase II and its relation to the arrest of the cell cycle in G(1) phase. Gene August 22, 2001;274(1):77–81.
[15] Johansen FE, Prywes R. Two pathways for serum regulation of the c-fos serum response element require specific sequence elements and a minimal domain of serum response factor. Mol Cell Biol September 1994; 14(9):5920–8.
[16] Gossen M, Bujard H. Tight control of gene expression in mammalian cells by tetracycline-responsive promoters. Proc Natl Acad Sci USA 1992;89(12):5547–51.
[17] Baron U, Bujard H. Tet repressor-based system for regulated gene expression in eukaryotic cells: principles and advances. Methods Enzymol 2000;327:401–21.
[18] Jia L, Mao Y, Ji Q, Dersh D, Yewdell JW, Qian S-B. Decoding mRNA translatability and stability from the 5′ UTR. Nat Struct Mol Biol July 27, 2020;27(9):814–21.
[19] Corbett AH. Post-transcriptional regulation of gene expression and human disease. In: Current opinion in cell biology, vol. 52. Elsevier Ltd; 2018. p. 96–104.
[20] Dassi E. Handshakes and fights: the regulatory interplay of RNA-binding proteins. Front Mol Biosci September 29, 2017;4(67):1–8.
[21] Hentze MW, Castello A, Schwarzl T, Preiss T. A brave new world of RNA-binding proteins. Nat Rev Mol Cell Biol January 17, 2018;19(5):327–41.
[22] Berka AJ. Discovery of RNA splicing and genes in pieces. In: Proceedings of the national academy of sciences of the United States of America. vol. 113. National Academy of Sciences; 2016. p. 801–5.
[23] Harvey SE, Cheng C. Methods for characterization of alternative RNA splicing. Long Non-Coding RNAs January 1, 2016;1402:229–41 [Chapter 18].
[24] Mignone F, Gissi C, Liuni S, Pesole G. Untranslated regions of mRNAs. Genome Biol February 28, 2002; 3(3):1–10.
[25] Casey JL, Hentze MW, Koeller DM, Caughman SW, Rouault TA, Klausner RD, et al. Iron-responsive elements: regulatory RNA sequences that control mRNA levels and translation. Science 1988;240(4854): 924–8.
[26] Hentze MW, Caughman SW, Rouault TA, Barriocanal JG, Dancis A, Harford JB, et al. Identification of the iron-responsive element for the translational regulation of human ferritin mRNA. Science 1987;238(4833): 1570–3.
[27] Groen AK. The pros and cons of gene expression analysis by microarrays. J Hepatol 2001;35:295–6.
[28] Sternburg EL, Karginov FV. Global approaches in studying RNA-binding protein interaction networks. Trends Biochem Sci July 1, 2020;45(7):593–603.

[29] Van Nostrand EL, Freese P, Pratt GA, Wang X, Wei X, Xiao R, et al. A large-scale binding and functional map of human RNA-binding proteins. Nature July 30, 2020;583(7818):711–9.

[30] Cherkasova VA, Hinnebusch AG. CLIP identifies nova-regulated RNA networks in the brain. Proc Natl Acad Sci USA 2003;302(5648):1212–5.

[31] Licatalosi DD, Mele A, Fak JJ, Ule J, Kayikci M, Chi SW, et al. HITS-CLIP yields genome-wide insights into brain alternative RNA processing. Nature November 27, 2008;456(7221):464–9.

[32] Hafner M, Katsantoni M, Köster T, Marks J, Mukherjee J, Staiger D, et al. CLIP and complementary methods. Nat Rev Methods Prim March 4, 2021;1(1):1–23.

[33] Van Nostrand EL, Pratt GA, Shishkin AA, Gelboin-Burkhart C, Fang MY, Sundararaman B, et al. Robust transcriptome-wide discovery of RNA-binding protein binding sites with enhanced CLIP (eCLIP). Nat Methods March 28, 2016;13(6):508–14.

CHAPTER

Epigenetic regulation of post-transcriptional machinery

4

Saddam Hussain, Anindita Dutta and Rohini Muthuswami
Chromatin Remodeling Laboratory, School of Life Sciences, JNU, New Delhi, India

Introduction

The term "epigenetics" was coined by Waddington in 1942 to explain the complex interplay between the genome and the environment that led to the manifestation of a particular phenotype. Over time, the term has come to imply heritable changes that do not involve changes in the DNA sequence of the genome.

Gene expression in eukaryotes is spatiotemporally regulated in response to environmental cues. The Spatio-temporal regulation occurs at many different levels: transcriptional, post-transcriptional, and post-translational. In this chapter, we will review the post-transcriptional machinery involved in the regulation of gene expression and understand how epigenetics plays a role in controlling alternative splicing.

Post-transcriptional regulation of gene expression

In eukaryotic cells, the production of mature mRNA involves multiple steps. In the nucleus, the mRNA transcript undergoes 5′-end-capping, splicing out of the introns, and 3′-end polyadenylation, after which it is exported out into the cytoplasm where they are translated into proteins [1]. Of these, the splicing of the introns is a critical step in generating diversity in the mRNA product.

Splicing out the introns

Splicing out of the introns is coupled to transcription and is mediated by the spliceosome complex. This was first shown by analysis of nascent RNA transcripts in Drosophila embryos using an electron microscope [2]. Using quantitative real-time PCR, it was shown that splicing of dystrophin mRNA is coupled to its transcription. Rapid sampling techniques combined with quantitative real-time PCR enabled Aitkens et al. to propose that splicing is most efficient and favored when it occurs simultaneously with transcription, that is, cotranscriptionally [3].

The spliceosome complex is believed to assemble when transcription of the target gene begins and the splicing event, or at least the commitment of removal of an intron, is completed before the release of the nascent pre-mRNA. It is a multistep process wherein the U1 snRNP first binds to the 5′-splice

Post-Transcriptional Gene Regulation in Human Disease, Volume 32. https://doi.org/10.1016/B978-0-323-91305-8.00013-2

site. This is followed by the binding of U2 snRNP to the 3′-splice site. Subsequently, U4, U5, and U6 snRNPs are recruited along with non-snRNPs, leading to cleavage of the intron [4,5]. Studies have also shown that the order of splicing is not strict also, that is, the intron transcribed first does not need to also be removed first, thus facilitating alternative splicing. For example, in the case of the Fibrinogen alpha gene, intron 3 was found to be spliced first. This is followed by splicing of intron 2, intron 4, and 1 [6]. Other studies, too, have shown that the order of splicing is not dependent on the order of transcription of the intron [7–9].

Alternative splicing

Alternative splicing was first proposed by Walter Gilbert [10]. It is estimated that >90% of the human genome undergoes alternative splicing and there are at least 7 mRNA variants per gene [11,12]. Alternative splicing includes exon skipping, mutually exclusive exons, alternative 5′-splice site, alternative 3′-splice site, and intron retention (Fig. 4.1) [13].

Alternative splicing was initially proposed to be regulated by cis-acting elements like enhancers and silencers as well as trans-acting factors such as SR (serine/arginine-rich) proteins and heterogeneous nuclear ribonuclear proteins (hnRNPs) [14–18]. The elongation rate of the RNA polymerase II

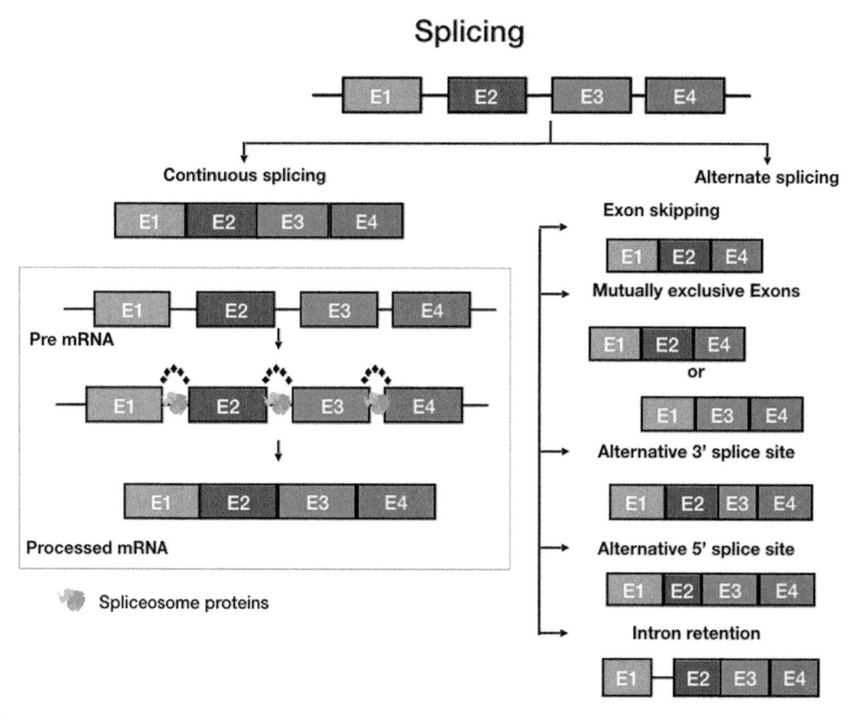

FIGURE 4.1

Schematic representation of constitutive and alternate splicing. During splicing, the spliceosome machinery assembles first on the pre-mRNA and ultimately forms the processed mRNA after skipping the introns and rejoining the exons.

(RNAPII) is yet another factor that is known to regulate this process [19,20]. The earliest evidence for the role of RNAPII came from an experiment involving different promoters. Using a minigene construct, Cramer et al. analyzed the inclusion of the ED1 exon of the Fibronectin (FN) gene under the control of different promoters. The inclusion of ED1 was high when a cytomegalovirus (CMV) or fibronectin (FN) promoter was used as opposed to a-1 globin promoter [21]. It was also shown that ED1 inclusion was inversely proportional to elongation efficiency. Thus, VP16, a transcriptional activator that enhances RNAPII elongation productivity, reduced ED1 inclusion [22]. Using a "slow" RNAPII also increases the inclusion of alternatively spliced exons [23,24].

Coupling alternative splicing to transcription

Two models, the kinetic model, and the recruitment model have been proposed to explain how transcription and splicing are coupled [25,26]. Although these are not mutually exclusive, the recruitment model is less supported by experimental evidence indicating that the kinetic model might be the favored method of coupling transcription with alternative splicing.

Kinetic model

This model proposes that the rate of RNAPII elongation is the major factor contributing to the coupling of transcription to splicing (Fig. 4.2).

DNA sequences or factors that induce RNAPII pausing promote exon inclusion, while factors like open chromatin conformation that allow RNAPII elongation increases the chance of exon skipping [27]. The exons 2 and 3 of the α-tropomyosin gene are mutually exclusive. The switch from exon 2 inclusion to exon 3 inclusion is mediated by cis-elements present in the flanking introns. It was further

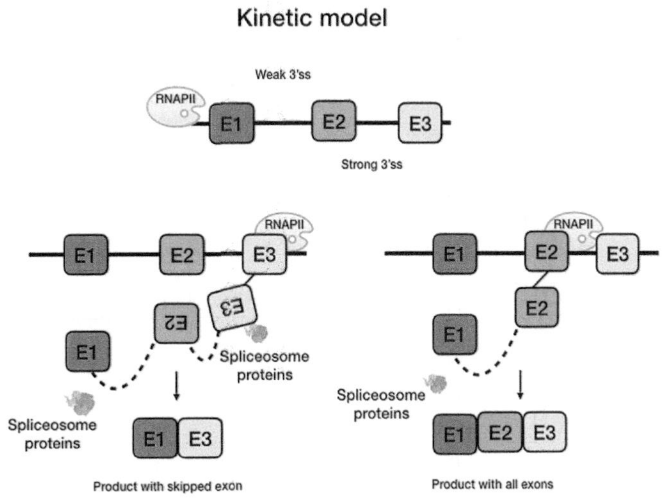

FIGURE 4.2

Kinetic model. The process of alternate splicing is regulated by the speed of RNAPII. Fast elongation rate (left panel) often results in exon skipping. The weak 3′-splicing site is generally skipped in this process. Slower elongation (right panel) consists of all the exons and forms a complete product of processed mRNA.

shown that the distance between the negative splicing regulatory element and the exon that it regulates is a critical factor in the exon exclusion. As the distance between the regulatory element and the exon increased, the exclusion of the exon decreased [28]. The importance of exon recognition by the spliceosome machinery was also demonstrated using a minigene construct containing human *Fibronectin* EDI exon and its flanking regions that were embedded in the third exon of the human α-globin gene [22]. DNA methylation, too, has been implicated to play a role in the kinetic model. The CCCTC-binding factor (CTCF) has been shown to regulate the inclusion of the alternative exon 45 in the case of the CD45 gene. CTCF binds to its cognate sequence when it is present in an unmethylated form and acts as a barrier to RNAPII progression, leading to the inclusion of exon 45 [29]. Downregulation of CTCF or methylation of the cognate sequence leads to decreased binding of CTCF and thus, increased exclusion of exon 45.

Recruitment model

The recruitment model proposes that the splicing machinery is recruited via protein–protein interaction with the transcription machinery to the transcription sites (Fig. 4.3).

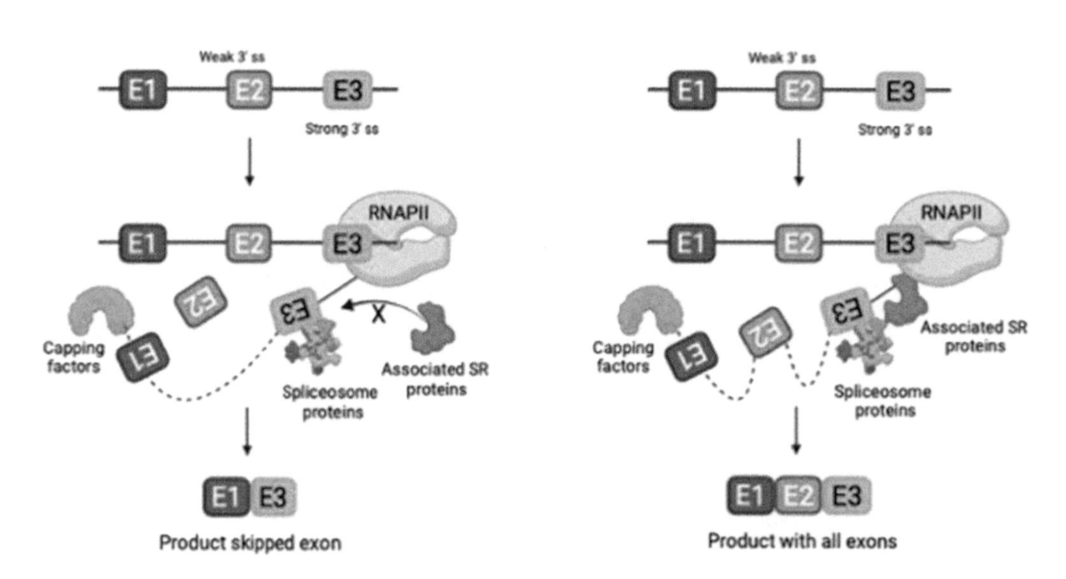

FIGURE 4.3

Recruitment model. This model proposes that the splicing machinery is recruited via protein-protein interaction with the transcription machinery to the transcription sites. The CTD of RNAPII loaded with splicing factors, capping proteins and associated adapter proteins determines the events of the splicing process. The SR proteins are a conserved family of proteins involved in RNA splicing. SR proteins are named because they contain a protein domain with long repeats of serine and arginine amino acid residues, whose standard abbreviations are "S" and "R," respectively. These proteins are one of the major regulators of the splicing process and can determine the accuracy of the process by influencing the skipping of particular exons.

The carboxy-terminal domain (CTD) of RNAPII contains 52 heptad repeats (consensus sequence: YSPTSPS) that are essential for transcriptional activation [27,30,31]. Using a mutant version of RNAPII containing only five repeats instead of the usual 52 repeats, Bentley's group showed that the CTD couples transcription to splicing [32]. They further showed that the CTD interacts with proteins mediated 5′ capping, polyadenylation, and splicing, giving rise to the concept of "transcription factories" [32,33]. Later experiments have shown that at least 22 tandem repeats are required for coupling transcription with splicing. However, this was not sufficient for the inclusion of an alternatively spliced exon in another pre-mRNA [34]. Using *Fibronectin* EDI minigene cassette Kornblihtt's group reported that at least 19 tandem repeats are needed for normal EDI splicing [35]. They further showed that the CTD of RNAPII recruits SRp20, a splicing factor belonging to the SR protein family [35]. The CTD-independent mechanism has also been reported. For example, the transcription coactivator, PGC-1, regulates alternative splicing by binding to the promoter of the gene [36]. The Mediator complex too regulates alternative splicing. MED23, a component of the Mediator complex, has been shown to regulate alternative splicing of a subset of hRNP L-targeted events by regulating occupancy of hRNP L on the coregulated genes [37]. Mechanistically, MED23 interacts with hRNP L and other splicing factors, thus, controlling alternative splicing.

Regulation of alternative splicing by chromatin

Nucleosome positioning

The DNA in every eukaryotic cell is wrapped around nucleosomes composed of an octamer of histone molecules—H2A, H2B, H3, and H4. The linker histone H1 connects the octamers and aids in higher folding. Genome-wide nucleosome positioning determined by MNase digestion has shown that nucleosomes are positioned in a nonrandom manner [38,39]. In yeast, it was shown that the transcription start site (TSS) and the 3′-polyadenylation site are nucleosome-depleted regions [40,41]. The nucleosome-depleted region around the TSS is surrounded by well-positioned +1 and −1 nucleosomes. However, nucleosome positioning becomes less precise beyond the +1 nucleosome. Studies now agree that the transcription start site is a nucleosome-depleted region in most eukaryotes [41,42]. Further, nucleosome enrichment at exons appears to be conserved through evolution. Interestingly, nucleosome enrichment is also observed at the intron−exon boundaries. Nucleosomes have long been known to act as barriers to RNAPII elongation [43,44]. Thus, the presence of the nucleosomes at the intron−exon boundaries can potentially reduce the rate of elongation and thus, aid alternative splicing.

The histones, both core, and variant are subject to post-translational modifications that have a major impact on nucleosome−DNA interaction. Further, ATP-dependent chromatin remodeling proteins can reposition nucleosomes. DNA methylation catalyzes the addition of methyl group on cytosine residues present as CpG dinucleotide, and thus, alters the chromatin architecture. Recent studies done in human H1 embryonic stem cells and IMR90 fetal fibroblasts have identified 11 chromatin modifications that differentially mark alternatively spliced exons depending on the level of exon inclusion, thus, indicating the importance of chromatin architecture in alternative splicing [45].

In the following sections, we will see how histone modifications, ATP-dependent chromatin remodeling proteins, and DNA methylation regulate and impact alternative splicing.

Histone modifications and alternative splicing

Analysis of histone modifications on alternatively spliced exons has illuminated the role of the modifications on the inclusion/exclusion of exons. Genome-wide studies using human CD4+ T-cells have shown that H3K36me3, H3K79me1, and H2BK5me1 were highly enriched over exons of highly expressed genes while H3K27me2 and H3K27me3 were enriched over exons of low expressed genes. These studies also showed that H3K36me3 facilitates exon inclusion [46]. Other genome-wide studies too have established the association of H3K36me3 with alternative splicing [47,48]. In addition, H3K4me3, H2BK12ac, and H4K5ac are also enriched around intron–exon boundaries [49].

The human fibroblast growth factor receptor 2 (FGFR2) gene undergoes alternative splicing wherein the exon IIIb is included in human prostate normal epithelium cells (PNT2s) while exon IIIc is included in human mesenchymal stem cells (hMSCs). The inclusion of exon IIIc in the case of hMSCs is dependent on polypyrimidine-tract binding protein (PTB). This protein binds to silencing elements around exon IIIb leading to its exclusion. ChIP-seq analysis showed that there was no difference in the occupancy of H3K4me2, H3K9ac, H3K27ac, and H4ac across these exons [50]. However, H3K36me3 and H3K4me1 modifications were enriched over the FGFR2 gene in hMSCs as compared to PNT2s. In contrast, H3K27me3, H3K4me3, and H3K9me1 marks were reduced, suggesting that H3K36me3 and H3K4me1 might be involved in the repression of exon IIIb inclusion in hMSCs. This was further confirmed in hMSCs where overexpression of SET2, the enzyme that catalyzes H3K36me3 modification, led to reduced inclusion of exon IIIb, while downregulation of SET2 led to increased inclusion of exon IIIb. This modification affects only the PTB-dependent exons and thus, has no effect on PTB-independent exons as well as constitutively spliced genes [50]. In contrast, in PNT2 cells, where exon IIIb is included, H3K27me3 and H3K9me3 were found to be enriched over the FGFR2 gene [50]. Analysis of FGFR2 locus in PNT2 cells showed enrichment of EZHZ and SUZ12, both components of the polycomb repressive complex (PRC) that mediates trimethylation of H3K27 [51]. In addition, enrichment of H3K36me3- and H3K9me3-binding proteins was also found. Overexpression of EZHZ in hMSC cells led to the recruitment of KDM2a (Histone methyltransferase) on the FGFR2 promoter, leading to increased H3K36me3 modification and inclusion of exon IIIb. These studies also uncovered the role of a long noncoding RNA (lncRNA) antisense to FGFR2, which was expressed when exon IIIb was included. This lncRNA helped in recruiting the PRC2 complex and KDM2a, thus, helping in establishing a specific chromatin signature for alternative splicing [51].

H3K9me3 has also been found to favor the inclusion of CD44 alternative exons. The CD44 gene contains two clusters of constant exons that frame a series of variant exons. H3K9me3 modification is enriched on the variant exons and downregulation of this modification by targeting the related/responsible histone methyltransferases involved in the catalytic process led to a reduction in the occupancy of H3K9me3 on the variant exons [52].

An elegant study showed that siRNA targeting gene sequences surrounding the ED1 alternative exon of *the Fibronectin* gene generates a closed, heterochromatin conformation involving the enrichment of H3K9me3 and H3K9me2. As a consequence, the RNAPII elongation rate reduces, leading to the inclusion of alternate exons [53].

The histone deacetylase (HDAC) activity has also been shown to regulate alternative splicing. Hu RNA-binding proteins interact directly with HDAC2 and inhibit its activity in a time- and dose-dependent manner. This inhibition promotes H3 acetylation of histones present over the alternately spliced exons, thus, regulating their inclusion [54]. Treatment with sodium butyrate, an inhibitor of

HDAC activity, was also found to alter the splicing of 683 human genes. The inhibition of HDAC activity was found to primarily affect H4 acetylation, thus, connecting this modification to the regulation of alternative splicing [55].

The role of histone modifications in regulating alternative splicing appears to be gene-dependent. The exon 23a of the Nf1 gene that encodes for neurofibromin is alternatively spliced in mouse cardiomyocytes. Elegant experiments showed that the inclusion of this exon decreases with an increase in KCl concentration. Elevated levels of KCl lead to an increase in Ca^{2+} influx via L-type voltage-gated calcium channels, and thus, Ca^{2+} regulates the inclusion of exon 23a. It was further observed that elevated levels of KCl in mouse cardiomyocytes lead to hyperacetylation of histones H3 and H4. ChIP experiments showed that H3 acetylation was indeed enriched over exon 23a of the Nf1 gene in the presence of KCl [56]. In contrast, enrichment was not observed on Capzb, a gene that contains a Ca^{2+} nonresponsive alternative exon [56].

Calcium-dependent epigenetic regulation has also been observed in the case of alternative splicing of the Neural cell adhesion molecule (ncam) gene. The exon 18 of the ncam gene is alternatively spliced, giving rise to two major isoforms: NCAM140, which is abundant in neuronal precursors, and NCAM180, which gradually increases as neurons differentiate. Once again, increasing of concentration of KCl induces depolarization resulting in exon 18 skipping. Exon 18 skipping again correlated with increased global H3 and H4 acetylation over the ncam locus [23]. Both these results suggest that alternative splicing is regulated by localized changes in the chromatin architecture mediated by the acetylation of histones.

Histone modifications-mediated epigenetic regulation of alternate splicing has also been linked to memory preservation. Neurexins (NRXNs) regulate synapse formation and are subjected to alternative splicing. In the case of Nrxn1, the inclusion of exon 22 was found to be dependent on the enrichment of H3K9me3 and H3K9me2 modifications, both of which are repressive marks, near this exon [57].

H2B123ub1 is known to regulate the methylation of H3K4, H3K36, and H3K79 methylation, all of which have been shown to regulate alternative splicing. A genome-wide map of *S. cerevisiae* has shown that H2B123 monoubiquitination (H2B123ub1) is enriched in the transcribed regions [58]. This study further analyzed the presence of this modification on the intron and exons of the ribosomal protein-coding genes and found that the level of this mark increased after the 5′-exon—intron boundary and peaked near the 3′-intron—exon boundary. This enrichment correlates with increased H3K4, H3K36, and H3K79 methylation at the intron—exon boundary. Depletion of H2B123ub1 led to decreased H3K36 methylation on the introns. In mammalian cells too, H2B monoubiquitination is enriched at exon boundaries [59]. Further, enrichment of H2B monoubiquitination marks was more on skipped exons indicating a role in alternative splicing.

Histone modifications, RNA binding proteins, and splicing

How do histone modifications regulate alternative splicing? Experimental evidence points to protein-protein interactions. The PWWP domain of Psip1 isoform, p52, has been shown to interact specifically with H3K36me3 modification and is enriched on the exons of active genes along with H3K36me3. Coimmunoprecipitation experiments show that both p52 isoform and H3K36me3 pull down the splicing factors, Srsf, Srsf2, and pTb. This interaction is dependent on Psip1 as downregulation of this gene results in reduced interaction of the splicing factors with H2K36me3, indicating that p52 might be bridging the histone modification with the splicing factors [60].

Alternative splicing is also regulated by Rbfox proteins that bind to (U)GCAUG sequences present in the pre-mRNA. In vitro cross-linking using HeLa nuclear extracts followed by mass, spectroscopy showed that Rbfox2 and Rbfox3 interacted with acetylated and methylated histones H2 and H4 [61]. The same study showed that the splicing factor proline and glutamine rich (SFPQ) is also a Rbfox3-binding partner, is bound to modified H2 and H3 [61]. In *S. cerevisiae*, H2B123ub1has been shown to genetically interact with components of U1snRNP and U2snRNP, both of which are involved in RNA splicing [58]. Thus, histone modifications appear to regulate alternative splicing via interactions with splicing factors.

Studies have also focused on the H3K36me3-binding protein, MORF-related gene 15 (MRG15). ChIP-seq experiments have shown that MRG15 occupancy on PTB-dependent exons mimics H3K36me3 enrichment. Mechanistically, MRG15 acts as an adaptor protein connecting histone modifications with a splicing regulator like PTB [50].

ATP-dependent chromatin remodeling proteins and alternative splicing

The ATP-dependent chromatin remodeling proteins use the energy released from ATP hydrolysis to reposition nucleosomes. These proteins are classified into 24 subfamilies [62]. The active DNA-dependent ATPase a domain inhibitor (ADAADi) has been identified as an inhibitor of these proteins [63,64]. This small molecule inhibits the ATPase activity of the ATP-dependent chromatin remodeling proteins specifically by binding to the protein and altering its conformation such that ATP hydrolysis is impeded [65]. RNA-sequencing showed that the expression of some of the splicing factors was altered upon ADAADi treatment. Evaluation of gene expression in Hela cells treated with ADAADi for 48 h using RNA-seq identified several differentially expressed genes involved in the biological processes of replication, transcription, translation, and DNA damage and repair. As these proteins are also involved in the alternative splicing, alternate transcript usage analysis of RNA-seq data was done. The data revealed that about 900 genes were switching to alternate transcripts due to ADAADi treatment, suggesting that the ATP-dependent chromatin remodeling proteins could be either directly or indirectly involved in alternative splicing.

Pre-mRNA splicing is carried out by the large ribonucleoprotein complexes (RNPs) known as the spliceosome. It is composed of core components such as five small nuclear RNPs (snRNPs) designated U1, U2, U4/U6, and U5, and many non-snRNP proteins such as U1/U2-related proteins, hnRNPs, and SR (serine/arginine-rich proteins) proteins. On treatment with ADAADi, the core components of U2 spliceosomal RNP, such as SF3a and SF3b, were found to be upregulated whereas U2-related subunit SR140, which is associated with splicing, alternative splicing, and nuclear export of mRNA, was found to be downregulated. The SF3b subunit is involved in the recognition and excision of the intron and 3′-end processing of the mRNA. SF3b not only splices pre-mRNA but also transports mRNA out of the nucleus through binding to 7-nt SF3b-binding motif, C/GAAGAAG. If the U2-snRNP complex is disrupted, the abundance of the SF3b increases over mRNA resulting in enhanced nuclear export [66]. Similarly, SR140 is found to be upregulated in many cancers. It interacts with calcium homeostasis endoplasmic reticulum protein (CHERP), another U2-related splicing protein, and mediates alternative splicing. In human colon cancer cells, SR140 and CHERP together radiate retention of exon 4 on UPF3A mRNA, thus maintaining an oncogenic phenotype in these cells [67].

Direct experimental evidence for the involvement of these proteins in alternative splicing has come from various studies. Brahma (Brm), one of the ATPases of the SWI/SNF chromatin remodeling complex, has been shown to regulate alternative splicing of genes encoding for E-cadherin, Bim, Cyclin D, and CD44 [68]. On the CD44 gene, ChIP studies showed that the protein is present both in the promoter region as well as in the coding region. The presence of the protein around the variant region correlated with increased enrichment of RNAPII, leading to the hypothesis that the presence of Brm might lead to decreased movement of the polymerases, thus, favoring alternative splicing. Brm interacts with the components of the spliceosome machinery including U1, which is involved in the 5′-splice-site recognition. The protein also interacts with the spliceosome-associated U5 snRNA and PRP6, a component of the U5 snRNP [68].

In Drosophila too, Brm has been shown to regulate alternative splicing [69]. Using ecdysone-inducible gene as the model system, Zraly and Dingwall showed that Brm, via SNR1 (SNF5), a component of the SWI/SNF complex, regulates alternative splicing by slowing down the RNAPII elongation [70]. Brm has also been shown to modulate the transcription elongation rate in Drosophila, thus, regulating alternative splicing. Both in *Drosophila melanogaster* and *Chironomus tentans*, Brm was found to be associated with nascent pre-mRNPs [71] In *S. cerevisiae*, deletion of Snf2, a homolog of Brm, resulted in enhanced splicing of the majority of the introns possibly due to redistribution of the spliceosome complex [72].

The SWI/SNF complexes in mammalian cells contain either Brg1 or Brm as the ATPase subunit [73]. Overexpression of Brg1 in C33A cells led to altered splicing of a subset of genes. Interestingly, this subset of genes could be further subdivided into two groups—one group of genes whose splicing was dependent on the ATPase activity of Brg1 and another group whose splicing was independent of the ATPase activity of the protein. However, the ATPase activity independent exons still were dependent on Brg1. Further, Brg1 and Brm can substitute for each other on some of the exons but not all suggesting specificity of the ATPase subunit for its target. Overexpression of Brg1 did not alter nucleosome spacing; instead, the protein when present as SWI/SNF complex appears to regulate splicing by interacting with the splicing machinery and helping in the recruitment of RNA binding proteins to the exons [74].

The imitation switch (ISWI) protein interacts with Williams syndrome transcription factor to form the WICH complex [75]. This complex has been shown to interact with many different nuclear proteins including SAP155, which is a component of U2 snRNP, thus, potentially linking this complex with splicing [76].

Chd1, an ATP-dependent chromatin remodeling protein possessing two chromodomains [77], has been shown to interact with splicing proteins like Srp20, SAF-B, and mKIAA0164, indicating a possible role in alternative splicing [78]. Chd1, via its chromodomain, recognizes H3K4me3 modification. Affinity purification of H3K4me3-associated factors from the nuclear extract of HeLa cells identified Chd1 as well as numerous proteins associated with pre-mRNA splicing, transcription elongation, and mRNA surveillance [79]. Chd1 is also a component of the Spt-Ada-Gcn5 acetyltransferase (SAGA) complex [80]. Studies have shown that Gcn5, a key component of the SAGA complex, helps in recruiting U2 snRNP to the exons and this recruitment is dependent on the histone acetyltransferase activity of Gcn5 [81]. It is hypothesized that the interaction between H3K4me3 and U2 snRNP is bridged by Chd1, bringing Gcn5, CHD1, and splicing machinery into one complex unit [79].

CHARGE syndrome, a severe developmental disorder, is characterized by coloboma of the eye, heart defects, atresia of choanae, retardation of growth, genital abnormalities, and ear anomalies. CHARGE syndrome has been associated with mutations in CHD7, encoding for a member of the

ATP-dependent chromatin remodeling protein family [82]. Chd7 interacts with Ago2 and Fam172a, an Ago2-binding protein. Further, Chd7 and Fam172a are both present on Ago2-regulated alternatively spliced exons of *CD44*. The underlying cause of CHARGE syndrome, therefore, appears to be the dysregulation of alternative splicing mediated by Chd7 and Fam172a [83,84].

Recently, Fun30 has also been shown to modulate splicing in yeast. Fun30 belongs to the Etl1 subfamily of ATP-dependent chromatin remodeling proteins [62] and has been shown to play a role in DNA double-strand end resection as well as transcription regulation [85–88]. In *S. cerevisiae*, fun30-depleted cells (*fun30D*) showed a decrease in splicing efficiency [89]. The protein was found to be localized on the intron-containing genes and the splicing efficiency was found to be dependent on the ATPase activity of the protein. Mutating the conserved lysine of the GKT motif (motif I) within the ATPase domain led to decreased splicing efficiency similar to fun30D cells. Fun30, like CHD1, also appears to recruit the spliceosome components, U1, U2, and U5 snRNPs to the splicing site [89].

Thus, the members of the ATP-dependent chromatin remodeling protein family appear to regulate alternative splicing by interacting with the modified histones and recruiting either the RNA binding proteins or the spliceosome machinery to the alternatively spliced exons.

DNA methylation and alternative splicing

DNA methylation mediated by DNA methyltransferases transfers a methyl group from S-adenosylmethionine to the 5′-position of cytosine residues that are present as CpG, dinucleotide in animal cells. In the case of plants, DNA methylation of cytosine can occur in CG, CHG, and CHH context where H is A, C, or T. In vertebrate genomes, more than 50% of the genes contain CpG islands that are 1 kb or longer. CpG dinucleotides are also present within the gene bodies. DNA methylation appears to have two distinct divergent functions: (1) methylation of CpG present in the promoter region leads to transcriptional repression; while (2) methylation of CpG present in the gene body correlates with increased transcription. The role of DNA methylation in intragenic regions has been less understood. For a long time, it was believed that methylation of CpG dinucleotides present in the gene body led to the repression of spurious transcription events. Genome-wide studies in the anemone (*Nematostella vectensis*) and the silkworm (*Bombyx mori*) showed a positive correlation between DNA methylation and gene expression. This, though, was found to be not universal. In honeybees, the majority of moderately transcribed genes were found to be more methylated as compared to low or highly expressed genes. This pattern appears to be true for many of the eukaryotes. Genome-wide studies performed in *Arabidopsis thaliana* showed that the nucleosome-bound DNA was enriched in DNA methylation [90]. Their study also demonstrated that nucleosomes, DNA methylation, and RNAPII are preferentially present on exons than on introns, leading the authors to hypothesize that DNA methylation could be playing a role in regulating splicing [90]. RNA-seq data combined with bisulfite-seq (BS-seq) data in honeybees showed that DNA methylation was higher on included exons as compared to excluded exons, once again indicating that this modification might play a role in alternative splicing [91]. BS-seq of human embryonic stem cells in various stages of differentiation showed that DNA methylation increased sharply at the 5′-splice site and decreased at the 3′-splice site of the intron–exon junction [92]. Genome-wide data analysis has shown preferential accumulation of DNA methylation over exons in human cells too [93,94]. Downregulation of DNMT3 using siRNA provided direct evidence that DNA methylation regulates alternative splicing in honeybees [95].

Evidence of the role of DNA methylation in regulating alternative splicing has come from many other studies. Alternately spliced events were found to be altered in IMR90 cells after treatment with 5′-aza-2′-deoxycytidine, an inhibitor of DNA methylation [93]. Using the *Fibronectin* EDI minigene construct, Yearim et al. provided the first causal evidence that exon inclusion is associated with high DNA methylation [94].

Intragenic DNA methylation can also dictate promoter selection and thus, regulate alternative splicing [96]. Genome-wide analysis showed that the methylation status of an intragenic CpG island present in SHANK3 as well as in Nfix genes was probably dictating the usage of the alternative promoter [96].

DNA methylation of the CpG islands present at the promoter or near the transcription start site can also affect alternative splicing. The IL-5RA gene encoding for the alpha chain of the IL-5 receptor undergoes alternative splicing producing multiple variants. The methylation status of the CpG island present near the transcription start site is important for the expression of IL-5RA transcripts as inhibition of DNA methylation by 5′-azacytidine altered the expression of the variants [97]. In prostate cancer cells, LNCaP and DU145, the switch from the production of the full-length transcript of *XAF* (XIAP-associated factor) mRNA to a shorter, alternatively spliced version are controlled by DNA methylation of the promoter region [98].

Many studies have shown that DNA methylation is altered in cancer cells. The CUG-BP and ETR-3 like factors (CELF) family of RNA-binding proteins are trans-activating factors that mediate alternate splicing. Interrogation of the genes encoding for these proteins showed that in many cancers cell lines, the CpG islands present in the promoters are hypermethylated, leading to dysregulated alternate splicing [99].

Mechanism of regulating alternative splicing by DNA methylation

Mechanistically, the regulation of alternative splicing by DNA methylation depends on three proteins that recognize methylated cytosine: CTCF, HP1, and MeCP2. MeCP2 occupancy over exons has been found to mirror DNA methylation in IMR90 cells. Downregulation of MeCP2 using RNAi technology in IMR90 cells led to alterations in the alternately spliced events, leading the authors to hypothesize a role for this methyl binding protein in regulating alternate splicing [93].

HP1 proteins recognize H3K9me3 and are a primary constituent of heterochromatin, thus, mediating transcriptional repression. However, its role is not limited to transcriptional repression as it also plays a role in transcriptional elongation, sister chromatid cohesion, DNA repair, and telomeric maintenance [100]. Mammalian cells have three isoforms: HP1a, Hp1b, and Hp1g [101]. RNA-seq experiments performed in mouse embryonic stem cells where each of the three isoforms and all three together were downregulated showed alterations in the splice variants indicating a role for these three proteins in alternate splicing [94]. Using the EDI minigene construct, the authors further established the link between HP1 and DNA methylation. EDI methylation recruits HP1, which in turn recruits splicing factor, SRSF3, and thus, connecting DNA methylation with splicing [94]. The H3K9me3 modification recruits HP1g to variant exons of CD44, which leads to slowing down of RNAPII, and thus favors the inclusion of these exons [52]. Recent studies have shown that HP1g is specifically targeted to the hexameric RNA motifs that are abundant in introns but absent in exons, thus, regulating alternative splicing events [102].

CTCF, a highly conserved zinc finger, mediates genomic imprinting, X-chromosome inactivation, functions as an insulator protein, as well as regulates transcription activation/repression [103]. ChIP-seq data has shown that approximately 40%–45% of CTCF binding sites are present intragenically [29]. CTCF binding has been shown to bind to exon 37a of Cacna 1b gene, promoting its inclusion [104] CTCF has also been shown to promote inclusion of exon 5 of the CD45 gene [29]. Analysis of ChIP-seq data showed that higher occupancy of CTCF on exon 5 correlates with RNAPII pausing, leading to the inclusion of this exon. DNA methylation of the CpG dinucleotides present on exon 5 precludes CTCF binding, and thus, affects the splicing process [29]. Mechanistically, it appears the CTCF might be promoting looping between promoter and exon, and thus, regulate alternate splicing [105].

Analysis of ChIP-seq data of histone marks H3K27me3, H3K9me2, H3K36me3, RNAPII, CTCF, and HP1a in MCF-7 and MCF-10 cells showed that CTCF and HP1a were present downstream of the alternative exon in relation to inclusion events, thus indicating a connection between these two proteins in relation to alternative splicing [106]. The analysis also showed cell-type specificity for CTCF was found to be significantly enriched in included exons and relatively depleted in excluded exons in MCF-7 cells, but not in MCF-10 cells [106].

Alternative splicing in cancer

Many cancers express alternative splice variants indicating alterations in the splicing process might be a contributing factor to the tumorigenesis process. Indeed, alternative splicing is now considered an emerging hallmark of cancer [107]. In many cancers, CD44 overexpression has been reported due to the usage of different splice sites [108]. A comparative study between tumor and adjacent normal tissues identified 15,818 alternate splicing events in 3955 annotated genes as significantly altered [109]. Another study involving 10,000 samples across 33 types of cancer identified more than 40,000 variant driver candidates indicating widespread deregulation in alternate splice events in cancer [110].

The expression of SETD2, an H3K36 methyltransferase, is downregulated in human colorectal cancers. Using SETD2 knockout mice, Yuan et al. showed that the expression of SETD2, as well as global H3K36 methylation, were reduced in these mice [111]. Concomitantly, there was an increased formation of adenomas in the gastrointestinal tract. Analysis of RNA-seq data showed that there was a widespread alteration in the mRNA splice variants, thus, connecting alternate splicing events with SETD2 expression and colorectal cancer [111].

The human KRAS transcript is alternately spliced to yield KRAS-4A and KRAS-4B variants. The ratio of KRAS-4A to KRAS-4B is altered in many cancers, including colorectal cancer. The enrichment of H3K27ac and H3K37me3 was found to be significantly altered around exon 4B between HCT116 and SW48, both of which are colorectal cancer cell lines. Decreased enrichment of H3K27ac and increased occupancy of H3K27me3 in SW48 cell line correlates with increased KRAS-4A/KRAS-4B ratio suggesting a role of histone modifications in the alternative splicing event [112].

DNA methylation-dependent regulation of alternate splicing has also been linked to colorectal cancer. In HCT116 cells, downregulation of DNMT3a and DNMT3b was found to activate the EMT pathway. Analysis of the RNA-seq data showed that downregulation of these DNA methyltransferases led to reduced recruitment of methyl CpG binding domain (MBD), HP1g, and H3K9me3 resulting in extensive reprogramming of the splicing events, thus providing a connection between alternate splicing, chromatin structure, and colorectal cancer [113].

Intragenic DNA methylation was also found to modulate splicing events during hypoxia in breast cancer. Pant et al. analyzed the publicly available datasets for differential isoform usage under hypoxic and normoxic conditions in breast cancer. They found that at least 821 genes showed differential transcript usage under hypoxic and normoxic conditions. They further correlated this differential usage to differences in intragenic DNA methylation. These changes in the isoform usage were found to correlate with survival rates indicating alternate splicing regulated by DNA methylation plays a role in breast cancer [114].

Alternative splicing events are also widespread in human papilloma-related oropharyngeal squamous cell carcinoma (HPV-OSCC). Approximately 109 alternatively spliced mRNAs have been identified from an RNA-seq experiment involving 46 HPV-OSCC and 25 normal tissue samples [115]. GenometriCorr (project statistics) analysis showed that these 109 alternatively spliced events significantly correlated with tumor-specific H3K27ac super-enhancer regions, suggesting a link between chromatin modifications and alternate splicing in this type of cancer [116].

In prostate cancers, PACE4, a proprotein convertase, is overexpressed due to alternate splicing caused by intraexonic DNA methylation. This altered splice variant is responsible for sustained growth properties of prostate cancer [117].

Summary

The "central dogma" is the process of accurate conversion of the basic DNA information to a final product, that is, proteins. Besides coding for proteins, DNA also acts as a functional unit to produce short-lived mRNA. Although all cells in the body contain the same DNA, cell types and functions differ due to qualitative and quantitative differences in gene expression. Therefore, the control of gene expression is the core of differentiation and development. The epigenetic processes regulate gene expression both via transcriptional and post-transcriptional regulation.

Chromatin remodeling mechanisms function as a scaffold to recruit the splicing machinery to the exon—intron boundary (Fig. 4.4). Studies have established crosstalk between histone modifications, ATP-dependent chromatin remodeling proteins, and DNA methylation in regulating chromatin architecture and thus, alternate splicing. The major lacunae are to understand how these chromatin remodeling proteins interact with the splicing machinery to regulate alternative splicing. Recent studies have shown that HDACs interact with spliceosome and ribonucleoprotein complexes to actively control the acetylation status of the splicing apparatus related to splicing factors, therefore, regulating splicing. Knockdown and coimmunoprecipitation studies have also revealed that HDACs can regulate alternative splicing by providing stability to the hnRNPs. Another study has investigated the interacting partners of U2 snRNP. Allemand et al. [118] have shown that U2 snRNP copurifies with HDAC2, CHD7, Sin3A, SMARCA5, and CHD5, indicating that protein—protein interactions might be the important factors connecting chromatin remodeling with alternative splicing. However, more studies need to be done to delineate the mechanisms involved in the regulation of alternative splicing by the chromatin remodeling factors.

Alternative splicing has been linked to cancer. Although studies have now established the connection between DNA methylation, alternative splicing, and cancer, the relationship of other chromatin remodeling with splicing and disease still needs to be explored.

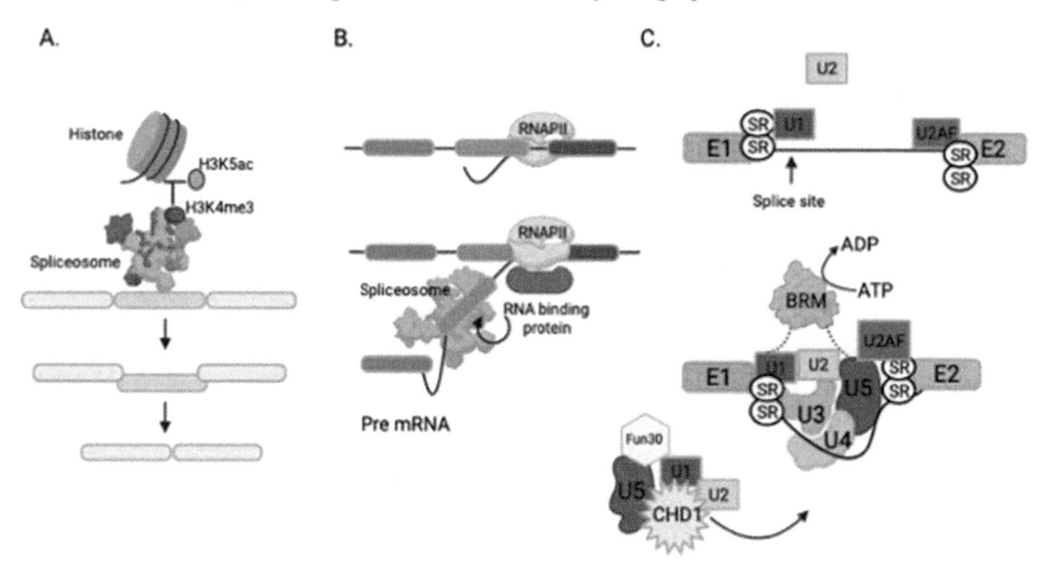

FIGURE 4.4

Regulation of alternative splicing by chromatin. (A). Histone modifiers like H3K4me3 and H3K5ac are found mainly at the exon—intron junction part and are associated with alternative splicing. (B). Different DNA-binding proteins bind to alternative exons (orange color) and can associate with the RNAPII to regulate splicing by incorporating RNA-binding proteins. They maintain the balance in the splicing process either by recognizing the pre-mRNA-binding motifs or by making the RNAPII elongation rate slower and favoring the insertion of the alternative exon. (C). ATP-dependent chromatin remodeler proteins can take part in the alternate splicing process by directly recruiting the spliceosome assembly factors like Fun30 and Chd1, can directly recruit and interact with U1, U2, and U5 of the spliceosome machinery, although the exact mode of action offered by the ATPase domain of the chromatin remodeler is not known.

References

[1] Moore MJ. From birth to death: the complex lives of eukaryotic mRNAs. Science September 2, 2005; 309(5740):1514—8.

[2] Beyer AL, Osheim YN. Splice site selection, rate of splicing, and alternative splicing on nascent transcripts. Genes Dev June 1988;2(6):754—65.

[3] Aitken S, Alexander RD, Beggs JD. Modelling reveals kinetic advantages of Co-transcriptional splicing. In: Guigo R, editor. PLoS Comput Biol, vol. 7; October 13, 2011. e1002215 (10).

[4] Will CL, Lührmann R. Spliceosome structure and function. Cold Spring Harbor Perspect Biol July 1, 2011; 3(7):a003707.

[5] Brow DA. Allosteric cascade of spliceosome activation. Annu Rev Genet 2002;36:333—60.

[6] Attanasio C, David A, Neerman-Arbez M. Outcome of donor splice site mutations accounting for congenital afibrinogenemia reflects order of intron removal in the fibrinogen alpha gene (FGA). Blood March 1, 2003;101(5):1851–6.
[7] Baurén G, Wieslander L. Splicing of Balbiani ring 1 gene pre-mRNA occurs simultaneously with transcription. Cell January 14, 1994;76(1):183–92.
[8] Kessler O, Jiang Y, Chasin LA. Order of intron removal during splicing of endogenous adenine phosphoribosyltransferase and dihydrofolate reductase pre-mRNA. Mol Cell Biol October 1993;13(10): 6211–22.
[9] LeMaire MF, Thummel CS. Splicing precedes polyadenylation during *Drosophila* E74A transcription. Mol Cell Biol November 1990;10(11):6059–63.
[10] Gilbert W. Why genes in pieces? Nature February 9, 1978;271(5645):501.
[11] Pan Q, Shai O, Lee LJ, Frey BJ, Blencowe BJ. Deep surveying of alternative splicing complexity in the human transcriptome by high-throughput sequencing. Nat Genet December 2008;40(12):1413–5.
[12] Croft L, Schandorff S, Clark F, Burrage K, Arctander P, Mattick JS. ISIS, the intron information system, reveals the high frequency of alternative splicing in the human genome. Nat Genet April 2000;24(4):340–1.
[13] Roy B, Haupt LM, Griffiths LR. Review: alternative splicing (AS) of genes as an approach for generating protein complexity. Curr Genom May 2013;14(3):182–94.
[14] Chasin LA. Searching for splicing motifs. Adv Exp Med Biol 2007;623:85–106.
[15] Long JC, Caceres JF. The SR protein family of splicing factors: master regulators of gene expression. Biochem J January 1, 2009;417(1):15–27.
[16] Han SP, Tang YH, Smith R. Functional diversity of the hnRNPs: past, present and perspectives. Biochem J September 15, 2010;430(3):379–92.
[17] House AE, Lynch KW. Regulation of alternative splicing: more than just the ABCs. J Biol Chem January 18, 2008;283(3):1217–21.
[18] Chen M, Manley JL. Mechanisms of alternative splicing regulation: insights from molecular and genomics approaches. Nat Rev Mol Cell Biol November 2009;10(11):741–54.
[19] Dujardin G, Lafaille C, Petrillo E, Buggiano V, Gómez Acuña LI, Fiszbein A, et al. Transcriptional elongation and alternative splicing. Biochim Biophys Acta January 2013;1829(1):134–40.
[20] Ip JY, Schmidt D, Pan Q, Ramani AK, Fraser AG, Odom DT, et al. Global impact of RNA polymerase II elongation inhibition on alternative splicing regulation. Genome Res March 2011;21(3):390–401.
[21] Cramer P, Pesce CG, Baralle FE, Kornblihtt AR. Functional association between promoter structure and transcript alternative splicing. Proc Natl Acad Sci USA October 14, 1997;94(21):11456–60.
[22] Nogues G, Kadener S, Cramer P, Bentley D, Kornblihtt AR. Transcriptional activators differ in their abilities to control alternative splicing. J Biol Chem November 8, 2002;277(45):43110–4.
[23] Schor IE, Rascovan N, Pelisch F, Alló M, Kornblihtt AR. Neuronal cell depolarization induces intragenic chromatin modifications affecting NCAM alternative splicing. Proc Natl Acad Sci USA March 17, 2009; 106(11):4325–30.
[24] Fong N, Kim H, Zhou Y, Ji X, Qiu J, Saldi T, et al. Pre-mRNA splicing is facilitated by an optimal RNA polymerase II elongation rate. Genes Dev December 1, 2014;28(23):2663–76.
[25] Lynch KW. Cotranscriptional splicing regulation: it's not just about speed. Nat Struct Mol Biol November 2006;13(11):952–3.
[26] Kornblihtt AR, Schor IE, Alló M, Dujardin G, Petrillo E, Muñoz MJ. Alternative splicing: a pivotal step between eukaryotic transcription and translation. Nat Rev Mol Cell Biol March 2013;14(3):153–65.
[27] Naftelberg S, Schor IE, Ast G, Kornblihtt AR. Regulation of alternative splicing through coupling with transcription and chromatin structure. Annu Rev Biochem 2015;84:165–98.
[28] Roberts G. Co-transcriptional commitment to alternative splice site selection. Nucleic Acids Res December 15, 1998;26(24):5568–72.

[29] Shukla S, Kavak E, Gregory M, Imashimizu M, Shutinoski B, Kashlev M, et al. CTCF-promoted RNA polymerase II pausing links DNA methylation to splicing. Nature November 3, 2011;479(7371):74–9.
[30] Muñoz MJ, de la Mata M, Kornblihtt AR. The carboxy terminal domain of RNA polymerase II and alternative splicing. Trends Biochem Sci September 2010;35(9):497–504.
[31] Gerber HP, Hagmann M, Seipel K, Georgiev O, West MA, Litingtung Y, et al. RNA polymerase II C-terminal domain required for enhancer-driven transcription. Nature April 13, 1995;374(6523):660–2.
[32] McCracken S, Fong N, Yankulov K, Ballantyne S, Pan G, Greenblatt J, et al. The C-terminal domain of RNA polymerase II couples mRNA processing to transcription. Nature January 23, 1997;385(6614): 357–61.
[33] McCracken S, Fong N, Rosonina E, Yankulov K, Brothers G, Siderovski D, et al. 5′-Capping enzymes are targeted to pre-mRNA by binding to the phosphorylated carboxy-terminal domain of RNA polymerase II. Genes Dev December 15, 1997;11(24):3306–18.
[34] Rosonina E, Blencowe BJ. Analysis of the requirement for RNA polymerase II CTD heptapeptide repeats in pre-mRNA splicing and 3′-end cleavage. RNA April 2004;10(4):581–9.
[35] de la Mata M, Kornblihtt AR. RNA polymerase II C-terminal domain mediates regulation of alternative splicing by SRp20. Nat Struct Mol Biol November 2006;13(11):973–80.
[36] Monsalve M, Wu Z, Adelmant G, Puigserver P, Fan M, Spiegelman BM. Direct coupling of transcription and mRNA processing through the thermogenic coactivator PGC-1. Mol Cell August 2000;6(2):307–16.
[37] Huang Y, Li W, Yao X, Lin Q-J, Yin J-W, Liang Y, et al. Mediator complex regulates alternative mRNA processing via the MED23 subunit. Mol Cell February 24, 2012;45(4):459–69.
[38] Yuan G-C, Liu Y-J, Dion MF, Slack MD, Wu LF, Altschuler SJ, et al. Genome-scale identification of nucleosome positions in *S. cerevisiae*. Science July 22, 2005;309(5734):626–30.
[39] Zhang Z, Pugh BF. High-resolution genome-wide mapping of the primary structure of chromatin. Cell January 21, 2011;144(2):175–86.
[40] Jiang C, Pugh BF. Nucleosome positioning and gene regulation: advances through genomics. Nat Rev Genet March 2009;10(3):161–72.
[41] Mavrich TN, Ioshikhes IP, Venters BJ, Jiang C, Tomsho LP, Qi J, et al. A barrier nucleosome model for statistical positioning of nucleosomes throughout the yeast genome. Genome Res July 2008;18(7): 1073–83.
[42] Barski A, Cuddapah S, Cui K, Roh T-Y, Schones DE, Wang Z, et al. High-resolution profiling of histone methylations in the human genome. Cell May 18, 2007;129(4):823–37.
[43] Bondarenko VA, Steele LM, Ujvári A, Gaykalova DA, Kulaeva OI, Polikanov YS, et al. Nucleosomes can form a polar barrier to transcript elongation by RNA polymerase II. Mol Cell November 3, 2006;24(3): 469–79.
[44] Gaykalova DA, Kulaeva OI, Volokh O, Shaytan AK, Hsieh F-K, Kirpichnikov MP, et al. Structural analysis of nucleosomal barrier to transcription. Proc Natl Acad Sci USA October 27, 2015;112(43):E5787–95.
[45] Agirre E, Oldfield AJ, Bellora N, Segelle A, Luco RF. Splicing-associated chromatin signatures: a combinatorial and position-dependent role for histone marks in splicing definition. Nat Commun December 2021;12(1):682.
[46] Andersson R, Enroth S, Rada-Iglesias A, Wadelius C, Komorowski J. Nucleosomes are well positioned in exons and carry characteristic histone modifications. Genome Res October 2009;19(10):1732–41.
[47] Kolasinska-Zwierz P, Down T, Latorre I, Liu T, Liu XS, Ahringer J. Differential chromatin marking of introns and expressed exons by H3K36me3. Nat Genet March 2009;41(3):376–81.
[48] Spies N, Nielsen CB, Padgett RA, Burge CB. Biased chromatin signatures around polyadenylation sites and exons. Mol Cell October 23, 2009;36(2):245–54.
[49] Shindo Y, Nozaki T, Saito R, Tomita M. Computational analysis of associations between alternative splicing and histone modifications. FEBS Lett March 1, 2013;587(5):516–21.

[50] Luco RF, Pan Q, Tominaga K, Blencowe BJ, Pereira-Smith OM, Misteli T. Regulation of alternative splicing by histone modifications. Science February 19, 2010;327(5968):996−1000.

[51] Gonzalez I, Munita R, Agirre E, Dittmer TA, Gysling K, Misteli T, et al. A lncRNA regulates alternative splicing via establishment of a splicing-specific chromatin signature. Nat Struct Mol Biol May 2015;22(5):370−6.

[52] Saint-André V, Batsché E, Rachez C, Muchardt C. Histone H3 lysine 9 trimethylation and HP1γ favor inclusion of alternative exons. Nat Struct Mol Biol March 2011;18(3):337−44.

[53] Alló M, Buggiano V, Fededa JP, Petrillo E, Schor I, de la Mata M, et al. Control of alternative splicing through siRNA-mediated transcriptional gene silencing. Nat Struct Mol Biol July 2009;16(7):717−24.

[54] Zhou H-L, Hinman MN, Barron VA, Geng C, Zhou G, Luo G, et al. Hu proteins regulate alternative splicing by inducing localized histone hyperacetylation in an RNA-dependent manner. Proc Natl Acad Sci USA September 6, 2011;108(36):E627−35.

[55] Hnilicová J, Hozeifi S, Dušková E, Icha J, Tománková T, Staněk D. Histone deacetylase activity modulates alternative splicing. PLoS One February 2, 2011;6(2):e16727.

[56] Sharma A, Nguyen H, Geng C, Hinman MN, Luo G, Lou H. Calcium-mediated histone modifications regulate alternative splicing in cardiomyocytes. Proc Natl Acad Sci USA November 18, 2014;111(46):E4920−8.

[57] Ding X, Liu S, Tian M, Zhang W, Zhu T, Li D, et al. Activity-induced histone modifications govern Neurexin-1 mRNA splicing and memory preservation. Nat Neurosci May 2017;20(5):690−9.

[58] Shieh GS, Pan C-H, Wu J-H, Sun Y-J, Wang C-C, Hsiao W-C, et al. H2B ubiquitylation is part of chromatin architecture that marks exon-intron structure in budding yeast. BMC Genom December 22, 2011;12:627.

[59] Jung I, Kim S-K, Kim M, Han Y-M, Kim YS, Kim D, et al. H2B monoubiquitylation is a 5′-enriched active transcription mark and correlates with exon−intron structure in human cells. Genome Res June 2012;22(6):1026−35.

[60] Pradeepa MM, Sutherland HG, Ule J, Grimes GR, Bickmore WA. Psip1/Ledgf p52 binds methylated histone H3K36 and splicing factors and contributes to the regulation of alternative splicing. PLoS Genet 2012;8(5):e1002717.

[61] Kim Y-E, Park C, Kim KE, Kim KK. Histone and RNA-binding protein interaction creates crosstalk network for regulation of alternative splicing. Biochem Biophys Res Commun April 30, 2018;499(1):30−6.

[62] Flaus A, Martin DMA, Barton GJ, Owen-Hughes T. Identification of multiple distinct Snf2 subfamilies with conserved structural motifs. Nucleic Acids Res 2006;34(10):2887−905.

[63] Muthuswami R, Mesner LD, Wang D, Hill DA, Imbalzano AN, Hockensmith JW. Phosphoaminoglycosides inhibit SWI2/SNF2 family DNA-dependent molecular motor domains. Biochemistry April 18, 2000;39(15):4358−65.

[64] Rakesh R, Chanana UB, Hussain S, Sharma S, Goel K, Bisht D, et al. Altering mammalian transcription networking with ADAADi: an inhibitor of ATP-dependent chromatin remodeling. PLoS One 2021;16(5):e0251354.

[65] Dutta P, Tanti GK, Sharma S, Goswami SK, Komath SS, Mayo MW, et al. Global epigenetic changes induced by SWI2/SNF2 inhibitors characterize neomycin-resistant mammalian cells. PLoS One 2012;7(11):e49822.

[66] Wang K, Yin C, Du X, Chen S, Wang J, Zhang L, et al. A U2-snRNP-independent role of SF3b in promoting mRNA export. Proc Natl Acad Sci USA April 16, 2019;116(16):7837−46.

[67] Wang Q, Wang Y, Liu Y, Zhang C, Luo Y, Guo R, et al. U2-related proteins CHERP and SR140 contribute to colorectal tumorigenesis via alternative splicing regulation. Int J Cancer November 15, 2019;145(10):2728−39.

[68] Batsché E, Yaniv M, Muchardt C. The human SWI/SNF subunit Brm is a regulator of alternative splicing. Nat Struct Mol Biol January 2006;13(1):22−9.

[69] Yu S, Waldholm J, Böhm S, Visa N. Brahma regulates a specific trans-splicing event at the mod(mdg4) locus of *Drosophila melanogaster*. RNA Biol February 2014;11(2):134–45.

[70] Zraly CB, Dingwall AK. The chromatin remodeling and mRNA splicing functions of the Brahma (SWI/SNF) complex are mediated by the SNR1/SNF5 regulatory subunit. Nucleic Acids Res July 2012;40(13):5975–87.

[71] Tyagi A, Ryme J, Brodin D, Östlund Farrants AK, Visa N. SWI/SNF associates with nascent pre-mRNPs and regulates alternative pre-mRNA processing. In: Akhtar A, editor. PLoS Genet, vol. 5; May 8, 2009. e1000470 (5).

[72] Venkataramanan S, Douglass S, Galivanche AR, Johnson TL. The chromatin remodeling complex Swi/Snf regulates splicing of meiotic transcripts in *Saccharomyces cerevisiae*. Nucleic Acids Res July 27, 2017;45(13):7708–21.

[73] Hargreaves DC, Crabtree GR. ATP-dependent chromatin remodeling: genetics, genomics and mechanisms. Cell Res March 2011;21(3):396–420.

[74] Zapater AG, Mackowiak SD, Guo Y, Jordan-Pla A, Friedländer MR, Visa N, et al. The SWI/SNF subunits BRG1 affects alternative splicing by changing RNA binding factor interactions with RNA. Online ahead of print Mol Genet Genom 2022. https://doi.org/10.1007/s00438-022-01863-9.

[75] Dirscherl SS, Krebs JE. Functional diversity of ISWI complexes. Biochem Cell Biol August 2004;82(4):482–9.

[76] Cavellán E, Asp P, Percipalle P, Farrants A-KO. The WSTF-SNF2h chromatin remodeling complex interacts with several nuclear proteins in transcription. J Biol Chem June 16, 2006;281(24):16264–71.

[77] Marfella CGA, Imbalzano AN. The Chd family of chromatin remodelers. Mutat Res May 1, 2007;618(1–2):30–40.

[78] Tai HH, Geisterfer M, Bell JC, Moniwa M, Davie JR, Boucher L, et al. CHD1 associates with NCoR and histone deacetylase as well as with RNA splicing proteins. Biochem Biophys Res Commun August 15, 2003;308(1):170–6.

[79] Sims RJ, Millhouse S, Chen C-F, Lewis BA, Erdjument-Bromage H, Tempst P, et al. Recognition of trimethylated histone H3 lysine 4 facilitates the recruitment of transcription postinitiation factors and pre-mRNA splicing. Mol Cell November 2007;28(4):665–76.

[80] Pray-Grant MG, Daniel JA, Schieltz D, Yates JR, Grant PA. Chd1 chromodomain links histone H3 methylation with SAGA- and SLIK-dependent acetylation. Nature January 2005;433(7024):434–8.

[81] Gunderson FQ, Johnson TL. Acetylation by the transcriptional coactivator Gcn5 plays a novel role in Co-transcriptional spliceosome assembly. In: Madhani HD, editor. PLoS genet, vol. 5; October 16, 2009. e1000682 (10).

[82] Hsu P, Ma A, Wilson M, Williams G, Curotta J, Munns CF, et al. CHARGE syndrome: a review. J Paediatr Child Health July 2014;50(7):504–11.

[83] Bélanger C, Bérubé-Simard F-A, Leduc E, Bernas G, Campeau PM, Lalani SR, et al. Dysregulation of cotranscriptional alternative splicing underlies CHARGE syndrome. Proc Natl Acad Sci USA January 23, 2018;115(4):E620–9.

[84] Bérubé-Simard F-A, Pilon N. Molecular dissection of CHARGE syndrome highlights the vulnerability of neural crest cells to problems with alternative splicing and other transcription-related processes. Transcription February 2019;10(1):21–8.

[85] Awad S, Ryan D, Prochasson P, Owen-Hughes T, Hassan AH. The Snf2 homolog Fun30 acts as a homodimeric ATP-dependent chromatin-remodeling enzyme. J Biol Chem March 26, 2010;285(13):9477–84.

[86] Costelloe T, Louge R, Tomimatsu N, Mukherjee B, Martini E, Khadaroo B, et al. The yeast Fun30 and human SMARCAD1 chromatin remodellers promote DNA end resection. Nature September 9, 2012;489:581–4. https://doi.org/10.1038/nature11353.

[87] Rowbotham SP, Barki L, Neves-Costa A, Santos F, Dean W, Hawkes N, et al. Maintenance of silent chromatin through replication requires SWI/SNF-like chromatin remodeler SMARCAD1. Mol Cell May 6, 2011;42(3):285−96.
[88] Lee J, Choi ES, Lee D. It's fun to transcribe with Fun30: a model for nucleosome dynamics during RNA polymerase II-mediated elongation. Transcription 2018;9(2):108−16.
[89] Niu Q, Wang W, Wei Z, Byeon B, Das AB, Chen B-S, et al. Role of the ATP-dependent chromatin remodeling enzyme Fun30/Smarcad1 in the regulation of mRNA splicing. Biochem Biophys Res Commun May 28, 2020;526(2):453−8.
[90] Chodavarapu RK, Feng S, Bernatavichute YV, Chen P-Y, Stroud H, Yu Y, et al. Relationship between nucleosome positioning and DNA methylation. Nature July 15, 2010;466(7304):388−92.
[91] Flores K, Wolschin F, Corneveaux JJ, Allen AN, Huentelman MJ, Amdam GV. Genome-wide association between DNA methylation and alternative splicing in an invertebrate. BMC Genom September 15, 2012;13:480.
[92] Laurent L, Wong E, Li G, Huynh T, Tsirigos A, Ong CT, et al. Dynamic changes in the human methylome during differentiation. Genome Res March 2010;20(3):320−31.
[93] Maunakea AK, Chepelev I, Cui K, Zhao K. Intragenic DNA methylation modulates alternative splicing by recruiting MeCP2 to promote exon recognition. Cell Res November 2013;23(11):1256−69.
[94] Yearim A, Gelfman S, Shayevitch R, Melcer S, Glaich O, Mallm J-P, et al. HP1 is involved in regulating the global impact of DNA methylation on alternative splicing. Cell Rep February 24, 2015;10(7):1122−34.
[95] Li-Byarlay H, Li Y, Stroud H, Feng S, Newman TC, Kaneda M, et al. RNA interference knockdown of DNA methyl-transferase 3 affects gene alternative splicing in the honey bee. Proc Natl Acad Sci USA July 30, 2013;110(31):12750−5.
[96] Maunakea AK, Nagarajan RP, Bilenky M, Ballinger TJ, D'Souza C, Fouse SD, et al. Conserved role of intragenic DNA methylation in regulating alternative promoters. Nature July 8, 2010;466(7303):253−7.
[97] Diniz SN, Pendeloski KP, Morgun A, Chepelev I, Gerbase-DeLima M, Shulzhenko N. Tissue-specific expression of IL-15RA alternative splicing transcripts and its regulation by DNA methylation. Eur Cytokine Netw December 2010;21(4):308−18.
[98] Fang X, Liu Z, Fan Y, Zheng C, Nilson S, Egevad L, et al. Switch to full-length of XAF1 mRNA expression in prostate cancer cells by the DNA methylation inhibitor. Int J Cancer May 15, 2006;118(10):2485−9.
[99] Piqué L, Martinez de Paz A, Piñeyro D, Martínez-Cardús A, Castro de Moura M, Llinàs-Arias P, et al. Epigenetic inactivation of the splicing RNA-binding protein CELF2 in human breast cancer. Oncogene November 7, 2019;38(45):7106−12.
[100] Zeng W, Ball AR, Yokomori K. HP1: heterochromatin binding proteins working the genome. Epigenetics May 16, 2010;5(4):287−92.
[101] Bosch-Presegué L, Raurell-Vila H, Thackray JK, González J, Casal C, Kane-Goldsmith N, et al. Mammalian HP1 isoforms have specific roles in heterochromatin structure and organization. Cell Rep November 21, 2017;21(8):2048−57.
[102] Rachez C, Legendre R, Costallat M, Varet H, Yi J, Kornobis E, et al. HP1γ binding pre-mRNA intronic repeats modulates RNA splicing decisions. EMBO Rep July 27, 2021:e52320.
[103] Phillips JE, Corces VG. CTCF: master weaver of the genome. Cell June 26, 2009;137(7):1194−211.
[104] López Soto EJ, Lipscombe D. Cell-specific exon methylation and CTCF binding in neurons regulate calcium ion channel splicing and function. Elife March 26, 2020:9.
[105] Ruiz-Velasco M, Kumar M, Lai MC, Bhat P, Solis-Pinson AB, Reyes A, et al. CTCF-mediated chromatin loops between promoter and gene body regulate alternative splicing across individuals. Cell Syst December 27, 2017;5(6):628−37. e6.
[106] Agirre E, Bellora N, Alló M, Pagès A, Bertucci P, Kornblihtt AR, et al. A chromatin code for alternative splicing involving a putative association between CTCF and HP1α proteins. BMC Biol May 2, 2015;13:31.

[107] Ladomery M. Aberrant alternative splicing is another hallmark of cancer. Int J Cell Biol 2013;2013:1–6.
[108] Kim Y-J, Kim H-S. Alternative splicing and its impact as a cancer diagnostic marker. Genomics Inform June 2012;10(2):74–80.
[109] Sun X, Tian Y, Wang J, Sun Z, Zhu Y. Genome-wide analysis reveals the association between alternative splicing and DNA methylation across human solid tumors. BMC Med Genom December 2020;13(1):4.
[110] Li Y, Sahni N, Pancsa R, McGrail DJ, Xu J, Hua X, et al. Revealing the determinants of widespread alternative splicing perturbation in cancer. Cell Rep October 2017;21(3):798–812.
[111] Yuan H, Li N, Fu D, Ren J, Hui J, Peng J, et al. Histone methyltransferase SETD2 modulates alternative splicing to inhibit intestinal tumorigenesis. J Clin Invest September 1, 2017;127(9):3375–91.
[112] Riffo-Campos ÁL, Gimeno-Valiente F, Rodríguez FM, Cervantes A, López-Rodas G, Franco L, et al. Role of epigenetic factors in the selection of the alternative splicing isoforms of human KRAS in colorectal cancer cell lines. Oncotarget April 17, 2018;9(29):20578–89.
[113] Batsché E, Yi J, Mauger O, Kornobis E, Hopkins B, Hanmer-Lloyd C, et al. CD44 alternative splicing senses intragenic DNA methylation in tumors via direct and indirect mechanisms. Nucleic Acids Res June 21, 2021;49(11):6213–37.
[114] Pant D, Narayanan SP, Vijay N, Shukla S. Hypoxia-induced changes in intragenic DNA methylation correlate with alternative splicing in breast cancer. J Biosci December 2020;45(1):3.
[115] Guo T, Sakai A, Afsari B, Considine M, Danilova L, Favorov AV, et al. A novel functional splice variant of AKT3 defined by analysis of alternative splice expression in HPV-positive oropharyngeal cancers. Cancer Res October 1, 2017;77(19):5248–58.
[116] Guo T, Zambo KDA, Zamuner FT, Ou T, Hopkins C, Kelley DZ, et al. Chromatin structure regulates cancer-specific alternative splicing events in primary HPV-related oropharyngeal squamous cell carcinoma. Epigenetics September 2020;15(9):959–71.
[117] Couture F, Sabbagh R, Kwiatkowska A, Desjardins R, Guay S-P, Bouchard L, et al. PACE4 undergoes an oncogenic alternative splicing switch in cancer. Cancer Res December 15, 2017;77(24):6863–79.
[118] Allemand E, Myers MP, Garcia-Bernardo J, Harel-Bellan A, Krainer AR, Muchardt C. A broad set of chromatin factors influences splicingEyras E, editor. PLoS Genet September 23, 2016;12(9):e1006318.

CHAPTER

Post-transcriptional regulation: a less explored territory in the world of neurodegenerative diseases

5

Ayeman Amanullah
University of Twente, Enschede, Netherlands

Introduction

In an organism, every cell encodes a set of instructions in the form of DNA, which is expressed into proteins by the process of transcription and translation. Unlike prokaryotes, in eukaryotes, the transcription and translation process is not coupled. The transcription site is in the nucleus, whereas the mature RNA translocates into the cytoplasm to get translated into proteins [1]. This segregation of transcription from translation permits eukaryotes to integrate an extensive post-transcriptional gene regulation step. This regulatory process involves recruiting numerous RNA binding proteins (RBPs) and other regulatory factors such as micro-RNAs. The interaction of nascent primary transcripts with RBPs and other regulatory factors decides the fate of primary transcripts that whether they would be translated into a protein or not [2,3]. The proteins thus synthesized then makes up the cellular proteome that largely determines the cell characteristics. Broadly, eukaryotic post-transcriptional regulation includes RNA splicing, capping, polyadenylation, nucleocytoplasmic export, RNA editing, and mRNA decay that would be discussed briefly in the following section.

In eukaryotes, the primary transcript RNA polymerase-II mediated transcription is subjected to various modification steps to make it suitable for its translation into proteins, done through the process of post-transcriptional modifications. These modifications provide necessary changes (such as removing noncoding regions) in immature mRNA so that the ribosomes may act on it for translation in the cytoplasm. The modification process is cotranscriptional, that is, it starts when the mRNA is still undergoing transcription [4]. The process begins with capping, where, the addition of 5′ m7G cap, N7-methylated guanosine, to the first nucleotide of pre-mRNA occurs. Capping helps in the splicing and export of nascent RNA from the nucleus. In the cytoplasm, the cap linked with the cap-binding complex serves as a recognition site for the translation initiation factors, which is crucial for the ribosomal action on pre-mRNA to translate it into proteins [5].

RNA splicing is the next step in the modification of primary mRNA which can be constitutive or alternative. In constitutive splicing, introns or noncoding regions of pre-mRNAs are removed initially, and then the resultant coding regions or exons are then ligated together to form mature mRNA. Whereas, in alternative splicing, exons that arise from splicing might or might not be skipped during the ligation step resulting in a diverse array of mature mRNAs from the individual pre-mRNA. Alternative splicing thus enhances the diversity of transcriptome obtained from the corresponding genome, consequently

Post-Transcriptional Gene Regulation in Human Disease, Volume 32. https://doi.org/10.1016/B978-0-323-91305-8.00001-6

increasing the overall complexity of the proteome [6,7]. A study on transcriptomes of 15 different human tissues and cell lines concluded ~94% of genes undergo alternative splicing [8]. Similarly, another study also estimated alternative splicing in ~95% of primary transcripts in human tissues [9]. The RNA splicing process is followed by the incorporation of an RNA stretch made up of 100–200 long adenine bases called polyadenylation tail to the 3′ end of nascent RNA. The addition of poly (A) tail protects the mRNA from degradation by phosphates and nucleases in the cytoplasm. Further, its role in nucleocytoplasmic export has also been elucidated in the past [10,11].

RNA editing is another mechanism that may provide additional diversity to the proteome. The process includes changes in RNA through insertion, deletion, or base substitution (conversion of one base to another). An estimated 85% of pre-mRNA reported in humans may undergo A - > I base substitution [12] indicating the significance of this process. Although not completely clear, the significant functions of RNA editing include altering the functions of protein, changing the splicing pattern, or cellular localization of RNA involved [13]. Increased understanding of eukaryotic post-transcriptional gene regulation in controlling gene expression has also gained interest in exploring molecular intricacies of mRNA decay. The process includes various pathways that degrade bulk (like normal protein-coding mRNAs) or specific mRNAs (e.g., defective mRNAs) with the help of endo or exonucleases along with several accessory proteins (e.g., RNA binding proteins) and micro-RNAs [14–16]. The mRNA decay process thus aids in regulating and surveillance of gene expression. This chapter's focus is to elucidate the importance of post-transcriptional gene regulation in neurodegenerative diseases (NDDs). Therefore, the following section provides an overview of recent advances in understanding the role of post-transcriptional gene regulation in a few commonly known NDDs.

Post-transcriptional gene regulation in neurodegenerative disorders

Neurodegenerative diseases encompass a group of complex and incurable disorders that exhibits a common feature of progressive loss of neurons. The resulting deterioration of the nervous system leads to the occurrence of mild to severe consequences. This may include mild cognitive impairment, loss of gait and balance, and dementia. The limitations in the understanding of the mechanistic aspect of the onset and development of neurodegenerative diseases can be realized from our current knowledge of protein functions that are frequently associated with these disorders. For instance, to date, functions of proteins like α-synuclein, amyloid-β, and huntingtin are not completely understood. Neurodegenerative diseases are observed frequently in older individuals. In 2016, neurological disorders were the second leading cause of death globally [17]. Among the neurodegenerative diseases, Alzheimer's disease (AD) is the most common, and an estimated 24 million people are affected by it globally [18]. Few other instances of neurodegenerative disorders include Amyotrophic lateral sclerosis (ALS), Parkinson's disease (PD), Spinocerebellar ataxia (SCA), and Huntington's disease (HD).

From the existing studies, role of disrupted proteostasis due to malfunctioning of protein folding and degradation machinery, mutations, oxidative damage, mitochondrial dysfunction, DNA damage, and inflammation has been suggested as contributing factors in the onset and progression of neurodegenerative disorders [19–22]. However, the post-transcriptional gene regulatory pathway seems to be comparatively less explored. Understanding the interplay between the post-transcriptional regulatory processes with the aforementioned factors might be crucial in terms of providing novel mechanistic insights into the etiology of neurodegenerative diseases. Fig. 5.1 provides a brief overview of mechanisms that are known to be crucial in the etiology of neurodegenerative diseases. The following sections will provide an overview of the current understanding of the role of post-transcriptional gene regulation in a few of the commonly known neurodegenerative disorders.

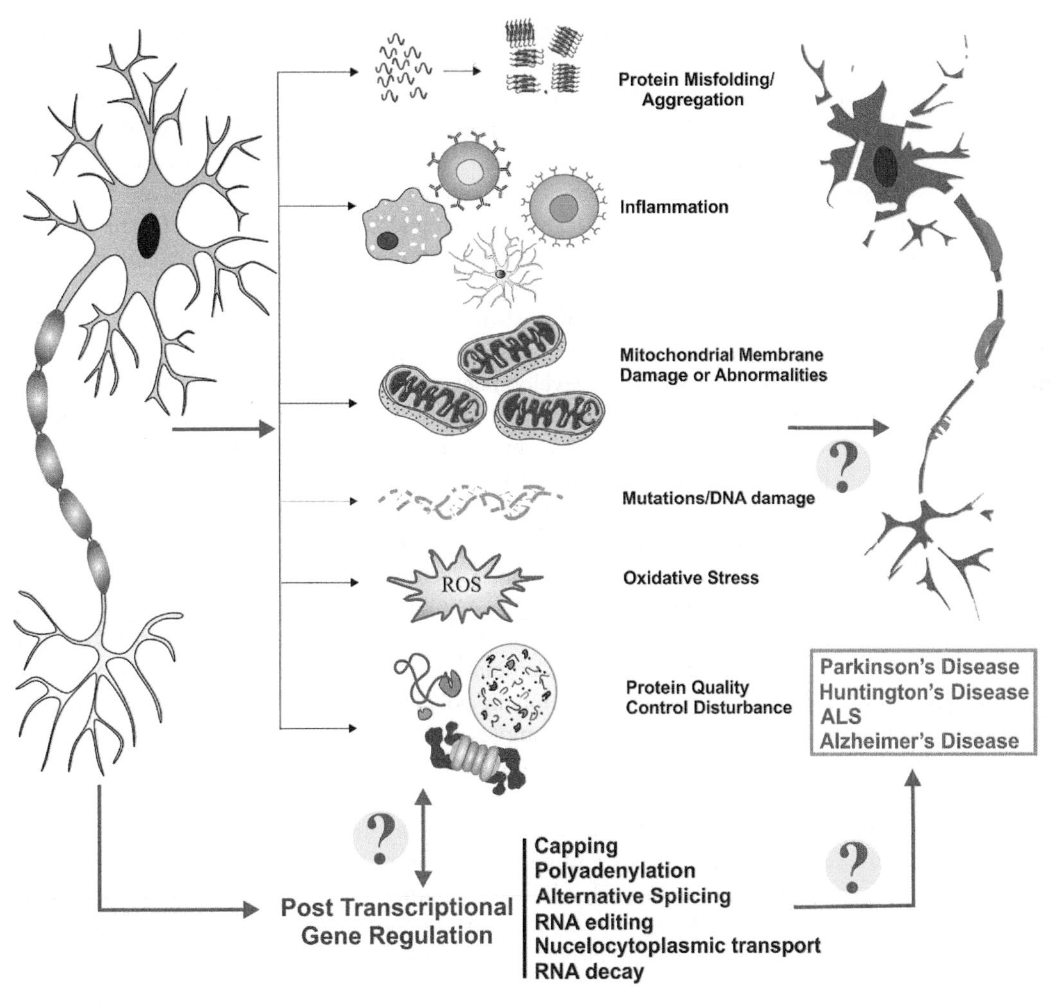

FIGURE 5.1

Overview of major mechanisms found to be crucial in neuronal health. Studies in the direction of understanding the role of post-transcriptional gene regulatory mechanisms may add useful insights in the hunt for gaining mechanistic insights into neurodegenerative diseases. Further understanding of possible interaction between other extensively studied pathological mechanisms with post-transcriptional gene regulation could be beneficial. The question sign indicates the links that need further attention in elucidating crucial aspects of this complex group of disorders.

Alzheimer's disease

According to the world health organization, 47 million people were affected by dementia in the year 2015; this is expected to increase up to 132 million in the year 2050 [23] due to an improved life expectancy globally. Alzheimer's disease is the most common neurodegenerative disorder and the primary cause of dementia in older adults. The incidence of Alzheimer's disease is 80% in individuals older than 85 years [24]. Alzheimer's disease is characterized by neuronal loss in parietal and temporal lobes, along with the hallmark presence of amyloid-β aggregates and τ protein neurofibrillary tangles. However, apart from indications of genetic mutations in a few instances; the underlying causal factors responsible for the disease are still not clear impeding the development of effective therapeutical solutions [25]. Mutations in amyloid precursor protein (APP) and Presenilin (γ-secretase that cuts APP) are considered to be crucial in AD pathology, as they are believed to be involved in the generation of neurotoxic amyloid-β aggregates [26–28]. Around 90% of cases of AD are sporadic; however, the neuropathological and cognitive outcomes are similar in both sporadic and familial forms of AD [29].

Recent evidence has suggested the role of post-transcriptional gene regulation in Alzheimer's disease pathology. Interestingly, the prevalence of alternative splicing and RNA editing is found to be highest in the brain, where it aids in neuronal development, synapse formation, and transmissions [30]. Therefore, understanding the potential involvement of various post-transcriptional gene regulators in neurodegenerative diseases becomes crucial. A recent study found an association of increased intron retention events due to altered alternative splicing with progressive aging and Alzheimer's disease [31]. Disrupted alternative splicing and RNA binding proteins have also been observed in various other studies associated with AD brain samples [32–34], depicting their crucial role in age-associated disease development. Micro-RNAs, another important constituent of post-transcriptional gene regulation was also found to be deregulated in AD patients. A study found deregulation of 41 micro-RNAs in late-onset AD patients and Tau hyperphosphorylation in neurons [35]. Micro-RNAs associated with cholesterol metabolism are also observed to be crucial in AD, as cholesterol levels are crucial in amyloid production [36,37]. Previously, reduction in polyadenylated mRNA in hippocampal and cerebellum regions was also observed in AD samples [38] further supporting the significance of post-transcriptional gene regulation in Alzheimer's disease.

Additionally, the role of deregulated RNA editing in AD pathology has also been reported in the past. RNA editing of glutamate receptor subunit (GluA2) in the hippocampus was shown to be reduced, this is important as altered RNA editing at Q/R site of GluA2 leads to excessive calcium influx resulting in neuronal cell death [39]. Similarly, another study involving AD patients brain samples found decreased RNA editing in the hippocampal region and to a lesser extent in frontal and temporal lobes in 22 genes at more than 30 target sites [40]. Interestingly, a study reported differential expression of seven miRNAs in plasma of Alzheimer's patients, further proving the influence of post-transcriptional gene regulation in neurodegenerative diseases [41]. Overall, despite little evidence, the additional insights obtained from these studies add crucial information to our current understanding of AD etiology.

Parkinson's disease

Parkinson's disease is a neurodegenerative disease affecting motor functions. The characteristic events include loss of dopaminergic neurons in substantia nigra and formation of Lewy bodies in neurons [42].

The selective vulnerability of certain neuronal subpopulations in PD is an interesting subject which is needed to be further addressed. It has been suggested that neuronal populations with certain characteristics such as long and thin axons in comparison to their somata, poorly myelinated or unmyelinated axons, and extensive axonal arborization are more susceptible to PD [42,43]. Parkinson's disease can be divided into six stages with the clinical possibility of late-phase assessment of the degenerative process [43]. Globally, PD is the second most commonly occurring neurodegenerative disease with an estimated 6.1 million people being affected in the year 2016 [44]. Gender-wise, the male population is diagnosed with PD more frequently than females. Clinically, rapid eye movement and rigidity are more common characteristics seen in the male population suffering from Parkinson's disease. Whereas, in PD-suffering females, dyskinesias and depression are more prevalent [45].

A meta-analysis found more than 1900 per 100,000 individuals of >80 year age group had PD compared to 41, 107, 428, and 1087 people in the fifth, sixth, seventh, and eighth decade of life respectively, showing its age-related prevalence [46]. The formation and accumulation of misfolded α-synuclein have largely been attributed to the occurrence of Parkinson's disease; however, the underlying triggers are still needed to be elucidated. Post-transcriptional gene regulation is one of the crucial links that may provide additional insights in this direction. A recent study found the change in differential transcript usage (i.e., relative influence of one transcript to the overall transcriptional output of a gene) in the PD brain [47]. The selective association of specific RNA transcript isoforms of synuclein having extended $3'$ untranslated region (aSynL) with Parkinson's disease pathology has also been reported earlier [48]. Deregulated isoform expression of genes such as synphilin-1 and synuclein α had also indicated the role of alternative splicing in PD [49]. In a separate study, the role of altered parkin isoforms was also reported in PD pathology [50]. Altogether, these studies provide crucial evidence supporting the involvement of changed alternative splicing patterns that may serve as a triggering factor in PD pathology. Similarly, to alternative splicing, altered micro-RNA expression levels have also been observed in PD.

A comparative miRNA profiling of PD prefrontal cortex samples with normal controls found 125 differentially expressed miRNAs [51]. In another study performed on putamen tissues of PD patients, altered expression of miRNAs associated with oxidative stress pathway was observed [52]. Recently, miR-153 and miR-223, two miRNAs involved in the regulation of α-synuclein levels were found to be significantly decreased in the saliva of PD patients. Consequently, this was proposed as a potential diagnostic biomarker for idiopathic PD [53]. A more recent study found deregulated miR-193b levels in PD peripheral blood mononuclear cells. The study concluded the role of miR-193b in PD development with the PGC-1α/FNDC5/BDNF pathway [54]. Overall, in the past decade, considerable interest has been shown in understanding the role of post-transcriptional gene regulatory mechanisms as a crucial factor in the development and progression of Parkinson's disease. Besides being a therapeutic target, deregulated micro-RNA levels in blood or saliva are also investigated as potential biomarkers in disease identification.

Spinocerebral ataxia

Spinocerebral ataxia is an umbrella term given to a subset of neurodegenerative disorders that share progressive ataxia as a prominent clinical manifestation [55]. To date, more than 40 different types of SCAs have been identified. These are denoted as SCA followed by the number in the order in which they were identified such as SCA 1 which was identified in the year 1993 [56]. Accumulated data from the past 3 decades reveals interesting information about the underlying disease-causing heterogenous genetic nature of SCA. Six of the SCAs, that is, SCA1, SCA2, SCA3, SCA6, SCA7, and SCA17 are classified as polyglutamine SCAs. In these SCAs, mutation leading to the generation of expanded Cytosine, Adenine, Guanine (CAG) nucleotide repeat sequences encodes abnormal polyglutamine (poly Q) stretches in the respective aberrant protein. The repeat expansion containing genes identified in SCA1, SCA2, SCA3, SCA6, SCA7, and SCA17 are ataxin-1 ataxin-2, ataxin-3, subunit-α voltage-dependent calcium channel, ataxin-7, and Thymine, Adenine, Thymine, Adenine (TATA)-box-binding protein, respectively [57]. Alternatively, SCAs with repeat expansions in noncoding regions and due to other types of mutations (insertions, deletions, etc.) have also been identified [58]. From the limited epidemiological data available, an average of 2.7 out of 100,000 individuals is estimated to be affected with hereditary ataxia [59]. In European children prevalence rate of ataxia is approximately 26/100,000 children and ataxic cerebral palsy was the main contributor out of all forms of ataxia [60]. Globally SCA 3 was found to be the most common of the autosomal dominant ataxias [61].

In the past few years, emerging pieces of evidence are beginning to provide novel insights on the role of post-transcriptional gene regulatory mechanisms in SCA disease pathology. A recent study conducted on the knockin SCA3 mouse model reported the potential role of alternative splicing in Spinocerebellar ataxia type 3 (SCA3). The study suggested that the presence of CAG repeat expansions promotes aberrant splicing leading to enhanced production of diseased protein aggregation-prone isoforms [62]. Interestingly, more recently, ataxin-2 has been reported to promote post-transcriptional polyadenylation of its target mRNA including TDP-43 that in turn aids in the activation of mRNA translation [63]. The study further suggests that this regulatory aspect of TDP-43 by ataxin-2 may be crucial in modifying neurodegeneration. The role of abnormal ribosomal RNA (rRNA) processing in SCA-2 and Huntington's disease has also been suggested in another study. The work concluded that the disrupted rRNA processing was mediated by abnormal interaction of expanded ATX-2 and HTT transcripts with RNA-binding proteins [64].

Previously, various studies involving the identification of deregulated levels of micro-RNAs in tissues of SCA disease models and human samples have also been performed having therapeutic and diagnosis significance. In a study involving serum samples of SCA3 patients, a cluster of miRNAs was found to be deregulated and the authors suggested four of the miRNAs viz. miR-125b miR-29a, miR-25, and miR-34b as biomarkers for SCA3/Machado-Joseph disease [65]. Similarly, various other studies have also identified differentially regulated miRNA levels in SCA types [66–68]. Additionally, the protective role of miRNA in neurodegenerative diseases is also being observed. It was found that reduction in miRNA processing resulted in increased polyQ toxicity of SCA-3 protein in drosophila and human cells [69]. Likewise, deregulated mir-29a/b levels have been reported to be associated with a common neuronal death mechanism in SCA-17 and other neurodegenerative diseases [70]. Altogether, the understanding of the role of post-transcriptional mechanism in SCA disease pathology is improving. However, still, the number of studies exploring the role of post-transcriptional processes in SCA is limited. More studies in this direction would be crucial in gaining further mechanistic insights into this complex group of neurodegenerative diseases.

Huntington's disease

Huntington's disease is classified as one of the rare forms of inherited neurodegenerative disorders caused by mutation in the huntingtin (HTT) gene that expresses the huntingtin protein. However, the possibility of sporadic forms of HD cannot be neglected [71]. The HTT gene mutation results in expansion of CAG repeats from normal 10–35 to 36–120 times in HD patients [72,73]. The disease involves cognitive, movement, and behavioral symptoms [74]. Mostly, the age of onset for Huntington's disease is around 40 years with a median of 15 years of survival postoccurrence of motor symptoms [75]. Interestingly, a study reported that though the disease duration (diagnosis to death period) is independent of CAG repeat length, the expansion mutation determines the age at death [76]. The basal ganglia and neocortex are the brain regions that are also highly connected show the most vulnerability to HD [77]. Epidemiological studies have found an increased occurrence of HD in the population of European descent [78]. Whereas, HD occurrence is less common in countries like Taiwan, Hongkong, and Japan [79]. Among the nine known hereditary polyglutamine neurodegenerative diseases, HD is the most common [80]. Available data from the western population shows that 10.3–13.7 per 100,000 individuals are afflicted by this disease [80,81]. Currently, although polyglutamine diseases are comparatively rare, a new study done on the general European population indicates higher proportions of individuals carrying intermediate to pathological ranges of polyglutamine disease-associated alleles [80].

Similar to most other neurodegenerative diseases no cure is available, and limited treatment options include medications for specific symptoms and slowing disease progression. Thus, studies directed toward an increase in understanding of mechanisms involved in disease progression can provide novel targets for therapeutic interventions. Interestingly, a recent mass spectrometry study found that the mutant HTT RNA captures majorly proteins involved in the spliceosome pathway and results in missplicing in HD cell models. The study also observed deregulated splicing in human HD samples and suggested the altered splicing as a critical mechanism for RNA-induced toxicity in HD [82]. Previously, the formation of pathogenic exon 1 protein has been shown to result from aberrant splicing of HTT [83]. With the progress in techniques such as next-generation sequencing, identification of global changes occurring in diseased conditions is becoming possible. Recently, a novel strategy based on such a sequencing approach showed a specific pathogenic mis-spliced signature in human striatal tissues of HD [84]. Additionally, another work using RNA sequencing had also reported aberrant alternative splicing in motor cortex samples of HD [85].

In the past, compromised RNA editing pattern of GluR-2 (glutamate ionotropic receptor α-amino-3-hydroxy-5-methyl-4-isoxazole propionic acid (AMPA)-type subunit 2) RNA in HD patients striatum region has also been shown. As GluR-2 is associated with calcium homeostasis or conductance, it was proposed that the resulting change in calcium permeability due to altered proportions of edited to unedited RNA may be a contributing factor in neuronal dysfunctions [86]. Interestingly, the relative abundance of HTT mRNA 3′ Untranslated region (UTR) isoforms in the human Huntington's disease brain was changed. In addition, knockdown of RNA-binding protein CNOT6 (CCR4-NOT Transcription Complex Subunit 6) in control fibroblast recapitulated some of the huntingtin isoform differences [87]. The role of miRNAs was also found to be crucial in HD pathology. In this regard, a study found downregulation of miRNA-128a in both monkey models and human patients of HD [88]. Similarly, various other studies have also reported disrupted miRNA levels in the brain and other peripheral tissues providing crucial pieces of evidence in support of the key role played by these entities in HD [89–91].

Amyotrophic lateral sclerosis

Amyotrophic lateral sclerosis, also known as motor neuron disease or Lou Gehrig's disease, is another disorder in the list of neurodegenerative diseases that affects voluntary motor function. Degeneration of upper and lower motor neurons results is known to be the cardinal factor involved in ALS pathology [92]. The early symptoms of weakness in arms or legs are referred to as "limb onset," while "bulbar onset" ALS is the term given when speaking and swallowing problems start appearing [93–95]. According to hopkinsmedicine.org, the prevalence of ALS is five times more than Huntington's disease and is estimated to be the cause of 5 deaths per 100,000 individuals of age 20 or more. Like other neurodegenerative diseases, there is no cure for ALS, and there is growing interest in understanding the role of post-transcriptional modifications in ALS pathology. The incidence rate of sporadic ALS cases is higher in comparison to familial ALS.

Approximately 60%–70% of the sporadic cases have a 30%–90% loss of EAAT2 (excitatory amino acid transporter 2) protein. One of the possible explanations for this loss was attributed to aberrant mRNA processing of EAAT2, which was observed only in neuropathologically affected brain regions of ALS patients [96]. RNA binding protein TDP-43 (transactivating response region DNA binding protein) aggregates and mutations are one of the pathological factors associated with ALS pathology. Interestingly, a recent study on ALS-linked TDP-43 mutations observed aberrant RNA splicing and adult-onset motor neuron disease [97]. However, no aggregation or loss of nuclear TPD-43 was observed indicating the critical role of disrupted RNA splicing in ALS pathology. Another recent work found similar aberrant splicing patterns in ALS models having fused in sarcoma (FUS) mutation; an RNA binding protein that is characteristically deposited in ALS brain regions [98]. Previously, another study involving the C9ORF72 (C9) gene with expanded GGGGCC hexanucleotide, a common causative factor of ALS also found disrupted splicing in ALS brains. The deregulation of splicing was associated with sequestration of splicing factor hnRNP H with the G-quadruplex structures formed by the expanded transcripts [99]. Further, reduced RNA editing in ALS motor neurons and its possible link with depleted Adenosine deaminases acting on RNA (ADAR2) activity is also suggested to be a contributing factor in motor neuron death seen in ALS [100].

Like other neurodegenerative diseases, in ALS also variations in miRNA profiles have been studied. A study reported 15 downregulated miRNAs in ALS patient cells, out of which two (miR-34a and miR504) were found to be crucial due to their involvement in neurodegenerative mechanisms [101]. Likewise, different studies have reported disrupted miRNA expression levels in different tissue types obtained from ALS patients [102–105]. In addition, new shreds of evidence are also beginning to elucidate the involvement of the nonsense-mediated mRNA decay (NMD) pathway in ALS pathology. It was observed that mutant FUS resulted in disrupted regulation of the NMD pathway [106]. Additionally, it was also shown recently that reactivation of nonsense-mediated mRNA decay by expression of UPF1, a core gene in the NMD pathway, protects against C9orf72 dipeptide-repeat neurotoxicity [107]. Overall, from the examples mentioned in the text post-transcriptional gene regulatory mechanisms can be considered as one of the key driving factors in some neurodegenerative diseases. Further, the association of several aberrantly functioning genes or proteins associated with post-transcriptional gene regulatory processes with neurodegenerative diseases makes this mRNA regulatory mechanism an important target for future studies. This in turn may provide an effective therapeutic solution against this highly complex group of disorders.

Discussion

Post-transcriptional regulation is the fundamental step in the expression of genes that systematically processes transcribed immature mRNA into the mature one. From previous and newly emerging evidence, altered post-transcriptional regulation can be considered as one of the contributing factors in neurodegenerative diseases. New studies on post-transcriptional gene regulatory mechanisms and their involvement in neurodegenerative diseases are continuously adding useful information. However, answers to some important questions are required. Mutations in TDP-43 and FUS, which are associated with post-transcriptional regulation, are known to be involved in ALS [108], but not all ALS instances have these mutations. The four major genes are known to be involved in familial ALS viz. C9orf72, SOD1, TDP43, and FUS accounts for 40%, 20%, 1%–5%, and 1%–5% cases, respectively [109]. This makes it difficult to get a clear perspective of such mutations in disease etiology. However, efforts like Project minE [110], which involves whole genome sequencing of large samples of ALS patients may be crucial in this direction. Similar other efforts like Global Parkinson's Genetics Program and Alzheimer's Disease Sequencing Project are also under progress, in order to understand the underlying genetic architecture of these complex neurodegenerative disorders [111,112]. Interestingly, as mentioned previously, studies have shown an association of aberrant C9orf72, TDP43, and FUS with altered splicing events. Further, a mutation in the RNU12 gene that is involved in editing more than 800 genes afflicting its RNA editing properties has also been reported to be linked with congenital cerebral ataxia [113]. Therefore, the possibility of mutations in specific genes affecting the proper functioning of the post-transcriptional regulation mechanism at a global scale cannot be neglected.

The presence of a single aberrant protein in cases with different neuropathological outcomes adds further complexity in understanding the underlying disease mechanisms involved. For instance, the pathogenic role of TDP43 has been observed in both ALS and frontotemporal dementia that is now being considered as overlapping disorders [55,114,115]. Similarly, expanded polyglutamine forms of another post-transcriptional regulator Ataxin-2(ATXN2) are implicated in neurodegenerative diseases like ALS and Spinocerebellar ataxia [116–118]. Interestingly, both SCA2 and ALS patients were reported clinically in a family with full CAG repeat expansions in ATXN2 [119]. Therapeutically, reduction of ataxin-2 levels using antisense oligonucleotide has been shown to improve motor performances in both ALS and SCA2 disease models [120]. Similarly, in Parkinson's disease mouse models antisense oligonucleotide against leucine-rich repeat kinase 2 (LRRK2) has been shown to reduce synuclein inclusions [121]. Likewise, results from clinical trials using antisense technology in other neurodegenerative diseases are also encouraging [122,123]. Antisense oligonucleotides are frequently looked as a therapeutic tool to block translation of aberrant proteins, but they can also be used in the toxic gain of function mutations where they can promote normal splicing around target mutations [124,125]. Spinraza (Nusinersen), an antisense technology-based drug that works on the splicing mechanism of survival motor neuron transcripts to increase survival motor neuron protein levels has been approved for spinal muscular atrophy treatment (a group of neuromuscular genetic disorder) recently [126,127].

Altogether, the studies discussed here provide an indication of typical biochemical hotspots, which should be investigated further for improving our understanding of these neurodegenerative diseases. The applications of deregulated levels of micro-RNAs (another component of Post-transcriptional regulation (PTR)) as a biomarker for various neurodegenerative diseases have been shown in various studies [128,129]. Unfortunately, antisense oligonucleotides as a therapeutic option for HD were not successful [130]. The current success of antisense oligonucleotides for therapeutic purposes shows the potential of this area but much work is needed to overcome the existing limitations. Future developments in understanding the role of post-transcriptional gene regulatory mechanisms may prove to be crucial in unraveling the puzzle of this complex group of disorders. The presence of abnormally functioning steps of the post-transcriptional gene regulatory mechanism opens an exciting area for further work to gain a better understanding of this disease. Further, it also provides an additional dimension that researchers must consider while addressing the problem of neurodegenerative diseases.

References

[1] Berg JM, Tymoczko JL, Stryer L, Stryer L. Biochemistry. 6th ed. New York: W.H. Freeman; 2007.
[2] Van Nostrand EL, Freese P, Pratt GA, Wang X, Wei X, Xiao R, et al. A large-scale binding and functional map of human RNA-binding proteins. Nature 2020;583:711–9.
[3] Nishtala S, Neelamraju Y, Janga SC. Dissecting the expression relationships between RNA-binding proteins and their cognate targets in eukaryotic post-transcriptional regulatory networks. Sci Rep 2016;6:25711.
[4] Bentley DL. Coupling mRNA processing with transcription in time and space. Nat Rev Genet 2014;15: 163–75.
[5] Jurado AR, Tan D, Jiao X, Kiledjian M, Tong L. Structure and function of pre-mRNA 5′-end capping quality control and 3′-end processing. Biochemistry 2014;53:1882–98.
[6] Ast G. How did alternative splicing evolve? Nat Rev Genet 2004;5:773–82.
[7] Graveley BR. Alternative splicing: increasing diversity in the proteomic world. Trends Genet TIG 2001;17: 100–7.
[8] Wang ET, Sandberg R, Luo S, Khrebtukova I, Zhang L, Mayr C, et al. Alternative isoform regulation in human tissue transcriptomes. Nature 2008;456:470–6.
[9] Pan Q, Shai O, Lee LJ, Frey BJ, Blencowe BJ. Deep surveying of alternative splicing complexity in the human transcriptome by high-throughput sequencing. Nat Genet 2008;40:1413–5.
[10] Eckner R, Ellmeier W, Birnstiel ML. Mature mRNA 3' end formation stimulates RNA export from the nucleus. EMBO J 1991;10:3513–22.
[11] Hilleren P, McCarthy T, Rosbash M, Parker R, Jensen TH. Quality control of mRNA 3′-end processing is linked to the nuclear exosome. Nature 2001;413:538–42.
[12] Costa Cruz PH, Kawahara Y. RNA editing in neurological and neurodegenerative disorders. Methods Mol Biol 2021;2181:309–30.
[13] Blow M, Futreal PA, Wooster R, Stratton MR. A survey of RNA editing in human brain. Genome Res 2004; 14:2379–87.
[14] Garneau NL, Wilusz J, Wilusz CJ. The highways and byways of mRNA decay. Nat Rev Mol Cell Biol 2007; 8:113–26.
[15] Long RM, McNally MT. mRNA decay: X (XRN1) marks the spot. Mol Cell 2003;11:1126–8.
[16] Iwakawa H-O, Tomari Y. The functions of MicroRNAs: mRNA decay and translational repression. Trends Cell Biol 2015;25:651–65.
[17] Collaborators GBDN. Global, regional, and national burden of neurological disorders, 1990–2016: a systematic analysis for the Global Burden of Disease Study 2016. Lancet Neurol 2019;18:459–80.
[18] Erkkinen MG, Kim M-O, Geschwind MD. Clinical neurology and epidemiology of the major neurodegenerative diseases. Cold Spring Harbor Perspect Biol 2018;10:a033118.
[19] Amanullah A, Upadhyay A, Joshi V, Mishra R, Jana NR, Mishra A. Progressing neurobiological strategies against proteostasis failure: challenges in neurodegeneration. Prog Neurobiol 2017;159:1–38.
[20] Amor S, Puentes F, Baker D, van der Valk P. Inflammation in neurodegenerative diseases. Immunology 2010;129:154–69.
[21] Sbodio JI, Snyder SH, Paul BD. Redox mechanisms in neurodegeneration: from disease outcomes to therapeutic opportunities. Antioxid Redox Signal 2019;30:1450–99.

[22] Madabhushi R, Pan L, Tsai L-H. DNA damage and its links to neurodegeneration. Neuron 2014;83: 266–82.
[23] Organization WH. Global action plan on the public health response to dementia 2017–2025. In: Global action plan on the public health response to dementia 2017–2025; 2017.
[24] Checkoway H, Lundin JI, Kelada SN. Neurodegenerative diseases. IARC scientific publications; 2011. p. 407–19.
[25] DeTure MA, Dickson DW. The neuropathological diagnosis of Alzheimer's disease. Mol Neurodegener 2019;14:32.
[26] O'Brien RJ, Wong PC. Amyloid precursor protein processing and Alzheimer's disease. Annu Rev Neurosci 2011;34:185–204.
[27] Shen J, Kelleher 3rd RJ. The presenilin hypothesis of Alzheimer's disease: evidence for a loss-of-function pathogenic mechanism. Proc Natl Acad Sci USA 2007;104:403–9.
[28] Tcw J, Goate AM. Genetics of β-amyloid precursor protein in Alzheimer's disease. Cold Spring Harb Perspect Med 2017;7:a024539.
[29] Ruberti F, Barbato C, Cogoni C. Post-transcriptional regulation of amyloid precursor protein by microRNAs and RNA binding proteins. Commun Integr Biol 2010;3:499–503.
[30] Kiebler M, Scheiffele P, Ule J. What, where, and when: the importance of post-transcriptional regulation in the brain. Front Neurosci 2013;7.
[31] Adusumalli S, Ngian ZK, Lin WQ, Benoukraf T, Ong CT. Increased intron retention is a post-transcriptional signature associated with progressive aging and Alzheimer's disease. Aging Cell 2019;18:e12928.
[32] Bai B, Hales CM, Chen P-C, Gozal Y, Dammer EB, Fritz JJ, et al. U1 small nuclear ribonucleoprotein complex and RNA splicing alterations in Alzheimer's disease. Proc Natl Acad Sci USA 2013;110: 16562–7.
[33] Tollervey JR, Wang Z, Hortobágyi T, Witten JT, Zarnack K, Kayikci M, et al. Analysis of alternative splicing associated with aging and neurodegeneration in the human brain. Genome Res 2011;21:1572–82.
[34] Apicco DJ, Zhang C, Maziuk B, Jiang L, Ballance HI, Boudeau S, et al. Dysregulation of RNA splicing in tauopathies. Cell Rep 2019;29:4377–88. e4374.
[35] Lau P, Bossers K, Janky R, Salta E, Frigerio CS, Barbash S, et al. Alteration of the microRNA network during the progression of Alzheimer's disease. EMBO Mol Med 2013;5:1613–34.
[36] Wang H, Kulas JA, Wang C, Holtzman DM, Ferris HA, Hansen SB. Regulation of beta-amyloid production in neurons by astrocyte-derived cholesterol. Proc Natl Acad Sci USA 2021;118. e2102191118.
[37] Yoon H, Flores LF, Kim J. MicroRNAs in brain cholesterol metabolism and their implications for Alzheimer's disease. Biochim Biophys Acta 2016;1861:2139–47.
[38] Harrison PJ, Barton AJ, Najlerahim A, McDonald B, Pearson RC. Regional and neuronal reductions of polyadenylated messenger RNA in Alzheimer's disease. Psychol Med 1991;21:855–66.
[39] Gaisler-Salomon I, Kravitz E, Feiler Y, Safran M, Biegon A, Amariglio N, et al. Hippocampus-specific deficiency in RNA editing of GluA2 in Alzheimer's disease. Neurobiol Aging 2014;35:1785–91.
[40] Khermesh K, D'Erchia AM, Barak M, Annese A, Wachtel C, Levanon EY, et al. Reduced levels of protein recoding by A-to-I RNA editing in Alzheimer's disease. RNA 2016;22:290–302.
[41] Kumar P, Dezso Z, MacKenzie C, Oestreicher J, Agoulnik S, Byrne M, et al. Circulating miRNA biomarkers for Alzheimer's disease. PLoS One 2013;8:e69807.
[42] Wong YC, Luk K, Purtell K, Burke Nanni S, Stoessl AJ, Trudeau L-E, et al. Neuronal vulnerability in Parkinson disease: should the focus be on axons and synaptic terminals? Mov Disord 2019;34:1406–22.
[43] Braak H, Ghebremedhin E, Rüb U, Bratzke H, Del Tredici K. Stages in the development of Parkinson's disease-related pathology. Cell Tissue Res 2004;318:121–34.

[44] Dorsey ER, Elbaz A, Nichols E, Abd-Allah F, Abdelalim A, Adsuar JC, et al. Global, regional, and national burden of Parkinson's disease, 1990–2016: a systematic analysis for the Global Burden of Disease Study 2016. Lancet Neurol 2018;17:939–53.
[45] Miller IN, Cronin-Golomb A. Gender differences in Parkinson's disease: clinical characteristics and cognition. Mov Disord Off J Mov Disord Soc 2010;25:2695–703.
[46] Pringsheim T, Jette N, Frolkis A, Steeves TD. The prevalence of Parkinson's disease: a systematic review and meta-analysis. Mov Disord Off J Mov Disord Soc 2014;29:1583–90.
[47] Dick F, Nido GS, Alves GW, Tysnes O-B, Nilsen GH, Dölle C, et al. Differential transcript usage in the Parkinson's disease brain. PLoS Genet 2020;16:e1009182.
[48] Rhinn H, Qiang L, Yamashita T, Rhee D, Zolin A, Vanti W, et al. Alternative α-synuclein transcript usage as a convergent mechanism in Parkinson's disease pathology. Nat Commun 2012;3. 1084-1084.
[49] Beyer K, Domingo-Sàbat M, Humbert J, Carrato C, Ferrer I, Ariza A. Differential expression of alpha-synuclein, parkin, and synphilin-1 isoforms in Lewy body disease. Neurogenetics 2008;9:163–72.
[50] Humbert J, Beyer K, Carrato C, Mate JL, Ferrer I, Ariza A. Parkin and synphilin-1 isoform expression changes in Lewy body diseases. Neurobiol Dis 2007;26:681–7.
[51] Hoss AG, Labadorf A, Beach TG, Latourelle JC, Myers RH. microRNA profiles in Parkinson's disease prefrontal cortex. Front Aging Neurosci 2016;8. 36-36.
[52] Nair VD, Ge Y. Alterations of miRNAs reveal a dysregulated molecular regulatory network in Parkinson's disease striatum. Neurosci Lett 2016;629:99–104.
[53] Cressatti M, Juwara L, Galindez JM, Velly AM, Nkurunziza ES, Marier S, et al. Salivary microR-153 and microR-223 levels as potential diagnostic biomarkers of idiopathic Parkinson's disease. Mov Disord 2020; 35:468–77.
[54] Baghi M, Yadegari E, Rostamian Delavar M, Peymani M, Ganjalikhani-Hakemi M, Salari M, et al. MiR-193b deregulation is associated with Parkinson's disease. J Cell Mol Med 2021;25:6348–60.
[55] Abramzon YA, Fratta P, Traynor BJ, Chia R. The overlapping genetics of amyotrophic lateral sclerosis and frontotemporal dementia. Front Neurosci 2020;14.
[56] Orr HT, Chung MY, Banfi S, Kwiatkowski Jr TJ, Servadio A, Beaudet AL, et al. Expansion of an unstable trinucleotide CAG repeat in spinocerebellar ataxia type 1. Nat Genet 1993;4:221–6.
[57] Paulson HL, Shakkottai VG, Clark HB, Orr HT. Polyglutamine spinocerebellar ataxias - from genes to potential treatments. Nat Rev Neurosci 2017;18:613–26.
[58] Klockgether T, Mariotti C, Paulson HL. Spinocerebellar ataxia. Nat Rev Dis Prim 2019;5:24.
[59] Ruano L, Melo C, Silva MC, Coutinho P. The global epidemiology of hereditary ataxia and spastic paraplegia: a systematic review of prevalence studies. Neuroepidemiology 2014;42:174–83.
[60] Musselman KE, Stoyanov CT, Marasigan R, Jenkins ME, Konczak J, Morton SM, et al. Prevalence of ataxia in children: a systematic review. Neurology 2014;82:80–9.
[61] Shakkottai VG, Fogel BL. Clinical neurogenetics: autosomal dominant spinocerebellar ataxia. Neurol Clin 2013;31:987–1007.
[62] Ramani B, Harris GM, Huang R, Seki T, Murphy GG, Costa MC, et al. A knockin mouse model of spinocerebellar ataxia type 3 exhibits prominent aggregate pathology and aberrant splicing of the disease gene transcript. Hum Mol Genet 2015;24:1211–24.
[63] Inagaki H, Hosoda N, Tsuiji H, Hoshino S-I. Direct evidence that Ataxin-2 is a translational activator mediating cytoplasmic polyadenylation. J Biol Chem 2020;295:15810–25.
[64] Li PP, Moulick R, Feng H, Sun X, Arbez N, Jin J, et al. RNA toxicity and perturbation of rRNA processing in spinocerebellar ataxia type 2. Mov Disord 2021;36:2519–29.
[65] Shi Y, Huang F, Tang B, Li J, Wang J, Shen L, et al. MicroRNA profiling in the serums of SCA3/MJD patients. Int J Neurosci 2014;124:97–101.

[66] Borgonio-Cuadra VM, Valdez-Vargas C, Romero-Córdoba S, Hidalgo-Miranda A, Tapia-Guerrero Y, Cerecedo-Zapata CM, et al. Wide profiling of circulating MicroRNAs in spinocerebellar ataxia type 7. Mol Neurobiol 2019;56:6106−20.

[67] Persengiev S, Kondova I, Otting N, Koeppen AH, Bontrop RE. Genome-wide analysis of miRNA expression reveals a potential role for miR-144 in brain aging and spinocerebellar ataxia pathogenesis. Neurobiol Aging 2011;32:2316.e2317−27.

[68] Koscianska E, Krzyzosiak WJ. Current understanding of the role of microRNAs in spinocerebellar ataxias. Cerebell & Ataxias 2014;1:7.

[69] Bilen J, Liu N, Burnett BG, Pittman RN, Bonini NM. MicroRNA pathways modulate polyglutamine-induced neurodegeneration. Mol Cell 2006;24:157−63.

[70] Roshan R, Ghosh T, Gadgil M, Pillai B. Regulation of BACE1 by miR-29a/b in a cellular model of Spinocerebellar ataxia 17. RNA Biol 2012;9:891−9.

[71] Alpaugh M, Cicchetti F. Huntington's disease: lessons from prion disorders. J Neurol 2021;268:3493−504.

[72] Testa CM, Jankovic J. Huntington disease: a quarter century of progress since the gene discovery. J Neurol Sci 2019;396:52−68.

[73] Myers RH. Huntington's disease genetics. NeuroRx 2004;1:255−62.

[74] Cardoso F. Chapter fifty - nonmotor symptoms in Huntington disease. In: Chaudhuri KR, Titova N, editors. International review of neurobiology. Academic Press; 2017. p. 1397−408.

[75] Chaganti SS, McCusker EA, Loy CT. What do we know about Late Onset Huntington's Disease? J Huntingtons Dis 2017;6:95−103.

[76] Keum JW, Shin A, Gillis T, Mysore JS, Abu Elneel K, Lucente D, et al. The HTT CAG-expansion mutation determines age at death but not disease duration in Huntington disease. Am J Hum Genet 2016;98:287−98.

[77] Blumenstock S, Dudanova I. Cortical and striatal circuits in Huntington's disease. Front Neurosci 2020;14. 82−82.

[78] Kay C, Hayden MR, Leavitt BR. Chapter 3 - epidemiology of Huntington disease. In: Feigin AS, Anderson KE, editors. Handbook of clinical neurology. Elsevier; 2017. p. 31−46.

[79] Ghosh R, Tabrizi SJ. Clinical features of Huntington's disease. Adv Exp Med Biol 2018;1049:1−28.

[80] Gardiner SL, Boogaard MW, Trompet S, de Mutsert R, Rosendaal FR, Gussekloo J, et al. Prevalence of carriers of intermediate and pathological polyglutamine disease-associated alleles among large population-based cohorts. JAMA Neurol 2019;76:650−6.

[81] Bates GP, Dorsey R, Gusella JF, Hayden MR, Kay C, Leavitt BR, et al. Huntington disease. Nat Rev Dis Prim 2015;1:15005.

[82] Schilling J, Broemer M, Atanassov I, Duernberger Y, Vorberg I, Dieterich C, et al. Deregulated splicing is a major mechanism of RNA-induced toxicity in Huntington's disease. J Mol Biol 2019;431:1869−77.

[83] Sathasivam K, Neueder A, Gipson TA, Landles C, Benjamin AC, Bondulich MK, et al. Aberrant splicing of HTT generates the pathogenic exon 1 protein in Huntington disease. Proc Natl Acad Sci USA 2013;110: 2366−70.

[84] Elorza A, Márquez Y, Cabrera JR, Sánchez-Trincado JL, Santos-Galindo M, Hernández IH, et al. Huntington's disease-specific mis-splicing unveils key effector genes and altered splicing factors. Brain 2021; 144:2009−23.

[85] Lin L, Park JW, Ramachandran S, Zhang Y, Tseng Y-T, Shen S, et al. Transcriptome sequencing reveals aberrant alternative splicing in Huntington's disease. Hum Mol Genet 2016;25:3454−66.

[86] Akbarian S, Smith MA, Jones EG. Editing for an AMPA receptor subunit RNA in prefrontal cortex and striatum in Alzheimer's disease, Huntington's disease and schizophrenia. Brain Res 1995;699:297−304.

[87] Romo L, Ashar-Patel A, Pfister E, Aronin N. Alterations in mRNA 3′ UTR isoform abundance accompany gene expression changes in human Huntington's disease brains. Cell Rep 2017;20:3057−70.

[88] Kocerha J, Xu Y, Prucha MS, Zhao D, Chan AWS. microRNA-128a dysregulation in transgenic Huntington's disease monkeys. Mol Brain 2014;7:46.
[89] Chang K-H, Wu Y-R, Chen C-M. Down-regulation of miR-9* in the peripheral leukocytes of Huntington's disease patients. Orphanet J Rare Dis 2017;12:185.
[90] Martí E, Pantano L, Bañez-Coronel M, Llorens F, Miñones-Moyano E, Porta S, et al. A myriad of miRNA variants in control and Huntington's disease brain regions detected by massively parallel sequencing. Nucleic Acids Res 2010;38:7219–35.
[91] Dong X, Cong S. The emerging role of microRNAs in polyglutamine diseases. Front Mol Neurosci 2019; 12.
[92] Ravits JM, La Spada AR. ALS motor phenotype heterogeneity, focality, and spread: deconstructing motor neuron degeneration. Neurology 2009;73:805–11.
[93] Morris J. Amyotrophic lateral sclerosis (ALS) and related motor neuron diseases: an overview. Neurodiagn J 2015;55:180–94.
[94] Chiò A, Logroscino G, Hardiman O, Swingler R, Mitchell D, Beghi E, et al. Prognostic factors in ALS: a critical review. Amyotroph Lateral Scler 2009;10:310–23.
[95] Walling AD. Amyotrophic lateral sclerosis: Lou Gehrig's disease. Am Fam Phys 1999;59:1489–96.
[96] Lin C-LG, Bristol LA, Jin L, Dykes-Hoberg M, Crawford T, Clawson L, et al. Aberrant RNA processing in a neurodegenerative disease: the cause for absent EAAT2, a glutamate transporter, in amyotrophic lateral sclerosis. Neuron 1998;20:589–602.
[97] Arnold ES, Ling S-C, Huelga SC, Lagier-Tourenne C, Polymenidou M, Ditsworth D, et al. ALS-linked TDP-43 mutations produce aberrant RNA splicing and adult-onset motor neuron disease without aggregation or loss of nuclear TDP-43. Proc Natl Acad Sci USA 2013;110:E736–45.
[98] Ito D, Taguchi R, Deguchi M, Ogasawara H, Inoue E. Extensive splicing changes in an ALS/FTD transgenic mouse model overexpressing cytoplasmic fused in sarcoma. Sci Rep 2020;10:4857.
[99] Conlon EG, Lu L, Sharma A, Yamazaki T, Tang T, Shneider NA, et al. The C9ORF72 GGGGCC expansion forms RNA G-quadruplex inclusions and sequesters hnRNP H to disrupt splicing in ALS brains. Elife 2016;5.
[100] Kwak S, Kawahara Y. Deficient RNA editing of GluR2 and neuronal death in amyotropic lateral sclerosis. J Mol Med 2005;83:110–20.
[101] Rizzuti M, Filosa G, Melzi V, Calandriello L, Dioni L, Bollati V, et al. MicroRNA expression analysis identifies a subset of downregulated miRNAs in ALS motor neuron progenitors. Sci Rep 2018;8:10105.
[102] Shioya M, Obayashi S, Tabunoki H, Arima K, Saito Y, Ishida T, et al. Aberrant microRNA expression in the brains of neurodegenerative diseases: miR-29a decreased in Alzheimer disease brains targets neurone navigator 3. Neuropathol Appl Neurobiol 2010;36:320–30.
[103] De Felice B, Annunziata A, Fiorentino G, Borra M, Biffali E, Coppola C, et al. miR-338-3p is overexpressed in blood, CFS, serum and spinal cord from sporadic amyotrophic lateral sclerosis patients. Neurogenetics 2014;15:243–53.
[104] Wakabayashi K, Mori F, Kakita A, Takahashi H, Utsumi J, Sasaki H. Analysis of microRNA from archived formalin-fixed paraffin-embedded specimens of amyotrophic lateral sclerosis. Acta Neuropathol Commun 2014;2:173.
[105] Benigni M, Ricci C, Jones AR, Giannini F, Al-Chalabi A, Battistini S. Identification of miRNAs as potential biomarkers in cerebrospinal fluid from amyotrophic lateral sclerosis patients. NeuroMolecular Med 2016; 18:551–60.
[106] Kamelgarn M, Chen J, Kuang L, Jin H, Kasarskis EJ, Zhu H. ALS mutations of FUS suppress protein translation and disrupt the regulation of nonsense-mediated decay. Proc Natl Acad Sci USA 2018;115: E11904–e11913.
[107] Xu W, Bao P, Jiang X, Wang H, Qin M, Wang R, et al. Reactivation of nonsense-mediated mRNA decay protects against C9orf72 dipeptide-repeat neurotoxicity. Brain 2019;142:1349–64.

[108] Colombrita C, Onesto E, Megiorni F, Pizzuti A, Baralle FE, Buratti E, et al. TDP-43 and FUS RNA-binding proteins bind distinct sets of cytoplasmic messenger RNAs and differently regulate their post-transcriptional fate in motoneuron-like cells *. J Biol Chem 2012;287:15635–47.
[109] van Es MA, Hardiman O, Chio A, Al-Chalabi A, Pasterkamp RJ, Veldink JH, et al. Amyotrophic lateral sclerosis. Lancet 2017;390:2084–98.
[110] Van Rheenen W, Pulit SL, Dekker AM, Al Khleifat A, Brands WJ, Iacoangeli A, et al. Project MinE: study design and pilot analyses of a large-scale whole-genome sequencing study in amyotrophic lateral sclerosis. Eur J Hum Genet 2018;26:1537–46.
[111] Global Parkinson's Genetics Program. GP2: The global Parkinson's genetics program. Mov Disord 2021;36:842–51.
[112] Nafikov RA, Nato Jr AQ, Sohi H, Wang B, Brown L, Horimoto AR, et al. Analysis of pedigree data in populations with multiple ancestries: strategies for dealing with admixture in Caribbean Hispanic families from the ADSP. Genet Epidemiol 2018;42:500–15.
[113] Elsaid MF, Chalhoub N, Ben-Omran T, Kumar P, Kamel H, Ibrahim K, et al. Mutation in noncoding RNA RNU12 causes early onset cerebellar ataxia. Ann Neurol 2017;81:68–78.
[114] Mackenzie IRA, Rademakers R. The role of transactive response DNA-binding protein-43 in amyotrophic lateral sclerosis and frontotemporal dementia. Curr Opin Neurol 2008;21:693–700.
[115] Lillo P, Matamala JM, Valenzuela D, Verdugo R, Castillo JL, Ibáñez A, et al. [Overlapping features of frontotemporal dementia and amyotrophic lateral sclerosis]. Rev Med Chile 2014;142:867–79.
[116] Watanabe R, Higashi S, Nonaka T, Kawakami I, Oshima K, Niizato K, et al. Intracellular dynamics of Ataxin-2 in the human brains with normal and frontotemporal lobar degeneration with TDP-43 inclusions. Acta Neuropathol Commun 2020;8:176.
[117] van den Heuvel DMA, Harschnitz O, van den Berg LH, Pasterkamp RJ. Taking a risk: a therapeutic focus on ataxin-2 in amyotrophic lateral sclerosis? Trends Mol Med 2014;20:25–35.
[118] Nkiliza A, Chartier-Harlin M-C. ATXN2 a culprit with multiple facets. Oncotarget 2017;8:34028–9.
[119] Tazen S, Figueroa K, Kwan JY, Goldman J, Hunt A, Sampson J, et al. Amyotrophic lateral sclerosis and spinocerebellar ataxia type 2 in a family with full CAG repeat expansions of ATXN2. JAMA Neurol 2013;70:1302–4.
[120] Crunkhorn S. Ataxin 2 reduction rescues motor defects. Nat Rev Drug Discov 2017;16:384–5.
[121] Zhao HT, John N, Delic V, Ikeda-Lee K, Kim A, Weihofen A, et al. LRRK2 antisense oligonucleotides ameliorate α-synuclein inclusion formation in a Parkinson's disease mouse model. Molecular therapy. Nucleic acids 2017;8:508–19.
[122] Miller TM, Pestronk A, David W, Rothstein J, Simpson E, Appel SH, et al. An antisense oligonucleotide against SOD1 delivered intrathecally for patients with SOD1 familial amyotrophic lateral sclerosis: a phase 1, randomised, first-in-man study. Lancet Neurol 2013;12:435–42.
[123] Bennett CF, Krainer AR, Cleveland DW. Antisense oligonucleotide therapies for neurodegenerative diseases. Annu Rev Neurosci 2019;42:385–406.
[124] Schneider SA, Alcalay RN. Precision medicine in Parkinson's disease: emerging treatments for genetic Parkinson's disease. J Neurol 2020;267:860–9.
[125] Martinovich KM, Shaw NC, Kicic A, Schultz A, Fletcher S, Wilton SD, et al. The potential of antisense oligonucleotide therapies for inherited childhood lung diseases. Mol Cell Pediatr 2018;5. 3-3.
[126] Hua Y, Vickers TA, Baker BF, Bennett CF, Krainer AR. Enhancement of SMN2 exon 7 inclusion by antisense oligonucleotides targeting the exon. PLoS Biol 2007;5. e73-e73.
[127] Neil EE, Bisaccia EK. Nusinersen: a novel antisense oligonucleotide for the treatment of spinal muscular atrophy. J Pediatr Pharmacol Therapeut 2019;24:194–203.

[128] Sheinerman KS, Toledo JB, Tsivinsky VG, Irwin D, Grossman M, Weintraub D, et al. Circulating brain-enriched microRNAs as novel biomarkers for detection and differentiation of neurodegenerative diseases. Alzheimer's Res Ther 2017;9:89.

[129] van den Berg MMJ, Krauskopf J, Ramaekers JG, Kleinjans JCS, Prickaerts J, Briedé JJ. Circulating microRNAs as potential biomarkers for psychiatric and neurodegenerative disorders. Prog Neurobiol 2020;185:101732.

[130] Kingwell K. Double setback for ASO trials in Huntington disease. Nat Rev Drug Discov 2021;20:412–3.

CHAPTER

Role of post-transcriptional gene regulation in hematological malignancies

Hafiz M. Ahmad

Department of Molecular Cell and Cancer Biology, University of Massachusetts Chan Medical School, Worcester, MA, United States

Introduction

Post-transcriptional gene regulation adds an additional layer of scrutiny for controlled gene expression. Cells use post-transcriptional gene regulation to synthesize the required amount of protein when required. They can also synthesize different variants of the same protein by utilizing alternate splicing as required for their normal proliferation and differentiation. Similarly, they can also increase or decrease the stability of mRNAs as per their need by using mRNA modifications, mRNA decay, and miRNA-mediated regulation. Cells can also utilize alternate translation initiation sites for the translation of proteins. Alternate translation sites are important for quick regulation of protein synthesis as well as are implicated in escaping shut down of translation by chemotherapeutic drugs. These and several other regulatory processes to control mRNA stability require interaction between RNA-binding proteins and their target mRNAs. RNA-binding proteins are crucial determinants in the post-transcriptional regulation of gene expression. Aberrant expression and mutations are reported in RNA-binding proteins in various hematological malignancies. Mutations in genes encoding splicing factors and altered expression of proteins that regulate modifications in mRNA are also frequent events in hematological disorders. We will learn about these factors in the context of normal hematopoiesis and their dysregulation in hematopoietic disorders as well as in chemotherapeutic resistance in this chapter.

IRES

5′-7 mG cap-dependent translation is the most frequent way of translation for most eukaryotic mRNAs. This mechanism involves the binding of translational apparatus to the 7-methyl guanosine cap using translation initiation factors and the 40S ribosome subunit. However, there are a lot of genes relevant to hematological malignancies that carry an internal ribosomal entry site (IRES) element. These genes include those involved in the characteristic translocations such as c-Myc, Cyclin D1, BCL2, BCL6, BCR/ABL1, and PML/RARα but their contribution to transformation and progression in hematological malignancies is not well understood. IRES is also implicated in drug resistance. In

Post-Transcriptional Gene Regulation in Human Disease, Volume 32. https://doi.org/10.1016/B978-0-323-91305-8.00007-7

response to chemotherapeutic drugs that target cap-dependent translation pathways, cancer cells can switch to IRES-dependent initiation of translation. For that, IRES interact with initiation factors and IRES trans-acting factors (ETFs) to drive cap-independent translation [1]. Inhibitors of the mTOR pathway and DNA-damaging drugs can obstruct cap-dependent translation in hematopoietic malignancies without affecting IRES-mediated translation. Oncogenes or chemoresistant genes can still be expressed using cap-independent translation. That is why it is important to target IRES-mediated translation that is bypassed in the chemotherapy-mediated global shutdown of protein translation.

Cap-independent translation is well documented in multiple myeloma, a disorder caused by a clonal increase in the number of plasmocytes. Broad-spectrum drugs, for example, Melphalan in combination with other drugs are used as a therapy for multiple myeloma. Proteosome-targeting drugs such as bortezomib are also used to treat multiple myeloma [2]. The immunomodulatory drug, thalidomide, is also increasingly used for treatments of multiple myeloma [3,4]. Thalidomide may inhibit the proliferative effect of multiple myeloma cells partly by interacting with the IRES of basic fibroblast growth factor (b-FGF) and blocking translation. However, a lot of focus has been on targeting c-Myc because multiple myeloma cells are c-Myc-dependent and disease progression is linked to the levels of c-Myc in the cells [5–7]. The c-Myc mRNA contains an IRES in its 5′-UTR [8,9]. Mutation of cytosine to thymidine at position 2796 of IRES interferes with the stability of its secondary structure that enhances translation of c-Myc mRNA [10]. The bone marrow of multiple myeloma patients showed an over-representation of C-T IRES mutation [11].

The interaction between IRES trans-acting factor (ITAF) proteins is also important for cap-independent translation. In various hematological malignancies, ITAFs are expressed in a dysregulated manner. For example, in chronic lymphocytic leukemia (CLL) and myeloid leukemia, nucleolin overexpression alters its subcellular localization [12,13]. Similarly, overexpression of SP1 is linked to a worse prognosis in several hematological malignancies, including acute leukemia [14,15]. IRES of SP1 mRNA is targeted by nucleolin. Overexpression of SP1 can cause resistance to chemotherapy in leukemia stem cells [15,16]. HuR is another IRES-binding protein that is upregulated in acute myeloid leukemia (AML), chronic myelogenous leukemia (CML), and ALL [17]. HuR and its implication in leukemia will be discussed in detail in the "RNA-binding proteins" section. HuR is known to block IRES-dependent translation of cyclin-dependent kinase inhibitor p27KIP1 [18]. Low expression of p27KIP1 is bad for AML prognosis [19]. Several ITAFs have been identified for c-Myc [20–23]. Two ITAFs, in particular, for c-Myc IRES, PTB1 and YB1 showed better binding to the mutated IRES of c-Myc. Cell culture-based studies have shown a correlation between the expression levels of PTB1 and YB1, and c-Myc in multiple myeloma-derived cell lines that carry these mutations [24].

MNK1 (MAP kinase-interacting serine/threonine-protein kinase 1) is an important player in the stimulation of the 5′-IRES of c-Myc mRNA during chemotherapy. MNK1 is essential for the interaction between ITAFs, hnRNPA1 and RPS25, and the 5′-IRES of c-Myc mRNA [23,25]. hnRNPA1 and RPS25 are shown to be necessary for cap-independent translation [26]. Pharmacological inhibition of hnRNPA1 also confirmed the interaction of hnRNPA1 to the c-Myc IRES upon induction of Endoplasmic reticulum (ER) stress [25]. Thus, targeting cap-independent inhibition along with cap-dependent translation can provide the exciting possibility to eliminate any escape route for malignant cells.

Modifications of RNA

Post-transcriptional modifications are reported both in coding and noncoding RNAs that widely impact gene expression. There are about 170 reported modifications. The most common modifications include N6-methyladenosine (m6A), A-to-I editing, pseudouridine (ψ), 5-methylcytosine (m5C), 5-hydroxymethylcytosine (hm5C), N1-methyladenosine (m1A), and N4-acetylcytidine (ac4C). These modifications affect the translation of mRNA and add an extra layer of epi-transcriptional regulation of gene expression. These modifications are found all over the mRNA; A-to-I editing is found in the exons of mRNA; 5-methylcytosine (m5C) and 7-methylguanosine (m7G) are present in the 5′-UTR of mRNAs; N6-methyladenosine (m6A) is present both in coding and noncoding regions of mRNAs.

m6A is a dynamic and reversible modification executed by a group of regulators. These regulators are categorized as writers that add m6A, erasers that selectively remove m6A, and readers that recognize m6A marks [27]. m6A modification affects mRNA stability, transport, degradation, and utilization by the translational machinery [28–30]. Strict regulation of m6A modification is critical for normal hematopoiesis [31,32]. Dysregulated expression levels of demethylase or methyltransferase may lead to misregulation of m6A modification, thus, leading to leukemogenesis [33]. m6A writers and erasers both are reported to be dysregulated in AML. Two major components of the m6A modulatory complex, METTL3 and METTL14 are critical for replenishing the hematopoietic pool of stem cells. The differentiation of hematopoietic stem and progenitor cells depends on low levels of METTL3 and METTL14. Overexpression of METTL3 and METTL14 are found in AML suggesting an oncogenic role of these two proteins [32,34]. METTL3 overexpression increases the survival of AML cells by stimulating the translation of c-Myc, Bcl-2, and Phosphatase and tensin homolog (PTEN) [31]. METTL14 also regulates c-Myc and MYB expression by regulating m6A marks. METTL14 is not only required for self-renewal of leukemia stem cells and progression of leukemia, but it also plays a significant role in the drug response of AML cells [32]. METTL16 has also been potentially implicated in AML [34].

Overexpression of an m6A eraser, fat mass, and obesity-associated demethylase (FTO), is reported in a variety of hematological malignancies, for example, AML with t(11q23)/MLL rearrangements, t(15; 17)/PML-RARA, FLT3-ITD (internal tandem duplication), or NPM1 mutations. FTO acts as an oncogene in these AMLs by post-transcriptionally regulating its target genes like ASB2 and RARA [35]. Similar to METTL3 and FTO, m6A reader YTHDF2 plays an important role in AML cells. Overexpression of YTHDF2 is linked to enhanced propagation of AML cells. YTHDF2 is not required for normal hematopoiesis, making it a suitable therapeutic target [36]. Insulin-like growth factor 2 mRNA-binding protein (IF2BP1–3) is another m6A reader protein that is also critical for oncogenic development by stabilizing m6A modified c-Myc oncogene [37]. This suggests that overexpression of the members of methyltransferase complex leading to dysregulation of the m6A modification might be a crucial contributor to leukemogenesis.

RNA-binding proteins

Expression from mRNAs is also regulated by another class of regulatory factors, RNA-binding proteins (RBPs). RBPs interact with mRNAs to produce a ribonucleoprotein complex (RNP), the formation of which is critical for the coordination of RNA processing and regulation of post-transcriptional gene expression. RBPs play an important role in the maturation of primary mRNA transcripts into mature mRNA. RBPs form nucleoprotein complexes that are needed for splicing, 5′-guanyl cap addition, polyadenylation, nucleocytoplasmic transport of mRNAs, and other organelles within the cell. Trans-acting regulatory RBPs bind to untranslated regions and coding regions to regulate the translation process. Alterations in the activities of RBPs have been implicated in various types of cancers including leukemia [38,39].

The RNA-binding proteins Musashi-1 (MSI1) and Musashi-2 (MSI2) regulate several important biological processes linked to the initiation of cancer, its progression, and the development of drug resistance. Musashi proteins play a critical role in the maintenance of hematopoietic stem cells. Overexpression of Musashi proteins was linked to several types of cancers including various types of leukemia. They regulate various oncogenic signaling pathways by binding and controlling the stability of mRNAs that are part of these signaling pathways, for example, c-Myc, c-MET, PTEN/mTOR, TGFβ/SMAD3, NUMB/Notch. By controlling the expression of these proteins, Musashi proteins regulate cancer progression, metastasis, and drug resistance [40]. Overexpression of MSI has been reported in AML, acute lymphoblastic leukemia (ALL), and chronic myelogenous leukemia-blastic phase (CML-BP) [41–43]. Inhibition of MSI proteins presents a novel therapeutic strategy in the treatment of hematological malignancies as well as in solid tumors. MSI2 targeting has been demonstrated recently by a small molecule inhibitor Ro 08-2750 (Ro). It interferes with the binding of MSI2 to its target mRNAs and increases apoptosis in myeloid leukemia cells [44]. (−)-Gossypol, a natural phenol extracted from cottonseed [45] and ω−9 monounsaturated fatty acids (e.g., oleic acid) [46] also show MSI1 inhibition, although further work is required for determining their specificity on MSI1 inhibition.

DDX3 is another RBP that regulates mRNA translation. DDX3 is dysregulated in hematological malignancies. It is also mutated in T-cell lymphoma and lymphocytic leukemia [47,48]. RK-33 is an inhibitor that has been developed to target DDX3 in lung cancer. RK-33 shows proapoptotic properties and increases the sensitivity of DDX3 overexpressing lung cancer cells to radiation [49,50]. This presents a promising outlook to develop a DDX3 inhibitor that can work in hematological malignancies.

HuR is another RBP that binds to Adenylate-uridylate (AU)-rich 3′-UTR and is essential for normal hematopoiesis and is implicated in malignant hematopoiesis. HuR stabilizes target mRNA by several possible mechanisms [51]. It can stabilize mRNAs by binding directly to their AU-rich elements [52]; it can work in association with miRNAs [53]; it can also affect mRNA stability through m6A interaction [54], and it can also affect target mRNA stability by transporting of them to the cytoplasm from the nucleus [55]. The activity of HuR is dysregulated in several types of leukemia [56–59] and can affect a broader aspect of leukemogenesis. HuR expression is increased in M4 acute myeloid leukemia (AML). Increased HuR levels regulate the stability of eukaryotic translation initiation factor 4E (elF4E) [60]. HuR expression also increases in CML when the disease progresses from chronic phase to blast crisis [61]. HuR expression is modulated through miR-16 [62]. HuR expression shows an inverse correlation with miR-16/miR-15 in chronic lymphocytic leukemia B-CLL [63]. More examples of RBPs providing chemo-resistance to AML cells are HuR and insulin-like growth factor mRNA-binding protein-3. These proteins are induced by IL-18 in chemo-resistant cells. IL-18 shows a positive correlation with chemo-resistance in AML cells. HuR and insulin-like growth factor mRNA-binding protein-3 stabilize antiapoptotic COX-2 mRNA that provides drug resistance to the AML cells [58].

Alternative splicing

Alternative splicing is an important process to make multiple mature isoforms from the same primary mRNA transcript in the nucleus. A core complex of ribonucleoproteins, the spliceosome is required for the splicing process. The spliceosome is composed of small nuclear riboproteins (snRNPs) and small nuclear RNA (snRNA). Splicing is a tightly regulated process controlled by cis- and trans-acting elements. Cis-acting elements are specific signals in the nucleotide sequence of primary mRNA transcript critical for proper splicing, while trans-acting factors are splicing factors. Trans-acting factors can make changes in splice site selection giving rise to alternative splicing [64–66]. mRNA modifications can also selectively affect the binding of trans-acting factors affecting splicing of target mRNA; for example, m6A modification increases binding of hnRNPs, A2B1 [67] or SRSF3 [68]. Splicing regulators play an important role in leukemia development and progression. TP53-2 (ASPP2) is a tumor suppressor required for TP53-mediated apoptosis. A splice variant of ASPP2, ASPP2-K is overexpressed in AML. ASPP2-K inhibits apoptosis and promotes the proliferation of AML cells [69]. There is a change in the splicing pattern of hnRNP A1 due to exon skipping when CML progresses from chronic phase to blast phase [70].

Alternative splicing factor 1 (SRSF1) is implicated in leukemia development. Protein arginine methyltransferase (PRMT5)-mediated methylation of SRSF1 modulates the splicing of several mRNAs related to leukemia development [71]. Its overexpression is also linked to the aggressive phenotype of leukemic cells [72]. PRMT5 inhibitors are undergoing clinical trials as a therapeutic for hematopoietic malignancies [73]. RNA-binding motif protein 39 (RBM39) is another RNA-binding protein that can regulate RNA splicing and is known to be critical for the proliferation of AML cells [74]. Proteasomal degradation of RBM39 by a sulfonamide drug (Indisulam, Tasisulam, E7820, and CQS) showed an anticancer effect [75].

Mutations are reported in splicing factor genes in myelodysplastic syndromes (MDS), chronic myelomonocytic leukemia (CMML), AML, and CLL [76]. The most common mutations are heterozygous point mutations in SF3B1, U2AF1, and SRSF2 [77]. Frequencies of mutations in LUC7L2, PRPF8, SF3B1, SRSF2, U2AF1, and ZRSR2 genes range between 40% and 85% in various subtypes of MDS, 5% in AML, and 10% in myeloproliferative neoplasms (MPN) [77]. Mutations in splicing genes are also linked to the prognosis of leukemia, for example, mutations in the SRSF2 gene are linked to poor survival outcome and increased AML progression whereas mutations in U2AF1 are linked with shorter survival as well as with lower remission rate [77]. Mutations in splicing factor SF3B are implicated in leukemia development and progression. A small-molecule inhibitor, H3B-8800 has been shown to selectively target mutant SF3B but not wildtype SF3B [78]. Somatic mutations in spliceosome genes are reported in almost 50% of the MDS patients. U2 small nuclear RNA auxiliary factor 1 (U2AF1) is a spliceosome component known to be mutated in MDS patients. AML cells including primary patient cells show high sensitivity to pharmacological inhibition by the drug Sudemycin, in vitro as well as in vivo [79]. Mutations in SF3B1 and SRSF2 result in hyperactivation of NF-kB signaling [80]. The presence of splicing mutations in MDS and other myeloid malignancies demonstrates the importance of the highly regulated splicing process for normal hematopoiesis.

MicroRNAs

Apart from coding mRNAs and long noncoding RNAs (lncRNAs), a group of small noncoding RNAs called microRNAs (miRNAs) have been identified as post-transcriptional regulators of eukaryotic gene expression. miRNAs can bind to specific mRNA and decisively regulate several biological processes. Their expression is frequently altered in solid cancers as well as hematological malignancies. Changes in the expression pattern of miRNAs can be used in diagnosis, prognosis, and treatment response in patients with various hematological malignancies including CLL, CML, acute lymphocytic leukemia (ALL), AML, and acute adult T-cell leukemia [81].

miRNAs in chronic lymphocytic leukemia: The commonly dysregulated miRNAs that are linked with leukemia progression and drug-resistant in CLL are miR-15/16 cluster, miR-34b/c, miR-29, miR-181b, miR-17/92, miR-150, and miR-155 [82]. miR-15a and miR-16-1 are frequently downregulated in CLL patients as well as miR-15a/miR-16-1 locus is frequently deleted in CLL [83]. miR-15a/miR-16-1 expression is also downregulated by histone deacetylases [84]. Downregulation of miR-15a/miR-16-1 is linked with resistance to apoptosis in B-CLL cells and therapeutic resistance [85]. Downregulation of miR-15a/miR-16-1 in CLL is correlated with a favorable prognosis [86]. Downregulation of the miR-29 family members, miR-29a, miR-29b, and miR-29c, is associated with unfavorable prognosis in a subset of CLL patients [86]. Genomic locus 11q that harbors miR-34 cluster is often deleted in CLL patients. miR-34 downregulation is linked with the inactivation of p53 leading to an aberrant response to DNA damage and repair [87]. miR-17/92 microRNA cluster is upregulated in a variety of lymphoid malignancies [88].

In addition, miR-150 is downregulated in CLL and reduces B-cell receptor signaling [89]. miR-155 is activated by STAT3 and upregulated in CLL. miR-155 expression in CLL is associated with poor response to treatment and clinical prognosis [82,90]. miR-181b is also commonly downregulated in CLL patients [91]. Downregulation of miR-181b is associated with poor prognosis [91].

miRNAs in chronic myeloid leukemia: miR-22, miR-10a, miR-17/92, miR-150, miR-203, and miR-328 are examples of commonly deregulated miRNAs in CML [92–94]. miR-17/92 is upregulated in the early chronic phase of CML and its expression is induced by breakpoint cluster region protein/Tyrosine-protein kinase ABL1 (BCR-ABL) and c-Myc, making it a potential therapeutic target [95]. Progression from chronic phase to blast stage is associated with downregulation of miR-328 [92]. Other miRNAs that are regulated by BCR-ABL are miR-130a and miR-130b [96]. miRNAs that downregulate oncogenic BCR-ABL are frequently silenced in leukemic cells either by deletion or epigenetic silencing, for example, miR-203 and miR-451 [97,98]. Expression of miR-150 and miR-10a is downregulated in CML [99,100]. miR-505-5p and miR-193b-3p are also proposed as imatinib response markers in CML patients [101].

miRNAs in acute lymphocytic leukemia: Based on 50 ALL samples, a common miRNA signature of miR-223, miR-19b, miR-20a, miR-92, miR-142–3p, miR-150, miR-93, miR-26a, miR-16, and miR-342 was identified [102]. Studies with the murine model have shown that miR-19b, miR-20a, miR-26a, miR-92, and miR-223 can act cooperatively with Notch for leukemia development by targeting tumor suppressors IKAROS, PTEN, BIM, PHF6, NF1, and FBXW7 [102].

miRNAs in acute myeloid leukemia: Several miRNAs including miR-29, miR-125, miR-142, miR-146, and miR-155 have been identified in AML. The expression of these miRNAs is perturbed and contributes to AML development and progression [103]. miR-29b regulates DNA methyltransferases DNMT3A and DNMT3B and downregulation of miR-29b results in hypermethylation in AML cells [104]. Altered expression of miR-29a and miR-29b in AML cells have also been implicated in aberrant apoptosis and cell cycle progression [105].

Expression of miR-125b is upregulated in the blast stage and is responsible for the malignant transformation of AML cells. Overexpression of miR-125 is implicated in an MPN like phenotype that later transformed into AML. miR-125b can also target tumor suppressor antagonists like Bak1 to inhibit apoptosis and enhance cell proliferation [106]. The genomic location of miR-125b on chromosome 21 makes it of particular interest in Down's syndrome (DS). Trisomy of chromosome 21 is linked with increased expression of miR-125b in DS as well as non-DS acute megakaryocytic leukemia [107].

Upregulation of let-7a-2-3p has been associated with the downregulation of oncogenic JDP2. It also favors a good prognosis and overall longer survival [108]. Upregulation of miR-181 is also associated with favorable prognosis in cytogenetically normal AML by downregulating toll-like receptors and interleukin- 1β [109]. miR-181b is down-regulated in relapsed and refractory AML cases [110]. Downregulation of miR-135a and miR-409−3p is identified in 238 intermediate-risk AML patients and their downregulation is correlated with increased chances of relapse [111]. miR-34 is linked with a favorable prognosis in AML [112].

Overall, microRNAs contribute to leukemia development, progression, and therapeutic resistance and could be used as a marker for disease development and progression.

Summary

Post-transcriptional gene regulation is a key process that controls the level of translated proteins in the cell according to their instant needs. Post-transcriptional regulation of gene expression has emerged as an important area in hematopoietic stem cell self-renewal, differentiation, and leukemogenesis. mRNA processing including alternate splicing, mRNA modifications, RNA-binding proteins, and feedback loops to transcriptional regulation represent opportunities for drug targeting, as well as immunotherapy. Besides the above-mentioned regulators of post-transcriptional gene expression, another important player is microRNAs. microRNAs are involved in the pathophysiology of almost every disease including hematological malignancies. MicroRNAs can be used as diagnostic and prognostic markers in leukemia. They can also be used as an indicator of drug response and clinical outcomes.

References

[1] Martinez-Salas E, Lozano G, Fernandez-Chamorro J, Francisco-Velilla R, Galan A, Diaz R. RNA-binding proteins impacting on internal initiation of translation. Int J Mol Sci 2013;14(11):21705−26.

[2] Lee AH, Iwakoshi NN, Anderson KC, Glimcher LH. Proteasome inhibitors disrupt the unfolded protein response in myeloma cells. Proc Natl Acad Sci USA 2003;100(17):9946−51.

[3] Olson KB, Hall TC, Horton J, Khung CL, Hosley HF. Thalidomide (N-phthaloylglutamimide) in the treatment of advanced cancer. Clin Pharmacol Ther 1965;6:292−7.

[4] Singhal S, Mehta J, Desikan R, Ayers D, Roberson P, Eddlemon P, et al. Antitumor activity of thalidomide in refractory multiple myeloma. N Engl J Med 1999;341(21):1565−71.

[5] Holien T, Vatsveen TK, Hella H, Waage A, Sundan A. Addiction to c-MYC in multiple myeloma. Blood 2012;120(12):2450–3.
[6] Chng WJ, Huang GF, Chung TH, Ng SB, Gonzalez-Paz N, Troska-Price T, et al. Clinical and biological implications of MYC activation: a common difference between MGUS and newly diagnosed multiple myeloma. Leukemia 2011;25(6):1026–35.
[7] Skopelitou A, Hadjiyannakis M, Tsenga A, Theocharis S, Alexopoulou V, Kittas C, et al. Expression of C-myc p62 oncoprotein in multiple myeloma: an immunohistochemical study of 180 cases. Anticancer Res 1993;13(4):1091–5.
[8] Nanbru C, Lafon I, Audigier S, Gensac MC, Vagner S, Huez G, et al. Alternative translation of the proto-oncogene c-myc by an internal ribosome entry site. J Biol Chem 1997;272(51):32061–6.
[9] Stoneley M, Paulin FE, Le Quesne JP, Chappell SA, Willis AE. C-Myc 5' untranslated region contains an internal ribosome entry segment. Oncogene 1998;16(3):423–8.
[10] Paulin FE, West MJ, Sullivan NF, Whitney RL, Lyne L, Willis AE. Aberrant translational control of the c-myc gene in multiple myeloma. Oncogene 1996;13(3):505–13.
[11] Chappell SA, LeQuesne JP, Paulin FE, deSchoolmeester ML, Stoneley M, Soutar RL, et al. A mutation in the c-myc-IRES leads to enhanced internal ribosome entry in multiple myeloma: a novel mechanism of oncogene de-regulation. Oncogene 2000;19(38):4437–40.
[12] Shen N, Yan F, Pang J, Wu LC, Al-Kali A, Litzow MR, et al. A nucleolin-DNMT1 regulatory axis in acute myeloid leukemogenesis. Oncotarget 2014;5(14):5494–509.
[13] Otake Y, Soundararajan S, Sengupta TK, Kio EA, Smith JC, Pineda-Roman M, et al. Overexpression of nucleolin in chronic lymphocytic leukemia cells induces stabilization of bcl2 mRNA. Blood 2007;109(7): 3069–75.
[14] Beishline K, Azizkhan-Clifford J. Sp1 and the 'hallmarks of cancer'. FEBS J 2015;282(2):224–58.
[15] Liu S, Wu LC, Pang J, Santhanam R, Schwind S, Wu YZ, et al. Sp1/NFkappaB/HDAC/miR-29b regulatory network in KIT-driven myeloid leukemia. Cancer Cell 2010;17(4):333–47.
[16] Zhang Y, Chen HX, Zhou SY, Wang SX, Zheng K, Xu DD, et al. Sp1 and c-Myc modulate drug resistance of leukemia stem cells by regulating survivin expression through the ERK-MSK MAPK signaling pathway. Mol Cancer 2015;14:56.
[17] Baou M, Norton JD, Murphy JJ. AU-rich RNA binding proteins in hematopoiesis and leukemogenesis. Blood 2011;118(22):5732–40.
[18] Kullmann M, Gopfert U, Siewe B, Hengst L. ELAV/Hu proteins inhibit p27 translation via an IRES element in the p27 5′UTR. Genes Dev 2002;16(23):3087–99.
[19] Yokozawa T, Towatari M, Iida H, Takeyama K, Tanimoto M, Kiyoi H, et al. Prognostic significance of the cell cycle inhibitor p27Kip1 in acute myeloid leukemia. Leukemia 2000;14(1):28–33.
[20] Jo OD, Martin J, Bernath A, Masri J, Lichtenstein A, Gera J. Heterogeneous nuclear ribonucleoprotein A1 regulates cyclin D1 and c-myc internal ribosome entry site function through Akt signaling. J Biol Chem 2008;283(34):23274–87.
[21] Evans JR, Mitchell SA, Spriggs KA, Ostrowski J, Bomsztyk K, Ostarek D, et al. Members of the poly (rC) binding protein family stimulate the activity of the c-myc internal ribosome entry segment in vitro and in vivo. Oncogene 2003;22(39):8012–20.
[22] Cobbold LC, Spriggs KA, Haines SJ, Dobbyn HC, Hayes C, de Moor CH, et al. Identification of internal ribosome entry segment (IRES)-trans-acting factors for the Myc family of IRESs. Mol Cell Biol 2008; 28(1):40–9.
[23] Shi Y, Frost PJ, Hoang BQ, Benavides A, Sharma S, Gera JF, et al. IL-6-induced stimulation of c-myc translation in multiple myeloma cells is mediated by myc internal ribosome entry site function and the RNA-binding protein, hnRNP A1. Cancer Res 2008;68(24):10215–22.

[24] Cobbold LC, Wilson LA, Sawicka K, King HA, Kondrashov AV, Spriggs KA, et al. Upregulated c-myc expression in multiple myeloma by internal ribosome entry results from increased interactions with and expression of PTB-1 and YB-1. Oncogene 2010;29(19):2884–91.
[25] Shi Y, Yang Y, Hoang B, Bardeleben C, Holmes B, Gera J, et al. Therapeutic potential of targeting IRES-dependent c-myc translation in multiple myeloma cells during ER stress. Oncogene 2016;35(8):1015–24.
[26] Hertz MI, Landry DM, Willis AE, Luo G, Thompson SR. Ribosomal protein S25 dependency reveals a common mechanism for diverse internal ribosome entry sites and ribosome shunting. Mol Cell Biol 2013; 33(5):1016–26.
[27] Shi H, Wei J, He C. Where, when, and how: context-dependent functions of RNA methylation writers, readers, and erasers. Mol Cell 2019;74(4):640–50.
[28] Frye M, Harada BT, Behm M, He C. RNA modifications modulate gene expression during development. Science 2018;361(6409):1346–9.
[29] Delaunay S, Frye M. RNA modifications regulating cell fate in cancer. Nat Cell Biol 2019;21(5):552–9.
[30] Barbieri I, Kouzarides T. Role of RNA modifications in cancer. Nat Rev Cancer 2020;20(6):303–22.
[31] Vu LP, Pickering BF, Cheng Y, Zaccara S, Nguyen D, Minuesa G, et al. The N(6)-methyladenosine (m(6) A)-forming enzyme METTL3 controls myeloid differentiation of normal hematopoietic and leukemia cells. Nat Med 2017;23(11):1369–76.
[32] Weng H, Huang H, Wu H, Qin X, Zhao BS, Dong L, et al. METTL14 inhibits hematopoietic stem/progenitor differentiation and promotes leukemogenesis via mRNA m(6)A modification. Cell Stem Cell 2018; 22(2):191–205. e9.
[33] Deng X, Su R, Feng X, Wei M, Chen J. Role of N(6)-methyladenosine modification in cancer. Curr Opin Genet Dev 2018;48:1–7.
[34] Barbieri I, Tzelepis K, Pandolfini L, Shi J, Millan-Zambrano G, Robson SC, et al. Promoter-bound METTL3 maintains myeloid leukaemia by m(6)A-dependent translation control. Nature 2017; 552(7683):126–31.
[35] Li Z, Weng H, Su R, Weng X, Zuo Z, Li C, et al. FTO plays an oncogenic role in acute myeloid leukemia as a N(6)-methyladenosine RNA demethylase. Cancer Cell 2017;31(1):127–41.
[36] Paris J, Morgan M, Campos J, Spencer GJ, Shmakova A, Ivanova I, et al. Targeting the RNA m(6)A reader YTHDF2 selectively compromises cancer stem cells in acute myeloid leukemia. Cell Stem Cell 2019;25(1): 137–48. e6.
[37] Huang H, Weng H, Sun W, Qin X, Shi H, Wu H, et al. Recognition of RNA N(6)-methyladenosine by IGF2BP proteins enhances mRNA stability and translation. Nat Cell Biol 2018;20(3):285–95.
[38] Pereira B, Billaud M, Almeida R. RNA-binding proteins in cancer: old players and new actors. Trends Cancer 2017;3(7):506–28.
[39] Hodson DJ, Screen M, Turner M. RNA-binding proteins in hematopoiesis and hematological malignancy. Blood 2019;133(22):2365–73.
[40] Kudinov AE, Karanicolas J, Golemis EA, Boumber Y. Musashi RNA-binding proteins as cancer drivers and novel therapeutic targets. Clin Cancer Res 2017;23(9):2143–53.
[41] Griner LN, Reuther GW. Aggressive myeloid leukemia formation is directed by the Musashi 2/Numb pathway. Cancer Biol Ther 2010;10(10):979–82.
[42] Ito T, Kwon HY, Zimdahl B, Congdon KL, Blum J, Lento WE, et al. Regulation of myeloid leukaemia by the cell-fate determinant Musashi. Nature 2010;466(7307):765–8.
[43] Kharas MG, Lengner CJ, Al-Shahrour F, Bullinger L, Ball B, Zaidi S, et al. Musashi-2 regulates normal hematopoiesis and promotes aggressive myeloid leukemia. Nat Med 2010;16(8):903–8.
[44] Minuesa G, Albanese SK, Xie W, Kazansky Y, Worroll D, Chow A, et al. Small-molecule targeting of MUSASHI RNA-binding activity in acute myeloid leukemia. Nat Commun 2019;10(1):2691.

[45] Lan L, Appelman C, Smith AR, Yu J, Larsen S, Marquez RT, et al. Natural product (-)-gossypol inhibits colon cancer cell growth by targeting RNA-binding protein Musashi-1. Mol Oncol 2015;9(7):1406–20.
[46] Clingman CC, Deveau LM, Hay SA, Genga RM, Shandilya SM, Massi F, et al. Allosteric inhibition of a stem cell RNA-binding protein by an intermediary metabolite. Elife 2014;3.
[47] Jiang L, Gu ZH, Yan ZX, Zhao X, Xie YY, Zhang ZG, et al. Exome sequencing identifies somatic mutations of DDX3X in natural killer/T-cell lymphoma. Nat Genet 2015;47(9):1061–6.
[48] Wang L, Lawrence MS, Wan Y, Stojanov P, Sougnez C, Stevenson K, et al. SF3B1 and other novel cancer genes in chronic lymphocytic leukemia. N Engl J Med 2011;365(26):2497–506.
[49] Bol GM, Vesuna F, Xie M, Zeng J, Aziz K, Gandhi N, et al. Targeting DDX3 with a small molecule inhibitor for lung cancer therapy. EMBO Mol Med 2015;7(5):648–69.
[50] Heerma van Voss MR, Vesuna F, Bol GM, Afzal J, Tantravedi S, Bergman Y, et al. Targeting mitochondrial translation by inhibiting DDX3: a novel radiosensitization strategy for cancer treatment. Oncogene 2018; 37(1):63–74.
[51] Suresh Babu S, Joladarashi D, Jeyabal P, Thandavarayan RA, Krishnamurthy P. RNA-stabilizing proteins as molecular targets in cardiovascular pathologies. Trends Cardiovasc Med 2015;25(8):676–83.
[52] Kim HH, Kuwano Y, Srikantan S, Lee EK, Martindale JL, Gorospe M. HuR recruits let-7/RISC to repress c-Myc expression. Genes Dev 2009;23(15):1743–8.
[53] Young LE, Moore AE, Sokol L, Meisner-Kober N, Dixon DA. The mRNA stability factor HuR inhibits microRNA-16 targeting of COX-2. Mol Cancer Res 2012;10(1):167–80.
[54] Wang Y, Li Y, Toth JI, Petroski MD, Zhang Z, Zhao JC. N6-methyladenosine modification destabilizes developmental regulators in embryonic stem cells. Nat Cell Biol 2014;16(2):191–8.
[55] Mukherjee N, Corcoran DL, Nusbaum JD, Reid DW, Georgiev S, Hafner M, et al. Integrative regulatory mapping indicates that the RNA-binding protein HuR couples pre-mRNA processing and mRNA stability. Mol Cell 2011;43(3):327–39.
[56] Annabi B, Currie JC, Moghrabi A, Beliveau R. Inhibition of HuR and MMP-9 expression in macrophage-differentiated HL-60 myeloid leukemia cells by green tea polyphenol EGCg. Leuk Res 2007;31(9): 1277–84.
[57] Ishimaru D, Ramalingam S, Sengupta TK, Bandyopadhyay S, Dellis S, Tholanikunnel BG, et al. Regulation of Bcl-2 expression by HuR in HL60 leukemia cells and A431 carcinoma cells. Mol Cancer Res 2009;7(8): 1354–66.
[58] Ko CY, Wang WL, Li CF, Jeng YM, Chu YY, Wang HY, et al. IL-18-induced interaction between IMP3 and HuR contributes to COX-2 mRNA stabilization in acute myeloid leukemia. J Leukoc Biol 2016;99(1): 131–41.
[59] Blackinton JG, Keene JD. Functional coordination and HuR-mediated regulation of mRNA stability during T cell activation. Nucleic Acids Res 2016;44(1):426–36.
[60] Topisirovic I, Siddiqui N, Orolicki S, Skrabanek LA, Tremblay M, Hoang T, et al. Stability of eukaryotic translation initiation factor 4E mRNA is regulated by HuR, and this activity is dysregulated in cancer. Mol Cell Biol 2009;29(5):1152–62.
[61] Radich JP, Dai H, Mao M, Oehler V, Schelter J, Druker B, et al. Gene expression changes associated with progression and response in chronic myeloid leukemia. Proc Natl Acad Sci USA 2006;103(8):2794–9.
[62] Xu F, Zhang X, Lei Y, Liu X, Liu Z, Tong T, et al. Loss of repression of HuR translation by miR-16 may be responsible for the elevation of HuR in human breast carcinoma. J Cell Biochem 2010;111(3):727–34.
[63] Calin GA, Cimmino A, Fabbri M, Ferracin M, Wojcik SE, Shimizu M, et al. MiR-15a and miR-16-1 cluster functions in human leukemia. Proc Natl Acad Sci USA 2008;105(13):5166–71.
[64] Fu XD, Ares Jr M. Context-dependent control of alternative splicing by RNA-binding proteins. Nat Rev Genet 2014;15(10):689–701.

[65] da Silva MR, Moreira GA, Goncalves da Silva RA, de Almeida Alves Barbosa E, Pais Siqueira R, Teixera RR, et al. Splicing regulators and their roles in cancer biology and therapy. BioMed Res Int 2015; 2015:150514.
[66] Wang E, Aifantis I. RNA splicing and cancer. Trends Cancer 2020;6(8):631−44.
[67] Alarcon CR, Goodarzi H, Lee H, Liu X, Tavazoie S, Tavazoie SF. HNRNPA2B1 is a mediator of m(6)A-dependent nuclear RNA processing events. Cell 2015;162(6):1299−308.
[68] Xiao W, Adhikari S, Dahal U, Chen YS, Hao YJ, Sun BF, et al. Nuclear m(6)A reader YTHDC1 regulates mRNA splicing. Mol Cell 2016;61(4):507−19.
[69] Schittenhelm MM, Walter B, Tsintari V, Federmann B, Bajrami Saipi M, Akmut F, et al. Alternative splicing of the tumor suppressor ASPP2 results in a stress-inducible, oncogenic isoform prevalent in acute leukemia. EBioMedicine 2019;42:340−51.
[70] Li SQ, Liu J, Zhang J, Wang XL, Chen D, Wang Y, et al. Transcriptome profiling reveals the high incidence of hnRNPA1 exon 8 inclusion in chronic myeloid leukemia. J Adv Res 2020;24:301−10.
[71] Radzisheuskaya A, Shliaha PV, Grinev V, Lorenzini E, Kovalchuk S, Shlyueva D, et al. PRMT5 methylome profiling uncovers a direct link to splicing regulation in acute myeloid leukemia. Nat Struct Mol Biol 2019; 26(11):999−1012.
[72] Shailesh H, Zakaria ZZ, Baiocchi R, Sif S. Protein arginine methyltransferase 5 (PRMT5) dysregulation in cancer. Oncotarget 2018;9(94):36705−18.
[73] Li X, Wang C, Jiang H, Luo C. A patent review of arginine methyltransferase inhibitors (2010−2018). Expert Opin Ther Pat 2019;29(2):97−114.
[74] Wang E, Lu SX, Pastore A, Chen X, Imig J, Chun-Wei Lee S, et al. Targeting an RNA-binding protein network in acute myeloid leukemia. Cancer Cell 2019;35(3):369−84. e7.
[75] Han T, Goralski M, Gaskill N, Capota E, Kim J, Ting TC, et al. Anticancer sulfonamides target splicing by inducing RBM39 degradation via recruitment to DCAF15. Science 2017;356(6336).
[76] Maciejewski JP, Padgett RA. Defects in spliceosomal machinery: a new pathway of leukaemogenesis. Br J Haematol 2012;158(2):165−73.
[77] Visconte V, Nakashima MO, Rogers HJ. Mutations in splicing factor genes in myeloid malignancies: significance and impact on clinical features. Cancers 2019;11(12).
[78] Seiler M, Yoshimi A, Darman R, Chan B, Keaney G, Thomas M, et al. H3B-8800, an orally available small-molecule splicing modulator, induces lethality in spliceosome-mutant cancers. Nat Med 2018;24(4): 497−504.
[79] Shirai CL, White BS, Tripathi M, Tapia R, Ley JN, Ndonwi M, et al. Mutant U2AF1-expressing cells are sensitive to pharmacological modulation of the spliceosome. Nat Commun 2017;8:14060.
[80] Lee SC, North K, Kim E, Jang E, Obeng E, Lu SX, et al. Synthetic lethal and convergent biological effects of cancer-associated spliceosomal gene mutations. Cancer Cell 2018;34(2):225−41. e8.
[81] Bouchie A. First microRNA mimic enters clinic. Nat Biotechnol 2013;31(7):577.
[82] Balatti V, Pekarky Y, Croce CM. Role of microRNA in chronic lymphocytic leukemia onset and progression. J Hematol Oncol 2015;8:12.
[83] Calin GA, Dumitru CD, Shimizu M, Bichi R, Zupo S, Noch E, et al. Frequent deletions and down-regulation of micro- RNA genes miR15 and miR16 at 13q14 in chronic lymphocytic leukemia. Proc Natl Acad Sci USA 2002;99(24):15524−9.
[84] Sampath D, Liu C, Vasan K, Sulda M, Puduvalli VK, Wierda WG, et al. Histone deacetylases mediate the silencing of miR-15a, miR-16, and miR-29b in chronic lymphocytic leukemia. Blood 2012;119(5): 1162−72.
[85] Cimmino A, Calin GA, Fabbri M, Iorio MV, Ferracin M, Shimizu M, et al. miR-15 and miR-16 induce apoptosis by targeting BCL2. Proc Natl Acad Sci USA 2005;102(39):13944−9.

[86] Marcucci G, Mrozek K, Radmacher MD, Bloomfield CD, Croce CM. MicroRNA expression profiling in acute myeloid and chronic lymphocytic leukaemias. Best Pract Res Clin Haematol 2009;22(2):239–48.
[87] Merkel O, Asslaber D, Pinon JD, Egle A, Greil R. Interdependent regulation of p53 and miR-34a in chronic lymphocytic leukemia. Cell Cycle 2010;9(14):2764–8.
[88] Xiao C, Srinivasan L, Calado DP, Patterson HC, Zhang B, Wang J, et al. Lymphoproliferative disease and autoimmunity in mice with increased miR-17-92 expression in lymphocytes. Nat Immunol 2008;9(4): 405–14.
[89] Mraz M, Chen L, Rassenti LZ, Ghia EM, Li H, Jepsen K, et al. miR-150 influences B-cell receptor signaling in chronic lymphocytic leukemia by regulating expression of GAB1 and FOXP1. Blood 2014;124(1): 84–95.
[90] Ferrajoli A, Shanafelt TD, Ivan C, Shimizu M, Rabe KG, Nouraee N, et al. Prognostic value of miR-155 in individuals with monoclonal B-cell lymphocytosis and patients with B chronic lymphocytic leukemia. Blood 2013;122(11):1891–9.
[91] Visone R, Veronese A, Rassenti LZ, Balatti V, Pearl DK, Acunzo M, et al. miR-181b is a biomarker of disease progression in chronic lymphocytic leukemia. Blood 2011;118(11):3072–9.
[92] Gordon JE, Wong JJ, Rasko JE. MicroRNAs in myeloid malignancies. Br J Haematol 2013;162(2):162–76.
[93] Vaz C, Ahmad HM, Bharti R, Pandey P, Kumar L, Kulshreshtha R, et al. Analysis of the microRNA transcriptome and expression of different isomiRs in human peripheral blood mononuclear cells. BMC Res Notes 2013;6:390.
[94] Vaz C, Ahmad HM, Sharma P, Gupta R, Kumar L, Kulshreshtha R, et al. Analysis of microRNA transcriptome by deep sequencing of small RNA libraries of peripheral blood. BMC Genom 2010;11:288.
[95] Venturini L, Battmer K, Castoldi M, Schultheis B, Hochhaus A, Muckenthaler MU, et al. Expression of the miR-17-92 polycistron in chronic myeloid leukemia (CML) $CD34^+$ cells. Blood 2007;109(10):4399–405.
[96] Suresh S, McCallum L, Lu W, Lazar N, Perbal B, Irvine AE. MicroRNAs 130a/b are regulated by BCR-ABL and downregulate expression of CCN3 in CML. J Cell Commun Signal 2011;5(3):183–91.
[97] Lopotova T, Zackova M, Klamova H, Moravcova J. MicroRNA-451 in chronic myeloid leukemia: miR-451-BCR-ABL regulatory loop? Leuk Res 2011;35(7):974–7.
[98] Bueno MJ, Perez de Castro I, Gomez de Cedron M, Santos J, Calin GA, Cigudosa JC, et al. Genetic and epigenetic silencing of microRNA-203 enhances ABL1 and BCR-ABL1 oncogene expression. Cancer Cell 2008;13(6):496–506.
[99] Agirre X, Jimenez-Velasco A, San Jose-Eneriz E, Garate L, Bandres E, Cordeu L, et al. Down-regulation of hsa-miR-10a in chronic myeloid leukemia $CD34^+$ cells increases USF2-mediated cell growth. Mol Cancer Res 2008;6(12):1830–40.
[100] Flamant S, Ritchie W, Guilhot J, Holst J, Bonnet ML, Chomel JC, et al. Micro-RNA response to imatinib mesylate in patients with chronic myeloid leukemia. Haematologica 2010;95(8):1325–33.
[101] Ramachandran SS, Muiwo P, Ahmad HM, Pandey RM, Singh S, Bakhshi S, et al. miR-505-5p and miR-193b-3p: potential biomarkers of imatinib response in patients with chronic myeloid leukemia. Leuk Lymphoma 2017;58(8):1981–4.
[102] Mavrakis KJ, Van Der Meulen J, Wolfe AL, Liu X, Mets E, Taghon T, et al. A cooperative microRNA-tumor suppressor gene network in acute T-cell lymphoblastic leukemia (T-ALL). Nat Genet 2011;43(7): 673–8.
[103] Khalaj M, Tavakkoli M, Stranahan AW, Park CY. Pathogenic microRNA's in myeloid malignancies. Front Genet 2014;5:361.
[104] Sandhu R, Rivenbark AG, Coleman WB. Loss of post-transcriptional regulation of DNMT3b by micro-RNAs: a possible molecular mechanism for the hypermethylation defect observed in a subset of breast cancer cell lines. Int J Oncol 2012;41(2):721–32.

[105] Garzon R, Heaphy CE, Havelange V, Fabbri M, Volinia S, Tsao T, et al. MicroRNA 29b functions in acute myeloid leukemia. Blood 2009;114(26):5331−41.

[106] Zhang H, Luo XQ, Feng DD, Zhang XJ, Wu J, Zheng YS, et al. Upregulation of microRNA-125b contributes to leukemogenesis and increases drug resistance in pediatric acute promyelocytic leukemia. Mol Cancer 2011;10:108.

[107] Klusmann JH, Li Z, Bohmer K, Maroz A, Koch ML, Emmrich S, et al. miR-125b-2 is a potential oncomiR on human chromosome 21 in megakaryoblastic leukemia. Genes Dev 2010;24(5):478−90.

[108] Jinlong S, Lin F, Yonghui L, Li Y, Weidong W. Identification of let-7a-2-3p or/and miR-188-5p as prognostic biomarkers in cytogenetically normal acute myeloid leukemia. PLoS One 2015;10(2):e0118099.

[109] Schwind S, Maharry K, Radmacher MD, Mrozek K, Holland KB, Margeson D, et al. Prognostic significance of expression of a single microRNA, miR-181a, in cytogenetically normal acute myeloid leukemia: a Cancer and Leukemia Group B study. J Clin Oncol 2010;28(36):5257−64.

[110] Lu F, Zhang J, Ji M, Li P, Du Y, Wang H, et al. miR-181b increases drug sensitivity in acute myeloid leukemia via targeting HMGB1 and Mcl-1. Int J Oncol 2014;45(1):383−92.

[111] Diaz-Beya M, Brunet S, Nomdedeu J, Tejero R, Diaz T, Pratcorona M, et al. MicroRNA expression at diagnosis adds relevant prognostic information to molecular categorization in patients with intermediate-risk cytogenetic acute myeloid leukemia. Leukemia 2014;28(4):804−12.

[112] Rucker FG, Russ AC, Cocciardi S, Kett H, Schlenk RF, Botzenhardt U, et al. Altered miRNA and gene expression in acute myeloid leukemia with complex karyotype identify networks of prognostic relevance. Leukemia 2013;27(2):353−61.

CHAPTER

Post-transcriptional gene regulation in solid tumors

7

Saba Tabasum[1,2] and Monika Yadav[3]

[1]*Department of Medical Oncology, Dana-Farber Cancer Institute, Harvard Medical School, Boston, MA, United States;* [2]*Melanoma Disease Center, Dana-Farber Cancer Institute, Harvard Medical School, Boston, MA, United States;* [3]*Cancer Biology Laboratory, School of Life Sciences, Jawaharlal Nehru University, New Delhi, India*

Abbreviations

ADAR	Adenosine deaminase acting on RNA
Akt/PKB	Protein kinase B
ANRIL	Antisense RNA in the INK4 locus
AS	Alternate splicing
ASO	Antisense oligonucleotide
BAX	BCL2-associated X
BCL	B-cell lymphoma gene family
Bcl-xL	Long form of BCL-X
BM11	RNA-binding motif protein 11
BMI1	B-cell-specific Moloney murine leukemia virus integration site 1
BRCA1	Breast cancer gene
c-Myc	cellular Myc
CD44	CD44 molecule (Indian blood group)
CDH1	Cadherin 1
CDR1	Cerebellar degeneration-related protein 1
CREB	cAMP response element-binding protein
CRISPR-Cas9	Clustered regularly interspaced short palindromic repeats and CRISPR-associated protein 9
CTNNB1	Catenin (cadherin-associated protein), beta 1
DCAF13	DDB1 and CUL4-associated factor 13
DCUN1D5	Defective in cullin neddylation 1 domain containing 5
DMTF1	Cyclin D binding Myb-like transcription factor 1
DTX3	Deltex E3 ubiquitin ligase 3
E2F1	E2F transcription factor 1
E2F8	E2F transcription factor 8
EMT	Epithelial-mesenchymal transition
EMT-TF	EMT-transcription factor
EPHX2	Epoxide hydrolase 2
ER2	Estrogen receptor 2
ERBB2	Erythroblastic oncogene B
ERα	Estrogen receptor alpha
ESR1	Estrogen receptor 1

Post-Transcriptional Gene Regulation in Human Disease, Volume 32. https://doi.org/10.1016/B978-0-323-91305-8.00011-9

ESRP1	Epithelial splicing regulatory protein 1
ESRP2	Epithelial splicing regulatory protein 2
FGF	Fibroblast growth factors
FGFR	Fibroblast growth factor receptor
FLT3	FMS-like tyrosine kinase 3
FOXP1	Forkhead box protein P1
GAS5	Growth arrest specific 5
HER2	Human epidermal growth factor receptor 2
HERC4	HECT And RLD domain containing E3 ubiquitin protein ligase 4
HGOC	High-grade ovarian carcinoma
HIPK1	Homeodomain interacting protein kinase 1
hnRNAs	Heterogeneous nuclear ribonucleic acid
hnRNP	Heterogeneous nuclear ribonucleoproteins
HOTAIR	HOX transcript antisense RNA
HRCT1	Histidine-rich carboxyl terminus 1
HSP27	Heat shock protein 27
HuR	Human antigen R
ITCH	Itchy E3 ubiquitin protein ligase
LARP1	La ribonucleoprotein 1, translational regulator
Let7a	Lethal-7
lncRNA	Long noncoding RNAs
LSD1	Lysine-specific histone demethylase 1A
m6A	N6-methyladenosine
MALAT1	Metastasis-associated lung adenocarcinoma transcript 1
MAPK	Mtogen-activated protein kinase
MCL-1	Myeloid cell leukemia 1
MCP1P1	Monocyte chemoattractant protein 1-induced protein 1
MD-1	Myeloid differentiation 1
MEG3	Maternally expressed 3
METTL3	N6-adenosine-methyltransferase 70 kDa subunit
MMP-2	Matrix metallopeptidase 2
mRNA	Messenger RNA
mRNP	complex mRNA-protein
MSI1	Musashi RNA binding protein 1
MSI2	Musashi RNA binding protein 2
NEAT2	Noncoding nuclear-enriched abundant transcript 2
NEIL	Nei endonuclease VIII-like gene
NEIL2	Nei-like DNA glycosylase 2
NF-kB	Nuclear factor-kappa B
NONO	Non-POU domain-containing octamer-binding protein
NOTCH4	Neurogenic locus notch homolog protein 4
NSCLC	Nonsmall-cell lung carcinoma
OSCC	Oral squamous cell carcinoma
p21	Potent cyclin-dependent kinase inhibitor
PABN1	Polyadenylate-binding nuclear protein 1
1PABPC1	Poly(A)-binding protein cytoplasmic 1
PD1	Programmed cell death protein 1

PDL1	Programmed death-ligand 1
PIK3R2	Phosphoinositide-3-kinase regulatory subunit 2
PIWI	P-element-induced WImpy testis in Drosophila
PKCA	Protein kinase C alpha
PLAGL2	Pleomorphic adenoma gene-like 2
PR	Progesterone receptor
PRC2	Polycomb repressive complex 2
PSF	Polypyrimidine tract-binding protein (PTB) associted splicing factor
PTENP1	Pseudogene of PTEN (phosphatase and tensin homolog) 1
PTK2	Protein tyrosine kinase 2
PVT1	Plasmacytoma variant translocation 1
QKI	KH domain-containing RNA binding
Rac1b	Ras-related C3 botulinum toxin substrate 1
Ras	Rat sarcoma virus
RASL-seq	RNA-mediated oligonucleotide annealing, selection, and ligation with next-generation sequencing
RISC	RNA-induced silencing complex
RNA	Ribonucleic acid
rRNA	Ribosomal ribonucleic acid
SCFD2	Sec1 family domain containing 2
SCLC	Small cell lung cancer
SF3B	Splicing factor 3B subunit 1
siRNAs	Small interfering RNA
SIRT1	Sirtuin 1
SKP2	S-phase kinase-associated protein 2
SLUG	SNAI2 gene
SMUG1	Single-strand-selective monofunctional uracil-DNA glycosylase 1
snoRNA	Small nucleolar RNAs
SNP	Single nucleotide polymorphism
snRNA	Small nuclear RNA
snRNPs	Small nuclear ribonucleoproteins
SOX4	SRY-box transcription factor 4
SqCC	Squamous cell carcinoma
SRF	Serum response factor
SRP	Signal recognition protein
SRPK1	SRSF protein kinase 1
SRSF1	Serine and arginine-rich splicing factor 1
SRSF3	Serine and arginine-rich splicing factor 3
SRSF5	Serine and arginine-rich splicing factor 5
SRSF6	Serine and arginine-rich splicing factor 6
STAT3	Signal transducer and activator of transcription 3
TCF21	Transcription factor 21
TCGA	The cancer genome atlas
TGF-β1	Transforming growth factor beta 1
TGFBR2	Transforming growth factor beta receptor 2
TNBC	Triple-negative breast cancer
TP53	Tumor protein P53

tRNA	Transfer ribonucleic acid
TUBB3	Tubulin beta 3 class III
U2AF1	U2 small nuclear RNA auxiliary factor 1
UHRF1	Ubiquitin-like with PHD and ring finger domains 1
UTRs	Untranslated regions
VEGF A	Vascular endothelial growth factor A
YB-1	Y box binding protein 1
YTHDF1	YTH N6-methyladenosine RNA-binding protein 1
YTHDF2	YTH N6-methyladenosine RNA-binding protein 2
ZEB1	Zinc finger E-box binding homeobox 1
ZFas1	ZNFX1 antisense RNA 1
ZNF217	Zinc-finger protein 217

Introduction

Every cell in the human body has the same genome. However, each cell type has its own functional identity due to the differential expression of genes. This differential gene expression is regulated at multiple levels beginning from gene transcription followed by post-transcriptional modifications that play important roles in providing a functional directive to the cell. The post-transcriptional events begin with the production of a primary transcript, which is processed in the nucleus by several mechanisms. Nuclear processing constitutes capping of the 5′-end by the addition of a 7-methylguanosine cap; splicing of the primary transcript, and 3′-end cleavage and addition of a poly-A tail. All these steps can be modulated by regulating the rate of capping/tailing and/or splicing to produce a myriad of different mature mRNAs. The processed mRNA then forms a messenger ribonucleoprotein complex (mRNP complex) and is transported to the cytoplasm via the nuclear pore complex [1,2]. The cytoplasmic end of the mRNP complex undergoes rework to orient the mRNA transport in the cytoplasm [3,4]. Once in the cytoplasm, there can be multiple fates of the mRNA apart from being translated. It can be stored, directed to a different subcellular compartment, or marked for degradation [5]. The mRNP complex is dynamic and the complement of proteins constituting it decides the fate of the mRNA [1]. At this stage, the mRNA's presence in the cytoplasm must be at the right place and at the right time for translation to achieve the desirable cellular function [6]. Once mRNA is localized correctly, the ribosomes come in contact to facilitate translation. During the end of its productive course of life, mRNA is passed on to definitive degradation pathways [7].

These post-transcriptional events are orchestrated by RNA-binding proteins (RBPs) and other processing factors associated with the primary transcript. This association begins as soon as the 5′-end of a nascent transcript emerges from the RNA polymerase II and continues until its degradation in the cytoplasm [1,8]. Interestingly, it has been reported by the analysis of gene expression data that only 6% of these RBPs are expressed in a tissue-specific fashion and the rest are universally expressed [9]. Defective tissue-specific RNA processing may result in a diseased phenotype due to the gain or loss of a tissue-specific protein function [10] (Fig. 7.1).

Cancer is defined as a group of diseases sharing hallmarks such as uncontrolled cell proliferation, resistance to growth-inhibitory signals and apoptosis, replicative immortality, metabolic reprogramming, angiogenesis, metastasis, and employing immune evasion mechanisms [11]. Underlying these hallmarks is an altered transcriptional impression that is not only responsible for the genesis of cancer

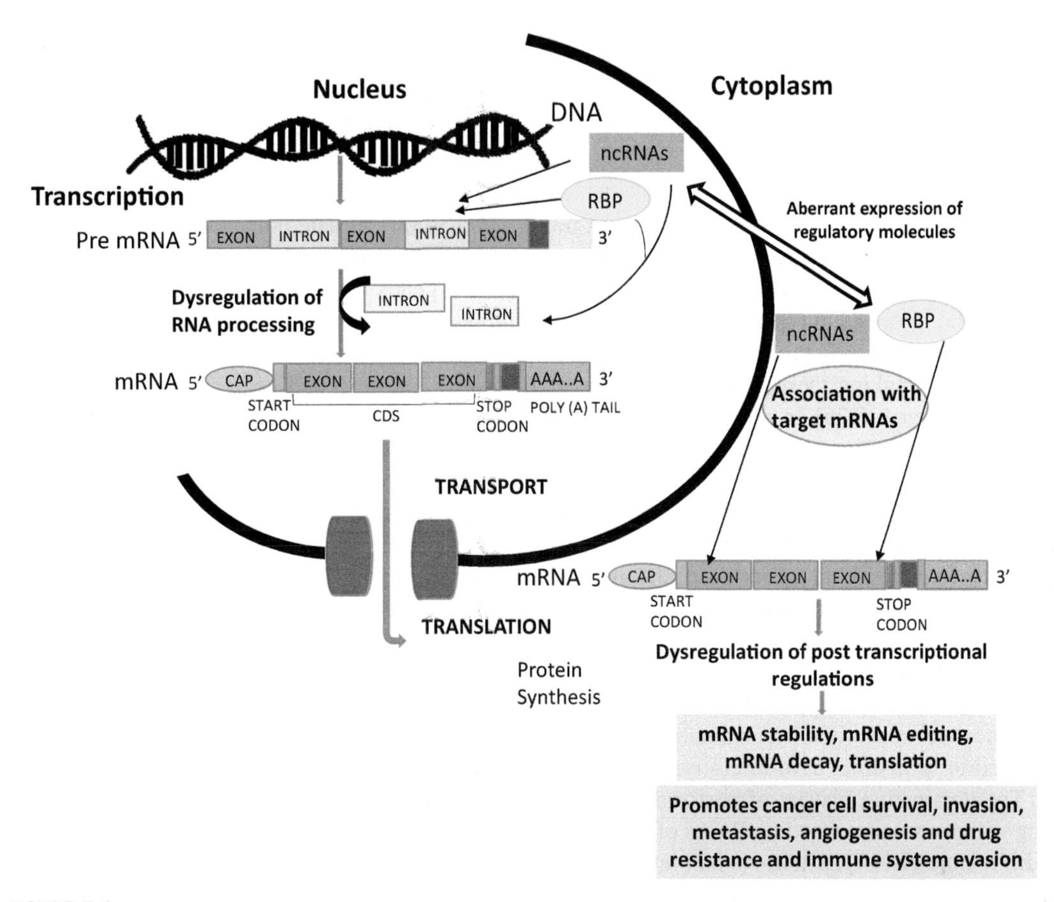

FIGURE 7.1

Representative flowchart depicts the various types of RNA regulators modulating post-transcriptional gene expression and the associated consequences of dysregulation in cancer. *DNA*, deoxy-ribonucleic acid; *mRNA*, messenger RNA; *ncRNA*, noncoding RNA; *pre-mRNA*, precursor mRNA; *RBP*, RNA binding protein.

but also results from it. The acquisition of control over transcriptional regulatory mechanisms allows the cancer cells to reenter the cell cycle, gain access to developmental pathways allowing self-renewal or invasive abilities. Thus, dysregulation of gene expression is at the core of carcinogenesis.

mRNA splicing involves the removal of some portions of the pre-mRNA known as introns and joining of other portions known as exons. The process of alternative splicing (AS) is regulated by cis and trans elements that distinguish the intron/exon boundaries and ensemble the splicing apparatus defined by 5 snRNPs (small nuclear ribonucleoprotein particles) and 100–200 non-snRNP proteins [12,13]. The splicing signature of malignant tissues varies widely as compared to the normal tissues [14–16]. This could be because of the following: mutations or single nucleotide polymorphism (SNP) at the acceptor or the donor sites or in regions responsible for enhancing or silencing the expression, thus changing the acceptor or donor site flags, or it could be due to abnormal expression or mutations leading to altered function of trans-regulatory elements [17,18]. The two protein families involved in facilitating splicing are Serine-rich (SR) protein family and the heterologous nuclear ribonuclear particle (hnRNP) family [19,20]. From the SR family, SRSF1 is the most studied splicing factor and is known to be overexpressed in breast [21,22] and lung cancer tissues [23].

Polyadenylation is another processing step that contributes to transcript variability in the human genome and approximately 30% of all mRNA transcripts harbor alternate polyadenylation sites. Like AS, the polyadenylation profile in malignant tissues is altered as compared to the normal tissues with proximal sites preferred [24]. In colorectal cancer, certain genes mediating cell cycle, nucleic acid binding, and processing elements prefer shorter 3′-untranslated regions (UTRs), a phenomenon also observed in particularly proliferative murine T lymphocytes. A shorter 3′-UTR is assumed to offer an advantage by making these transcripts inaccessible for miRNA-mediated gene regulation and thus, favor their surge [25,26]. In addition to the above-mentioned steps of regulation, mammalian mRNA is also post-transcriptionally regulated via RNA editing catalyzed by a group of enzymes known as adenosine and cytidine deaminases. These enzymes deaminate A to I or C to U in a cell type-specific or microenvironment-driven manner resulting in a change of the resulting protein sequence and thus adding another layer of complexity in the context of cancer transcriptomics [27]. The consequences of recoding dependence on the region of the mRNA being edited; if the pre-miRNA is edited it may alter its ability to identify its target or change its biogenesis [28–30]. Editing in the 3′-UTR region of mRNA may disrupt its identification by regulating miRNA or lncRNA [29,31,32]. Similarly, RNA decay may undergo manipulation as a result of RNA editing, for example, the RBP human antigen R (HuR) undergoes editing by adenosine deaminase acting on RNA (ADAR1) fostering binding of HuR to its desired transcripts by augmenting its stability [33]. Thus, the RNA editing fingerprints were reported to be distinct in different types of tumors [31,32].

The dominating players of post-transcriptional regulation involve RNA-based machinery, beginning with the small noncoding RNAs of approximately 20–30 nucleotides that are further divided into miRNA (microRNA), siRNAs (small interfering RNA), and piRNA (PIWI-interacting RNA). miRNAs bind to the mRNA by complementarity and bring about a variety of consequences including recruiting the RISC at the 3′-UTR [34], calling up proteins to flag mRNA for degradation, deadenylation, or termination [35,36]. Apart from this, miRNA may also interact with the 5′-UTR or the coding sequences to modulate translation. The malignant transformations in tissues have been reported to have their characteristic miRNA fingerprints as compared to the normal tissues that may act as a tumor suppressor or oncogene [37]. Among the first reports in 2002 in chronic lymphocytic leukemia, 70% of all cases were found to have low or no expression of miR-15a and miR-16-1, both known to regulate apoptosis by regulating BCL-2 [38,39]. miRNAs downregulated in cancer are considered tumor suppressors [40,41] whereas other miRNAs known as oncomiRs are overexpressed in cancer and thus, considered to be drivers. miR-125a or miR-125b are known to target HER2 and HER3 oncogenes that are associated with poor survival in breast cancer [42,43]. However, upregulating the expression of miR-155 and miR-21 in mice leads to lymphomagenesis [44,45]. Similarly, the let-7 family modulates genes associated with Ras oncogene as well as central pathways in the cell cycle and cell multiplication [46]. In ovarian cancer, the polymorphism of SNP34091 generates a binding site for miR-191, an indicator of chemosensitivity [47]. Thus, the miRNA profiles of different cancers may serve as an ideal tool to classify or define a subtype of tumors and, at the same time, help design the treatments based on chemosensitive or resistance indicative miRNAs in tumors.

The other important players of RNA regulatory machinery are noncoding RNAs that are shorter or longer than 200 nucleotides known as short ncRNAs and long ncRNAs, respectively [48,49]. Short ncRNAs comprised rRNA (ribosomal RNAs), tRNAs (transfer RNA), snRNAs (small nuclear RNA), and snoRNA (small nucleolar RNA). Long noncoding RNAs are also cataloged as NATs (natural antisense transcripts), long intronic ncRNAs, and lincRNAs (long intergenic RNA) based on their location as compared to the coding sequences in the genome [50]. As mentioned previously for miRNAs, lncRNAs also exhibit a tissue-specific profile that is altered in cancer and can define hallmarks of cancer contributing from initiation, progression to metastasis [51–53]. Among the most reported and well-studied lncRNAs are HOTAIR (HOX transcript antisense RNA) known to interact with PRC2 and LSD1, which couples these two repressive complexes to facilitate epigenetic silencing of HOX genes in liver, breast, and lung tumors [54–56]. GAS5 acts as a tumor suppressor by halting cell proliferation and increasing the sensitivity of the cells to chemotherapeutic agents [57,58]. Similarly, H19 [59], MALAT1(metastasis-associated lung adenocarcinoma transcript 1) [60–62], MEG3 [63,64], pseudogene of PTEN (phosphatase and tensin homolog) 1 (PTENP1) [65–67], and ZFas1 [68,69] are notably reported to be involved in carcinogenesis. Profiling lncRNA will provide a means of precisely diagnosing and predicting the course of the disease and identifying therapeutic targets.

A recently recognized class of RNA regulators is circRNA (circular RNA), produced from pre-mRNA by back splicing of the exons. They are acknowledged for carrying out a spectrum of roles, from acting as RNA sponges, to regulating transcription and splicing, and serving as scaffolds for protein or templates for protein translation [70]. CDR1as was reported to interact with miR-7 acting as an RNA sponge and inhibiting miR-7 [71]. CircPABN1 derived from the PABN1 gene competes with HuR to bind to PABN1 mRNA and thus inhibiting its translation [72]. CircRNAs have been added to the list of possible clinical biomarkers and therapeutic targets as they are reported to be involved in multiple steps of carcinogenesis.

A fast-growing tumor needs a continuous supply of nutrients and oxygen, to facilitate angiogenesis. The formation of new blood vessels is normally suppressed in adult tissues; however, in cancer, the angiogenic switch is turned on. VEGF is the most common signaling molecule studied in this context. Interestingly, AS of the VEGF primary transcript gives rise to two sets of isoforms of VEGF, VEGFxxx, and VEGFxxxb. Here, xxx denotes the number of amino acid residues in the resulting isoform. The two isoform sets, VEGFxxx and VEGFxxxb, are distinguished by picking off a far-end splice site present in the exon 8b, the last exon before the 3′-UTR. The resulting protein has a dissimilar sequence of six amino acids at the C-terminal compared to its counterpart resulting in contrasting functions [73]. VEGFxxx isoforms are angiogenic while VEGFxxxb isoforms are antiangiogenic, both expressed in most normal adult tissues [73,74]. It is the balance of the anti-angiogenic and angiogenic isoforms of VEGF that maintains the healthy state of tissues. A decrease in the endogenous levels of VEGxxxb has been reported in renal cancer compared to the nondiseased conditions [75], and in invasive prostate cancers as compared to localized prostate tumors [76].

Epithelial to mesenchymal transition (EMT) is a mechanism by which cancer cells gain access to motility and invasion. In physiology, EMT is involved in embryogenesis and repair [77]. miRNAs are involved in post-transcriptional regulation of EMT function at multiple levels by modulating EMT-transcription factors (EMT-TF) to directly manipulate EMT or by targeting signaling pathways indirectly involved in EMT. The hallmark of EMT induction is the loss of the epithelial marker E-cadherin. miR-9 directly targets CDH1 gene transcript to suppress its expression in breast cancer, leading to E-cadherin loss [78]. Similarly, miR-200 and miR-34 directly modulate the transcription factors involved in EMT and uphold the EMT state by establishing dual-negative feedback loops [79–81]. TGF-β-induced EMT involves upregulation of Slug expression that is targeted by miR-1 and miR-200b but TGF-β signaling suppresses both these miRNAs to establish EMT [82].

In the current chapter, we review six of the most frequently diagnosed solid tumors and elucidate the regulation of post-transcriptional gene expression involved in the development and progression of these solid tumors.

Breast cancer

According to the GLOBOCAN 2020 data, breast cancer is the most frequently diagnosed cancer with an estimated 2.3 million new cases and the fifth leading cause of global cancer mortality [83]. On the basis of expression of specific protein markers-estrogen receptor α (ERα), progesterone receptor (PR), ERBB2 (HER2); breast tumors are classified into the following: luminal A [ERα (+), PR (+), HER2 (−)], luminal B [ERα (+), PR (−), HER2 (+)], HER2 [ERα (−), PR (−), HER2 (+)], and triple-negative breast cancer (TNBC) [ERα (−), PR (−), HER2 (−)] with varied aggressiveness and therapeutic responses [84,85]. Although advancements in early diagnosis and treatment have shown substantial improvement in the 5-year patient survival rate; recurrence, metastasis, and drug resistance remain the primary reasons and consequences of dysregulated gene expression profiles observed in breast cancer deaths. Therefore, in view of understanding the molecular causes associated with the abnormal gene expression patterns, several layers of post-transcriptional regulations have been studied to decode their significance in breast cancer research.

AS of the Bcl-x gene results in the formation of two isoforms with contrasting effects in apoptosis regulation. The short splice variant Bcl-x stimulates apoptosis, while the long splice variant Bcl-xL subdues apoptosis. Consequently, overexpression of Bcl-xL is connected with metastasis risk in breast cancer [86]. Similarly, splicing shift from CD44 variant splice isoforms (CD44v) to CD44 standard splice isoforms (CD44s) can result in stemness and metastasis in breast cancer [87–89]. Growing evidence suggests the role of hypoxia as a key driver of AS in breast cancer [90,91]. Using a transcriptome-wide analysis, Wang et al. established a prognostic model to detect HERC4 (exon 23), EPHX2 (exon 7), and DDX39B as a potential target in invasive breast cancer [92]. Oh et al. conducted a global analysis of AS events and demonstrated that a decreased skipping of DCUN1D5 exon 4 is associated with an increase of total DCUN1D5 gene expression, which is further correlated with poor overall and relapse-free survival in aggressive breast cancer cases [93]. Bellanger et al. for the first time highlighted the significance of the novel Exon 4-Skipping Isoform of the ZNF217 Oncogene (ZNF217-ΔE4 transcripts) in breast cancer, where ectopic overexpression of ZNF217-ΔE4 transcripts promotes cellular aggressiveness and increases the expression levels of wild-type oncogene (ZNF217-WT). Thus, the assessment of expression levels of these two isoforms may serve as a novel prognostic biomarker for breast cancer patients [94]. Likewise, several vital splicing events targeting important tumorigenesis factors (BRCA1, TP53, HER2, FGFR, KLF6, surviving, and DMTF1) have been recognized in breast cancer. Overall, these AS events present potential therapeutic biomarkers and candidate targets that could be used for guiding clinical approaches [95].

Similarly, RNA-binding protein non-POU domain-containing octamer-binding protein (NONO) and polypyrimidine tract-binding protein (PTB)-associated splicing factor (PSF) stimulate breast cancer progression via post-transcriptional regulation of SKP2 and E2F8; and ESR1 and SCFD2, respectively [96,97]. Moreover, Chen et al. demonstrated the antioncogenic function of MCPIP1 (a zinc finger RBP) in TNBC [98]. Musashi (MSI1 and MSI2) is also considered as an oncogenic RBP involved in the development of several cancers. In TNBC, the knockdown of Musashi RBP promotes cell invasion and migration [99]. Additionally, the overexpression of miR-125b decreases the expression of MSI1 and hence modulates the expression of epithelial markers in breast cancer [100]. Further, increased expression of another novel RBP, DCAF13 is associated with post-transcriptional decay of DTX3 mRNA and therefore, activation of NOTCH4 signaling in TNBC [101]. Dysregulation of splicing factor SRSF1 is associated with the transformation of mammary cells [22,102]. Altogether, RBPs are critical regulators of post-transcriptional gene expression and aberrant RBP levels can lead to cancer progression.

Out of the 170 discovered post-transcriptional modifications, N6-methyladenosine (m6A) is the most ubiquitously present modification, which is placed along with the coding sequences of mRNA and 5′/3′-UTRs. The process of m6A modification is regulated by a group of complex interplayers called “writers” (methyltransferases), “readers” (binding proteins), and “erasers” (demethylases). Fang et al. reviewed the dysregulated effects of m6A modification and its corresponding proteins in the development of breast cancer [103]. Using RBP-focused CRISPR-Cas9 screening, Einstein et al. identified that the inhibition of m6A reader YTHDF2 could selectively trigger apoptosis in MYC-driven breast cancer; thus underlining the therapeutic potential of RBPs in countering the global increment of mRNA synthesis in MYC-driven breast cancers [104]. Recently, noncoding RNAs (ncRNAs) also emerged as key regulators of gene expression in breast cancer. For instance, the small nucleolar RNAs SNORD50A/B via mediating TRIM21-GMPS interaction works as an oncogene in p53 wild-type breast cancer [105]. Similarly, Das et al. for the first time showed that miR-20a and Nucleolin act as potent post-transcriptional regulators of MMP2 mRNA in breast cancer cell lines [106]. Venkatesh et al. highlighted the potential of lncRNA-microRNA axes as a novel therapeutic target in breast cancer [107]. Further, genetic variation in 3′UTRs of base excision repair glycosylases SMUG1 and Nei-like DNA glycosylase 2 (NEIL2) activated by miRNAs was found to modulate the prognosis of hormonally treated breast cancer [108]. Altogether, post-transcriptional gene regulation is a crucial process, which allows cancer cells to modify their proteome as per the cellular requirement enabling them to survive in extremities. Therefore, understanding the complex mechanism behind such regulations is important to open new avenues in studies of breast cancer progression and management.

Lung cancer

Lung cancer is the second most frequent cause of cancer morbidity and the leading cause of cancer deaths worldwide, with 2.2 million new diagnoses and 1.8 million mortalities reported in 2020 [83]. The 5-year survival rate of lung cancer patients depends on the cancer stage and regional differences and ranges from 4% to 17% because the diagnosis is often at advanced cancer stages [109]. It is a heterogeneous disease, and the two prime types of lung cancers are nonsmall cell lung cancer (NSCLC) and small cell lung cancer (SCLC) [110,111]. NSCLC accounts for 85% of all lung cancers, and SCLC for 15%. NSCLC is further classified into adenocarcinoma, squamous cell carcinoma (SqCC), and large cell carcinoma [110]. The underlying causes of lung cancer include both genetic and environmental factors leading to cellular transformation and tumorigenesis. Some risk factors for lung cancer include tobacco consumption usually by cigarette smoking, radiation exposure, and toxins, for example, asbestos, arsenic, chromium, etc.

Recent research in 22 NSCLC cell lines, aimed at collecting full-length transcriptomic profiles exhibiting genomic mutations and aberrations, identified many aberrant splicing isoforms. These included aberrations like unannotated exon, exon skipping, exon shuffling, intron retention, alternative first exon, alternative last exon, alternative 5′-splice site, and alternative 3′-splice site [112]. Disrupted expression of these splicing regulatory proteins directly affects the splicing events as discussed earlier. For example, SRSF1 is a member of the SR family and is observed to be upregulated in small cell lung cancer, where it is associated with tumorigenicity [23]. S34F mutation in the splicing factor U2AF1 renders it to modulate the splicing of introns containing the trinucleotide CAG at the 3′-splice site. The U2AF1 S34F enables alternative splicing of many genes, including the CTNNB1 protooncogene, and promotes EMT and tumor cell invasion in lung adenocarcinoma [113,114]. Another splicing factor SF3B1 is often mutated in several cancers, including lung cancer [115,116]. Mutations in SF3B1 and SUGP1 splicing factors were shown to be associated with the use of cryptic 3′-splice sites, leading to an increase in aberrant transcripts in lung adenocarcinoma [117].

The miRNA profiles are distinct for different subtypes of lung cancers, miR-205-5p is expressed only in lung squamous cell carcinoma but not in adenocarcinoma [118] miRNAs also play a crucial role in metabolic reprogramming of lung cancer cells to aid in survival, growth, and proliferation. One such example can be seen with miRNA-144, which is downregulated in NSCLC tissue [119,120]. Liu et al. elucidated that the downregulation of miRNA-144 induces the expression of glucose transporter 1 (GLUT1), which in turn leads to an increase in glucose uptake and lactate secretion [120]. This increased glucose uptake causes enhanced glycolysis that promotes and supports the continuous proliferation of lung cancer cells.

LncRNAs may act as suppressors or promoters of lung carcinogenesis. Some oncogenic lncRNAs upregulated in lung cancer include MALAT1 and H19. MALAT1 or NEAT2 (noncoding nuclear-enriched abundant transcript 2) is highly expressed in NSCLC cells and promotes metastasis. The lncRNA targets and downregulates miR-206, thereby inducing Akt/mTOR signaling, which then facilitates EMT, cell migration, and invasion [121]. MALAT1 also promotes cancer progression and metastasis by repressing miR-204 and activates the downstream SLUG pathway in lung adenocarcinoma [122]. LncRNA H19 is shown to be overexpressed in NSCLC cells and tissue, and this overexpression is driven by c-Myc. Accumulated H19 binds and represses miR-107 to promote cell cycle progression in NSCLC cell lines [123]. H19 could also promote NSCLC development, migration, and invasion by targeting and sponging miR-17 that leads to an upregulation of signal transducers and activators of transcription (STAT3) [124].

Circular RNAs have been recently associated with the initiation, progression, and suppression of lung cancer [125]. In lung cancer, circular RNAs can promote or inhibit cancer progression through various cellular pathways and mechanisms. For example, a recent study reported the role of circFOXP1 in lung adenocarcinoma where it is overexpressed. It has been shown that circFOXP1 directly interacts with miR-185-5p and modulates the miR-185-5p/Wnt1 signaling pathway to favor cell proliferation and inhibit apoptosis [126]. Another circular RNA circ-SOX4 promotes tumor initiation, proliferation, invasion, and migration in lung adenocarcinoma by sponging miR-1270 and activating the Wnt pathway via PLAGL2 accumulation [127]. Moreover, circular RNA circ_0001946 promotes cell growth by activation of the Wnt/β-catenin pathway. In this case, circ_0001946 repressed miR-135a-5p to upregulate Sirtuin1 (SIRT1) [128]. Conversely, several circular RNAs inactivate the Wnt pathway and inhibit cell proliferation to suppress tumorigenesis. For instance, circ_0043256 was shown to hamper cell proliferation and promote apoptosis in NSCLC, when overexpressed under the influence of cinnamaldehyde. The mechanism of action of hsa_circ_0043256 was found to be sequestration of miR-1252 and downstream upregulation of Itchy E3 ubiquitin-protein ligase to inhibit the canonical Wnt pathway [129]. Many circular RNAs target the mitogen-activated protein kinase (MAPK) signaling pathway to modulate tumorigenesis in lung cancer cells. For example, hsa_circ_0007580 & hsa_circRNA_101237 are upregulated in NSCLC cells where they promote cell proliferation and invasion. Hsa_circ_0007580 binds to repress miR-545-3p and upregulates the downstream protein kinase C alpha (PKCA), thereby inducing p38/MAPK signaling [130], whereas hsa_circRNA_101237 modulates MAPK1 signaling through miRNA-490-3p downregulation [131]. Circular RNAs may affect tumorigenesis through the NF-kB signaling pathway as well, as is seen in the case of circ_cMras. It was shown by Zhou et al. that circ_cMras is downregulated in lung adenocarcinoma tissue and cells. Overexpression of circ_cMras inhibits cell proliferation, migration, invasion, and tumor growth via the NF-kB signaling pathway [132]. Another example of circular RNAs acting as tumor suppressors is hsa_circ_100395, which arrests cell cycle progression in NSCLC and inhibits cell proliferation, migration, and invasion. The mechanism of action of hsa_circ_100395 involves sequestration of miR-1228, leading to upregulation of transcription factor 21 (TCF21) and activation of the TCF21 pathway [133].

Circular RNAs can also target EMT as is seen with hsa_circ_0007059 that inhibits Wnt/β-catenin and ERK1/2 pathways by sequestering miR-378 leading to suppression of cell proliferation and EMT [134]. On the other hand, circular RNA hsa_circ_00008305 (circPTK2) is downregulated in NSCLC that causes inhibition of TGF-β-induced EMT and cell invasion, accompanied by an accumulation of miR-429/miR-200b-3p [135]. Other than the mechanisms mentioned here, circular RNAs participate in lung cancer proliferation and apoptosis through the myriad of different pathways [136]. The extensive involvement of noncoding RNAs in lung cancer exhibits a promising approach for biomarker development and target-specific treatment.

Colorectal cancer

As per GLOBOCAN, there were 1.9 million cases of colorectal cancer (CRC) and 935,000 deaths worldwide in 2020. It is the second deadliest cancer [83]. About 20% of cases of CRC are diagnosed at an advanced stage while the other 40% of those treated at early stages experience relapse. The 5-year survival rate for advanced CRC is 20%, thus resonating a poor prognosis [137]. Surgery and systemic chemotherapy often are therapeutic options. The optimal first-line chemotherapeutic regimen involves a combination of molecularly targeted drugs along with anti-VEGF (bevacizumab) or anti-epidermal growth factor (cetuximab) agents [138]. Thus, novel drugs are urgently needed to improve the prognosis of CRC.

Histological examination of lymph nodes from normal, metastatic, and CRC patients has confirmed differential lncRNA profiles. A study reported that 14 lncRNAs were found to be over-expressed and 5 lncRNAs were downregulated in metastatic lymph nodes as compared to normal and tumor tissues [139]. Another profiling study found 2636 lncRNAs out of which POU6F2-AS1,

RAB6C-AS1, DDP10-AS1, HOXA11-AS, LINC00944, and FEZF1-AS1 were found to be associated with liver metastasis [140]. A recent study confirmed that FEZF1-AS1 associates with LSD1, the histone demethylase, in CRC to repress the tumor suppressor KLF2 (Krüppel-like factor) regulating cell growth and motility [141]. The search to find the lncRNA signature associated with metastatic CRC led to finding a new lncRNA Gastric adenocarcinoma predictive long intergenic noncoding (GAPLINC) RNA associated with tumor size, advancement of the disease in terms of the stage (T stage), and lymph node stage (N stage) and overall survival in CRC. GAPLINC was found to mediate the mentioned effects by collaborating with NONO and PSF [142].

Zhang et al. analyzed CRC data available in the cancer genome atlas (TCGA) for differential expression of RBPs and reported that out of the 242 differentially expressed, 200 were overexpressed, and 42 were downregulated in tumors. Eight of the differentially expressed RBPs, that is, RRS1, PABPC1L, TERT, SMAD6, UPF3B, RP9, NOL3, and PTRH1 were associated with CRC and demonstrated prognostic marker potential [143]. A recent finding demonstrated RBM10, a tumor suppressor is known to facilitate apoptosis via a negative feedback loop of MDM2-p53, Bcl2, Bax, and repressing tumor growth by Rapa/AKT/cAMP response element-binding protein (CREB) and Notch signaling, is frequently mutated in CRC [144]. Some of the most notable RBPs in CRC include Lin28, IGF2BPs/IMPs, Musashi, HuR, RBM3, Mex3A, and CELF1 and are involved in intestinal homeostasis, wound healing, and neoplastic transformations [145].

The miRNA studies in CRC have reported tumor suppressive as well as promoting effects of miRNAs. miR-21 overexpression in CRC decreases the expression of PDCD4, SPRY2 (sprouty homolog 2), TIAM1, transforming growth factor beta receptor 2 (TGFBR2), and CDC25A (cell division cycle 25A), thus regulating cell growth, apoptosis, motility, invasiveness, and plasticity of associated stem cells [146]. Similarly, miR-92a from the miR-17-92 oncomiR cluster is reported to drive tumor progression in CRC, angiogenesis, evasion of apoptosis signals, and cell proliferation [147]. miR-92a is also secreted in stool and plasma and, thus, may serve as a diagnostic marker for CRC [148]. The other oncomiRs reported in CRC are miR-96, miR-135, miR-155, and miR-224 [149]. Lethal 7a (let7a) consists of a family of 12 miRNA members facilitating EMT by regulating key oncogenes, that is, RAS, c-Myc, CDC34 (cell division cycle 34), CDC25A, CDK6 (cyclin-dependent kinase 6), HMGA2 (high mobility group AT-hook 2), LIN28 (lin-28 homolog), and LIN28B, which are downregulated in CRC [150]. Masanobu et al. reported the intratumoral expression of let7a in CRC samples positively correlated with patient survival [151].

Many profiling studies have emphasized the differential expression of circRNAs in CRC. Zhou et al. found 31,557 circRNAs in eight pairs of CRC samples out of which the majority (81%) originated from protein-coding sequences and the rest from intronic (5%) and other sequences [128]. The circRNA hsa_circ_0014717 [152], circHIPK3 [153], and hsa_circ_0009361 [154] are among the most deregulated circRNA in CRC. circRNAs have been found to have diagnostic, prognostic, as well as therapeutic potential in CRC.

Gastric cancer

Gastric cancer remains one of the most leading causes of cancer worldwide with 1.1 million new cases and 0.77 million deaths in 2020. The incidence and mortality rates in men are twice that in women [83]. Gastric cancer is divided into two categories based on topography, namely, cardia gastric cancer and noncardia gastric cancer. Cardia cancer arises in the upper part of the stomach adjacent to the oesophageal—gastric junction, while noncardia cancer arises in the middle and distal stomach areas [155,156]. The common risk factors for both types of gastric cancers include familial disposition, low fruit intake, foods preserved by salting, alcohol intake, and tobacco smoking. Cardia cancer is associated with obesity and gastroesophageal reflux as the primary causes, while noncardia cancer is strongly caused by *Helicobacter pylori* infection [157]. Histologically, gastric cancer is divided into intestinal gastric cancer (well-differentiated, glandular with intercellular junctions) and diffused (poorly differentiated, highly metastatic, without intercellular junctions) [158]. The overall survival rate of gastric cancer patients remains 25% worldwide [159]; furthermore, intestinal cancer has a better prognosis and survival rate than diffused cancer [156,160].

By comparing the RNA editing landscape between normal gastric tissues and gastric cancer tissues, Chan et al. observed an ADAR-mediated RNA misediting phenotype in gastric cancer cells. The ADAR1/2 dysregulation occurs due to genomic gain of ADAR1 and loss of ADAR2 gene. ADAR1 acts as the oncogene and ADAR2 acts as the tumor suppressor for gastric cancer [161]. Moreover, poly(A)-binding protein C1 (PABPC1), which controls mRNA deadenylation and participates in mRNA decay, behaves as an oncogenic protein in gastric cancer cells [162]. PABPC1 is upregulated in gastric cancer tissues and contributes toward poor survival. It was shown that PABPC1 downregulates miR-34c to promote the growth and survival of gastric cancer cells [162].

Alternative splicing has been shown as one of the important factors governing gastric cancer development and progression [163]. A mutation in exon 10 of the Nei endonuclease VIII-like gene (NEIL) gene leads to the alternatively spliced transcript and shortened mRNA that produces a truncated protein with a lower expression level. The truncated NEIL protein was shown to lack nuclear localization and base excision repair activity and promote gastric carcinogenesis [164].

The occurrence and progression of gastric cancer are closely associated with miRNA dysregulation in gastric cancer stem cells. Alteration in the expression of miRNAs responsible for maintaining gastric cancer stem cells can perturb the balance between self-renewal and differentiation and, thus, induce cellular transformation and tumor development [165]. Golestaneh et al. profiled miRNAs from gastric cancer cells and gastric cancer stem cells and reported that miR-21 and miR-302 expression was higher in gastric cancer stem cells, whereas the expression of let-7a, miR-372, miR-373, and miR-502c-5p was more in the case of gastric cancer cells [165]. The involvement of several miRNAs in the cellular transformation from gastric cancer stem cells to gastric cancer cells has been validated since then. For instance, miRNA-144 could inhibit tumorsphere formation and chemoresistance in gastric cells by downregulating CD44. Gastric cancer cells highly enriched in cancer stem cells exhibited downregulation of miR-145 and overexpression of CD44 [166]. Another tumor-suppressive miRNA is miR-71, which was shown to downregulate CD44 expression and inhibit EMT in gastric cells [167]. Many miRNAs also regulate the main signaling pathways involved in cancer stemness regulation. miR-17-92 cluster members miR-19b, miR-20a and miR-92a were observed to be downregulated during gastric cancer stem cell differentiation. The miR-17-92 cluster was involved in the self-renewal of gastric cancer stem cells and the progression of proliferation in gastric cancer cells. The mechanism of action was described as miR-17-92 targeting E2F1 and HIPK1 proteins, thereby inhibiting Wnt/β-catenin signaling [168].

Various lncRNAs participate in the suppression or progression of gastric cancer. Duan et al. associated the presence of SNPs within three lncRNAs with a reduced risk of developing gastric cancer [169]. The lncRNA HOTAIR was observed to be overexpressed in cisplatin-resistant gastric cancer cells where it promotes cell cycle, G1/S transition, and cell proliferation and inhibits apoptosis. Yan et al. elucidated the mechanism of action in which HOTAIR was shown to target and sponge miR-126, which leads to the overexpression of VEGFA and PIK3R2, and activation of downstream PI3K/AKT/MRP1 pathway [170]. Furthermore, Cheng explained that HOTAIR confers cisplatin resistance in gastric cancer cells by targeting miR-34, and cisplatin resistance can be reduced by the knockdown of HOTAIR [171]. The lncRNA MALAT1 was found to be upregulated in gastric cancer tissue that caused chemoresistance due to autophagy. The lncRNA acts by sponging miR-23b-3p, which accumulates the regulator of autophagy, ATG12 [172]. Another mechanism of induction of autophagy-related chemoresistance by MALAT1 is by sponging and downregulating miR-30b and overexpressing the downstream effector ATG5 [173].

Circular RNAs have been implicated to have important roles in promoting or suppressing gastric cancer. For example, circNRIP1 favors cell proliferation, migration, and invasion in gastric cancer cells, which is brought about by repression of miR-149-5p and overexpression of the downstream effector AKT1 [174]. Another circular RNA hsa_circ_0001368 acts as a tumor suppressor of gastric cancer by sponging miR-6506e5p and inducing the overexpression of FOXO3. hsa_circ_0001368 is found to be downregulated in gastric cancer cells, and its knockdown promoted cell viability and motility and accelerated cell proliferation and tumor growth in vivo [175]. circRNA0047905 functions as a sponge for miR-4516 and miR-1227-5p and promotes the accumulation of SERPINB5 and MMP11, which in turn leads to activation of Akt/CREB signaling pathway and progression of gastric cancer [176]. CircCACTIN is overexpressed in gastric cancer tissues and cell lines, and its knockdown inhibits cell proliferation, migration, invasion, and EMT. CircCACTIN upregulation promotes tumor growth and EMT in vivo. Mechanistically, CircCACTIN assists in gastric cancer progression by inhibition of miR-331-3p, upregulation of TGFBR1 [177]. The tumor suppressor role of circular RNA circGRAMD1B was elucidated by Dai et al. in 2019. circGRAMD1B works by inhibition of miR-130a-3p and regulation of PTEN and p21 expression, which eventually promotes proliferation, migration, and invasion of gastric cancer cells [178].

Being one of the most common malignancies and leading causes of death worldwide, the molecular mechanisms governing the development and progression of gastric cancer are crucial for improved diagnosis and therapy.

Oral squamous cell carcinoma

Cancer of the oral cavity is the most frequent type of head and neck cancer, estimated to account for about 354,864 new cases and 177,384 deaths globally [179]. Oral squamous cell carcinoma (OSCC) contributes to over 90% of oral tumors emerging from the basal layer of oral epithelium and results from either transformed stem cells or by dedifferentiation of early phase differentiated cells. The key causative agents of OSCC include smoking, tobacco, alcohol consumption, and human papillomavirus infections. Cancer metastasis, recurrence, and drug resistance remain the major shortcomings to a 5-year survival rate of less than 65% which further reduces to 30% in case of advanced disease [180,181]. Therefore, deducing the key mechanisms of the malignant spread of OSCC is vital for the development of efficient therapeutics and the extension of patient survival.

Aberrant splicing of TGFBR2 is associated with early exposition, tumor-specific manifestation, and prognostic significance in OSCC [182]. Expression of splicing factor SRSF3 is found to be positively correlated with OSCC carcinogenesis [183]. Jia et al. also confirmed the oncogenic role of SRSF3 in OSCC where both the expression and AS of SRSF3 exon 4 was found to be modulated by high expression of HnRNP L in OSCC tissues [184]. Differential expression of antiangiogenic VEGF isoforms in oral cancer was supported by splicing regulatory factors SRSF1, SRSF6, SRSF5, and SRPK1 [185]. SRSF5 was identified as a novel oncogene overexpressed by SRSF3 in oral cancer cells [186]. Osada et al. showed a positive link between AS machinery and EMT phenotype in OSCC. An autocrine FGF−FGFR (IIIc) signaling supports mesenchymal-like phenotypes in OSCC by sustained expression of ZEB1/2 that in return regulates AS via isoform switching of the fibroblast growth factor receptors IIIc-isoform [187]. Zhang et al. identified seven prognostic AS signatures in OSCC through univariate and multivariate Cox regression analysis of OSCC data available in TGCA [188]. Overall, the splicing machinery and its components together decide tumor growth and development and, thus, serve as diagnostic markers and therapeutic targets.

RBPs are key regulators of migration, invasion, and metastatic processes in OSCC [189]. The RBPs, ESRP1, and ESRP2, were reported to regulate the AS events related to the epithelial phenotype of cells and the expression of these two proteins was found to be reduced during EMT. In 2014, Ishii et al. observed that ESRPs suppress the motility of cancer cells via a different mechanism in OSCC. The overexpression of ESRPs was identified during carcinogenesis while downregulation was perceived in invasive fronts. ESRP1 downregulates the Rac1b isoform expression, thus, altering the actin cytoskeleton dynamics whereas ESRP2 suppresses EMT-inducing transcription factors [190]. Another RBP Quaking (QKI) confirms its tumor-suppressive role by decreasing SOX2 expression and consequently impairing the stemness and tumorigenic potential in oral cancer [191].

RNA modification enzyme methyltransferase-like 3 (METTL3) was upregulated and associated with OSCC pathogenesis through m6A methylation of BMI1 mRNA [192]. Correspondingly, METTL3 was found to enhance c-Myc stability through YTHDF1-mediated m6A modification and hence, facilitates OSCC tumorigenesis [193]. Recently, interactions between N6-methyladenosine (m6A) modifications and noncoding RNAs have become an attention point of epigenetic investigations in cancer [194]. For example, m6A regulation of circCUX1 affirms radio-resistance by the caspase 1 pathway [195]. LINC00668 modulated VEGFA signaling by suppressing miR-297 [196]. A significant connection was observed between lncRNA SNHG20/miR-197/LIN28 axis, oncogenesis, and stemness in the case of OSCC [197]. Li et al. identified that c-Myc-induced overexpression of small nucleolar RNA host gene 16 (SNHG16) escalated OSCC progression [198]. A novel circRNA, circUHRF1 promotes OSCC tumorigenesis through circUHRF1/miR-526b-5p/c-Myc/TGF-β1/ESRP1 feedback loop [199]. H19 transcript variants promote EMT and oncogenesis by enhancing ZEB1 protein expression in oral cancer [200]. Altogether, several post-transcriptional modifications have been successfully deduced in oral cancer cells that have been shown to enhance the key oncogenic processes and, thus, remain helpful in guiding clinical therapeutics. Importantly, all earlier studies discussed the role of post-transcriptional regulation in human cancer at the bulk tumor level, which involves most of the tumor cells collected with other types of cells that might influence tumor development and therapeutic response. Therefore, more studies are required to dissect the connection of post-transcriptional complexity present within tumors at the single-cell level that could be beneficial for directing future cancer therapies.

Ovarian cancer

According to the GLOBOCAN 2020, there were 313,959 cases of ovarian cancer cases worldwide [83]. Among gynecological cancers, ovarian cancer is the second most diagnosed and deadliest cancer [201]. For a long time, it has been associated with a poor 5-year survival rate of 27% for stage III, and at stage IV it falls to 13% because 58% of the cases are diagnosed when it has already spread beyond the pelvis [202]. It is a heterogeneous disease classified based on the cell type that becomes cancerous; the most common type being epithelial ovarian cancer that will be discussed [203]. Interestingly, ovarian cancer is distinguished from other cancers for having a subtle mutational profile. The most-reported subtype of ovarian cancer, high-grade serous ovarian cancer (HGSOC) accounts for only 9 non-synonymous gene mutations in TCGA data. Mutations in TP53 were found to be present in 96% of the cases while BRCA was mutated in 22% of cases [204]. Ovarian cancers with a low mutational load show poor response toward the checkpoint inhibitors targeting PD1 and PDL1 [205]. As ovarian cancers are mostly diagnosed at advanced stages, chemotherapy is the most used treatment with pseudoalkylating agent carboplatin and paclitaxel, the microtubule spindle poison, being the two most popular choices.

Ovarian cancer demonstrates an increased genomic mutability and involves deregulation of 17 miRNAs in HGSOC out of the 34 known to be generically deregulated in cancer. The miRNAs known to be associated with the resistance toward cisplatin are enlisted as let-7e, let-7i, miR-214, and miR-30a-5p, and those associated with paclitaxel resistance are miR-663, miR-622, and miR-130b [206–209]. There are also miRNAs associated with sensitization of the cells toward drugs such as miR-200c that modulates TUBB3. The miRNAs known to be modulating EMT in ovarian cancer include miR-135a-3p, miR-200c, miR-216a, and miR-340 [210].

RBPs may work in collaboration with miRNA or may compete by binding and modifying the mRNA of interest and secluding them from interacting with miRNAs. RNA-binding motif protein 3 (RBP3) is known to be overexpressed in several cancers including ovarian and is correlated with favorable outcomes. The mRNA and protein levels of RBM3 were found to be in collation with the results in tissues poised from 163 to 151 patients of ovarian cancer, respectively. RBP3 interacts with the mRNA of BCL-2 and BAX, as well as genes involved in maintaining the integrity of DNA. Increased levels of RBP3 are linked with platinum sensitivity and, thus, a good prognosis [211]. Recently, RBM11 was also reported to be an oncoprotein that can be used as a prognostic biomarker in ovarian cancer [212]. The ubiquitously expressed RBP, HuR is known to stabilize and/or endorse the mRNAs with AU-rich elements in their 3′-end region for translation in response to stress, is normally expressed in low copy numbers [213,214].

HuR cooperates with NEAT1, a lncRNA over-expressed in advanced ovarian cancer and linked to bad prognosis [215]. Chai et al. reported that overexpression of NEAT1 in ovarian cancer cells gave them a proliferative and invasive advantage. They also found that HuR secures NEAT1 by binding and overexpressing it while miR124-3p competes with HuR to destabilize NEAT1 [216]. The other lncRNAs with which HuR interacts include HOTAIR and MALAT1, and these interactions enhance their ability to act as miRNA sponges [217,218]. Both HOTAIR and MALAT1 are known to be overexpressed in ovarian cancers and drive cell division as well as invasion. HOTAIR downregulates Rab22a, a member of the Ras oncogene family, by acting as a sponge for miR-373 [170,219]. Nonetheless, MALAT1 downregulates the expression of iASPP, a cell death inhibitor, by acting as a sponge for miR-506 in ovarian cancer [220]. Thus, HuR is an important post-transcriptional regulator in ovarian cancer-driving oncogenesis as well as drug resistance. Y-Box binding protein 1 (YB-1) is an oncogene linked with Ovarian cancer manifesting RNA-binding abilities due to the two RNA-binding domains present in its cold shock region [221,222]. The higher expression levels of YB-1 indicate poor prognosis as well as drug resistance [223]. The nuclear localization of YB-1 is associated with MD-1 binding at the CCAAT elements that lead to the expression of P-glycoprotein. Kumara et al. reported the expression of YB-1 in the nucleus of 30% of the ovarian cancer patient samples associated with the worst survival prognosis [224]. Expression levels of another lncRNA, LINC01969, have been correlated with poor prognosis in ovarian cancer. LINC01969, found primarily in the cytoplasm, sponges miR-144-5p to upregulate the expression of LARP1 promoting an invasive phenotype in ovarian cancer [225]. LARP1 is known to modulate approximately 300 mRNAs in cancer responsible for survival signaling, RNA biogenesis proteins and flags a survival advantage in ovarian cancer [226]. lncRNA panels may serve as diagnostic and prognostic biomarkers addressing the main hurdle in the early detection of ovarian cancer. For instance, the combined expression of SPRY4-IT1, NEAT1, and ANRIL served as an effective diagnostic biomarker rather than individually assessing them in NSCLC [227]. Similarly, an integrated expression of PVT1, lnc-SERTD2-3, lnc-SOX4-1, and lnc-HRCT1 in stage I epithelial ovarian cancer patient's tumor biopsy samples was associated with the risk of relapse [228].

Summary

More than ever before, now is the time that post-transcriptional regulation of gene expression should take a center stage in clinical investigations to better understand cancer. The discussed studies emphasize the complexity of deregulation of post-transcriptional events seen in different types of solid tumors at different stages demanding inventorying the panels of miRNA, lncRNA, RBPs, and epigenetic modulators involved in tumorigenesis. An atlas comprising the global panels of altered post-transcriptional regulators in the context of tumor heterogeneity will be instrumental in identifying targets, diagnosis, and the selection of a chemotherapeutic regimen. Cancer cells also adapt by tapping the post-transcriptional regulation to adjust to therapies and, thus, targeting it serves the dual purpose of inhibiting the growth of the tumor as well as precision medicine.

Although initially considered undruggable, RNA targeting has now been realized as a potential avenue and several drugs have been approved to be used in the clinic for diseases other than cancer. Therapies being tested in preclinical studies employ antisense oligonucleotides (ASO), small interfering RNA (siRNAs), short hairpin RNA (shRNAs), antimicro-RNA (antimiRs), miRNA mimics, miRNA sponges, therapeutic circRNAs, and CRISPR-Cas9-based gene editing [229–231]. RNA-based drugs for solid tumors are still in preclinical trials like siG12D-LODER (NCT01188785; NCT01676259), (siRNA) for advanced pancreatic tumor-targeting G12D-mutated KRAS mRNA; Danvatirsen (ASO) (NCT03819465, NCT03794544, NCT02983578) for early and advanced NSCLC, pancreatic cancer, mismatch repair-deficient colorectal cancer that targets STAT3 mRNA; and Apatorsen (ASO) for SCLC and nonsquamous NSCLC, advanced bladder and prostate cancer and urinary tract tumors targeting HSP27 mRNA [232]. Although most of the candidates in clinical trials are siRNAs or ASOs, targeting miRNA has its benefits. The advantage of targeting miRNAs lies in the fact that these are naturally occurring RNA moieties and, thus, can be cleared from the system on their own, unlike other synthetic agents. Additionally, these are involved in every cancer ever studied and regulate numerous genes within the same pathways. For example, multiple apoptotic factors are regulated by miR-15/miR-16 cluster including BCL-2 and MCL-1 [39,233] implying broader consequences. Almost every year, new RNA-based drugs are being approved to be used in clinics for other diseases. With our ever-increasing understanding of post-transcriptional gene regulation in solid tumors, novel, promising, and creative developments will soon advance to the clinic.

References

[1] Singh G, Pratt G, Yeo GW, Moore MJ. The clothes make the mRNA: past and present trends in mRNP fashion. Annu Rev Biochem 2015;84:325–54.
[2] Corbett AH. Post-transcriptional regulation of gene expression and human disease. Curr Opin Cell Biol June 2018;52:96–104.
[3] Oeffinger M, Zenklusen D. To the pore and through the pore: a story of mRNA export kinetics. Biochim Biophys Acta June 2012;1819(6):494–506.
[4] Stewart M. Ratcheting mRNA out of the nucleus. Mol Cell February 9, 2007;25(3):327–30.
[5] Xing L, Bassell GJ. mRNA localization: an orchestration of assembly, traffic and synthesis. Traffic Cph Den January 2013;14(1):2–14.
[6] Kapur M, Monaghan CE, Ackerman SL. Regulation of mRNA translation in neurons-A matter of life and death. Neuron November 1, 2017;96(3):616–37.
[7] Garneau NL, Wilusz J, Wilusz CJ. The highways and byways of mRNA decay. Nat Rev Mol Cell Biol February 2007;8(2):113–26.
[8] Dassi E. Handshakes and fights: the regulatory interplay of RNA-binding proteins. Front Mol Biosci 2017; 4:67.
[9] Gerstberger S, Hafner M, Ascano M, Tuschl T. Evolutionary conservation and expression of human RNA-binding proteins and their role in human genetic disease. Adv Exp Med Biol 2014;825:1–55.
[10] Carey KT, Wickramasinghe VO. Regulatory potential of the RNA processing machinery: implications for human disease. Trends Genet TIG April 2018;34(4):279–90.
[11] Hanahan D, Weinberg RA. Hallmarks of cancer: the next generation. Cell March 4, 2011;144(5):646–74.
[12] McManus CJ, Graveley BR. RNA structure and the mechanisms of alternative splicing. Curr Opin Genet Dev August 2011;21(4):373–9.
[13] Castle JC, Zhang C, Shah JK, Kulkarni AV, Kalsotra A, Cooper TA, et al. Expression of 24,426 human alternative splicing events and predicted cis regulation in 48 tissues and cell lines. Nat Genet December 2008;40(12):1416–25.
[14] Venables JP, Klinck R, Koh C, Gervais-Bird J, Bramard A, Inkel L, et al. Cancer-associated regulation of alternative splicing. Nat Struct Mol Biol June 2009;16(6):670–6.
[15] Xi L, Feber A, Gupta V, Wu M, Bergemann AD, Landreneau RJ, et al. Whole genome exon arrays identify differential expression of alternatively spliced, cancer-related genes in lung cancer. Nucleic Acids Res November 2008;36(20):6535–47.
[16] Armero VES, Tremblay M-P, Allaire A, Boudreault S, Martenon-Brodeur C, Duval C, et al. Transcriptome-wide analysis of alternative RNA splicing events in Epstein-Barr virus-associated gastric carcinomas. PLoS One 2017;12(5):e0176880.
[17] Silipo M, Gautrey H, Tyson-Capper A. Deregulation of splicing factors and breast cancer development. J Mol Cell Biol October 2015;7(5):388–401.
[18] Srebrow A, Kornblihtt AR. The connection between splicing and cancer. J Cell Sci July 1, 2006;119(Pt 13): 2635–41.

[19] Venables JP, Koh C-S, Froehlich U, Lapointe E, Couture S, Inkel L, et al. Multiple and specific mRNA processing targets for the major human hnRNP proteins. Mol Cell Biol October 2008;28(19):6033–43.
[20] Oltean S, Bates DO. Hallmarks of alternative splicing in cancer. Oncogene November 13, 2014;33(46): 5311–8.
[21] Anczuków O, Rosenberg AZ, Akerman M, Das S, Zhan L, Karni R, et al. The splicing factor SRSF1 regulates apoptosis and proliferation to promote mammary epithelial cell transformation. Nat Struct Mol Biol January 15, 2012;19(2):220–8.
[22] Anczuków O, Akerman M, Cléry A, Wu J, Shen C, Shirole NH, et al. SRSF1-Regulated alternative splicing in breast cancer. Mol Cell October 1, 2015;60(1):105–17.
[23] Jiang L, Huang J, Higgs BW, Hu Z, Xiao Z, Yao X, et al. Genomic landscape survey identifies SRSF1 as a key oncodriver in small cell lung cancer. PLoS Genet April 2016;12(4):e1005895.
[24] Lin Y, Li Z, Ozsolak F, Kim SW, Arango-Argoty G, Liu TT, et al. An in-depth map of polyadenylation sites in cancer. Nucleic Acids Res September 1, 2012;40(17):8460–71.
[25] Sandberg R, Neilson JR, Sarma A, Sharp PA, Burge CB. Proliferating cells express mRNAs with shortened 3' untranslated regions and fewer microRNA target sites. Science June 20, 2008;320(5883):1643–7.
[26] Morris AR, Bos A, Diosdado B, Rooijers K, Elkon R, Bolijn AS, et al. Alternative cleavage and polyadenylation during colorectal cancer development. Clin Cancer Res Off J Am Assoc Cancer Res October 1, 2012;18(19):5256–66.
[27] Baysal BE, Sharma S, Hashemikhabir S, Janga SC. RNA editing in pathogenesis of cancer. Cancer Res July 15, 2017;77(14):3733–9.
[28] Axtell MJ, Westholm JO, Lai EC. Vive la différence: biogenesis and evolution of microRNAs in plants and animals. Genome Biol 2011;12(4):221.
[29] Ghildiyal M, Zamore PD. Small silencing RNAs: an expanding universe. Nat Rev Genet February 2009; 10(2):94–108.
[30] Fire A, Xu S, Montgomery MK, Kostas SA, Driver SE, Mello CC. Potent and specific genetic interference by double-stranded RNA in *Caenorhabditis elegans*. Nature February 19, 1998;391(6669):806–11.
[31] Tomaselli S, Galeano F, Alon S, Raho S, Galardi S, Polito VA, et al. Modulation of microRNA editing, expression and processing by ADAR2 deaminase in glioblastoma. Genome Biol January 13, 2015;16:5.
[32] Zhang L, Yang C-S, Varelas X, Monti S. Altered RNA editing in 3′ UTR perturbs microRNA-mediated regulation of oncogenes and tumor-suppressors. Sci Rep March 16, 2016;6:23226.
[33] Bartel DP. MicroRNAs: genomics, biogenesis, mechanism, and function. Cell January 23, 2004;116(2): 281–97 [cited 2017 April 10].
[34] Bartel DP. MicroRNAs: target recognition and regulatory functions. Cell January 23, 2009;136(2):215–33 [cited 2017 June 6].
[35] Huntzinger E, Izaurralde E. Gene silencing by microRNAs: contributions of translational repression and mRNA decay. Nat Rev Genet February 2011;12(2):99–110.
[36] Vinchure OS, Sharma V, Tabasum S, Ghosh S, Singh RP, Sarkar C, et al. Polycomb complex mediated epigenetic reprogramming alters TGF-β signaling via a novel EZH2/miR-490/TGIF2 axis thereby inducing migration and EMT potential in glioblastomas. Int J Cancer September 2019;145(5):1254–69.
[37] Malumbres M. miRNAs versus oncogenes: the power of social networking. Mol Syst Biol February 14, 2012;8:569.
[38] Calin GA, Sevignani C, Dumitru CD, Hyslop T, Noch E, Yendamuri S, et al. Human microRNA genes are frequently located at fragile sites and genomic regions involved in cancers. Proc Natl Acad Sci U S A March 2, 2004;101(9):2999–3004.
[39] Cimmino A, Calin GA, Fabbri M, Iorio MV, Ferracin M, Shimizu M, et al. miR-15 and miR-16 induce apoptosis by targeting BCL2. Proc Natl Acad Sci U S A September 27, 2005;102(39):13944–9.

[40] Bucay N, Sekhon K, Yang T, Majid S, Shahryari V, Hsieh C, et al. MicroRNA-383 located in frequently deleted chromosomal locus 8p22 regulates CD44 in prostate cancer. Oncogene May 11, 2017;36(19):2667−79.
[41] Volinia S, Galasso M, Costinean S, Tagliavini L, Gamberoni G, Drusco A, et al. Reprogramming of miRNA networks in cancer and leukemia. Genome Res May 2010;20(5):589−99.
[42] Wiseman SM, Makretsov N, Nielsen TO, Gilks B, Yorida E, Cheang M, et al. Coexpression of the type 1 growth factor receptor family members HER-1, HER-2, and HER-3 has a synergistic negative prognostic effect on breast carcinoma survival. Cancer May 1, 2005;103(9):1770−7.
[43] Scott GK, Goga A, Bhaumik D, Berger CE, Sullivan CS, Benz CC. Coordinate suppression of ERBB2 and ERBB3 by enforced expression of micro-RNA miR-125a or miR-125b. J Biol Chem January 12, 2007;282(2):1479−86.
[44] Costinean S, Zanesi N, Pekarsky Y, Tili E, Volinia S, Heerema N, et al. Pre-B cell proliferation and lymphoblastic leukemia/high-grade lymphoma in E(mu)-miR155 transgenic mice. Proc Natl Acad Sci U S A May 2, 2006;103(18):7024−9.
[45] Medina PP, Nolde M, Slack FJ. OncomiR addiction in an in vivo model of microRNA-21-induced pre-B-cell lymphoma. Nature September 2, 2010;467(7311):86−90.
[46] Johnson SM, Grosshans H, Shingara J, Byrom M, Jarvis R, Cheng A, et al. RAS is regulated by the let-7 microRNA family. Cell March 11, 2005;120(5):635−47.
[47] Acunzo M, Romano G, Wernicke D, Croce CM. MicroRNA and cancer−a brief overview. Adv Biol Regul January 2015;57:1−9.
[48] Kapranov P, Willingham AT, Gingeras TR. Genome-wide transcription and the implications for genomic organization. Nat Rev Genet June 2007;8(6):413−23.
[49] Morris KV, Mattick JS. The rise of regulatory RNA. Nat Rev Genet June 2014;15(6):423−37.
[50] Shao J, Chen H, Yang D, Jiang M, Zhang H, Wu B, et al. Genome-wide identification and characterization of natural antisense transcripts by strand-specific RNA sequencing in *Ganoderma lucidum*. Sci Rep July 18, 2017;7(1):5711.
[51] Brunner AL, Beck AH, Edris B, Sweeney RT, Zhu SX, Li R, et al. Transcriptional profiling of long non-coding RNAs and novel transcribed regions across a diverse panel of archived human cancers. Genome Biol August 28, 2012;13(8):R75.
[52] Rinn JL, Chang HY. Genome regulation by long noncoding RNAs. Annu Rev Biochem 2012;81.
[53] Gutschner T, Diederichs S. The hallmarks of cancer: a long non-coding RNA point of view. RNA Biol June 2012;9(6):703−19.
[54] Gupta RA, Shah N, Wang KC, Kim J, Horlings HM, Wong DJ, et al. Long non-coding RNA HOTAIR reprograms chromatin state to promote cancer metastasis. Nature April 15, 2010;464(7291):1071−6.
[55] Bhan A, Mandal SS. LncRNA HOTAIR: a master regulator of chromatin dynamics and cancer. Biochim Biophys Acta August 2015;1856(1):151−64.
[56] Zhang J, Zhang P, Wang L, Piao H, Ma L. Long non-coding RNA HOTAIR in carcinogenesis and metastasis. Acta Biochim Biophys Sin January 2014;46(1):1−5.
[57] Pickard MR, Williams GT. The hormone response element mimic sequence of GAS5 lncRNA is sufficient to induce apoptosis in breast cancer cells. Oncotarget March 1, 2016;7(9):10104−16.
[58] Mourtada-Maarabouni M, Pickard MR, Hedge VL, Farzaneh F, Williams GT. GAS5, a non-protein-coding RNA, controls apoptosis and is downregulated in breast cancer. Oncogene January 15, 2009;28(2):195−208.
[59] Berteaux N, Aptel N, Cathala G, Genton C, Coll J, Daccache A, et al. A novel H19 antisense RNA overexpressed in breast cancer contributes to paternal IGF2 expression. Mol Cell Biol November 2008;28(22):6731−45.

[60] Tripathi V, Ellis JD, Shen Z, Song DY, Pan Q, Watt AT, et al. The nuclear-retained noncoding RNA MALAT1 regulates alternative splicing by modulating SR splicing factor phosphorylation. Mol Cell September 24, 2010;39(6):925–38.
[61] Jadaliha M, Zong X, Malakar P, Ray T, Singh DK, Freier SM, et al. Functional and prognostic significance of long non-coding RNA MALAT1 as a metastasis driver in ER negative lymph node negative breast cancer. Oncotarget June 28, 2016;7(26):40418–36.
[62] Gutschner T, Hämmerle M, Diederichs S. MALAT1 – a paradigm for long noncoding RNA function in cancer. J Mol Med Berl Ger July 2013;91(7):791–801.
[63] Sun L, Li Y, Yang B. Downregulated long non-coding RNA MEG3 in breast cancer regulates proliferation, migration and invasion by depending on p53's transcriptional activity. Biochem Biophys Res Commun September 9, 2016;478(1):323–9.
[64] Zhou Y, Zhang X, Klibanski A. MEG3 noncoding RNA: a tumor suppressor. J Mol Endocrinol June 2012; 48(3):R45–53.
[65] Poliseno L, Marranci A, Pandolfi PP. Pseudogenes in human cancer. Front Med 2015;2:68.
[66] Poliseno L, Salmena L, Zhang J, Carver B, Haveman WJ, Pandolfi PP. A coding-independent function of gene and pseudogene mRNAs regulates tumour biology. Nature June 24, 2010;465(7301):1033–8.
[67] Poliseno L, Haimovic A, Christos PJ, Vega Y Saenz de Miera EC, Shapiro R, Pavlick A, et al. Deletion of PTENP1 pseudogene in human melanoma. J Invest Dermatol December 2011;131(12):2497–500.
[68] Liu F, Gao H, Li S, Ni X, Zhu Z. Long non-coding RNA ZFAS1 correlates with clinical progression and prognosis in cancer patients. Oncotarget September 22, 2017;8(37):61561–9.
[69] Askarian-Amiri ME, Crawford J, French JD, Smart CE, Smith MA, Clark MB, et al. SNORD-host RNA Zfas1 is a regulator of mammary development and a potential marker for breast cancer. RNA N Y N May 2011;17(5):878–91.
[70] Liu J, Zhang X, Yan M, Li H. Emerging role of circular RNAs in cancer. Front Oncol 2020;10:663.
[71] Guo JU, Agarwal V, Guo H, Bartel DP. Expanded identification and characterization of mammalian circular RNAs. Genome Biol July 29, 2014;15(7):409.
[72] Abdelmohsen K, Panda AC, Munk R, Grammatikakis I, Dudekula DB, De S, et al. Identification of HuR target circular RNAs uncovers suppression of PABPN1 translation by CircPABPN1. RNA Biol March 4, 2017;14(3):361–9.
[73] Ladomery MR, Harper SJ, Bates DO. Alternative splicing in angiogenesis: the vascular endothelial growth factor paradigm. Cancer Lett May 8, 2007;249(2):133–42.
[74] Bevan HS, van den Akker NMS, Qiu Y, Polman JAE, Foster RR, Yem J, et al. The alternatively spliced anti-angiogenic family of VEGF isoforms VEGFxxxb in human kidney development. Nephron Physiol 2008; 110(4):p57–67.
[75] Bates DO, Cui T-G, Doughty JM, Winkler M, Sugiono M, Shields JD, et al. VEGF165b, an inhibitory splice variant of vascular endothelial growth factor, is down-regulated in renal cell carcinoma. Cancer Res July 15, 2002;62(14):4123–31.
[76] Woolard J, Wang W-Y, Bevan HS, Qiu Y, Morbidelli L, Pritchard-Jones RO, et al. VEGF165b, an inhibitory vascular endothelial growth factor splice variant: mechanism of action, in vivo effect on angiogenesis and endogenous protein expression. Cancer Res November 1, 2004;64(21):7822–35.
[77] Behbahani GD, Ghahhari NM, Javidi MA, Molan AF, Feizi N, Babashah S. MicroRNA-mediated post-transcriptional regulation of epithelial to mesenchymal transition in cancer. Pathol Oncol Res January 2017;23(1):1–12.
[78] Ma L, Young J, Prabhala H, Pan E, Mestdagh P, Muth D, et al. miR-9, a MYC/MYCN-activated microRNA, regulates E-cadherin and cancer metastasis. Nat Cell Biol March 2010;12(3):247–56.
[79] Díaz-López A, Moreno-Bueno G, Cano A. Role of microRNA in epithelial to mesenchymal transition and metastasis and clinical perspectives. Cancer Manag Res 2014;6:205–16.

[80] Brabletz S, Brabletz T. The ZEB/miR-200 feedback loop–a motor of cellular plasticity in development and cancer? EMBO Rep September 2010;11(9):670–7.

[81] Hill L, Browne G, Tulchinsky E. ZEB/miR-200 feedback loop: at the crossroads of signal transduction in cancer. Int J Cancer February 15, 2013;132(4):745–54.

[82] Liu Y-N, Yin JJ, Abou-Kheir W, Hynes PG, Casey OM, Fang L, et al. MiR-1 and miR-200 inhibit EMT via Slug-dependent and tumorigenesis via Slug-independent mechanisms. Oncogene January 17, 2013;32(3): 296–306.

[83] Sung H, Ferlay J, Siegel RL, Laversanne M, Soerjomataram I, Jemal A, et al. Global cancer statistics 2020: GLOBOCAN estimates of incidence and mortality worldwide for 36 cancers in 185 countries. CA Cancer J Clin 2021;71(3):209–49 [cited 2021 August 15].

[84] Petri BJ, Klinge CM. Regulation of breast cancer metastasis signaling by miRNAs. Cancer Metastasis Rev September 2020;39(3):837–86.

[85] Canu V, Donzelli S, Sacconi A, Lo Sardo F, Pulito C, Bossel N, et al. Aberrant transcriptional and post-transcriptional regulation of SPAG5, a YAP-TAZ-TEAD downstream effector, fuels breast cancer cell proliferation. Cell Death Differ May 2021;28(5):1493–511.

[86] Mercatante DR, Bortner CD, Cidlowski JA, Kole R. Modification of alternative splicing of Bcl-x pre-mRNA in prostate and breast cancer cells: analysis OF apoptosis and cell death. J Biol Chem May 11, 2001;276(19):16411–7 [cited 2021 August 21].

[87] Xu Y, Gao XD, Lee J-H, Huang H, Tan H, Ahn J, et al. Cell type-restricted activity of hnRNPM promotes breast cancer metastasis via regulating alternative splicing. Genes Dev June 1, 2014;28(11):1191–203.

[88] Zhang F-L, Cao J-L, Xie H-Y, Sun R, Yang L-F, Shao Z-M, et al. Cancer-associated MORC2-mutant M276I regulates an hnRNPM-mediated CD44 splicing switch to promote invasion and metastasis in triple-negative breast cancer. Cancer Res October 15, 2018;78(20):5780–92 [cited 2021 August 21].

[89] Zhang H, Brown RL, Wei Y, Zhao P, Liu S, Liu X, et al. CD44 splice isoform switching determines breast cancer stem cell state. Genes Dev February 1, 2019;33(3–4):166–79 [cited 2021 August 21].

[90] Han J, Li J, Ho JC, Chia GS, Kato H, Jha S, et al. Hypoxia is a key driver of alternative splicing in human breast cancer cells. Sci Rep June 22, 2017;7(1):4108 [cited 2021 August 21].

[91] Liu Z, Sun L, Cai Y, Shen S, Zhang T, Wang N, et al. Hypoxia-induced suppression of alternative splicing of MBD2 promotes breast cancer metastasis via activation of FZD1. Cancer Res March 1, 2021;81(5): 1265–78 [cited 2021 August 21].

[92] Wang L, Wang Y, Su B, Yu P, He J, Meng L, et al. Transcriptome-wide analysis and modelling of prognostic alternative splicing signatures in invasive breast cancer: a prospective clinical study. Sci Rep October 5, 2020;10(1):16504 [cited 2021 August 22].

[93] Oh J, Pradella D, Shao C, Li H, Choi N, Ha J, et al. Widespread alternative splicing changes in metastatic breast cancer cells. Cells April 2021;10(4):858 [cited 2021 August 21].

[94] Bellanger A, Le DT, Vendrell J, Wierinckx A, Pongor LS, Solassol J, et al. Exploring the significance of the exon 4-skipping isoform of the ZNF217 oncogene in breast cancer. Front Oncol 2021:11 [cited 2021 August 21].

[95] Martínez-Montiel N, Anaya-Ruiz M, Pérez-Santos M, Martínez-Contreras RD. Alternative splicing in breast cancer and the potential development of therapeutic tools. Genes October 5, 2017;8(10):217 [cited 2021 August 22].

[96] Iino K, Mitobe Y, Ikeda K, Takayama K-I, Suzuki T, Kawabata H, et al. RNA-binding protein NONO promotes breast cancer proliferation by post-transcriptional regulation of SKP2 and E2F8. Cancer Sci January 2020;111(1):148–59.

[97] Mitobe Y, Iino K, Takayama K, Ikeda K, Suzuki T, Aogi K, et al. PSF promotes ER-positive breast cancer progression via posttranscriptional regulation of ESR1 and SCFD2. Cancer Res June 1, 2020;80(11): 2230–42 [cited 2021 August 23].

[98] Chen F, Wang Q, Yu X, Yang N, Wang Y, Zeng Y, et al. MCPIP1-mediated NFIC alternative splicing inhibits proliferation of triple-negative breast cancer via cyclin D1-Rb-E2F1 axis. Cell Death Dis April 6, 2021;12(4):1−16 [cited 2021 August 23].

[99] Troschel FM, Minte A, Ismail YM, Kamal A, Abdullah MS, Ahmed SH, et al. Knockdown of Musashi RNA-binding proteins decreases radioresistance but enhances cell motility and invasion in triple-negative breast cancer. Int J Mol Sci March 21, 2020;21(6):E2169.

[100] Forouzanfar M, Lachinani L, Dormiani K, Nasr-Esfahani MH, Ghaedi K. Increased expression of MUSASHI1 in epithelial breast cancer cells is due to down regulation of miR-125b. BMC Mol Cell Biol February 4, 2021;22(1):10 [cited 2021 August 23].

[101] Liu J, Li H, Mao A, Lu J, Liu W, Qie J, et al. DCAF13 promotes triple-negative breast cancer metastasis by mediating DTX3 mRNA degradation. Cell Cycle December 16, 2020;19(24):3622−31 [cited 2021 August 23].

[102] Du J-X, Luo Y-H, Zhang S-J, Wang B, Chen C, Zhu G-Q, et al. Splicing factor SRSF1 promotes breast cancer progression via oncogenic splice switching of PTPMT1. J Exp Clin Cancer Res May 15, 2021;40(1): 171 [cited 2021 August 23].

[103] Fang R, Ye L, Shi H. Understanding the roles of N6-methyladenosine writers, readers and erasers in breast cancer. Neoplasia N Y N June 2021;23(6):551−60.

[104] Einstein JM, Perelis M, Chaim IA, Meena JK, Nussbacher JK, Tankka AT, et al. Inhibition of YTHDF2 triggers proteotoxic cell death in MYC-driven breast cancer. Mol Cell August 5, 2021;81(15). 3048.e9−3064.e9.

[105] Su X, Feng C, Wang S, Shi L, Gu Q, Zhang H, et al. The noncoding RNAs SNORD50A and SNORD50B-mediated TRIM21-GMPS interaction promotes the growth of p53 wild-type breast cancers by degrading p53. Cell Death Differ August 2021;28(8):2450−64 [cited 2021 August 25].

[106] Das S, De S, Sengupta S. Post-transcriptional regulation of MMP2 mRNA by its interaction with miR-20a and Nucleolin in breast cancer cell lines. Mol Biol Rep March 1, 2021;48(3):2315−24 [cited 2021 August 25].

[107] Venkatesh J, Wasson M-CD, Brown JM, Fernando W, Marcato P. LncRNA-miRNA axes in breast cancer: novel points of interaction for strategic attack. Cancer Lett July 1, 2021;509:81−8 [cited 2021 August 25].

[108] Cumova A, Vymetalkova V, Opattova A, Bouskova V, Pardini B, Kopeckova K, et al. Genetic variations in 3′UTRs of SMUG1 and NEIL2 genes modulate breast cancer risk, survival and therapy response. Mutagenesis June 7, 2021:geab017 [cited 2021 August 25].

[109] Hirsch FR, Scagliotti GV, Mulshine JL, Kwon R, Curran WJ, Wu Y-L, et al. Lung cancer: current therapies and new targeted treatments. Lancet Lond Engl January 21, 2017;389(10066):299−311.

[110] Inamura K. Lung cancer: understanding its molecular pathology and the 2015 WHO classification. Front Oncol August 28, 2017;7:193 [cited 2021 September 2].

[111] Tabasum S, Singh RP. Fisetin suppresses migration, invasion and stem-cell-like phenotype of human non-small cell lung carcinoma cells via attenuation of epithelial to mesenchymal transition. Chem Biol Interact April 2019;303:14−21.

[112] Oka M, Xu L, Suzuki T, Yoshikawa T, Sakamoto H, Uemura H, et al. Aberrant splicing isoforms detected by full-length transcriptome sequencing as transcripts of potential neoantigens in non-small cell lung cancer. Genome Biol December 2021;22(1):9 [cited 2021 September 2].

[113] Esfahani MS, Lee LJ, Jeon Y-J, Flynn RA, Stehr H, Hui AB, et al. Functional significance of U2AF1 S34F mutations in lung adenocarcinomas. Nat Commun December 2019;10(1):5712 [cited 2021 September 2].

[114] The Cancer Genome Atlas Research Network. Comprehensive molecular profiling of lung adenocarcinoma. Nature July 31, 2014;511(7511):543−50 [cited 2021 September 2].

[115] Zhou Z, Gong Q, Wang Y, Li M, Wang L, Ding H, et al. The biological function and clinical significance of SF3B1 mutations in cancer. Biomark Res 2020;8:38.

[116] Imielinski M, Berger AH, Hammerman PS, Hernandez B, Pugh TJ, Hodis E, et al. Mapping the hallmarks of lung adenocarcinoma with massively parallel sequencing. Cell September 14, 2012;150(6):1107–20.

[117] Alsafadi S, Dayot S, Tarin M, Houy A, Bellanger D, Cornella M, et al. Genetic alterations of SUGP1 mimic mutant-SF3B1 splice pattern in lung adenocarcinoma and other cancers. Oncogene January 7, 2021;40(1): 85–96 [cited 2021 September 2].

[118] Bishop JA, Benjamin H, Cholakh H, Chajut A, Clark DP, Westra WH. Accurate classification of non–small cell lung carcinoma using a novel MicroRNA-based approach. Clin Cancer Res January 15, 2010;16(2): 610–9 [cited 2021 September 2].

[119] Chen Y-J, Guo Y-N, Shi K, Huang H-M, Huang S-P, Xu W-Q, et al. Down-regulation of microRNA-144-3p and its clinical value in non-small cell lung cancer: a comprehensive analysis based on microarray, miRNA-sequencing, and quantitative real-time PCR data. Respir Res December 2019;20(1):48 [cited 2021 September 2].

[120] Liu M, Gao J, Huang Q, Jin Y, Wei Z. Downregulating microRNA-144 mediates a metabolic shift in lung cancer cells by regulating GLUT1 expression. Oncol Lett June 2016;11(6):3772–6.

[121] Tang Y, Xiao G, Chen Y, Deng Y. LncRNA MALAT1 promotes migration and invasion of non-small-cell lung cancer by targeting miR-206 and activating Akt/mTOR signaling. Anticancer Drug September 2018; 29(8):725–35 [cited 2021 September 2].

[122] Li J, Wang J, Chen Y, Li S, Jin M, Wang H, et al. LncRNA MALAT1 exerts oncogenic functions in lung adenocarcinoma by targeting miR-204. Am J Cancer Res 2016;6(5):1099–107.

[123] Cui J, Mo J, Luo M, Yu Q, Zhou S, Li T, et al. c-Myc-activated long non-coding RNA H19 downregulates miR-107 and promotes cell cycle progression of non-small cell lung cancer. Int J Clin Exp Pathol 2015; 8(10):12400–9.

[124] Huang Z, Lei W, Hu H, Zhang H, Zhu Y. H19 promotes non-small-cell lung cancer (NSCLC) development through STAT3 signaling via sponging miR-17. J Cell Physiol October 2018;233(10):6768–76 [cited 2021 September 2].

[125] Ng WL, Mohd Mohidin TB, Shukla K. Functional role of circular RNAs in cancer development and progression. RNA Biol 2018;15(8):995–1005.

[126] Li O, Kang J, Zhang J-J, Wang J, Hu L-W, Li L, et al. Circle RNA FOXP1 promotes cell proliferation in lung cancer by regulating miR-185-5p/Wnt1 signaling pathway. Eur Rev Med Pharmacol Sci June 2020; 24(12):6767–78.

[127] Gao N, Ye B. Circ-SOX4 drives the tumorigenesis and development of lung adenocarcinoma via sponging miR-1270 and modulating PLAGL2 to activate WNT signaling pathway. Cancer Cell Int December 2020; 20(1):2 [cited 2021 September 2].

[128] Yao Y, Hua Q, Zhou Y, Shen H. CircRNA has_circ_0001946 promotes cell growth in lung adenocarcinoma by regulating miR-135a-5p/SIRT1 axis and activating Wnt/β-catenin signaling pathway. Biomed Pharmacother Biomedecine Pharmacother March 2019;111:1367–75.

[129] Tian F, Yu CT, Ye WD, Wang Q. Cinnamaldehyde induces cell apoptosis mediated by a novel circular RNA hsa_circ_0043256 in non-small cell lung cancer. Biochem Biophys Res Commun November 2017;493(3): 1260–6 [cited 2021 September 2].

[130] Chen S, Lu S, Yao Y, Chen J, Yang G, Tu L, et al. Downregulation of hsa_circ_0007580 inhibits non-small cell lung cancer tumorigenesis by reducing miR-545-3p sponging. Aging July 18, 2020;12(14):14329–40.

[131] Zhang Z, Gao X, Ma M, Zhao C, Zhang Y, Guo S. CircRNA_101237 promotes NSCLC progression via the miRNA-490-3p/MAPK1 axis. Sci Rep December 2020;10(1):9024 [cited 2021 September 2].

[132] Zhou Q, Sun Y. Circular RNA cMras suppresses the progression of lung adenocarcinoma through ABHD5/ ATGL Axis using NF-κB signaling pathway. Cancer Biother Radiopharm August 19, 2020:3709 [cited 2021 September 2];cbr.2020.

[133] Chen D, Ma W, Ke Z, Xie F. CircRNA hsa_circ_100395 regulates miR-1228/TCF21 pathway to inhibit lung cancer progression. Cell Cycle August 18, 2018;17(16):2080–90 [cited 2021 September 2].

[134] Gao S, Yu Y, Liu L, Meng J, Li G. Circular RNA hsa_circ_0007059 restrains proliferation and epithelial-mesenchymal transition in lung cancer cells via inhibiting microRNA-378. Life Sci September 15, 2019; 233:116692.

[135] Wang ET, Sandberg R, Luo S, Khrebtukova I, Zhang L, Mayr C, et al. Alternative isoform regulation in human tissue transcriptomes. Nature November 2008;456(7221):470–6 [cited 2021 September 2].

[136] Chen H-H, Zhang T-N, Wu Q-J, Huang X-M, Zhao Y-H. Circular RNAs in lung cancer: recent advances and future perspectives. Front Oncol 2021;11:664290.

[137] Kahi CJ, Boland CR, Dominitz JA, Giardiello FM, Johnson DA, Kaltenbach T, et al. Colonoscopy surveillance after colorectal cancer resection: recommendations of the US multi-society task force on colorectal cancer. Gastrointest Endosc March 2016;83(3). 489.e10–498.e10.

[138] Ooki A, Shinozaki E, Yamaguchi K. Immunotherapy in colorectal cancer: current and future strategies. J Anus Rectum Colon January 28, 2021;5(1):11–24.

[139] Han J. Screening of lymph nodes metastasis associated lncRNAs in colorectal cancer patients. World J Gastroenterol 2014;20(25):8139.

[140] Chen D, Sun Q, Cheng X, Zhang L, Song W, Zhou D, et al. Genome-wide analysis of long noncoding RNA (lnc RNA) expression in colorectal cancer tissues from patients with liver metastasis. Cancer Med July 2016;5(7):1629–39.

[141] Tian Y, Zhou J, Zou Y, Luo B, Liu Q, Cao X. Upregulated long noncoding RNAs LINC02163 and FEZF1-AS1 exert oncogenic roles in colorectal cancer. Anti Cancer Drugs January 2021;32(1):66–73.

[142] Yang P, Chen T, Xu Z, Zhu H, Wang J, He Z. Long noncoding RNA GAPLINC promotes invasion in colorectal cancer by targeting SNAI2 through binding with PSF and NONO. Oncotarget July 5, 2016;7(27): 42183–94.

[143] Zhang Z, Wang L, Wang Q, Zhang M, Wang B, Jiang K, et al. Molecular characterization and clinical relevance of RNA-binding proteins in colorectal cancer. Front Genet October 16, 2020;11:580149.

[144] Giannakis M, Mu XJ, Shukla SA, Qian ZR, Cohen O, Nishihara R, et al. Genomic correlates of immune-cell infiltrates in colorectal carcinoma. Cell Rep April 2016;15(4):857–65.

[145] Chatterji P, Rustgi AK. RNA-binding proteins in intestinal epithelial biology and colorectal cancer. Trends Mol Med May 2018;24(5):490–506.

[146] Thomas J, Ohtsuka M, Pichler M, Ling H. MicroRNAs: clinical relevance in colorectal cancer. Int J Mol Sci November 25, 2015;16(12):28063–76.

[147] Mogilyansky E, Rigoutsos I. The miR-17/92 cluster: a comprehensive update on its genomics, genetics, functions and increasingly important and numerous roles in health and disease. Cell Death Differ December 2013;20(12):1603–14.

[148] Chang P-Y, Chen C-C, Chang Y-S, Tsai W-S, You J-F, Lin G-P, et al. MicroRNA-223 and microRNA-92a in stool and plasma samples act as complementary biomarkers to increase colorectal cancer detection. Oncotarget March 1, 2016;7(9):10663–75.

[149] Ding L, Lan Z, Xiong X, Ao H, Feng Y, Gu H, et al. The dual role of MicroRNAs in colorectal cancer progression. Int J Mol Sci September 17, 2018;19(9):2791.

[150] Takahashi M, Sung B, Shen Y, Hur K, Link A, Boland CR, et al. Boswellic acid exerts antitumor effects in colorectal cancer cells by modulating expression of the let-7 and miR-200 microRNA family. Carcinogenesis December 1, 2012;33(12):2441–9.

[151] Saridaki Z, Weidhaas JB, Lenz H-J, Laurent-Puig P, Jacobs B, De Schutter J, et al. A let-7 microRNA-binding site polymorphism in KRAS predicts improved outcome in patients with metastatic colorectal cancer treated with salvage Cetuximab/Panitumumab monotherapy. Clin Cancer Res September 1, 2014;20(17):4499–510.
[152] Wang F, Wang J, Cao X, Xu L, Chen L. Hsa_circ_0014717 is downregulated in colorectal cancer and inhibits tumor growth by promoting p16 expression. Biomed Pharmacother February 2018;98:775–82.
[153] Zeng K, Chen X, Xu M, Liu X, Hu X, Xu T, et al. CircHIPK3 promotes colorectal cancer growth and metastasis by sponging miR-7. Cell Death Dis April 2018;9(4):417.
[154] Geng Y, Zheng X, Hu W, Wang Q, Xu Y, He W, et al. Hsa_circ_0009361 acts as the sponge of miR-582 to suppress colorectal cancer progression by regulating APC2 expression. Clin Sci May 31, 2019;133(10):1197–213.
[155] Mukaisho K, Nakayama T, Hagiwara T, Hattori T, Sugihara H. Two distinct etiologies of gastric cardia adenocarcinoma: interactions among pH, *Helicobacter pylori*, and bile acids. Front Microbiol May 11, 2015:6 [cited 2021 September 1].
[156] Sabarwal A, Agarwal R, Singh RP. Fisetin inhibits cellular proliferation and induces mitochondria-dependent apoptosis in human gastric cancer cells: Fisetin targets mitochondria for ROS-mediated apoptosis. Mol Carcinog February 2017;56(2):499–514.
[157] Colquhoun A, Arnold M, Ferlay J, Goodman KJ, Forman D, Soerjomataram I. Global patterns of cardia and non-cardia gastric cancer incidence in 2012. Gut December 2015;64(12):1881–8 [cited 2021 September 1].
[158] Correa P. Gastric cancer. Gastroenterol Clin North Am June 2013;42(2):211–7 [cited 2021 September 1].
[159] Rugge M, Fassan M, Graham DY. Epidemiology of gastric cancer. In: Strong VE, editor. Gastric cancer. Cham: Springer International Publishing; 2015. p. 23–34 [cited 2021 September 1].
[160] Petrelli F, Berenato R, Turati L, Mennitto A, Steccanella F, Caporale M, et al. Prognostic value of diffuse versus intestinal histotype in patients with gastric cancer: a systematic review and meta-analysis. J Gastrointest Oncol February 2017;8(1):148–63 [cited 2021 September 1].
[161] Chan THM, Qamra A, Tan KT, Guo J, Yang H, Qi L, et al. ADAR-mediated RNA editing predicts progression and prognosis of gastric cancer. Gastroenterology October 2016;151(4) [cited 2021 September 1] 637.e10–650.e10.
[162] Zhu J, Ding H, Wang X, Lu Q. PABPC1 exerts carcinogenesis in gastric carcinoma by targeting miR-34c. Int J Clin Exp Pathol 2015;8(4):3794–802.
[163] Li Y, Yuan Y. Alternative RNA splicing and gastric cancer. Mutat Res Mutat Res July 2017;773:263–73 [cited 2021 September 1].
[164] Shinmura K. Inactivating mutations of the human base excision repair gene NEIL1 in gastric cancer. Carcinogenesis June 24, 2004;25(12):2311–7 [cited 2021 September 1].
[165] Golestaneh AF, Atashi A, Langroudi L, Shafiee A, Ghaemi N, Soleimani M. miRNAs expressed differently in cancer stem cells and cancer cells of human gastric cancer cell line MKN-45: MIRNAS IN cancer stem cells and cancer cells. Cell Biochem Funct July 2012;30(5):411–8 [cited 2021 September 2].
[166] Zeng J-F, Ma X-Q, Wang L-P, Wang W. MicroRNA-145 exerts tumor-suppressive and chemo-resistance lowering effects by targeting CD44 in gastric cancer. World J Gastroenterol 2017;23(13):2337 [cited 2021 September 2].
[167] Xiao W, Li D, Tang Y, Chen Y, Deng W, Chen J, et al. Inhibition of epithelial-mesenchymal transition in gastric cancer cells by miR-711-mediated downregulation of CD44 expression. Oncol Rep September 4, 2018 [cited 2021 September 2].
[168] Wu Q, Yang Z, Wang F, Hu S, Yang L, Shi Y, et al. MiR-19b/20a/92a regulates the self-renewal and proliferation of gastric cancer stem cells. J Cell Sci January 1, 2013 [cited 2021 September 2]; jcs.127944.

[169] Duan F, Jiang J, Song C, Wang P, Ye H, Dai L, et al. Functional long non-coding RNAs associated with gastric cancer susceptibility and evaluation of the epidemiological efficacy in a central Chinese population. Gene March 2018;646:227−33 [cited 2021 September 2].

[170] Yan J, Dang Y, Liu S, Zhang Y, Zhang G. LncRNA HOTAIR promotes cisplatin resistance in gastric cancer by targeting miR-126 to activate the PI3K/AKT/MRP1 genes. Tumor Biol December 2016;37(12): 16345−55 [cited 2021 September 2].

[171] Cheng C, Qin Y, Zhi Q, Wang J, Qin C. Knockdown of long non-coding RNA HOTAIR inhibits cisplatin resistance of gastric cancer cells through inhibiting the PI3K/Akt and Wnt/β-catenin signaling pathways by up-regulating miR-34a. Int J Biol Macromol February 2018;107:2620−9 [cited 2021 September 2].

[172] YiRen H, YingCong Y, Sunwu Y, Keqin L, Xiaochun T, Senrui C, et al. Long noncoding RNA MALAT1 regulates autophagy associated chemoresistance via miR-23b-3p sequestration in gastric cancer. Mol Cancer December 2017;16(1):174 [cited 2021 September 2].

[173] Xi Z, Si J, Nan J. LncRNA MALAT1 potentiates autophagy-associated cisplatin resistance by regulating the microRNA-30b/autophagy-related gene 5 axis in gastric cancer. Int J Oncol October 26, 2018 [cited 2021 September 2].

[174] Zhang X, Wang S, Wang H, Cao J, Huang X, Chen Z, et al. Circular RNA circNRIP1 acts as a microRNA-149-5p sponge to promote gastric cancer progression via the AKT1/mTOR pathway. Mol Cancer December 2019;18(1):20 [cited 2021 September 2].

[175] Lu J, Zhang P, Li P, Xie J, Wang J, Lin J, et al. Circular RNA hsa_circ_0001368 suppresses the progression of gastric cancer by regulating miR-6506−5p/FOXO3 axis. Biochem Biophys Res Commun April 2019; 512(1):29−33 [cited 2021 September 2].

[176] Lai Z, Yang Y, Wang C, Yang W, Yan Y, Wang Z, et al. Circular RNA 0047905 acts as a sponge for microRNA4516 and microRNA1227-5p, initiating gastric cancer progression. Cell Cycle July 18, 2019; 18(14):1560−72 [cited 2021 September 2].

[177] Zhang L, Song X, Chen X, Wang Q, Zheng X, Wu C, et al. Circular RNA CircCACTIN promotes gastric cancer progression by sponging MiR-331-3p and regulating TGFBR1 expression. Int J Biol Sci 2019;15(5): 1091−103 [cited 2021 September 2].

[178] Dai X, Guo X, Liu J, Cheng A, Peng X, Zha L, et al. Circular RNA circGRAMD1B inhibits gastric cancer progression by sponging miR-130a-3p and regulating PTEN and p21 expression. Aging November 13, 2019;11(21):9689−708 [cited 2021 September 2];.

[179] Bray F, Ferlay J, Soerjomataram I, Siegel RL, Torre LA, Jemal A. Global cancer statistics 2018: GLOBOCAN estimates of incidence and mortality worldwide for 36 cancers in 185 countries. CA Cancer J Clin 2018;68(6):394−424.

[180] Ling Z, Cheng B, Tao X. Epithelial-to-mesenchymal transition in oral squamous cell carcinoma: challenges and opportunities. Int J Cancer 2021;148(7):1548−61 [cited 2021 September 1].

[181] Yadav M, Pradhan D, Singh RP. Integrated analysis and identification of nine-gene signature associated to oral squamous cell carcinoma pathogenesis. 3 Biotech May 2021;11(5):215.

[182] Sivadas VP, Gulati S, Varghese BT, Balan A, Kannan S. The early manifestation, tumor-specific occurrence and prognostic significance of TGFBR2 aberrant splicing in oral carcinoma. Exp Cell Res September 10, 2014;327(1):156−62 [cited 2021 September 1].

[183] Peiqi L, Zhaozhong G, Yaotian Y, Jun J, Jihua G, Rong J. Expression of SRSF3 is correlated with carcinogenesis and progression of oral squamous cell carcinoma. Int J Med Sci June 30, 2016;13(7):533−9 [cited 2021 September 1].

[184] Jia R, Zhang S, Liu M, Zhang Y, Liu Y, Fan M, et al. HnRNP L is important for the expression of oncogene SRSF3 and oncogenic potential of oral squamous cell carcinoma cells. Sci Rep November 3, 2016;6:35976.

[185] Biselli-Chicote PM, Biselli JM, Cunha BR, Castro R, Maniglia JV, Neto D de S, et al. Overexpression of antiangiogenic vascular endothelial growth factor isoform and splicing regulatory factors in oral, laryngeal and pharyngeal squamous cell carcinomas. Asian Pac J Cancer Prev APJCP August 27, 2017;18(8): 2171–7.
[186] Yang S, Jia R, Bian Z. SRSF5 functions as a novel oncogenic splicing factor and is upregulated by oncogene SRSF3 in oral squamous cell carcinoma. Biochim Biophys Acta Mol Cell Res September 2018;1865(9): 1161–72.
[187] Osada AH, Endo K, Kimura Y, Sakamoto K, Nakamura R, Sakamoto K, et al. Addiction of mesenchymal phenotypes on the FGF/FGFR axis in oral squamous cell carcinoma cells. PLoS One November 4, 2019; 14(11):e0217451 [cited 2021 September 1].
[188] Zhang S, Wu X, Diao P, Wang C, Wang D, Li S, et al. Identification of a prognostic alternative splicing signature in oral squamous cell carcinoma. J Cell Physiol 2020;235(5):4804–13 [cited 2021 September 1].
[189] Weiße J, Rosemann J, Krauspe V, Kappler M, Eckert AW, Haemmerle M, et al. RNA-binding proteins as regulators of migration, invasion and metastasis in oral squamous cell carcinoma. Int J Mol Sci September 17, 2020;21(18):E6835.
[190] Ishii H, Saitoh M, Sakamoto K, Kondo T, Katoh R, Tanaka S, et al. Epithelial splicing regulatory proteins 1 (ESRP1) and 2 (ESRP2) suppress cancer cell motility via different mechanisms. J Biol Chem October 3, 2014;289(40):27386–99.
[191] Lu W, Feng F, Xu J, Lu X, Wang S, Wang L, et al. QKI impairs self-renewal and tumorigenicity of oral cancer cells via repression of SOX2. Cancer Biol Ther September 1, 2014;15(9):1174–84 [cited 2021 September 2].
[192] Liu L, Wu Y, Li Q, Liang J, He Q, Zhao L, et al. METTL3 promotes tumorigenesis and metastasis through BMI1 m6A methylation in oral squamous cell carcinoma. Mol Ther October 7, 2020;28(10):2177–90 [cited 2021 September 2].
[193] Zhao W, Cui Y, Liu L, Ma X, Qi X, Wang Y, et al. METTL3 facilitates oral squamous cell carcinoma tumorigenesis by enhancing c-Myc stability via YTHDF1-mediated m6A modification. Mol Ther Nucleic Acids June 5, 2020;20:1–12 [cited 2021 September 2].
[194] Chen Y, Lin Y, Shu Y, He J, Gao W. Interaction between N6-methyladenosine (m6A) modification and noncoding RNAs in cancer. Mol Cancer May 22, 2020;19(1):94. [cited 2021 September 2].
[195] Wu P, Fang X, Liu Y, Tang Y, Wang W, Li X, et al. N6-methyladenosine modification of circCUX1 confers radioresistance of hypopharyngeal squamous cell carcinoma through caspase1 pathway. Cell Death Dis March 19, 2021;12(4):1–13 [cited 2021 September 2].
[196] Zhang C-Z. Long intergenic non-coding RNA 668 regulates VEGFA signaling through inhibition of miR-297 in oral squamous cell carcinoma. Biochem Biophys Res Commun August 5, 2017;489(4):404–12.
[197] Wu J, Zhao W, Wang Z, Xiang X, Zhang S, Liu L. Long non-coding RNA SNHG20 promotes the tumorigenesis of oral squamous cell carcinoma via targeting miR-197/LIN28 axis. J Cell Mol Med 2019; 23(1):680–8 [cited 2021 September 1].
[198] Li S, Zhang S, Chen J. c-Myc induced upregulation of long non-coding RNA SNHG16 enhances progression and carcinogenesis in oral squamous cell carcinoma. Cancer Gene Ther November 2019;26(11): 400–10 [cited 2021 September 1].
[199] Zhao W, Cui Y, Liu L, Qi X, Liu J, Ma S, et al. Splicing factor derived circular RNA circUHRF1 accelerates oral squamous cell carcinoma tumorigenesis via feedback loop. Cell Death Differ March 2020;27(3): 919–33 [cited 2021 September 1].
[200] Zhou W, Wang X-Z, Fang B-M. A variant of H19 transcript regulates EMT and oral cancer progression. Oral Dis December 3, 2020.

[201] Obermayr E, Reiner A, Brandt B, Braicu EI, Reinthaller A, Loverix L, et al. The long-term prognostic significance of circulating tumor cells in ovarian cancer—a study of the OVCAD consortium. Cancers May 26, 2021;13(11):2613.
[202] Menon U, Gentry-Maharaj A, Burnell M, Singh N, Ryan A, Karpinskyj C, et al. Ovarian cancer population screening and mortality after long-term follow-up in the UK Collaborative Trial of Ovarian Cancer Screening (UKCTOCS): a randomised controlled trial. Lancet June 2021;397(10290):2182–93.
[203] Momenimovahed Z, Tiznobaik A, Taheri S, Salehiniya H. Ovarian cancer in the world: epidemiology and risk factors. Int J Womens Health April 2019;11:287–99.
[204] The Cancer Genome Atlas Research Network. Integrated genomic analyses of ovarian carcinoma. Nature June 30, 2011;474(7353):609–15.
[205] Rizvi NA, Hellmann MD, Snyder A, Kvistborg P, Makarov V, Havel JJ, et al. Mutational landscape determines sensitivity to PD-1 blockade in non–small cell lung cancer. Science April 3, 2015;348(6230): 124–8.
[206] Zong C, Wang J, Shi T-M. MicroRNA 130b enhances drug resistance in human ovarian cancer cells. Tumour Biol J Int Soc Oncodevelopmental Biol Med December 2014;35(12):12151–6.
[207] Ahn WS, Kim Y-W, Liu J-L, Kim H, Kim EY, Jeon D, et al. Differential microRNA expression signatures and cell type-specific association with Taxol resistance in ovarian cancer cells. Drug Des Devel Ther February 2014:293.
[208] Liu J, Wu X, Liu H, Liang Y, Gao X, Cai Z, et al. Expression of microRNA-30a-5p in drug-resistant and drug-sensitive ovarian cancer cell lines. Oncol Lett September 2016;12(3):2065–70.
[209] Cai J, Yang C, Yang Q, Ding H, Jia J, Guo J, et al. Deregulation of let-7e in epithelial ovarian cancer promotes the development of resistance to cisplatin. Oncogenesis October 2013;2(10). e75–e75.
[210] Ghafouri-Fard S, Shoorei H, Taheri M. miRNA profile in ovarian cancer. Exp Mol Pathol April 2020;113: 104381.
[211] Ehlén Õ, Nodin B, Rexhepaj E, Brändstedt J, Uhlén M, Alvarado-Kristensson M, et al. RBM3-Regulated genes promote DNA integrity and affect clinical outcome in epithelial ovarian cancer. Transl Oncol August 2011;4(4). 212-IN1.
[212] Fu C, Yuan M, Sun J, Liu G, Zhao X, Chang W, et al. RNA-binding motif protein 11 (RBM11) serves as a prognostic biomarker and promotes ovarian cancer progression. Dis Markers 2021;2021:3037337.
[213] Hinman MN, Lou H. Diverse molecular functions of Hu proteins. Cell Mol Life Sci October 2008;65(20): 3168–81.
[214] Lal A, Kawai T, Yang X, Mazan-Mamczarz K, Gorospe M. Antiapoptotic function of RNA-binding protein HuR effected through prothymosin α. EMBO J May 18, 2005;24(10):1852–62.
[215] Li Z, Wei D, Yang C, Sun H, Lu T, Kuang D. Overexpression of long noncoding RNA, NEAT1 promotes cell proliferation, invasion and migration in endometrial endometrioid adenocarcinoma. Biomed Pharmacother December 2016;84:244–51.
[216] Chai Y, Liu J, Zhang Z, Liu L. HuR-regulated lncRNA NEAT1 stability in tumorigenesis and progression of ovarian cancer. Cancer Med July 2016;5(7):1588–98.
[217] Xu C-Z, Jiang C, Wu Q, Liu L, Yan X, Shi R. A feed-forward regulatory loop between HuR and the long noncoding RNA HOTAIR promotes head and neck squamous cell carcinoma progression and metastasis. Cell Physiol Biochem 2016;40(5):1039–51.
[218] Latorre E, Carelli S, Raimondi I, D'Agostino V, Castiglioni I, Zucal C, et al. The ribonucleic complex HuR-MALAT1 represses CD133 expression and suppresses epithelial–mesenchymal transition in breast cancer. Cancer Res May 1, 2016;76(9):2626–36.
[219] Zou A, Liu R, Wu X. Long non-coding RNA MALAT1 is up-regulated in ovarian cancer tissue and promotes SK-OV-3 cell proliferation and invasion. Neoplasma 2016;63(06):865–72.

[220] Li Q, Zhang C, Chen R, Xiong H, Qiu F, Liu S, et al. Disrupting MALAT1/miR-200c sponge decreases invasion and migration in endometrioid endometrial carcinoma. Cancer Lett December 1, 2016;383(1): 28–40.

[221] Sobočan M, Bračič S, Knez J, Takač I, Haybaeck J. The communication between the PI3K/AKT/mTOR pathway and Y-box binding protein-1 in gynecological cancer. Cancers January 14, 2020;12(1):E205.

[222] Oda Y, Sakamoto A, Shinohara N, Ohga T, Uchiumi T, Kohno K, et al. Nuclear expression of YB-1 protein correlates with P-glycoprotein expression in human osteosarcoma. Clin Cancer Res Off J Am Assoc Cancer Res September 1998;4(9):2273–7.

[223] Jiang M-P, Xu W-X, Hou J-C, Xu Q, Wang D-D, Tang J-H. The emerging role of the interactions between circular RNAs and RNA-binding proteins in common human cancers. J Cancer 2021;12(17):5206–19.

[224] Kamura T, Yahata H, Amada S, Ogawa S, Sonoda T, Kobayashi H, et al. Is nuclear expression of Y box-binding protein-1 a new prognostic factor in ovarian serous adenocarcinoma? Cancer June 1, 1999;85(11): 2450–4.

[225] Chen J, Li X, Yang L, Zhang J. Long non-coding RNA LINC01969 promotes ovarian cancer by regulating the miR-144-5p/LARP1 Axis as a competing endogenous RNA. Front Cell Dev Biol 2020;8:625730.

[226] Hopkins TG, Mura M, Al-Ashtal HA, Lahr RM, Abd-Latip N, Sweeney K, et al. The RNA-binding protein LARP1 is a post-transcriptional regulator of survival and tumorigenesis in ovarian cancer. Nucleic Acids Res February 18, 2016;44(3):1227–46.

[227] Hu X, Bao J, Wang Z, Zhang Z, Gu P, Tao F, et al. The plasma lncRNA acting as fingerprint in non-small-cell lung cancer. Tumour Biol J Int Soc Oncodevelopmental Biol Med March 2016;37(3):3497–504.

[228] Martini P, Paracchini L, Caratti G, Mello-Grand M, Fruscio R, Beltrame L, et al. lncRNAs as novel indicators of patients' prognosis in stage I epithelial ovarian cancer: a retrospective and multicentric study. Clin Cancer Res Off J Am Assoc Cancer Res May 1, 2017;23(9):2356–66.

[229] Ling H, Fabbri M, Calin GA. MicroRNAs and other non-coding RNAs as targets for anticancer drug development. Nat Rev Drug Discov November 2013;12(11):847–65.

[230] Rupaimoole R, Slack FJ. MicroRNA therapeutics: towards a new era for the management of cancer and other diseases. Nat Rev Drug Discov March 2017;16(3):203–22.

[231] van Rooij E, Olson EN. MicroRNA therapeutics for cardiovascular disease: opportunities and obstacles. Nat Rev Drug Discov November 2012;11(11):860–72.

[232] Winkle M, El-Daly SM, Fabbri M, Calin GA. Noncoding RNA therapeutics - challenges and potential solutions. Nat Rev Drug Discov August 2021;20(8):629–51.

[233] Calin GA, Cimmino A, Fabbri M, Ferracin M, Wojcik SE, Shimizu M, et al. MiR-15a and miR-16-1 cluster functions in human leukemia. Proc Natl Acad Sci U S A April 1, 2008;105(13):5166–71.

CHAPTER

8

A comprehensive view of the prostate cancer metastasis and role of androgen receptor splice variants

Yashika Jawa[1], Sangeeta Kumari[1], Gargi Bagchi[2] and Rakesh K. Tyagi[1]

[1]*Special Centre for Molecular Medicine, Jawaharlal Nehru University, New Delhi, India;* [2]*Amity Institute of Biotechnology, Amity University Haryana, Amity Education Valley, Gurgaon, Haryana, India*

Prostate cancer: origin and metastasis

Prostate cancer (PCa) affects around 1.5 million men worldwide and causes 375,000 deaths annually. It is reported as one of the leading cancers worldwide in terms of morbidity and mortality. PCa features as the most prevalent cancer in men in 112 countries and the second leading mediator of cancer-related mortality ([1], Fig. 8.1). The prostate is a male accessory gland located under the urinary bladder, which contributes to the secretion of semen. The prostate gland cells often become neoplastic during the mid or late stages of life [2]. In humans, the prostate gland is divided into three zones, viz., central, peripheral, and transition zones of which peripheral zones cover over 70% of glandular tissue in young adults. Interestingly, this zone is also the most common site of neoplasm origin, contributing to nearly 80 % of all tumors in the aged prostate gland.

The cells in the normal and neoplastic prostate express high levels of androgen receptor (AR) that promotes cell proliferation and hormone dependency of the cancer cells. These cells also secrete a serine protease called prostate-specific antigen or PSA, which is often used as a marker in PCa patients. A seminal work in 1941 established PCa as a hormone-dependent disease [3]. Huggins' group investigated whether PCa responds to the level of androgens (male sex hormones) by using the enzyme prostatic acid phosphatase (PAP) as a biomarker. This enzyme produced by prostate epithelial cells is expressed highly in PCa patients. Their work revealed that PAP activity was elevated in serum samples in 21 of 47 PCa cases, and the frequency was higher in serum of patients with bone metastasis. They also demonstrated that castration in eight of these patients caused their PAP levels to reduce swiftly, indicating a role for androgens in PCa and paving the path for antiandrogen therapy in PCa patients for years to come. This established the role of AR as an oncogene in the development of prostate cancer. For demonstrating the role of endocrine manipulation in the treatment of metastatic PCa, Huggins received the Nobel Prize in medicine in 1966.

Post-Transcriptional Gene Regulation in Human Disease, Volume 32. https://doi.org/10.1016/B978-0-323-91305-8.00010-7

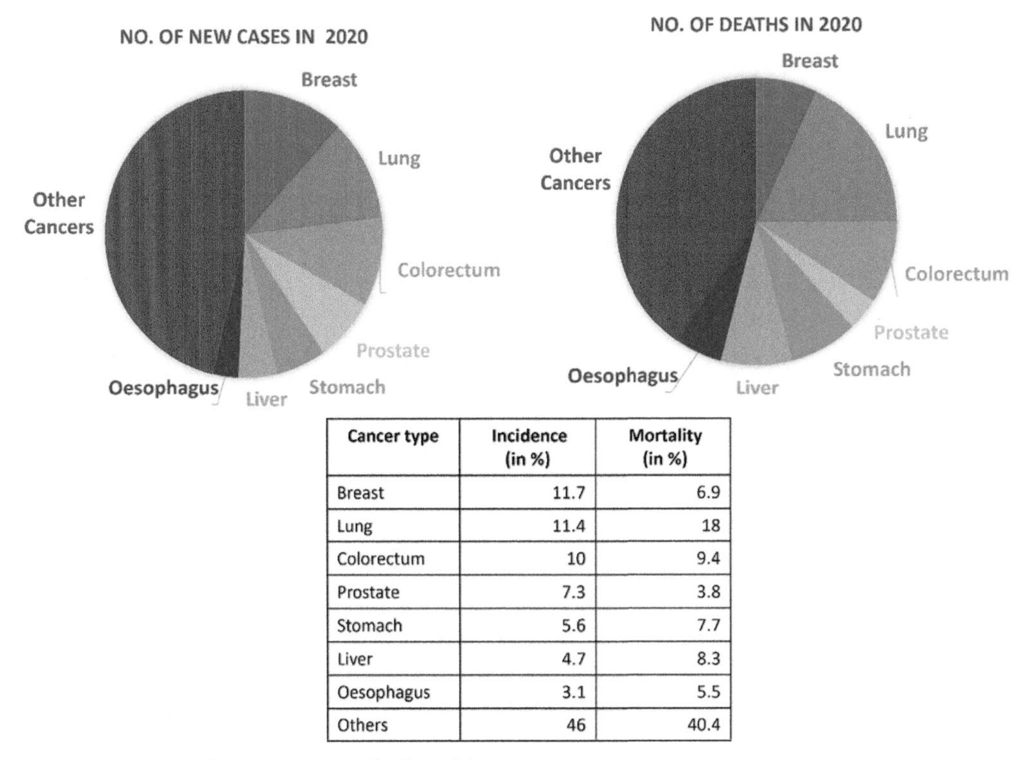

Cancer type	Incidence (in %)	Mortality (in %)
Breast	11.7	6.9
Lung	11.4	18
Colorectum	10	9.4
Prostate	7.3	3.8
Stomach	5.6	7.7
Liver	4.7	8.3
Oesophagus	3.1	5.5
Others	46	40.4

FIGURE 8.1 Cancer incidence and mortality in 2020.

The pie charts represent GLOBOCAN data on new cancer cases and deaths based on cancer types. The highest incidence rates are observed in breast cancers, while mortality is highest in lung cancers.

Metastatic prostate cancer

In the early stages of advanced PCa, cells shed from their site of origin migrate to distant sites invading blood vessels [4]. This process of establishing a secondary tumor is known as metastasis, which occurs through a series of sequential events that begins with proliferative inflammatory atrophy of normal epithelial cells leading to complex interactions between the cancer cells and the milieu. The tumor cells escaping into the vasculature and lymphatics further lead to extravasation at the secondary site and further growth ([5]; Fig. 8.2). Metastatic PCa is often incurable and manifests in progressive enervation of the patient, bone invasion, compression of the spinal cord, loss of bladder outflow, and renal failure. The process of metastasis is inefficient as not all tumor cells have the capacity to metastasize and only a fraction of cells that leave the primary site can colonize at a secondary spot and develop into a secondary neoplasm [5].

Androgen receptor

The central role of male hormones (androgens) and AR in PCa was established in 1941 [3]. Since then, androgen deprivation therapy or ADT has been the primary option in PCa treatment. Androgens or

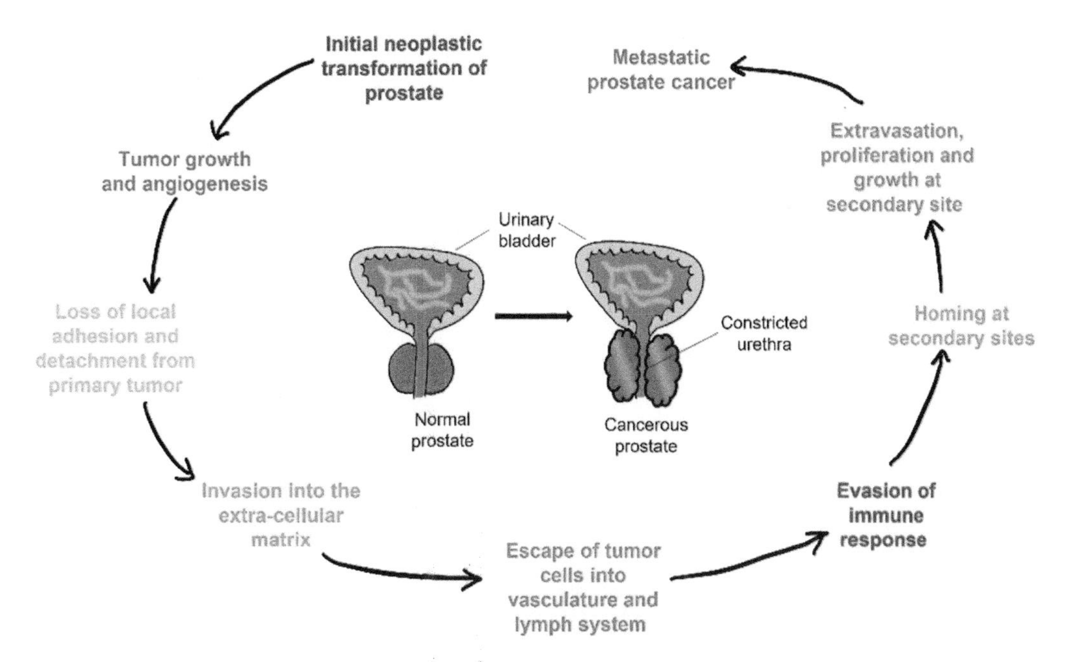

FIGURE 8.2 The metastatic cascade of prostate cancer.

The onset of cancer in the prostate starts with the initial neoplastic transformation of prostate tissue. The progression may further lead to metastasis, reduced bladder outflow, renal failure, etc.

Created from the concept adapted from Arya M, Bott SR, Shergill IS, Ahmed HU, Williamson M, Patel HR. The metastatic cascade in prostate cancer. Surg Oncol 2006;15(3):117–28.

male sex hormones, namely, testosterone and 5α-dihydrotestosterone (DHT) are responsible for promoting reproduction and development of secondary sexual features in males [6]. Most endogenous androgen is produced by stimulating the hypothalamus-pituitary-Leydig circuit cells, although a minute amount is also produced by adrenal glands [7]. The primary male hormone testosterone (T) is converted to a more active form DHT by the 5α-reductase enzyme. Both T and DHT mediate their hormonal actions by binding and activating AR [6].

The AR, a member of the nuclear receptor (NR) superfamily, forms the backbone of the androgen signaling axis. NRs are ligand-modulated transcription factors that regulate various physiological processes in the body such as metabolism, reproduction, inflammation, and circadian rhythm [8]. The sex steroid receptors such as estrogen and androgen receptors are established therapeutic targets and therefore have been explored to greater depths. These receptors are known to play vital roles in cancers, endocrine disorders, aging, and age-related disorders [9,10]. The basic structure of AR can be divided into a large NTD or N-terminal domain for transactivation (exon 1), a DBD corresponding to DNA-binding domain (exons 2–3), a C-terminal domain (CTD), or C-terminal domain for ligand-binding (LBD, exons 4–8), and a hinge region located between the DBD and LBD that facilitates nuclear entry and subsequent degradation [11,12]. The basic structure of NRs and AR is depicted in Fig. 8.3A and B.

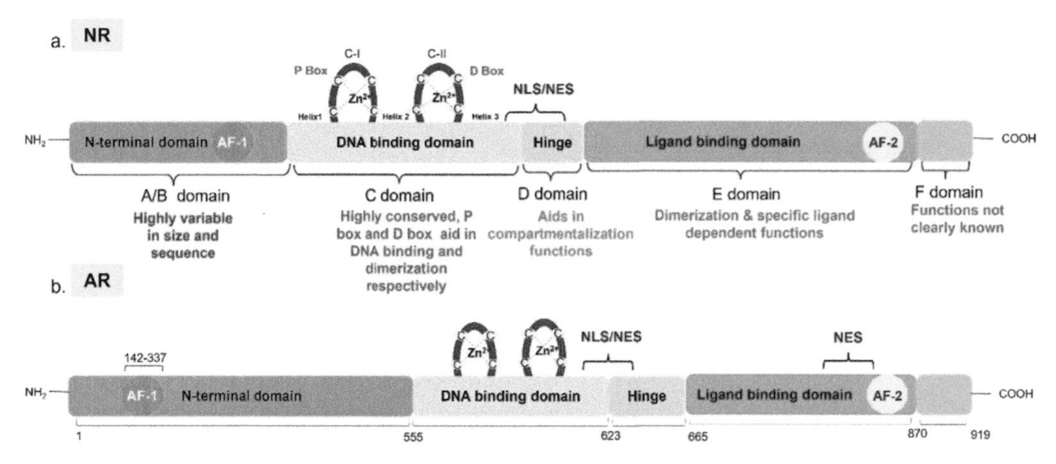

FIGURE 8.3 The modular structures of (A) nuclear receptor and (B) androgen receptor.

The basic structure of NR consists of an N-terminal (AF-1 containing) domain, a DNA binding domain, a hinge, a ligand-binding domain, and an F-domain. The functions of these domains are depicted in the figure. The nuclear localization signal (NLS) and nuclear export signals (NES) are present in DBD and Hinge region. An additional NES has also been observed in the AR LBD region. The size of each region is depicted by numbers denoting amino acids.

Before androgen binding, the inactive apo-AR is maintained as a heterocomplex with several molecular chaperones including HSP70 and HSP90. The binding of HSP90 to the AR helps in presenting the AR in a conformation that allows androgen-binding with high affinity [13]. Phosphorylation of HSP90 by activation of protein kinase A (PKA) in the cells by androgens or another agent, releases the AR allowing it to bind to the androgens [14]. The binding of T or DHT promotes a conformational change in the position of helix 12 residing in AR LBD that stabilizes ligand binding and creates a hydrophobic cleft for interaction with coactivator proteins [11]. The ligand-bound AR subsequently homodimerizes and translocates into the nucleus to bind its regulatory sequences upstream of target genes known as androgen response element (ARE). AR binding to ARE is usually dependent upon prior binding of FOXA1, another transcription factor that helps in unwinding the chromatin near the ARE, making it accessible for AR [15]. Two transcription activation domains, viz, AF1 and AF2 are predicted to be present in AR. Of these AF-1 is located in the NTD of the receptor and functions independent of ligand binding. Contrary to AF-1, AF-2 is a ligand-dependent activation domain. However, AF-2 that lies within the LBD is not well defined in the case of AR and was predicted by sequence comparison with steroid hormone receptors [16]. Nuclear receptors modulate the rate of transcription initiation by binding to basal transcription factors and altering chromatin organization within the promoter of the target genes. The transcription modulation is known to depend on the integrity of AF-2 and the recruitment of cofactors such as histone acetyltransferases or histone acetylases [17].

Studies performed using cell lines and mouse xenografts have demonstrated that AR signaling is critical for both normal as well as neoplastic prostate [18, 19]. In most cases, PCa begins in an androgen-dependent manner, where cells are completely reliant on androgens for their growth and proliferation. Upon androgen ablation, many of the normal and cancer cells die off, though, a few cancer cells can induce changes in their normal signaling mechanisms that allow them to activate the AR axis despite the removal of androgens [20]. This is when the tumor relapses resulting in a more aggressive disease that is mostly untreatable and is known as Castration-resistant PCa or CRPC [11]. Scientists have identified different mechanisms adopted by cells to rekindle the AR axis; these include AR gene mutation, gene amplification, alterations in coregulatory protein expression, and ligand-independent pathways known as outlaw pathways [20]. One of the common strategies by which AR tries to compensate for the loss of androgen signaling upon androgen ablation is by overexpressing itself, both at RNA and protein levels. Studies reveal that 25 %–30 % of androgen-independent prostate cancer (AIPC) contains gene amplification [21]. AR, which is typically a single copy gene that maps to q11-12 of the human X chromosome, can be duplicated in hormone-refractory prostate cancer and is found at another amplification site Xq11-13 [22].

Aberrant activation of AR signaling could also occur by gene mutations [20]. Usually, at early stages, the mutation frequency is low and rises 10%–20% in aggressive AIPC [23]. The first reported mutation was observed in androgen-dependent LNCaP cells, which are derived from a lymph node biopsy. The LNCaP cells contain a missense mutation at codon 877 that replaces amino acid residue "Threonine" by "Alanine," within the LBD of AR [24]. This change in single amino acid residue reduces the specificity of the AR LBD for androgens and allows it to bind to other steroid hormones such as estrogens or progesterone as well as antiandrogens. The change is quite beneficial to PCa cells as they no longer need androgen for stimulation and can utilize any other circulating hormone or anti-androgen to promote AR signaling.

The primary function of AR is to act as a transcription factor modulating the transcription of a variety of genes. Over 170 proteins are known to modulate the transcriptional activity of AR and hence are known as AR coregulators. The proteins either activate or suppress AR-signaling, hence are called AR coactivators or corepressors, respectively. Changes in the expression of any of these coregulators could alter AR signaling. Gregory et al. found that elevated levels of SRC1 and TIF2 expression in AIPC samples caused an increase in AR signaling [25]. On the contrary, the expression of corepressors NCoR and SMRT, which are histone deacetylases is downregulated, impacting chromatin condensation and increasing AR transcriptional activity [26].

Outlaw and bypass pathways

Apart from androgens and other steroid hormones (in the case of AR mutants), AR can also be activated by ligand-independent processes. These mostly result due to nongenomic activation of AR and are known as outlaw pathways [20,27]. AR also gets activated by growth factors, cytokines, and protein kinases [28]. Indirect activation of AR also occurs by activation of PKA by forskolin [29], or via activation of membrane G-Protein Coupled Receptors (GPCRs) such as GPR56 [30].

Post-transcriptional regulation of AR

The expression of AR at both RNA and protein levels is regulated by many factors [31]. Androgens themselves regulate the level of AR mRNA and protein, though the effect may vary from cell to cell [32]. As the upstream region of the AR promoter contains SP1 binding sites instead of a typical TATA box, AR transcription is influenced by the SP1 transcription factor that recruits Transcription Factor

IID (TFIID) and TFIID-binding proteins in TATA-less promoters. In fact, in a study inhibition of SP1 activity markedly reduced AR protein level in LNCaP cells [33]. Another important regulatory element for AR mRNA expression is CRE or cAMP response element, which is located within the human AR gene. Cells treated with a cAMP analog dibutyryl-cAMP demonstrated enhanced AR transcription by 4-to 6-folds [34]. This established that AR transcription has a profound impact on prostate cell proliferation.

Additionally, AR also undergoes post-transcriptional regulation, that is, regulation after transcription of a primary RNA transcript, such as alternate splicing, or effect on RNA stability. Both RNA splicing and RNA stability are important events that can regulate gene expression and play essential roles in PCa [35].

AR splice variants

Multiple splice variants of AR (ARSV) have been detected in PCa cell lines, tissues, and xenografts. One of the earliest known ARSV, AR45 was detected in placental tissue [36]. This particular variant of AR is inhibitory to wild-type AR full length (AR-FL) in the heart, but increases AR activity in the prostate, indicating that the function of ARSVs may be tissue-specific. Most of the other ARSVs were detected in PCa cell lines, particularly the CWR22RV1 cell line that is derived from a xenograft serially propagated in mice [37].

Studies show that the formation of ARSVs can occur through two different mechanisms, that is, genome rearrangements and alternate splicing. Li et al. have demonstrated that there is a 35 kb tandem duplication in CRPC 22Rv1 that encompasses the AR exon 3 [38]. This rearrangement causes increased mRNA and protein expression of truncated ARSVs, AR V7 or AR3, and allows robust growth of the cells under castrate conditions. These cells continue to grow upon treatment with bicalutamide or enzalutamide. The second mechanism of ARSV generation is alternate splicing. Differential splicing of pre-mRNA is a mechanism frequently used by malignant cells to generate protein variants with oncogenic activity [39].

As mentioned earlier, the AR can be structurally divided into NTD, DBD, Hinge region, and CTD. Structural analysis reveals that the NTD is coded by exon 1, the highly conserved DBD is coded by exons 2 and 3, the LBD is coded by exons 4–8, and the hinge region is encoded by exons 3 and 4 [40]. Also, the AR NTD is relatively long, comprising 538 amino acids, encompassing the two transcription activation units TAU1 and TAU5 [41].

The earliest reported ARSVs were AR45 and AR 23 [36,42]. The AR45 was identified by rapid amplification of cDNA ends and weighed around 45 kDa. It is an NTD-truncated isoform that contains the DBD, hinge region, and LBD. In addition, it also contains a novel N-terminal peptide (7 amino acids long) that is encoded by exon 1b, situated approximately 22.1 kb downstream of exon 1 in AR. The mRNA for the AR45 isoform was detected in the heart, uterus, breast, prostate, and skeletal muscle. The AR45 isoform is thought to be a dominant-negative that suppresses AR-Fl function. The AR23 that was detected in a bone metastasis results by aberrant splicing of intron 2 sequence, leading to an in-frame insertion of 23 amino acid sequences between the two zinc fingers of the AR-DBD [42].

The ARSVs came into prominence in 2008 due to the discovery of LBD-lacking variants that had the potential to migrate into the nucleus constitutively and mediate AR signaling [43]. A study by Kumar et al. showed that mutant AR lacking the LBD (denoted as AR-ΔLBD) is predominantly

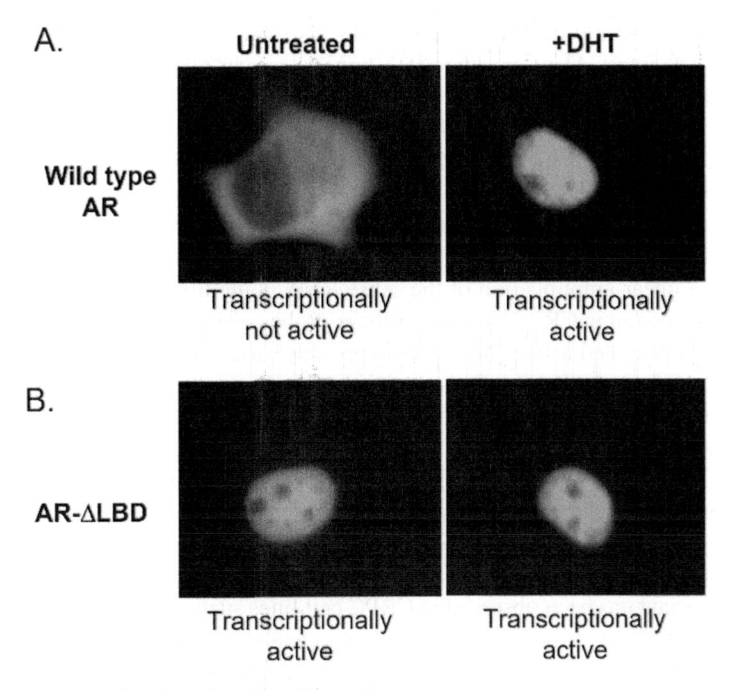

FIGURE 8.4 Intracellular localization of androgen receptor.

(A) GFP- labeled wild-type AR is predominantly cytoplasmic in the absence of ligand but translocates to the nucleus on treatment with DHT (dihydrotestosterone) while (B) AR-ΔLBD mutant is nuclear-localized both in presence and absence of ligand and is transcriptionally active [44].

localized in the nucleus. It was also observed to organize into "nuclear foci" and showed weak binding to mitotic chromatin. All these properties were similar to the wild type of AR albeit observed to be ligand-independent.Fig. 8.4 depicts the subcellular localization patterns of wild type and AR-ΔLBD mutant tagged with Green Fluorescent Protein (GFP) [44]. In 2008, Dehm et al. identified a new exon, exon 2b, in the CWR22Rv1 cell line by performing 3′-RACE [43]. This resulted in the generation of two truncated AR variants AR1/2/2b and AR1/2/3/2b that exhibited ligand-independent AR activity.

Hu et al. identified more AR cryptic exons in 2009 in established cell lines and clinical samples [45]. They identified three cryptic exons (CE1, CE2, CE3) in intron 3. CE4 was identical to exon 2b mentioned earlier. Splicing of these exons generated seven ARSVs (AR-V1 to AR V7), that lacked LBD due to the occurrence of stop codons within the transcribed introns. The variants AR-V1 and AR-V7 have been often detected in PCa specimens, with almost 20-fold higher expression levels in CRPC as compared to early-stage prostate tumors [45].

Subsequently, Guo et al., discovered variants AR3, AR4, and AR5 in both androgens sensitive as well as androgen-independent cell lines using 3′ RACE [46]. Immunofluorescence analysis was performed to demonstrate the presence of AR3 in both cytoplasm and nucleus in 22Rv1 and CWR-R1 cells. When compared the sequences of AR3, AR4 and AR5 were similar to AR-V7, AR-V1, and AR-4 as mentioned earlier. Notably, the expression of the ARSVs was significantly higher in androgen-independent PCa cell lines. Also, specific AR3 knockdown in CWR-R1 or CWR22Rv1 revealed that gene-set expression was modulated by the splice variant [46].

These ARSVs were confirmed by 3′-RACE followed by next-generation sequencing by Watson et al. in 2010 [47]. By this technique, they also detected four more ARSVs, ARV8 to ARV11 in VCaP cells derived from vertebral metastasis lesions. They also found that both ARSVs and AR-Fl were increased at mRNA and protein level by castration whereas administration of testosterone suppressed their expression [47]. For detecting transcribed AR sequences in an unbiased manner, Hu et al. used SLASR or selective linear amplification of the sense RNA, a modified RNA amplification approach. To analyze RNA expression in CRPC samples, 60-mer probes overlapping across the human AR locus were used [48]. This provided a snapshot of all expressed variants and identified three new ARSVs, viz., AR-V12, AR-V13, and AR-V14. This study identified sequences downstream of exon 8 (now called exon 9) as sequences that potentially participate in AR splicing [48].

A novel AR transcript was identified by Sun et al. while investigating AR isoforms in LnCaP PCa xenografts [49]. This AR variant called ARV567es was generated by skipping exons 57 and retaining exons 1−4. In this variant, the open reading frame encounters an early stop codon after 29 nucleotides in exon 8, thus covering the complete hinge region of AR encoded by exons 3 and 4 [49].

A membrane-associated variant of AR called AR8 has also been identified by RACE [50]. Higher expression of this variant has been observed in CRPC cell lines such as C4-2, C4-2b, and CWR22Rv1. AR8 has no DBD, LBD, or transactivating function as analyzed by reporter assays. It likely forms complexes with AR-Fl or EGFR and likely mediates Src-induced activation of AR [50].

The most distinguishing feature of many of the ARSVs is their ability to constitutively translocate into the nucleus. Nuclear translocation is certainly a prerequisite step for genomic AR signaling. In AR-Fl, upon androgen binding, a NLS encompassing the C-terminal part of DBD, and hinge region are exposed that interacts with importin proteins helping AR to migrate through the nuclear pore complex [51]. While AR-V7 (also called AR3) and AR-12 (also called ARV567es) demonstrate constitutive nuclear translocation [47], in absence of androgens, other AR variants have variable translocation capability. Studies have shown that AR-V1, AR-V9, and AR-V13 are primarily cytoplasmic probably due to the lack of basic amino acid stretch typical of a bipartite NLS. Notably, the genomic signaling of these variants is not always dependent upon their location. The variants AR-V1 and AR-V9 demonstrate ligand-induced activity in LNCaP cells but no activity in PC3 cells. These variants are therefore known as conditionally active [48]. To understand the constitutive migration of AR-V7, Chan et al. analyzed its unique C-terminal sequence. Their studies revealed that this sequence resembles the truncated AR NLS. Site-directed mutagenesis of AR-V7, to replace K629 and R631 by Alanine, resulted in a mutated variant that was distributed in both cytoplasm and nucleus instead of being localized completely in the nucleus [52]. Fig. 8.5 depicts all the splice variants discussed above.

Role of AR splice variants in CRPC

As mentioned earlier, PCa patients are subjected to ADT as the first line of treatment, yet, almost invariably, they progress to androgen-independent PCa or CRPC [35]. Research conducted using cell lines, xenografts, and patient samples demonstrate that the androgen signaling axis remains activated

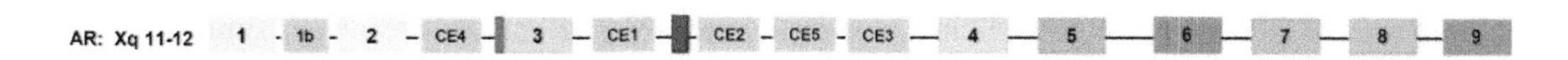

b.

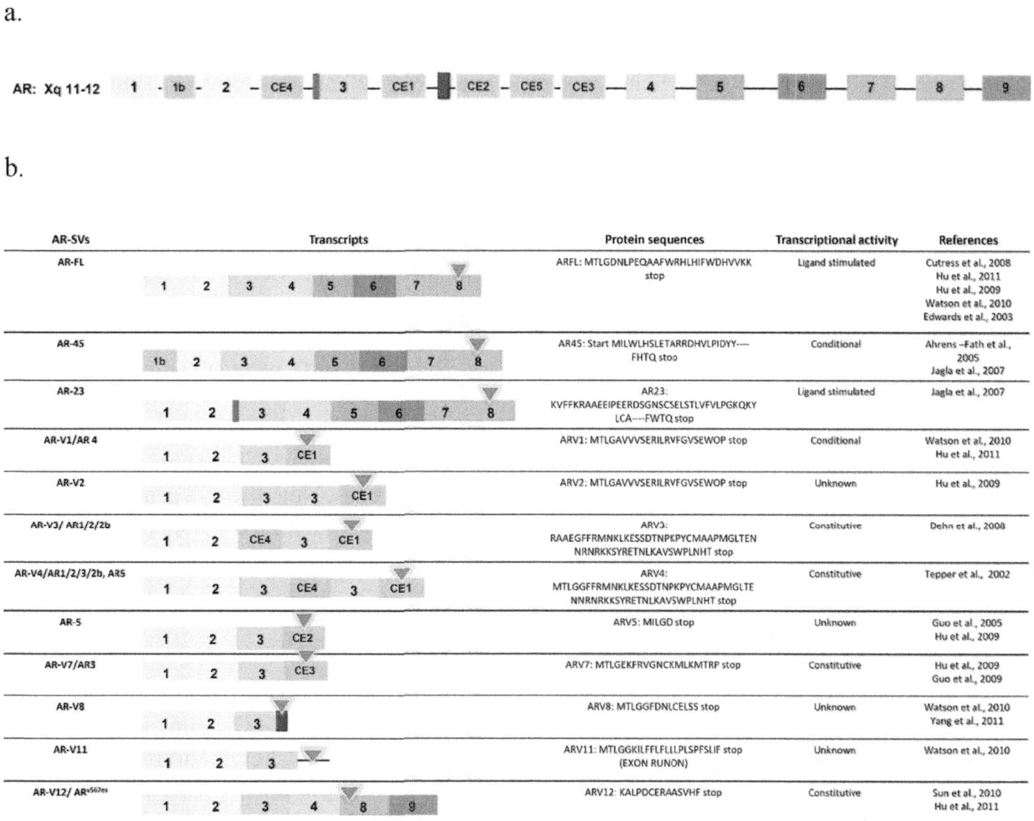

AR-SVs	Transcripts	Protein sequences	Transcriptional activity	References
AR-FL	1 - 2 - 3 - 4 - 5 - 6 - 7 - 8	ARFL: MTLGDNLPEQAAFWRHLHIFWDHVVKK stop	Ligand stimulated	Cutress et al., 2008 Hu et al., 2011 Hu et al., 2009 Watson et al., 2010 Edwards et al., 2003
AR-45	1b - 2 - 3 - 4 - 5 - 6 - 7 - 8	AR45: Start MILWLHSLETARRDHVLPIDYY----FHTQ stoo	Conditional	Ahrens –Fath et al., 2005 Jagla et al., 2007
AR-23	1 - 2 - 3 - 4 - 5 - 6 - 7 - 8	AR23: KVFFKRAAEEIPEERDSGNSCSELSTLVFVLPGKQKY LCA----FWTQ stop	Ligand stimulated	Jagla et al., 2007
AR-V1/AR 4	1 - 2 - 3 - CE1	ARV1: MTLGAVVVSERILRVFGVSEWOP stop	Conditional	Watson et al., 2010 Hu et al., 2011
AR-V2	1 - 2 - 3 - 3 - CE1	ARV2: MTLGAVVVSERILRVFGVSEWOP stop	Unknown	Hu et al., 2009
AR-V3/ AR1/2/2b	1 - 2 - CE4 - 3 - CE1	ARV3: RAAEGFFRMNKLKESSDTNPKPYCMAAPMGLTEN NRNRKKSYRETNLKAVSWPLNHT stop	Constitutive	Dehn et al., 2008
AR-V4/AR1/2/3/2b, AR5	1 - 2 - 3 - CE4 - 3 - CE1	ARV4: MTLGGFFRMNKLKESSDTNPKPYCMAAPMGLTE NNRNRKKSYRETNLKAVSWPLNHT stop	Constitutive	Tepper et al., 2002
AR-5	1 - 2 - 3 - CE2	ARV5: MILGD stop	Unknown	Guo et al., 2005 Hu et al., 2009
AR-V7/AR3	1 - 2 - 3 - CE3	ARV7: MTLGEKFRVGNCKMLKMTRP stop	Constitutive	Hu et al., 2009 Guo et al., 2009
AR-V8	1 - 2 - 3	ARV8: MTLGGFDNLCELSS stop	Unknown	Watson et al., 2010 Yang et al., 2011
AR-V11	1 - 2 - 3	ARV11: MTLGGKILFFLFLLLPLSPFSLIF stop (EXON RUNON)	Unknown	Watson et al., 2010
AR-V12/ AR^{v567es}	1 - 2 - 3 - 4 - 8 - 9	ARV12: KALPDCERAASVHF stop	Constitutive	Sun et al., 2010 Hu et al., 2011

FIGURE 8.5 AR splice variants.

Twenty two AR splice variants have so far been reported in the literature. The figure depicts important AR splice variants, their nomenclature, exon composition, peptide sequences, functional annotation, etc. (A) AR gene structure as per GRCh37/hg19 assembly with canonical splice sites, cryptic exons and (B) Nomenclature, exonic organization, and splice variant-specific peptide sequences. Corresponding exonic regions in AR splice variants have been color-matched to (A) the full-length AR gene.

despite androgen removal. These findings resulted in the development of second-generation new drugs including enzalutamide, abiraterone acetate, darolutamide, and apalutamide that can target both androgen-sensitive and well as androgen-independent PCa. Surprisingly, within a year of starting treatment, resistance to these drugs is observed, possibly due to the appearance of constitutively active ARSVs [53]. The AR-V7 is considered as most significant among these due to its constitutively active nature and the ability to drive PCa cell growth in cells undergoing AR-directed treatments.

Homo- and heterodimerization of AR-V7 influences AR signaling cascade. Because of its ability to migrate into the nucleus in the absence of androgens, it can bind to AREs in androgen-independent cells undergoing ADT and trigger the activation of a transcriptome that supports tumor formation [55].

Also, the forced expression of AR-V7 in transgenic mice causes them to display a protumorigenic phenotype. Bromodomain inhibitors and inhibitors of extra-terminal proteins were identified to suppress the formation of AR-V7 by mRNA splicing. However, as these target multiple pathways simultaneously, their use raised many concerns [56].

Sharp et al., have performed extensive studies on the expression and role of AR-V7 in primary PCa and CRPC [35]. They analyzed the location of the AR-V7 protein critically by developing an AR-V7 specific antibody. Interestingly, their work with castration-sensitive PCa (CSPC) and primary PCa cell lines established that AR-V7 is only marginally expressed in these cells. However, the expression of AR-V7 increased significantly when the patient progressed from CSPC to CRPC. Moreover, biopsy samples of CRPC patients also showed a marked increase in AR-V7 levels after Abiraterone acetate or Enzalutamide treatment. The levels of AR-V7 were comparatively higher in lymph node metastases compared to bone metastases. Similar results were obtained with mouse VCaP xenografts.

A positive correlation was observed between mRNA and protein expression levels of AR-V7 using 41 metastatic biopsies from 24 men in a cohort study. This correlation distribution was shifted to significantly higher values showing a strong correlation of 487 genes. Notably, 407 genes from a pool of 487 were upregulated, whereas 80 were downregulated. This study identified a unique transcriptome regulated by the AR-V7 protein.

AR variants as clinical biomarkers

Among the known 22 AR splice variants reported in the literature [57], AR-V7 is selectively overexpressed at mRNA and protein levels in CRPC patients. The variant AR-V7 is linked with disease progression, and the levels are correlated to poor clinical outcomes and shorter survival in these patients [46,58]. AR-V7 levels are also correlated with enzalutamide resistance in CRPC patients [59]. Of the patients treated with enzalutamide, 57 % showed detectable levels of AR-V7. The disease progression was observed within 4 months of treatment while the patients who were successfully responding to enzalutamide treatment for more than 6 months did not show AR-V7 expression. Thus, indicating toward resistance to enzalutamide possibly leading to AR-V7 expression. Detection of AR-V7 in circulating tumor cells provides a prognostic tool and is correlated with resistance development in mCRPC patients treated with the second generation of antiandrogens [53]. These patients showed a lack of PSA, response to treatment and overall survival was also reduced and vice-versa. Moreover, enzalutamide or abiraterone-treated patients showed elevated levels of AR-V7. Notingly, the patients treated with both the drugs in either sequence had higher levels of AR-V7, suggesting a role for this variant in prostate cancer progression.

A study by Kohli et al. (2017) reported the cooccurrence of AR-V9 along with AR-V7 in CRPC model systems. Increased AR-V9 mRNA in CRPC metastases before therapy appeared indicative of subsequent resistance toward abiraterone acetate. Interestingly, both AR-V7 and AR-V9 contain CE3 exonic sequence at their 3′-region, and both are constitutively active splice forms. Because of an extended 3′-UTR, AR-V9 is longer than AR-V7 [54]. Therefore, these splice variants that lack certain domains or contain extra stretches of nucleotides play an important role in modulating AR-FL activity by interacting with it, thus making a functional impact in prostate cancer progression.

AR variants as therapeutic targets

Although multiple treatment options are available for mCRPC, the mortality rates are still high. The second line of antiandrogen drugs, including enzalutamide and abiraterone, often provides better survival for prostate cancer patients but eventually fails and leads to secondary resistance, highlighting the need for new therapeutic strategies to combat it. The presence of AR-Vs in cancer cells and their mechanisms of action appears to be one crucial method of escape in mCRPC. Since AR-Vs lack ligand-binding domains, they do not respond to enzalutamide or abiraterone acetate. Among these AR-Vs, AR-V7 has emerged as a potential predictive marker of resistance to most of the current antiandrogen therapies in CRPC. Thus, therapeutic agents that counter and account for both, AR and AR-V7 specific signaling are being developed. One such alternative is provided by mechanisms of RNA-based regulations or nucleic-acid-based therapy. These are stemming as complements and alternatives to current treatment regimens (Fig. 8.6).

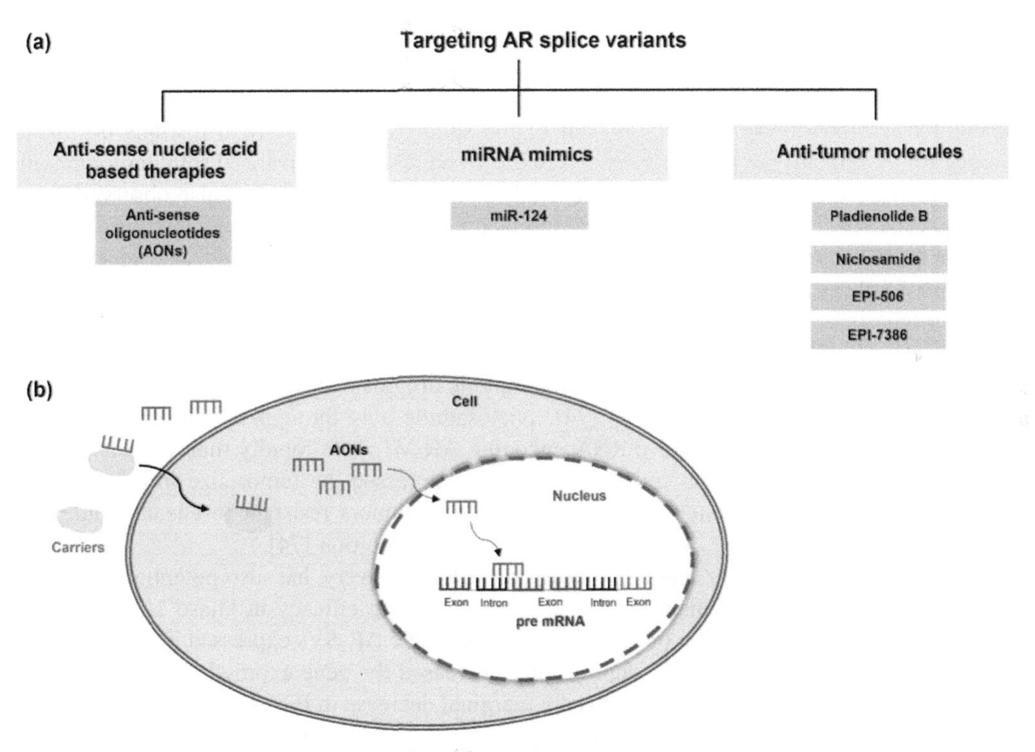

FIGURE 8.6 Therapeutic strategies targeting AR splice variants.

RNA-based regulations are emerging as key alternatives to the current therapies. (A) Strategies under development for targeting AR splice variants, (B) Mechanism of action of RNA-based regulations. The oligonucleotide or micro-RNAs (miRNAs) bind to carrier proteins to enter the cell and further modulate the gene expressions by binding to specific pre-mRNAs.

Nucleic acid-based therapies have recently emerged as promising candidates for the treatment of CRPC patients. The single-stranded antisense oligonucleotides (AONs) downregulate AR-V7 expression through base pairing with AR-V7 cryptic splice sites and therefore block aberrant splicing events [60,61]. A recent study has shown that rational designing of AONs that specifically target AR-V7 mRNA synthesis blocks prostate cancer cell proliferation [62]. An antisense morpholino was developed in another study. The synthetic oligomer molecule contained DNA bases on a methylene morpholino backbone. These morpholino compounds target the splice sites both in AR pre-mRNA and in oncogenes associated with prostate cancer progression (e.g., ERG) [63]. Hence, AONs are attractive candidates as an alternative treatment for advanced stages of prostate cancer.

Besides AONs, miRNA mimics are also emerging as a promising alternative. These are still in the clinical development stage but have proven successful in other cancers as well [64–66]. The role of miRNAs is relatively established in the development and progression of complex cancers like Head and Neck cancer. The dysregulated miRNAs are shown to alter the epigenetic mechanisms in cancer. Thus providing the much-needed space for therapeutic intervention as biomarkers and drug targets [67]. In PCa, miR-124 inhibits prostate cancer progression by downregulating the expression of AR splice variants (ARV4/7) and other oncogenic targets (EZH2 and Src). The mimics of miR-124 have also been shown to repress AR-Vs in animal models [68], and therefore appear like another strategy to combat AR-V7 expression.

An antitumor macrolide, Pladienolide B, has been shown to inhibit spliceosome machinery [69]. Pladienolide B binds to the core component of the spliceosome, SF3B1and disrupts downstream splicing [70]. A derivative of pladienolides, E7107, showed restrained advanced solid tumor growth in approximately 30% of patients in two clinical trials. However, vision loss was also observed in some patients [71] and [72]. Pladienolide B strongly targeted high SF3B2 expressing aggressive cancer cells and suppressed the tumor growth in the CRPC animal model system [73]. It also modulated the expression of full-length and oncogenic splice variants. Thus, splicing inhibitors can efficiently target aggressive tumors exhibiting modified splicing patterns. Therefore, these may provide yet another potential treatment of CRPC where the traditional treatments such as enzalutamide are not effective.

Other than RNA-based approaches, an antihelminthic drug named niclosamide has emerged as an unexpected success in targeting AR-V7 [74]. Niclosamide acts through a proteasome-dependent pathway and induces degradation of mRNA encoding AR-V7 more rapidly than AR-FL. The drug targeted AR-V7 in various CRPC cell lines and also suppressed the tumor size in animal studies. Niclosamide alone showed antitumor activity not only in tumors resistant to enzalutamide but in combination with enzalutamide, it showed maximal tumor inhibition [74].

Another class of therapeutic agents, prodrug-based drug delivery, has also potentiated PCa treatment. EPI-506 is an AR-NTD inhibitor that showed promising efficacy in Phase 1/2 trials [76]. It inhibits the transcriptional function of AR and other oncogenic AR-SVs expressed in advanced stages of prostate cancer. Although EPI-506 significantly decreased the gene expression and tumor size in AR-SV expressing xenograft model, it showed a marginal decrease in PSA levels of mCRPC patients in the phase 1 trial [75], revealing the need for more potent and metabolically stable NTD inhibitors. EPI-7386 represents the second generation of NTD inhibitors that are more active and metabolically more stable than EPI-506. The EPI-7386 controlled tumor growth and induced tumor regressions in several CRPC xenografts, including AR-V7-driven 22Rv1 and enzalutamide resistant LNCaP95 models, as well as the VCaP model. It showed on-target activity in AR-FL and AR-V7 driven models and demonstrates a high therapeutic index in preclinical models [75].

Summary

RNA-based mechanisms and other strategies mentioned above have shown unprecedented potential in treating prostate cancer. Recent advances in nucleic acid-based therapeutics have garnered broad interest in targeting the previously known "undruggable" targets that manipulate the gene expression and hence regulate protein function. Future efforts to identify the trans-acting factors and cis-elements regulating these mechanisms will play a pivotal role in making these novel therapies successful. However, more in-depth studies on their potentials, mechanisms of action, and efficacy are still awaited.

Acknowledgments

The RKT laboratory has been supported by research grants from the Government of India including CSIR, DBT, DST, ICMR, and ICAR-NASF. Central grants from ICMR-CAR, UGC-SAP, and DST-PURSE are also acknowledged. The work in the GB laboratory was supported by SERB-DST and DBT grants.

References

[1] Sung H, Ferlay J, Siegel RL, Laversanne M, Soerjomataram I, Jemal A, Bray F. Global cancer statistics 2020: GLOBOCAN estimates of incidence and mortality worldwide for 36 cancers in 185 countries. CA A Cancer J Clin 2021;71(3):209−49.

[2] Rebello RJ, Oing C, Knudsen KE, et al. Prostate cancer. Nat Rev Dis Prim 2021;7:9.

[3] Huggins C, Stevens RE, Hodges CV. Studies on prostatic cancer: II. The effects of castration on advanced carcinoma of the prostate gland. Arch Surg 1941;43(2):209−23.

[4] Wong SK, Mohamad NV, Giaze TR, Chin KY, Mohamed N, Ima-Nirwana S. Prostate cancer and bone metastases: the underlying mechanisms. Int J Mol Sci 2019;20(10):2587.

[5] Arya M, Bott SR, Shergill IS, Ahmed HU, Williamson M, Patel HR. The metastatic cascade in prostate cancer. Surg Oncol 2006;15(3):117−28.

[6] Davey RA, Grossmann M. Androgen receptor structure, function and biology: from bench to bedside. Clin Biochem Rev 2016;37(1):3−15.

[7] Radmayr C, Lunacek A, Schwentner C, Oswald J, Klocker H, Bartsch G. 5-alpha-reductase and the development of the human prostate. Ind J Urol 2008;24(3):309.

[8] Huang P, Chandra V, Rastinejad F. Structural overview of the nuclear receptor superfamily: insights into physiology and therapeutics. Annu Rev Physiol 2010;72:247−72.

[9] Zhao L, Zhou S, Gustafsson JÅ. Nuclear receptors: recent drug discovery for cancer therapies. Endocr Rev 2019;40(5):1207−49.

[10] Bagchi G, Dash AK, Kumar S, Shoulei J, Ahn SC, Chatterjee B, Tyagi RK. In: Rath PC, editor. Sex steroids, cognate receptors, and aging in models, molecules and mechanisms in biogerontology. Singapore: Springer, Springer Science; 2019. p. 265−96.

[11] Yuan X, Cai C, Chen S, Chen S, Yu Z, Balk SP. Androgen receptor functions in castration-resistant prostate cancer and mechanisms of resistance to new agents targeting the androgen axis. Oncogene 2014;33(22): 2815−25.

[12] Roy AK, Tyagi RK, Song CS, Lavrovsky Y, Ahn SC, Oh TS, Chatterjee B. Androgen receptor: structural domains and functional dynamics after ligand-receptor interaction. Ann N Y Acad Sci 2001;949:44−57.

[13] Kumar S, Saradhi M, Chaturvedi NK, Tyagi RK. Intracellular localization and nucleocytoplasmic trafficking of steroid receptors: an overview. Mol Cell Endocrinol 2006;246(1–2):147–56.
[14] Dagar M, Singh JP, Dagar G, Tyagi RK, Bagchi G. Phosphorylation of HSP90 by protein kinase A is essential for the nuclear translocation of androgen receptor. J Biol Chem 2019;294(22):8699–710.
[15] He HH, Meyer CA, Shin H, Bailey ST, Wei G, Wang Q, Zhang Y, Xu K, Ni M, Lupien M, Mieczkowski P, Lieb JD, Zhao K, Brown M, Liu XS. Nucleosome dynamics define transcriptional enhancers. Nat Genet 2010;42(4):343–7.
[16] Beato M, Truss M, Chávez S. Control of transcription by steroid hormones. Ann N Y Acad Sci 1996;784:93–123.
[17] Shang Y, Myers M, Brown M. Formation of the androgen receptor transcription complex. Mol Cell 2002;9(3):601–10.
[18] Wang Q, Li W, Zhang Y, Yuan X, Xu K, Yu J, Chen Z, Beroukhim R, Wang H, Lupien M, Wu T, Regan MM, Meyer CA, Carroll JS, Manrai AK, Jänne OA, Balk SP, Mehra R, Han B, Chinnaiyan AM, Brown M. Androgen receptor regulates a distinct transcription program in androgen-independent prostate cancer. Cell 2009;138(2):245–56.
[19] Lonergan PE, et al. Androgen receptor signaling in prostate cancer development and progression. J Carcinog 2011. https://doi.org/10.4103/1477-3163.83937.
[20] Saraon P, Drabovich AP, Jarvi KA, Diamandis EP. Mechanisms of androgen-independent prostate cancer. EJIFCC 2014;25(1):42.
[21] Koivisto P, Kononen J, Palmberg C, Tammela T, Hyytinen E, Isola J, Trapman J, Cleutjens K, Noordzij A, Visakorpi T, Kallioniemi OP. Androgen receptor gene amplification: a possible molecular mechanism for androgen deprivation therapy failure in prostate cancer. Cancer Res 1997;57(2):314–9.
[22] Linja MJ, Savinainen KJ, Saramäki OR, Tammela TL, Vessella RL, Visakorpi T. Amplification and overexpression of androgen receptor gene in hormone-refractory prostate cancer. Cancer Res 2001;61(9):3550–5.
[23] Taplin ME, Bubley GJ, Shuster TD, Frantz ME, Spooner AE, Ogata GK, Keer HN, Balk SP. Mutation of the androgen-receptor gene in metastatic androgen-independent prostate cancer. N Engl J Med 1995;332(21):1393–8.
[24] Wilding G, Chen M, Gelmann EP. Aberrant response in vitro of hormone-responsive prostate cancer cells to anti-androgens. Prostate 1989;14(2):103–15.
[25] Gregory CW, He B, Johnson RT, Ford OH, Mohler JL, French FS, Wilson EM. A mechanism for androgen receptor-mediated prostate cancer recurrence after androgen deprivation therapy. Cancer Res 2001;61(11):4315–9.
[26] Liao G, Chen LY, Zhang A, Godavarthy A, Xia F, Ghosh JC, Li H, Chen JD. Regulation of androgen receptor activity by the nuclear receptor corepressor SMRT. J Biol Chem 2003;278(7):5052–61.
[27] Kumar S, Kumar S, Thakur K, Kumar S, Baghchi G, Tyagi RK. Androgen receptor signaling by Growth factors in androgen independent Prostate cancer: recent advances and emerging perspectives. Publication Cell press, Banaras Hindu University; 2016.
[28] Culig Z, Hobisch A, Cronauer MV, Radmayr C, Trapman J, Hittmair A, Bartsch G, Klocker H. Androgen receptor activation in prostatic tumor cell lines by insulin-like growth factor-I, keratinocyte growth factor, and epidermal growth factor. Cancer Res 1994;54(20):5474–8.
[29] Nazareth LV, Weigel NL. Activation of the human androgen receptor through a protein kinase A signaling pathway. J Biol Chem 1996;271(33):19900–7.
[30] Singh JP, Dagar M, Dagar G, Kumar S, Rawal S, Sharma RD, Tyagi RK, Bagchi G. Activation of GPR56, a novel adhesion GPCR, is necessary for nuclear androgen receptor signaling in prostate cells. PLoS One 2020;15(9):e0226056.

[31] Lee MM, Gomez SL, Chang JS, Wey M, Wang RT, Hsing AW. Soy and isoflavone consumption in relation to prostate cancer risk in China. Cancer Epidemiol Biomark Prev 2003;12(7):665–8.
[32] Wolf DA, Herzinger T, Hermeking H, Blaschke D, Hörz W. Transcriptional and post-transcriptional regulation of human androgen receptor expression by androgen. Mol Endocrinol 1993;7(7):924–36.
[33] Faber PW, Van Rooij HC, Schipper HJ, Brinkmann AO, Trapman J. Two different, overlapping pathways of transcription initiation are active on the TATA-less human androgen receptor promoter. The role of Sp1. J Biol Chem 1993;268(13):9296–301.
[34] Mizokami A, Yeh SY, Chang C. Identification of 3', 5'-cyclic adenosine monophosphate response element and other cis-acting elements in the human androgen receptor gene promoter. Mol Endocrinol 1994;8(1): 77–88.
[35] Sharp A, Coleman I, Yuan W, Sprenger C, Dolling D, Rodrigues DN, et al. Androgen receptor splice variant-7 expression emerges with castration resistance in prostate cancer. J Clin Invest 2019;129(1):192–208.
[36] Ahrens-Fath I, Politz O, Geserick C, Haendler B. Androgen receptor function is modulated by the tissue-specific AR45 variant. FEBS J 2005;272(1):74–84.
[37] Sprenger CC, Plymate SR. The link between androgen receptor splice variants and castration-resistant prostate cancer. Hormones Cancer 2014;5(4):207–17.
[38] Li Y, Chan SC, Brand LJ, Hwang TH, Silverstein KA, Dehm SM. Androgen receptor splice variants mediate enzalutamide resistance in castration-resistant prostate cancer cell lines. Cancer Res 2013;73(2):483–9.
[39] Chen M, Manley JL. Mechanisms of alternative splicing regulation: insights from molecular and genomics approaches. Nat Rev Mol Cell Biol 2009;10(11):741–54.
[40] Hu R, Denmeade SR, Luo J. Molecular processes leading to aberrant androgen receptor signaling and castration resistance in prostate cancer. Expet Rev Endocrinol Metabol 2010;5(5):753–64.
[41] Jenster G, van der Korput HA, Trapman J, Brinkmann AO. Identification of two transcription activation units in the N-terminal domain of the human androgen receptor. J Biol Chem 1995;270(13):7341–6.
[42] Jagla M, Fève M, Kessler P, Lapouge G, Erdmann E, Serra S, Bergerat JP, Céraline J. A splicing variant of the androgen receptor detected in a metastatic prostate cancer exhibits exclusively cytoplasmic actions. Endocrinology 2007;148(9):4334–43.
[43] Dehm SM, Schmidt LJ, Heemers HV, Vessella RL, Tindall DJ. Splicing of a novel androgen receptor exon generates a constitutively active androgen receptor that mediates prostate cancer therapy resistance. Cancer Res 2008;68(13):5469–77.
[44] Kumar S, Tyagi RK. Androgen receptor association with mitotic chromatin - analysis with introduced deletions and disease-inflicting mutations. FEBS J 2012;279(24):4598–614.
[45] Hu R, Dunn TA, Wei S, Isharwal S, Veltri RW, Humphreys E, Han M, Partin AW, Vessella RL, Isaacs WB, Bova GS, Luo J. Ligand-independent androgen receptor variants derived from splicing of cryptic exons signify hormone-refractory prostate cancer. Cancer Res 2009;69(1):16–22.
[46] Guo Z, Yang X, Sun F, Jiang R, Linn DE, Chen H, Chen H, Kong X, Melamed J, Tepper CG, Kung HJ, Brodie AM, Edwards J, Qiu Y. A novel androgen receptor splice variant is up-regulated during prostate cancer progression and promotes androgen depletion-resistant growth. Cancer Res 2009;69(6):2305–13.
[47] Watson PA, Chen YF, Balbas MD, Wongvipat J, Socci ND, Viale A, Kim K, Sawyers CL. Constitutively active androgen receptor splice variants expressed in castration-resistant prostate cancer require full-length androgen receptor. Proc Natl Acad Sci USA 2010;107(39):16759–65.
[48] Hu R, Isaacs WB, Luo J. A snapshot of the expression signature of androgen receptor splicing variants and their distinctive transcriptional activities. Prostate 2011;71(15):1656–67.
[49] Sun S, Sprenger CC, Vessella RL, Haugk K, Soriano K, Mostaghel EA, Page ST, Coleman IT, Nguyen HM, Sun H, Nelson P, Plymate SR. Castration resistance in human prostate cancer is conferred by a frequently occurring androgen receptor splice variant. J Clin Invest 2010;120(8):2715–30.

[50] Yang X, Guo Z, Sun F, Li W, Alfano A, Shimelis H, Chen M, Brodie A, Chen H, Xiao Z, Veenstra TD, Qiu Y. Novel membrane-associated androgen receptor splice variant potentiates proliferative and survival responses in prostate cancer cells. J Biol Chem 2011;286(41):36152–60.

[51] Cutress ML, Whitaker HC, Mills IG, Stewart M, Neal DE. Structural basis for the nuclear import of the human androgen receptor. J Cell Sci 2008;121(7):957–68.

[52] Chan SC, Li Y, Dehm SM. Androgen receptor splice variants activate androgen receptor target genes and support aberrant prostate cancer cell growth independent of canonical androgen receptor nuclear localization signal. J Biol Chem 2012;287(23):19736–49.

[53] Antonarakis ES, Lu C, Wang H, Luber B, Nakazawa M, Roeser JC, et al. AR-V7 and resistance to enzalutamide and abiraterone in prostate cancer. N Engl J Med 2014;371(11):1028–38.

[54] Kohli M, et al. Androgen Receptor Variant AR-V9 Is Coexpressed with AR-V7 in Prostate Cancer Metastases and Predicts Abiraterone Resistance. Clin. Cancer Res 2017. https://doi.org/10.1158/1078-0432.CCR-17-0017.

[55] Xu D, Zhan Y, Qi Y, Cao B, Bai S, Xu W, Gambhir SS, Lee P, Sartor O, Flemington EK, Zhang H, Hu CD, Dong Y. Androgen receptor splice variants dimerize to transactivate target genes. Cancer Res 2015;75(17): 3663–71.

[56] Asangani IA, Wilder-Romans K, Dommeti VL, Krishnamurthy PM, Apel IJ, Escara-Wilke J, Plymate SR, Navone NM, Wang S, Feng FY, Chinnaiyan AM. BET bromodomain inhibitors enhance efficacy and disrupt resistance to AR antagonists in the treatment of prostate cancer. Mol Cancer Res 2016;14(4):324–31.

[57] Robinson D, Van Allen EM, Wu YM, Schultz N, Lonigro RJ, Mosquera JM, Montgomery B, Taplin ME, Pritchard CC, Attard G, Beltran H, Abida W, Bradley RK, Vinson J, Cao X, Vats P, Kunju LP, Hussain M, Feng FY, Tomlins SA, et al. Integrative clinical genomics of advanced prostate cancer. Cell 2015;161(5): 1215–28.

[58] Qu Y, Dai B, Ye D, Kong Y, Chang K, Jia Z, Yang X, Zhang H, Zhu Y, Shi G. Constitutively active AR-V7 plays an essential role in the development and progression of castration-resistant prostate cancer. Sci Rep 2015;5:7654.

[59] Efstathiou E, Titus M, Wen S, Hoang A, Karlou M, Ashe R, Tu SM, Aparicio A, Troncoso P, Mohler J, Logothetis CJ. Molecular characterization of enzalutamide-treated bone metastatic castration-resistant prostate cancer. Eur Urol 2015;67(1):53–60.

[60] Sazani P, Kole R. Therapeutic potential of antisense oligonucleotides as modulators of alternative splicing. J Clin Invest 2003;112(4):481–6.

[61] Havens MA, Hastings ML. Splice-switching antisense oligonucleotides as therapeutic drugs. Nucleic Acids Res 2016;44(14):6549–63.

[62] Luna Velez MV, Verhaegh GW, Smit F, Sedelaar JPM, Schalken JA. Suppression of prostate tumor cell survival by antisense oligonucleotide-mediated inhibition of AR-V7 mRNA synthesis. Oncogene 2019;38: 3696–709.

[63] Van Etten JL, Nyquist M, Li Y, Yang R, Ho Y, Johnson R, Ondigi O, Voytas DF, Henzler C, Dehm SM. Targeting a single alternative polyadenylation site coordinately blocks expression of androgen receptor mRNA splice variants in prostate cancer. Cancer Res 2017;77(19):5228–35.

[64] Trang P, Wiggins JF, Daige CL, Cho C, Omotola M, Brown D, Weidhaas JB, Bader AG, Slack FJ. Systemic delivery of tumor suppressor microRNA mimics using a neutral lipid emulsion inhibits lung tumors in mice. Mol Ther 2011;19(6):1116–22.

[65] Daige CL, Wiggins JF, Priddy L, Nelligan-Davis T, Zhao J, Brown D. Systemic delivery of a miR34a mimic as a potential therapeutic for liver cancer. Mol Cancer Therapeut 2014;13(10):2352–60.

[66] Pramanik D, Campbell NR, Karikari C, Chivukula R, Kent OA, Mendell JT, Maitra A. Restitution of tumor suppressor microRNAs using a systemic nanovector inhibits pancreatic cancer growth in mice. Mol Cancer Therapeut 2011;10(8):1470–80.

[67] Jawa Y, Yadav P, Gupta S, Mathan SV, Pandey J, Saxena AK, Kateriya S, Tiku AB, Mondal N, Bhattacharya J, Ahmad S, Chaturvedi R, Tyagi RK, Tandon V, Singh RP. Current insights and advancements in Head and Neck cancer: emerging biomarkers and therapeutics with cues from single cell and 3D model omics profiling. Front Oncol 2021;11:676948.

[68] Shi XB, Ma AH, Xue L, Li M, Nguyen HG, Yang JC, Tepper CG, Gandour-Edwards R, Evans CP, Kung HJ, deVere White RW. miR-124 and androgen receptor signaling inhibitors repress prostate cancer growth by downregulating androgen receptor splice variants, EZH2, and Src. Cancer Res 2015;75(24):5309−17.

[69] Bonnal S, Vigevani L, Valcárcel J. The spliceosome as a target of novel antitumor drugs. Nat Rev Drug Discov 2012;11(11):847−59.

[70] Effenberger KA, Urabe VK, Prichard BE, Ghosh AK, Jurica MS. Interchangeable SF3B1 inhibitors interfere with pre-mRNA splicing at multiple stages. RNA (New York, NY) 2016;22(3):350−9.

[71] Eskens FA, Ramos FJ, Burger H, O'Brien JP, Piera A, de Jonge MJ, Mizui Y, Wiemer EA, Carreras MJ, Baselga J, Tabernero J. Phase I pharmacokinetic and pharmacodynamic study of the first-in-class spliceosome inhibitor E7107 in patients with advanced solid tumors. Clin Cancer Res 2013;19(22):6296−304.

[72] Hong DS, Kurzrock R, Naing A, Wheler JJ, Falchook GS, Schiffman JS, Faulkner N, Pilat MJ, O'Brien J, LoRusso P. A phase I, open-label, single-arm, dose-escalation study of E7107, a precursor messenger ribonucleic acid (pre-mRNA) splicesome inhibitor administered intravenously on days 1 and 8 every 21 days to patients with solid tumors. Invest N Drugs 2014;32(3):436−44.

[73] Kawamura N, Nimura K, Saga K, Ishibashi A, Kitamura K, Nagano H, Yoshikawa Y, Ishida K, Nonomura N, Arisawa M, Luo J, Kaneda Y. SF3B2-Mediated RNA splicing drives human prostate cancer progression. Cancer Res 2019;79(20):5204−17.

[74] Liu C, Lou W, Zhu Y, Nadiminty N, Schwartz CT, Evans CP, Gao AC. Niclosamide inhibits androgen receptor variants expression and overcomes enzalutamide resistance in castration-resistant prostate cancer. Clin Cancer Res 2014;20(12):3198−210.

[75] Ronan RL, et al. EPI-7386 is a novel N-terminal domain androgen receptor inhibitor for the treatment of prostate cancer. Ann oncol 2019.

[76] Myung JK, Banuelos CA, Fernandez JG, Mawji NR, Wang J, Tien AH, Yang YC, Tavakoli I, Haile S, Watt K, McEwan IJ, Plymate S, Andersen RJ, Sadar MD. An androgen receptor N -terminal domain antagonist for treating prostate cancer. J Clin Invest 2013;123(7):2948−60.

CHAPTER

9 Post-transcriptional gene regulation in Cardiorenal syndrome

Ramandeep Singh[1], Anupam Mittal[2], Ajay Bahl[1] and Madhu Khullar[3]

[1]*Department of Cardiology, PGIMER, Chandigarh, India;* [2]*Department of Translational and Regenerative Medicine, PGIMER, Chandigarh, India;* [3]*Department of Experimental Medicine and Biotechnology, PGIMER, Chandigarh, India*

Introduction

Cardiorenal syndrome (CRS) encompasses clinical conditions involving coexisted dysfunctions of the heart and kidney. The interorgan crosstalk maintains homeostasis between various organs on multiple levels such as physiological, cellular, and molecular levels, which implies that imbalance in the functioning of one organ can have negative repercussions on the functioning of other organ systems. Much research and literature on the crosstalk between organs are available, but the fundamental basis has not been fully unraveled. Multiple factors are known to be contributing components in the manifestation of the CRS. Risk factors in developing a heart or kidney disease include diabetes, pre-existing heart or kidney problems, and high blood pressure. On cellular levels, immune dysregulation, oxidative stress, uremia, renin-angiotensin system activation, and inflammation are known to conduce the phenotype of the CRS; however, the complex pathways have not been fully elucidated yet. Apart from the conventional causes of the syndrome, new causative agents have been suggested. These include post-transcriptional modifications of the RNA, epigenetic mechanisms, small noncoding RNAs, prenatal programming, extracellular vesicles, and transcription factors [1–4].

The syndrome has been classified into five major subtypes as follows:

CRS Type 1: Acute cardiorenal syndrome: Acute cardiac disorder followed by acute kidney injury.

CRS Type 2: Chronic cardiorenal syndrome: Chronic cardiac dysfunction followed by Chronic kidney disease.

CRS Type 3: Acute renocardiac syndrome: Acute kidney dysfunction causing acute cardiac injury.

CRS Type 4: Chronic renocardiac syndrome: Chronic kidney disease-causing chronic cardiac injury.

CRS Type 5: Secondary cardiorenal syndrome: Simultaneous dysfunction of both kidney and heart due to a systemic condition.

The patients suffering from heart and kidney disease move from one classification to another with the progression of the disease [1,4,5].

Post-Transcriptional Gene Regulation in Human Disease, Volume 32. https://doi.org/10.1016/B978-0-323-91305-8.00012-0

Lately, due consideration has been endowed to the nonconventional and proposed causative agents of the CRS. Epigenomic changes are the structural changes in the DNA packaging controlling the gene expression by not altering the DNA sequence itself. Epigenomics can amend the transcriptional outcomes of the nucleotide sequences, which can curtail the gene expression of certain genes. These changes may arise due to some intercellular, intracellular, or other external factors such as drugs. Several epigenetic mechanisms like histone modifications, DNA methylation, chromatin structure organization, miRNA, short noncoding RNA (snRNA) are linked with the regulation of gene expression. Epigenetic modifications are known to cause phenotypic changes in the various cell types. Recent studies suggest that epigenetic changes are crucial in cellular responses generated from cardiovascular diseases and kidney injuries [1,2]. Epigenetic modifications can be categorized into several distinct processes as follows:

Histone modifications

The multi-level organized chromatin structure is tightly held together by the electrostatic forces of attraction generated between negatively charged DNA and positively charged histone proteins. The histone proteins are acetylated at lysine residues rendering them devoid of positive charge and more susceptible to transcriptional changes. Similarly, deacetylation of the histone proteins makes the chromatin more condensed and less susceptible to transcription. All the core histone proteins holding the nucleosome structure are acetylated. The histone proteins are acetylated rapidly in response to the signals generated within cells and serve as a docking site for many transcription factors to facilitate transcription [6,7]. Interestingly, a study showed that a histone acetyltransferase, PCAF (p300/CBP-associated factor), is overexpressed in renal injuries and results in the upregulation of certain inflammation genes such as VCAM-1 and ICAM-1. Supporting the earlier study, an in vitro study also revealed that knockdown of PCAF gene in human renal proximal tubule epithelial cells resulted in the downregulation of the previously mentioned inflammatory genes [8]. Researchers have shown that composite signaling systems like crosstalk between heart and kidney can lead to dysfunction, which may be a consequence of the remodeling processes at cellular and subcellular levels. These systems are altered with histone modifications which result in the structural alterations of chromatin. Evidence has been found that these histone proteins get released on damage to the tubular epithelial cells, and on encounter with Toll-like receptors, proinflammatory cytokines are released. In this way, alterations in the histone proteins can result in acute kidney injury with its proinflammatory effects [9]. In some experimental animals, histone acetylation and deacetylation have been linked to cardiac hypertrophy. Some transcription coactivators like CRB (CREB-binding protein), when overexpressed, can lead to hypertrophic growth in the cardiomyocytes culminating in hypertrophic cardiomyopathy [7].

Apart from the acetylation, methylation can also lead to the manifestation of cardiorenal syndrome (Fig. 9.1). A histone methylation study of heart tissues in heart failure revealed altered methylation patterns of histone proteins H3H4 and H3K9 [7]. A group of researchers showed that acetylation, dimethylation, and phosphorylation of H3 histone protein get elevated on kidney failure and the expression of mRNA of multiple genes associated with cardiomyopathy increases [10]. Irregular DNA methylation patterns can contribute to the development of atherosclerosis by regulating the associated genes, and uremia is also known to alter DNA methylation, which suggests that epigenetic modifications can give rise to cardiorenal syndrome [2].

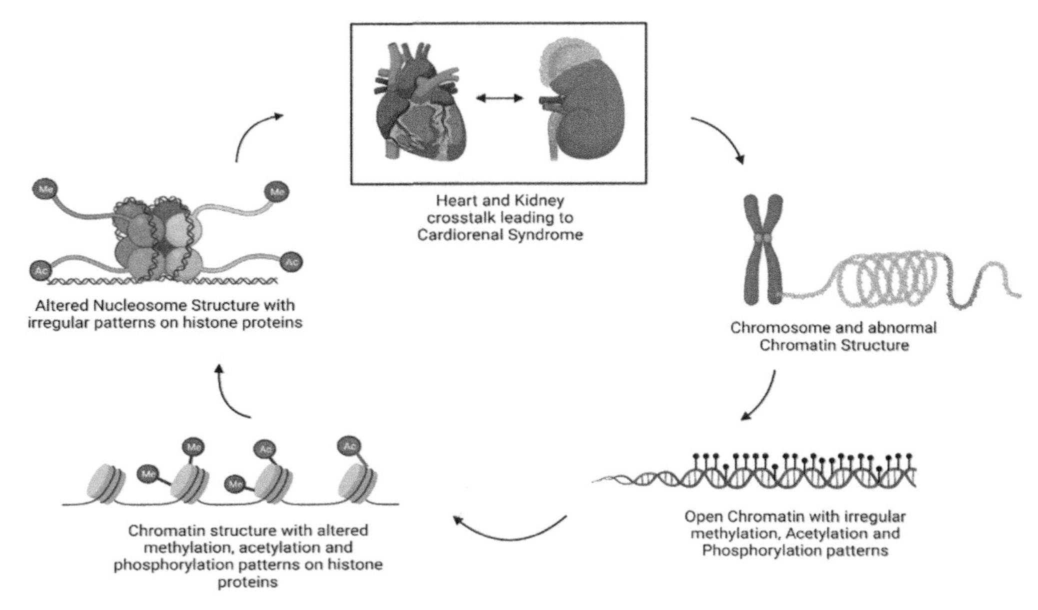

FIGURE 9.1

Pictorial illustration of irregular methylation and acetylation patterns leading to molecular and structural changes in chromatin and nucleosome structures giving rise to cardiorenal syndrome.

RNA-based modifications

Small noncoding RNAs are known to control the genetic expression in both cardiac and renal cells and play an imperative role in maintaining homeostasis [11]. snRNA and miRNA altogether have been the center of the contemporary research that function on the post-transcriptional level to control genetic expression by upregulation and downregulation of the genes. They perform a wide range of functions in different realms such as transcriptional regulation, mRNA destruction, and the dynamics of chromosomes (Fig. 9.2). Similarly, the long noncoding RNAs are an important key factor in controlling epigenetic modifications [7,12].

A recent focus in gene regulation by small noncoding RNAs is through micro-RNAs (miRNA) with an average length of 22 nucleotides. Most of the miRNA is transcribed from the DNA as primary miRNAs and are edited through site-specific modifications yielding diverse products and, therefore, providing various actions. They mainly function in RNA splicing and post-transcriptional gene expression regulation. There are more than 1000 different types of miRNAs described in humans. Overexpression of certain miRNAs, for instance, miR-24, miR-23a, and miR-23b induce hypertrophic cardiomyopathy in neonatal cardiomyocytes; however, overexpression of particular miRNAs such as miR-133 is known to inhibit cardiac hypertrophy.

The silencing of RNA is achieved through miRNA incorporated in RNA-induced silencing complex. The miRNA is divided into two separate strands post unwinding, and mature miRNA is retained in the complex. The complex scavenges for the target mRNA, which is degraded [13].

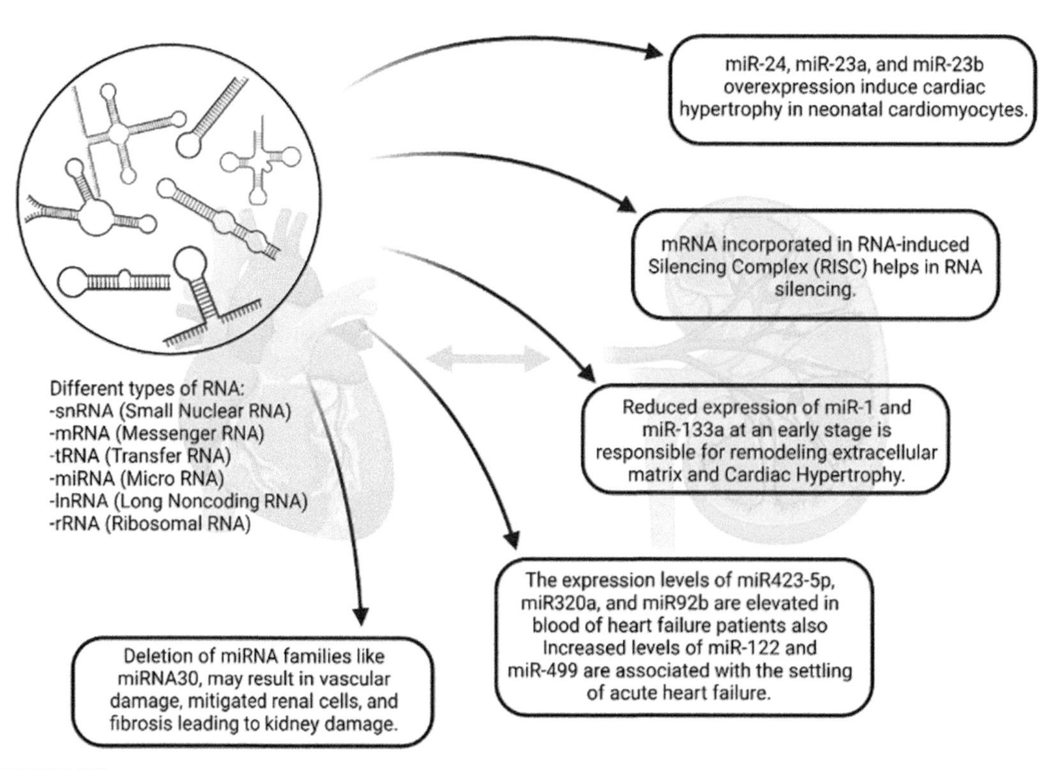

FIGURE 9.2

snRNA and miRNA perform a wide range of functions in different realms like transcriptional regulation, mRNA destruction, and the dynamics of chromosomes and their escalated or attenuated expressions may result in organ damage.

miRNAs are known to regulate the programming and gene expression of various types of cells that participate in the development of atherosclerosis. They also play a crucial role in lipid metabolism, cellular differentiation, and inflammation [7].

Researchers have investigated the expression levels of miRNA in a transgenic mouse with progressing hypertrophic cardiomyopathy from early to end-stage with increasing severity. Chronological analysis showed mitigated expression levels of miR-1 and miR-133a at an early stage, whereas increased expression of genes responsible for remodeling extracellular matrix and cardiac hypertrophy. While the expression of miR-31 was upregulated at the end stage of hypertrophic cardiomyopathy, implying the particular role of this miRNA toward the final stages of the disease [7,14].

The miR423-5p levels have been known to be elevated in the blood of heart failure patients, and its extent is increased with the progression of the disease. Several other miRNAs such as miR320a and miR92b are also found with elevated levels in the blood of heart failure patients. Increased levels of miR-122 and miR-499 are linked to the settling of acute heart failure.

The miRNAs expressed in the kidney are miR192, miR194, miR215, and miR216, which are associated with the proliferation and migration of renal cells. Deletion of several miRNA families such as miRNA30 may result in vascular damage, mitigated renal cells, and fibrosis [1].

DNA-based modifications

One of the imperative DNA-based modifications in the nucleotide sequence is the methylation of the base cytosine, which results in epigenetic changes. This methylation of the cytosine base is carried out by DNA methyltransferases transferring a methyl group to the cytosine ring structure. Nucleotide sequences of the cardiac genome exhibit distinct DNA methylation patterns at various progressive levels of cardiomyopathy. These epigenetic modifications control the functioning of the genes, which are known to be expressed in myocardial stress-related conditions. They also pointed out three molecules that were angiogenic in nature, that is, ARHGAP24 (rho GTPase-activating protein 24), AMOTL2 (angiomotin-like2), PECAM1 (platelet/endothelial cell adhesion molecule 1). It has been proposed that tracking the methylation patterns and the disease's progression may help in the earlier prognosis of myocardial infarction [7,15,16].

Studies have shown that the methylation patterns are associated with various levels of pluripotency in different cell types. A group of researchers performed genome-wide mapping of the cytosine methylation patterns of fibroblasts from two sources, that is, mammalian embryonic stem cells and fibroblasts isolated from the fetus. They also performed transcriptomic analysis of the RNA, histone modifications, and DNA–protein interaction sites. Vast differences were found in the methylation patterns of both sources. It was also observed that close to one-fourth of all the methylation patterns found in the embryonic stem cells was non-CG methylation. This finding suggests that a different mechanism might control the upregulation and downregulation of the genes in embryonic stem cells. The differentiation of the embryonic stem cells mitigated the non-CG methylation patterns, whereas the differentiation of induced pluripotent stem cells reinstated the non-CG methylation patterns (Fig. 9.3) [17]. Impaired DNA methylation at the different development stages of the embryo may culminate into congenital heart diseases with an increased risk of cardiovascular diseases in the future [7].

Researchers have shown that altered DNA methylation patterns can be helpful in the identification of the diseases much before the phenotypic signs of the disease are visible. A combination of different techniques like southern blotting, PCR, DNA fingerprinting was used to detect hyper and hypomethylation in peripheral blood mononuclear cells and aorta of mutant mice even before the appearance of any atherosclerotic lesion. They also revealed that some atherogenic lipoproteins assist in the hypermethylation of DNA in a monocyte cell line [18].

SUMOylation

SUMOylation is involved in the regulation of various biomolecular and cellular processes such as apoptosis, transcription, translation, stability of the protein, and stress response by attachment and detachment of a family of small proteins called SUMO proteins to other cellular proteins. Moreover, the SUMO protein conjugation pathways are involved in cardiac development which functions by modulating signal transduction pathways. Researchers have explored the activity of a deSUMOylating

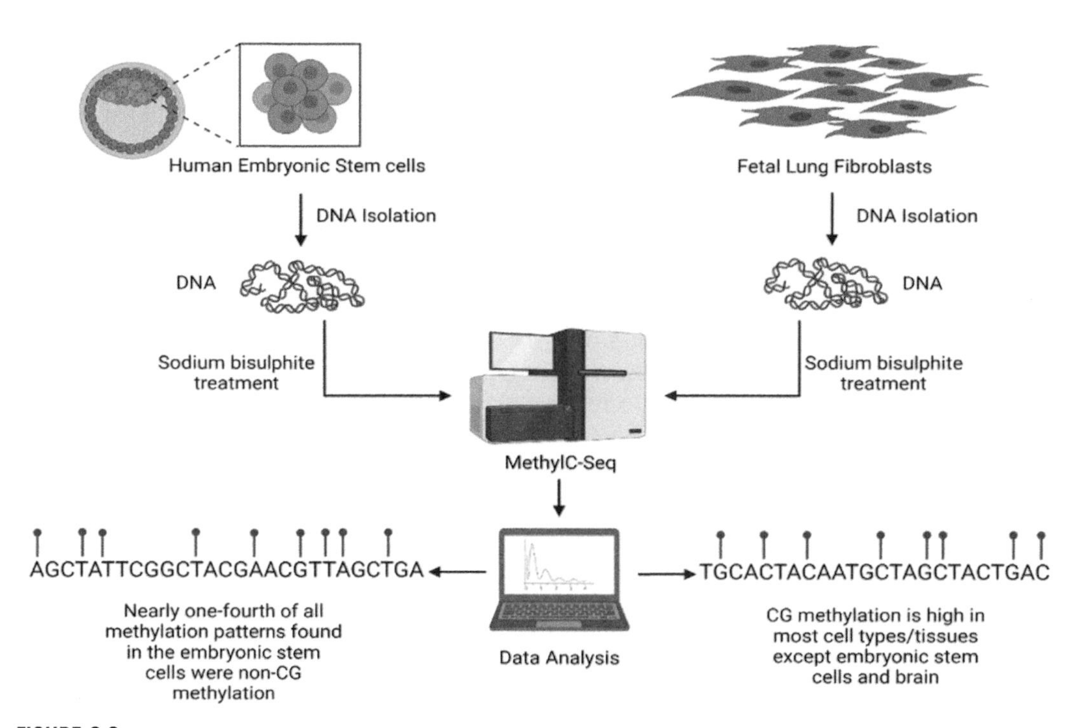

FIGURE 9.3

Schematic representation of the mapping of methylation patterns in cells from two different sources. Human embryonic stem cells showed non-CG methylation patterns in nearly one-fourth of all the patterns whereas fetal lung fibroblasts expressed CG methylation.

enzyme under the promoter of myosin heavy chain in cardiomyocytes of rodents and found that more than 50% of the first-generation population died within 1 week of birth suggesting that the balance of SUMOylation and deSUMOylation is essential for the normal physiology of the cardiac system. SUMOylation is also involved in the transactivation of the cardiogenic genes in the fibroblasts, which otherwise is poorly exhibited by myocardin. SUMOylation by the protein SUMO1 increases the efficiency of myocardin to activate the genes involved in the smooth muscle activity, for instance, myosin heavy chain and cardiac actin. Ion channel activity of cardiomyocytes is also moderated by SUMO conjugation However, it still needs to be investigated if defects in SUMOylation pathways of ion channels lead to any cardiac disorders in humans. Apart from maintaining the cellular processes their role in cardiac disorders has been unraveled. SUMO2/3 protein conjugation is induced by heart failure. Increased modifications by SUMO2/3 protein can induce cardiomyocytes death culminating in cardiac disorders. Researchers have shown that overinvolvement of SUMO proteins in the cellular processes can lead to degradation of the proteins. It was observed that dilated cardiomyopathy can be induced by overexpression of certain proteins like SENP5. Satiated evidence advocates the involvement of SUMO proteins in the pathophysiology of cardiac disorders which in turn can give rise to cardiorenal syndrome [19–21].

Summary

Despite enormous literature available on the crosstalk between heart and kidney, there is still a lot of missing information on epigenetic regulation of cardiorenal syndrome, which forbids fully comprehending the multifactorial interactions in the disease. This void calls for diligent research for understanding the basics and magnitude of the complexity of interaction between two organs. The nonorthodox causative agents have provided a new representation of communication in interorgan crosstalk. There may be innumerable causative agents for establishing the injury or disease in one organ but to comprehend the transmission of disease from one organ to another needs thorough investigation. This may also lead to the discovery of some new biomarkers for the disease even before the appearance of the phenotype, which can drastically ameliorate the clinical outcomes. The genetic elements as the cause of the disorder discussed in this chapter remain primarily unexplored and possess the potential to appear appealing targets for the betterment of patients with heart and kidney diseases.

References

[1] Virzì GM, Clementi A, Brocca A, de Cal M, Ronco C. Molecular and genetic mechanisms involved in the pathogenesis of cardiorenal cross talk. Pathobiology 2016;83(4):201−10.
[2] Kingma JG, Simard D, Rouleau JR, Drolet B, Simard C. The physiopathology of cardiorenal syndrome: a review of the potential contributions of inflammation. J Cardiovasc Dev Dis 2017;4(4):21.
[3] Bongartz LG, Cramer MJ, Doevendans PA, Joles JA, Braam B. The severe cardiorenal syndrome: 'Guyton revisited. Eur Heart J 2005;26(1):11−7.
[4] Shah BN, Greaves K. The cardiorenal syndrome: a review. Int J Nephrol 2010;2011.
[5] Hadjiphilippou S, Kon SP. Cardiorenal syndrome: review of our current understanding. J R Soc Med 2016; 109(1):12−7.
[6] Bomsztyk K, Denisenko O. Epigenetic alterations in acute kidney injury. Semin Nephrol 2013;33(4): 327−40. WB Saunders.
[7] Cao Y, Lu L, Liu M, Li XC, Sun RR, Zheng Y, Zhang PY. Impact of epigenetics in the management of cardiovascular disease: a review. Eur Rev Med Pharmacol Sci 2014;18(20):3097−104.
[8] Huang J, Wan D, Li J, Chen H, Huang K, Zheng L. Histone acetyltransferase PCAF regulates inflammatory molecules in the development of renal injury. Epigenetics 2015;10(1):62−71.
[9] Virzì GM, Day S, de Cal M, Vescovo G, Ronco C. Heart−kidney crosstalk and role of humoral signaling in critical illness. Crit Care 2014;18(1):1−11.
[10] Gaikwad AB, Sayyed SG, Lichtnekert J, Tikoo K, Anders HJ. Renal failure increases cardiac histone h3 acetylation, dimethylation, and phosphorylation and the induction of cardiomyopathy-related genes in type 2 diabetes. Am J Pathol 2010;176(3):1079−83.
[11] Clementi A, Virzì GM, Brocca A, de Cal M, Pastori S, Clementi M, Granata A, Vescovo G, Ronco C. Advances in the pathogenesis of cardiorenal syndrome type 3. Oxid Med Cell Longev 2015;2015.
[12] Rodríguez-Romo R, Berman N, Gómez A, Bobadilla NA. Epigenetic regulation in the acute kidney injury to chronic kidney disease transition. Nephrology 2015;20(10):736−43.
[13] Esteller M. Non-coding RNAs in human disease. Nat Rev Genet 2011;12(12):861−74.
[14] Bagnall RD, Tsoutsman T, Shephard RE, Ritchie W, Semsarian C. Global microRNA profiling of the mouse ventricles during the development of severe hypertrophic cardiomyopathy and heart failure. 2012.

[15] Movassagh M, Choy MK, Knowles DA, Cordeddu L, Haider S, Down T, Siggens L, Vujic A, Simeoni I, Penkett C, Goddard M. Distinct epigenomic features in end-stage failing human hearts. Circulation 2011;124(22):2411–22.
[16] Virzì GM, Clementi A, Brocca A, de Cal M, Ronco C. Epigenetics: a potential key mechanism involved in the pathogenesis of cardiorenal syndromes. J Nephrol 2018;31(3):333–41.
[17] Lister R, Pelizzola M, Dowen RH, Hawkins RD, Hon G, Tonti-Filippini J, Nery JR, Lee L, Ye Z, Ngo QM, Edsall L. Human DNA methylomes at base resolution show widespread epigenomic differences. Nature 2009;462(7271):315–22.
[18] Lund G, Andersson L, Lauria M, Lindholm M, Fraga MF, Villar-Garea A, Ballestar E, Esteller M, Zaina S. DNA methylation polymorphisms precede any histological sign of atherosclerosis in mice lacking apolipoprotein E. J Biol Chem 2004;279(28):29147–54.
[19] Wang J, Schwartz RJ. Sumoylation and regulation of cardiac gene expression. Circ Res 2010;107(1):19–29.
[20] Li XC, Zeng Y, Sun RR, Liu M, Chen S, Zhang PY. SUMOylation in cardiac disorders-a review. Eur Rev Med Pharmacol Sci 2017;21(7):1583–7.
[21] Wang J, Li A, Wang Z, Feng X, Olson EN, Schwartz RJ. Myocardin sumoylation transactivates cardiogenic genes in pluripotent 10T1/2 fibroblasts. Mol Cell Biol 2007;27(2):622–32.

CHAPTER

Noncoding RNAs as modulators of post-transcriptional changes and their role in CVDs 10

Swati Sharma, Shankar Chanchal, Yasir Khan and Zahid Ashraf
Department of Biotechnology, Jamia Millia Islamia, New Delhi, India

Introduction

Cardiovascular diseases (CVDs) are considered a global epidemic as they inflict major stress on global health and societal costs. They account for the leading cause of death around the globe. CVDs such as ischemic heart disease, stroke, and thrombosis account for more than 80% of deaths worldwide. Even with the advancement in the prevention and management of CVDs, they remain a major public health issue in various countries including India. The average death rate worldwide due to CVD is 235 per 100,000 deaths, whereas in India it is 272 per 100,000 deaths in India [1]. In the Indian population, increased fatality along with early disease onset of CVDs is observed. CVDS are a group of multifactorial diseases, and there are various risk factors (inherited and acquired) that have been associated with the pathogenesis of the disease. The major hurdle in the treatment and prevention of CVDs is that variability in the risk of getting CVDs among the different populations cannot be fully explained based on known genetic and environmental risk factors. Various studies have addressed the pathophysiology, gene polymorphism, epidemiology, genetic linkage maps, and effect of environmental stresses on the CVDs, leading to a better understanding of the diseases and improved treatment. Still, the link that connects the genetic predisposition with the environmental stresses in the progression of CVD is missing. As inherited genetic risk factors are not the sole contributors and environmental factors including lifestyle play a crucial role in deciding the outcome of the disease, it is pertinent to look into post-transcriptional changes (PTC) mediated by epigenetic control to bridge the gap in our knowledge. PTC or modifications are the "heritable changes in the genome that does not include changes in DNA sequences." This involves several molecular mechanisms mediated by DNA methylation, demethylation, histone modifications, and ncRNAs-mediated alterations. Various PTC is mediated by environmental stimuli, as they serve as a mechanism to respond to changing environmental stress quickly. There are various lifestyle/environment-related risk factors such as smoking, nutrition, circadian rhythm, obesity, and pollution that are linked with the PTC changes and associated modification in epigenetic biomarkers. Thus, examination of these changes can help in the early detection and prevention of CVDs and better therapeutic interventions. In the Indian scenario, the optimal therapeutic interventions are not being received by patients belonging to lower socioeconomic backgrounds. This leads to poorer outcomes of treatment. In the proximity of poorer outcomes, such multifactorial

Post-Transcriptional Gene Regulation in Human Disease, Volume 32. https://doi.org/10.1016/B978-0-323-91305-8.00015-6

diseases can only be combated by developing the protocols that involve the restructuring of the health care system for better and effective implementation of evidence-based treatment, early detection, therapeutic interventions with both conventional and innovative techniques. Among these protocols, the studies and work related to noncoding RNAs (ncRNAs) such as miRNA, aptamers, and RNA binding proteins (RBPs) are also being worked on so that they can be used as a biomarker for early detection or can be used in treatment for better outcome among patients [2].

Therefore, the focus of this chapter is to acquaint the medical community dealing with CVDs with the advent field of ncRNAs, how they regulate PTC, and its role in cardiovascular medicine. The chapter consists of a brief introduction of ncRNAs, PTC/epigenetic mechanisms mediated by them, and scientific evidence linking them with CVD. The main area of the thrust of this chapter would be on ncRNAs as a regulator of CVDs to decipher their clinical reality and translational potential in cardiovascular medicine.

Noncoding RNAs

Although all the cells of a human body possess the same genetic, that is, DNA blueprint, still each cell type has distinct functional properties because of differential expression of the genes in each cell type. This differential expression of genes is majorly achieved by gene-specific transcription. But various post-transcriptional changes also influence the cell or stimuli specific expression of the genes. Much of these PTCs are dictated by ncRNAs (Table 10.1).

The deep sequencing analysis of mammalian genomes showed that the protein-coding genes account for only 2% of the whole human genome, while the rest is transcribed RNA that lacks protein-coding information and is never translated. These are termed ncRNAs. ncRNAs are generally classified on the size basis into two using an arbitrary threshold of 200 nucleotides (nt) namely small ncRNAs (snRNAs) having less than 200 nt and lncRNA with more than 200 nt. Short ncRNAs comprise microRNAs (miRNAs), small interfering RNAs (siRNAs), and P-element-induced wimpy testis interacting RNAs (piRNAs), which generally regulate gene expression negatively. In contrast, lncRNAs comprise a diverse class of RNAs such as circular RNA (circRNA), enhancer RNAs (eRNAs), long intergenic RNA (lincRNA), sense or antisense transcripts (AS), natural antisense transcripts (NATs), and promoter-associated long RNA (pRNA) [3].

ncRNA in CVD

miRNAs: miRNAs are the most extensively studied ncRNAs that play an important function in maintaining various cell processes such as apoptosis, autophagy, cell metabolism, and proliferation [21,22]. Any abnormalities in their function and biogenesis would directly or indirectly lead to several pathogeneses of cardiovascular diseases and its risk factors such as atherosclerosis, arrhythmia, diabetes, myocardial infarction, heart failure, hypertension, coronary artery disease, and many more. They are identified in many cardiac cell types like cardiomyocytes, endothelial cells, fibroblasts, and cardiac tissues at all stages of development [23,24]. Many miRNAs such as miR-1, miR-16, miR-27b, miR-30d, miR-126, miR-133, miR-143, miR-208, and the let-7 family in a healthy adult heart [25]. miR expressed in fetal heart includes miR-21, miR-29a, miR-129, miR-210, miR-211, miR-320, miR-

Table 10.1 Key features of ncRNAs and their occurrence in various CVDs.

	Small ncRNAs				Long ncRNAs
Characteristics	**miRNAs**	**siRNAs**	**piRNAs**	**Circular RNAs**	**lncRNAs**
Size (base pair)	20–25 nt	20–24 nt	21–35 nt	Varies	>200 nt
Source	ssRNA	dsRNA	ssRNA	Multiple ways according to its-site (exonic, intronic, intergenic, UTRs)	Multiple ways according to its-site (exonic, intronic, intergenic)
Discovery	*C. elegans* in 1993	Plants in 1999	*D. melanogaster* in 2001	RNA viruses in 1979	Mammals in the 1990s (XIST)
Interaction with other ncRNAs	Trigger lncRNAs decay	Act as lncRNA decay	Target lncRNAs	Act as miRNAs sponge.	Act as miRNAs decay/sponge.
Role in CVDs	miR-106, miR-197, miR-223, miR-1, miR-21, miR-29b, miR-34a—AMI with ischemia related HF [4–6]. miR-1,miR-21-HF [7] miR-150-PAH [8]	siRNAs in atherosclerosis, thrombosis, etc [9,10]	piR-823—changes property of endothelial cells [11] piR_2106027-MI [12] piR-006426, piR-020009-HF [13]	cANRIL—atherosclerosis [14] cZNF292—angiogenesis [15] HRCR—hypertrophy and heart failure [16]	ANRIL, MALAT1-atherosclerosis [17,18] CRRL—cardiomyocytes proliferation and repair [19] CCRR—antiarrhythmic [20]

Abbreviations used: cANRIL, *circular ANRIL;* CCRR, *cardiac conduction regulatory RNA;* CRRL, *cardiomyocyte regeneration-related lncRNA;* dsRNA, *double -stranded RNA;* HF, *heart failure;* HRCR, *heart-related circRNA;* MI, *myocardial infarction;* PAH, *pulmonary arterial hypertension; ssRNA, single-stranded RNA.*

423, and let-7c [26]. Using a mouse model of heart failure compared to control has shown a correlation of miR-25 to cardiac contractility. It depicts inhibition of miR-25 using antagomiR technology has restored cardiac contractility and improved its survival [27]. The human genome encodes approximately 1000 miRNAs, out of which more than 100 miRNAs have been located in the serum of healthy individuals and termed as circulating miRNAs [28]. The presence of miRNA in human plasma and serum was detected in 2008 by Mitchell and the group [29]. There is enormous evidence that circulating miRNAs are endogenous, single-stranded, more stable, and resistant to degradation by RNase activity as they reside in microvesicles such as apoptotic bodies, exosomes, and microparticles [30–32]. Circulating miRNAs has the potential to be used as diagnostic biomarkers in various CVD such as Acute myocardial infarction (AMI), heart failure (HF), atrial fibrillation, and hypertension. In patients of AMI with ischemia related HF, there are changes in the circulating miRNAs levels to name some decreases in miR-106, miR-197, and miR-223 and increase in miR-1, miR-21, miR-29b, miR-34a, miR-126,miR-133, miR-134, miR-208, miR-23, and miR-499 [4–6]. In symptomatic HF, miR-1 is alleviated and miR-21 is elevated [7].

Several circulating miRNAs have been linked to pulmonary arterial hypertension development and its progression. A pilot study by Rhodes and colleagues showed miR-150 is downregulated in peripheral blood of patients with pulmonary arterial hypertension [8].

Circular RNAs: Many recent studies have suggested the essential role of circRNAs in the initiation and development of cardiovascular diseases. The first circRNAs expression profiling by RNA sequence done by Jakobi et al. [33] on adult mouse hearts identified 575 cardiac circRNAs [33]. Later, a different group identified approximately more than 2400 circRNAs [34] in human blood that can be a good diagnostic biomarker, and subsequent research discovered approximately 1000 circRNAs from human sera in exosomes [35] and during pregnancy, 19 circRNAs detected in human plasma [36]. Circular ANRIL (cANRIL), one of the first identified circRNAs is associated with single nucleotide polymorphisms (SNPs) on chromosome 9p21.3. These SNPs alter cANRIL splicing and cause INK4A/ARF locus repression, which is associated with atherosclerotic vascular disease [14]. Boeckel et al. [15] showed proangiogenic activities of circRNAs cZNF292 in vitro, which is regulated by hypoxia and control angiogenesis. Circular RNAs circRNAs_000203, from the Myo9A gene, are upregulated in diabetic mouse hearts than the control. Mechanistically these circRNAs sponge miR-26b-5p and enhance the expression of Acta2, Col1a2, and Col1a3 and subsides the expression of Col1a2 and ctgf in cardiac fibroblasts [16]. CircRNA derived from the PWWP domain containing 2A (PWWP2A) gene inhibits cardiac hypertrophy, heart failure, and hypertrophy in cardiomyocytes by sponging miR-223 by targeting apoptosis repressor with CARD domain [37]. It is seen that circAmortl1 levels are linked to cardiomyocytes' survival and its inhibition leads to decreased survival and increased apoptosis. Its overexpression acts as cardioprotective in a doxorubicin-induced cardiomyopathy mouse model by decreasing collagen deposition decreased apoptotic cells and importantly decreased ventricular dilation [38].

piRNAs: According to previous studies, piRNAs are generally play role in the reproductive system, brain, nervous system, and in some form of cancers [39–42]. But interestingly recent studies showed that the presence of piRNAs in heart and vascular tissues plays a regulatory role in the initiation and development of CVD [43]. The research by Zhao and colleagues has demonstrated the role of piR-823 in DNA methylation by DNMT3B to show its carcinogenic effect in esophageal cancer [44]. Thus, this work indicates the role of piRNAs in DNA methylation and opens a wide scope in research to the CVD scenario. Another work by Li et al. has shown the presence of piR-823 in multiple myeloma-derived extracellular vesicles that potentially changes the biology of endothelial cells and enhances its capillary structure formation, proliferation, and migration [11]. Ranjan et al. has shown the circulating piRNAs from human blood samples and their differential expression in induced myocardial hypertrophy from in vitro and in vivo samples compared with the control 394. They have demonstrated the role of piR_2106027 in patients with and without MI and its direct relation with the level of Troponin-I [12]. In 2018, Yang et al. through RNA sequence and bioinformatic analysis have shown the piRNAs in the human serum of healthy and HF patients. This study has identified 585 piRNAs upregulated and 4623 piRNAs as downregulated piRNAs, in which hsa-piR-006426 and hsa-piR-020009 were most differentially expressed [13]. Currently, the research and the understanding of piRNAs are still in their preliminary stage and need great attention on their potential use as a diagnostic biomarker in CVD.

LncRNAs: Gene regulation by lncRNAs has a great impact on the pathogenesis of inflammation and various CVD. Recent work has demonstrated the involvement of lncRNAs such as antisense noncoding RNA in the INK4 locus (ANRIL), metastasis-associated lung adenocarcinoma transcript 1 (MALAT1) in the occurrence of atherosclerosis and observed its expression in various atherogenic cells like endothelial cells, monocytes/macrophages, and vascular smooth muscle cells (VSMCs) and involved in vascular biology, inflammatory diseases, and various cellular activities like autophagy [17,18]. LncRNAs ANRIL [17], X-inactive specific transcript (XIST) [45], small nucleolar RNA host gene 1 (SNHG1) [46] and nuclear enriched abundant transcript 1 [47], functional intergenic repeating RNA element (FIRRE) [48], growth arrest-specific 5 (GAS5) [48], and MALAT1 [18] regulates NLRP3 inflammasome and its downstream molecules such as caspase-1 and various cytokines such as 1L-1β and IL-18. It is well established the involvement of NLRP3 inflammasome in cardiovascular diseases and various inflammatory diseases [49]. LncRNA CRRL (cardiomyocyte regeneration related lncRNA) is identified in cardiac tissues and its inhibition increases cardiomyocytes proliferation and repair as well as improves cardiac function in the MI rat model by suppressing the activity of miR-199a-3p thereby increasing its target Hopx expression [19]. Zhang et al. has recognized antiarrhythmic lncRNA known as cardiac conduction regulatory RNA whose suppression leads to decreases cardiac conduction and increases arrhythmogenicity in mice. This is achieved through the degradation of gap junction protein connexin-43 [20]. MALAT1 expression in the heart has shown its role in angiogenesis, cardiac fibrosis, and stress remodeling. In ApoE and MALAT1 deficient mice model has indicated its role in plaque size and infiltration of CD45+ cells enhancing atherosclerotic plaque formation by sponging miR-503 [50].

siRNAs: The silencing of JAK2 by siRNA inhibits Pam3 CSK4-induced factor III (tissue factor) expression, which plays a crucial role in coagulation and is involved in the progression of atherosclerosis [9]. Wang et al. under low shear stress in HUVECs, has demonstrated that the knockdown of platelet endothelial cell adhesion molecule 1 (PECAM-1) by siRNA decreases p-JNK and VCAM-1 levels. JNK is crucial in atherosclerosis by PECAM-1-mediated mechanisms altering NF-kB activity and VCAM-1 expression [51]. The LncRNAs lincRNA-p21 inhibition by siRNA causes neointima hyperplasia by modulating p53 activity in the carotid artery injury model [52]. Use of hypoxia-inducible factor 1-alpha (HIF-1α) siRNAs showed a reduction in HIF-1α mRNA and subsequently decreases the expression of the NLRP3 inflammasome, cytokine 1L-1β, and caspase-1, which subsequently decreases thrombus formation. Similarly, NLRP3 siRNAs indicated a significant reduction in prothrombin fragment 1 + 2 and D-dimer, which also alters thrombus formation. SiRNA is a very

Table 10.2 Regulation of gene expression through histone modification by ncRNAs in vascular smooth muscle cells (VSMCs).

Noncoding RNA	PTC (histone modification)	Gene influenced	Effect exerted
miR-125b	H3K9me3	MCP-1	Inhibits monocyte adherence
LincRNA-21	HAT/p53 interaction	p53	Inhibits VSMCs proliferation
FENDRR	HEK27Me3	–	Inhibits VSMCs proliferation
Giver	HEK27Me3	–	Inhibits VSMCs proliferation
miR-2861	HDAC5	RunX2	Induces VSMCs
TUG1	EZH2	F-actin	Induces VSMCs
miR-22	HDAC4	PCNA	Inhibits VSMCs proliferation

efficient and powerful technology whose application can be for a very specific target to study disease pathology and mechanistic approach. It has certain limitations such as a very short half-life in vivo and probe to enzymatic and thermal degradation (Table 10.2).

RNA-binding protein and CVD

The conversion of pre-mRNA to mature mRNA undergoes substantial processing and its associated RBPs play a crucial role in determining its destiny in cells [10]. RBPs are involved in coordinating various aspects of pathophysiological RNA processing such as stabilization or destabilization, splicing, cellular localization, and rate of translation in response to extracellular signals or stress [53,54].

There are approximately 729 RBPs encoded by the human genome [44] such as Quaking, HuR, Muscleblind, Tristetraprolin (TTP), monocyte chemoattractant protein-1 (MCP)-induced protein 1 (MCPIP1), Drosha and DGCR8, and SRSF1. It interacts with pre-mRNA at 5′-untranslated region (UTRs), 3′-UTRs along with intronic and exonic regions and influences the event it catalyzes [55].

TTP is an mRNA destabilizing protein that binds to AU-rich elements (ARE) and promotes their deadenylation and degradation. It is also known as ZFP36, G0S24, Nup475, and TIS11 having tandem CCCH zinc fingers [56]. It facilitates degradation of mRNA by binding to the components of its machinery for example mRNA decapping enzyme DCP1A and DCP2, the deadenylase CCR4-NOT transcription complex subunit 6 (CNOT6), exosome complex endonuclease RRP45, the 5−3′-exoribonuclease 1, an argonaute protein component of the RNAi silencing complex argonaute 2 (AGO2) [57,58]. Sato and colleagues have demonstrated that TTP is induced by iron deficiency and is essential for maintaining cardiac function by showing that its deletion in mice leads to cardiac dysfunction with low iron [59]. TTP is minimally expressed in the vascular endothelium of normal mice but significantly expressed in mice and humans with atherosclerosis [60].

Hu-antigen R (**HuR**) is a ubiquitous RBP, which is also known as Elavl1. It binds to ARE in 3′-UTR of target mRNA and prevents its RNAse-mediated degradation and increases mRNA stability against its degradation by competing for AREs position [61,62]. Researchers have demonstrated that HuR is crucial for B-cell antibody response, T-cell selection, hematopoietic progenitor cell survival, and chemotaxis in tissue-specific $HuR^{-/-}$ models [63,64]. HuR enhances the stability of β(2)-adrenergic receptor mRNAs, angiotensin receptors, vascular endothelial growth factor, toll-like receptor 4 (TLR-4), iNOS, cyclooxygenase (COX-2), tumor necrosis factor (TNF-α), and cell adhesion molecules such as ICAM and V-CAM mRNAs. It is well established that these molecules have a significant role in different cardiovascular diseases such as myocardial infarction, heart failure, and hypertension [65–68]. HuR binds to the 3′-UTR of sarco/endoplasmic reticulum Ca^{2+} pump (SERCA2b) leading to VSMC-mediated vasoconstriction [69]. The new and modern delivery methods [70] for RNA-based therapeutics [71] give the fascinating possibility of changing the transcriptome by modulating RBP expression or its activity, or targeting specific RBP-mediated events, making it easy to direct molecular pathways playing role in disease pathogenesis. Quaking (QKI) is an RBPs belonging to the Signal Transduction and Activator of RNA metabolism family and an hnRNP K homology type family. Generally, there are three main isoforms of QKI namely QKI-5, QKI-6, and QKI-7, which are produced by alternative splicing [72]. It functions in various aspects of RNA metabolism such as mRNA stability [73], alternative splicing [74], and miRNAs biogenesis [75]. QKI-5 and QKI-6 inhibit the FOXO1, proapoptotic transcription factor by binding to the 3′-UTR of FOXO1 mRNA and directly decreasing its stability following ischemic reperfusion in neonatal cardiomyocytes and adult rat hearts [76,77]. Quaking is expressed by endothelial cells and maintains endothelial barrier function in humans and mice by interacting with 3′UTRs of mature VE-cadherin and β-catenin mRNA [78]. QKI plays a pivotal role in human vasculature and is significantly expressed in neointimal VSMCs of human coronary restenotic lesions, but not in healthy vessels. The enhanced expression of QKI leads to interaction with Myocardin pre-mRNA causing alternative splicing changing the balance of myocardin protein isoform (Myocd_v3/Myocd_v1), which activates proliferative gene expression profiles in SMCs [79]. QKI RBPs' role in CVD suggests that QKI may be a possible early biomarker for its development and deserve further attention as a therapeutic target.

Summary

ncRNAs are complex regulatory components responsible for differential gene expression. The discovery of novel ncRNAs and their use in prognosis and diagnosis is highly appreciable. Various ncRNAs especially miRNAs have emerged as diagnostic biomarkers for several diseases, as they offer several advantages over other therapeutics such as small size, stability in the body fluids, and ease of delivery. Thus, ncRNAs have a promising future, but a lot more is yet to be done for elucidating their potential. For the implementation of ncRNAs in therapeutics, new strategies need to be worked out for better delivery, their stability, off-target effects, inefficient endocytosis by target cells, or the immunogenicity of delivery vehicles. Cardiovascular diseases will significantly improve the understanding of their involvement. Thus, the ncRNAs approach has a remarkable potential to identify new biomarkers that can be used in the early diagnosis of CVDs. This will lead to new avenues for targeted therapies along with improvisation in the treatment and outcome of the disease. However, more studies are warranted that will aid in deciphering the complex link connecting PTC mediated by ncRNAs and CVDs.

Acknowledgment

This work has been supported by DST-SERB.

References

[1] de Bruin RG, Rabelink TJ, van Zonneveld AJ, Van der Veer EP. Emerging roles for RNA-binding proteins as effectors and regulators of cardiovascular disease. Eur Heart J 2016;38(18):1380–8.
[2] Adams BD, Parsons C, Walker L, Zhang WC, Slack FJ. Targeting noncoding RNAs in disease. J Clin Invest 2017;127(3):761–71.
[3] Lee RC, Feinbaum RL, Ambros V. The *C. elegans* heterochronic gene lin-4 encodes small RNAs with antisense complementarity to lin-14. Cell 1993;75:843–54. https://doi.org/10.1016/0092-8674(93)90529-Y.
[4] Bauters C, Kumarswamy R, Holzmann A, Bretthauer J, Anker SD, Pinet F, et al. Circulating mir-133a and mir-423-5p fail as biomarkers for left ventricular remodeling after myocardial infarction. Int J Cardiol 2013; 168:1837–40.
[5] Grabmaier U, Clauss S, Gross L, Klier I, Franz WM, Steinbeck G, et al. Diagnostic and prognostic value of mir-1 and mir-29b on adverse ventricular remodeling after acute myocardial infarction–the SITAGRAMI-miR analysis. Int J Cardiol 2017;244:30–6.
[6] Colpaert RMW, Calore M. MicroRNAs in cardiac diseases. Cells 18 Jul. 2019;8(7):737. https://doi.org/10.3390/cells8070737.
[7] Sygitowicz G, Tomaniak M, Blaszczyk O, Koltowski L, Filipiak KJ, Sitkiewicz D. Circulating micro-ribonucleic acids mir-1, mir-21 and mir-208a in patients with symptomatic heart failure: preliminary results. Arch Cardiovasc Dis 2015;108:634–42.
[8] Rhodes CJ, Wharton J, Boon RA, Roexe T, Tsang H, Wojciak-Stothard B, et al. Reduced microRNA-150 is associated with poor survival in pulmonary arterial hypertension. Am J Respir Crit Care Med 2013;187: 294–302.
[9] Park DW, Lyu JH, Kim JS, Chin H, Bae YS, Baek SH. Role of JAK2-STAT3 in TLR2-mediated tissue factor expression. J Cell Biochem 2013b;114(6):1315–21.
[10] Gupta N, Sahu A, Prabhakar A, Chatterjee T, Tyagi T, Kumari B, et al. Activation of NLRP3 inflammasome complex potentiates venous thrombosis in response to hypoxia. Proc Natl Acad Sci USA 2017;114:4763–8. https://doi.org/10.1073/pnas.1620458114.
[11] Li B, Hong J, Hong M, Wang Y, Yu T, Zang S, Wu Q. piRNA-823 delivered by multiple myeloma-derived extracellular vesicles promoted tumorigenesis through re-educating endothelial cells in the tumor environment. Oncogene 2019;38(26):5227–38.
[12] Rajan KS, Velmurugan G, Gopal P, Ramprasath T, Babu DDV, Krithika S, et al. Abundant and altered expression of PIWI-interacting RNAs during cardiac hypertrophy. Heart Lung & Circ 2016. https://doi.org/10.1016/j.hlc.2016.02.015.
[13] Yang J, Xue FT, Li YY, Liu W, Zhang S. Exosomal piRNA sequencing reveals differences between heart failure and healthy patients. Eur Rev Med Pharmacol Sci 2018;22(22):7952–61. https://doi.org/10.26355/eurre v_201811_16423.
[14] Burd CE, Jeck WR, Liu Y, Sanoff HK, Wang Z, Sharpless NE. Expression of linear and novel circular forms of an INK4/ARFassociated non-coding RNA correlates with atherosclerosis risk. PLoS Genet 2010;6: e1001233. https://doi.org/10.1371/journal.pgen.1001233.
[15] Boeckel J-N, Jaé N, Heumüller AW, Chen W, Boon RA, Stellos K, et al. Identification and characterization of hypoxia-regulated endothelial circular RNA. Circ Res 2015;117:884–90. https://doi.org/10.1161/CIRCRESAHA.115.306319.

[16] Tang C-M, et al. CircRNA_000203 enhances the expression of fibrosis-associated genes by derepressing targets of miR-26b-5p, Col1a2 and CTGF, in cardiac fibroblasts. Sci Rep 2017;7:40342.

[17] Holdt LM, Beutner F, Scholz M, Gielen S, Gäbel G, Bergert H, et al. ANRIL expression is associated with atherosclerosis risk at chromosome 9p21. Arterioscler Thromb Vasc Biol 2010;30:620–7. https://doi.org/10.1161/ATVBAHA.109.196832.

[18] Yu S-Y, Dong B, Tang L, Zhou S-H. LncRNA MALAT1 sponges miR-133 to promote NLRP3 inflammasome expression in ischemia-reperfusion injured heart. Int J Cardiol 2017;254(2018):50. https://doi.org/10.1016/j.ijcard.10.071.

[19] Chen G, et al. Loss of long non-coding RNA CRRL promotes cardiomyocyte regeneration and improves cardiac repair by functioning as a competing endogenous RNA. J Mol Cell Cardiol 2018;122:152–64 [PubMed: 30125571].

[20] Zhang Y, et al. Long non-coding RNA CCRR controls cardiac conduction via regulating intercellular coupling. Nat Commun 2018;9(1):4176 [PubMed: 30301979] This is one of the first study implicating lncRNAs in the regulation of cardiac conduction.

[21] Li Q, Xie J, Li R, Shi J, Sun J, Gu R, et al. Overexpression of microRNA-99a attenuates heart remodelling and improves cardiac performance after myocardial infarction. J Cell Mol Med 2014;18:919–28.

[22] Li X, Zeng Z, Li Q, Xu Q, Xie J, Hao H, et al. Inhibition of microRNA-497 ameliorates anoxia/reoxygenation injury in cardiomyocytes by suppressing cell apoptosis and enhancing autophagy. Oncotarget 2015;6:18829–44.

[23] Yang H, Qin X, Wang H, Zhao X, Liu Y, Wo HT, et al. An in vivo miRNA delivery system for restoring infarcted myocardium. ACS Nano 2019;13(9):9880–94 [PubMed: 31149806].

[24] Bang C, Batkai S, Dangwal S, et al. Cardiac fibroblast-derived microRNA passenger strand-enriched exosomes mediate cardiomyocyte hypertrophy. J Clin Invest 2014;124:2136–46.

[25] Thum T, Catalucci D, Bauersachs J. MicroRNAs: novel regulators in cardiac development and disease. Cardiovasc Res 2008;79:562–70.

[26] Thum T, Galuppo P, Wolf C, et al. MicroRNAs in the human heart: a clue to fetal gene reprogramming in heart failure. Circulation 2007;116:258–67.

[27] Wahlquist C, Jeong D, Rojas-Munoz A, et al. Inhibition of miR-25 improves cardiac contractility in the failing heart. Nature 2014;508:531–5.

[28] Chen X, Ba Y, Ma L, Cai X, Yin Y, Wang K, et al. Characterization of microRNAs in serum: a novel class of biomarkers for diagnosis of cancer and other diseases. Cell Res 2008;18:997–1006.

[29] Mitchell PS, Parkin RK, Kroh EM, et al. Circulating microRNAs as stable blood-based markers for cancer detection. Proc Natl Acad Sci USA 2008;105:10513–8.

[30] Tsui NB, Ng EK, Lo YM. Stability of endogenous and added RNA in blood specimens, serum, and plasma. Clin Chem 2002;48:1647–53.

[31] Wang K, Zhang S, Marzolf B, Troisch P, Brightman A, Hu Z, et al. Circulating microRNAs, potential biomarkers for drug-induced liver injury. Proc Natl Acad Sci USA 2009;106:44027.

[32] Zampetaki A, Willeit P, Drozdov I, Kiechl S, Mayr M. Profiling of circulating microRNAs: from single biomarkers to re-wired networks. Cardiovasc Res 2012;93:555–62.

[33] Jakobi T, Czaja-Hasse LF, Reinhardt R, Dieterich C. Profiling and validation of the circular RNA repertoire in adult murine hearts. Dev Reprod Biol 2016;14:216–23. https://doi.org/10.1016/j.gpb.2016.02.003.

[34] Memczak S, Papavasileiou P, Peters O, Rajewsky N. Identification and characterization of circular RNAs as a new class of putative biomarkers in human blood. PLoS One 2015;10:e0141214.

[35] Li Y, et al. Circular RNA is enriched and stable in exosomes: a promising biomarker for cancer diagnosis. Cell Res 2015;25:981–4.

[36] Koh W, et al. Noninvasive in vivo monitoring of tissue-specific global gene expression in humans. Proc Natl Acad Sci USA 2014;111:7361–6.

[37] Wang K, et al. A circular RNA protects the heart from pathological hypertrophy and heart failure by targeting miR-223. Eur Heart J 2016;37:2602–11.
[38] Zeng Y, et al. A circular RNA binds to and activates AKT phosphorylation and nuclear localization reducing apoptosis and enhancing cardiac repair. Theranostics 2017;7:3842–55.
[39] Schulze M, Sommer A, Plötz S, Farrell M, Winner B, Grosch J, et al. Sporadic Parkinson's disease derived neuronal cells show disease-specific mRNA and small RNA signatures with abundant deregulation of piRNAs. Acta Neuropathol Commun 2018;6(1):58. https://doi.org/10.1186/s40478-018-0561-x.
[40] Qiu W, Guo X, Lin X, Yang Q, Zhang W, Zhang Y, et al. Transcriptome-wide piRNA profiling in human brains of Alzheimer's disease. Neurobiol Aging 2017;57:170–7. https://doi.org/10.1016/j.neurobiolaging.2017.05.020.
[41] Taborska E, Pasulka J, Malik R, Horvat F, Jenickova I, JelićMatošević Z, et al. Restricted and non-essential redundancy of RNAi and piRNA pathways in mouse oocytes. PLoS Genet 2019;15(12):e1008261. https://doi.org/10.1371/journal.pgen.1008261.
[42] Liu Y, Dou M, Song X, Dong Y, Liu S, Liu H, et al. The emerging role of the piRNA/piwi complex in cancer. Mol Cancer 2019;18(1):123. https://doi.org/10.1186/s12943-019-1052-9.
[43] Rajan KS, Velmurugan G, Pandi G, Ramasamy S. miRNA and piRNA mediated Akt pathway in heart: antisense expands to survive. Int J Biochem Cell Biol 2014;55:153–6. https://doi.org/10.1016/j.biocel.2014.09.001.
[44] Su J-F, Zhao F, Gao Z-W, Hou Y-J, Li Y-Y, Duan L-J, et al. piR-823 demonstrates tumor oncogenic activity in esophageal squamous cell carcinoma through DNA methylation induction via DNA methyltransferase 3B. Pathol Res Pract 2020. https://doi.org/10.1016/j.prp.2020.152848.
[45] Liu J, Yao L, Zhang M, Jiang J, Yang M, Wang Y. Downregulation of LncRNA-XIST inhibited development of non-small cell lung cancer by activating miR-335/SOD2/ROS signal pathway mediated pyroptotic cell death. Aging 2019;11(18).
[46] Cao B, Wang T, Qu Q, Kang T, Yang Q. Long noncoding RNA SNHG1 promotes neuroinflammation in Parkinson's disease via regulating miR-7/NLRP3 pathway. Neuroscience 2018. https://doi.org/10.1016/j.neuroscience.2018.07.019.
[47] Zhang P, Cao L, Zhou R, Yang X, Wu M. The lncRNA Neat1 promotes activation of inflammasomes in macrophages. Nat Commun 2019;10:1495. https://doi.org/10.1038/s41467-019-09482-6.
[48] Zang Y, et al. LncRNA FIRRE/NF-kB feedback loop contributes to OGD/R injury of cerebral microglial cells. Biochem Biophys Res Commun 2018. https://doi.org/10.1016/j.bbrc.2018.04.194.
[49] Chanchal S, Mishra A, Singh MK, Ashraf MZ. Understanding inflammatory responses in the manifestation of prothrombotic phenotypes. Front Cell Dev Biol February 14, 2020;8:73. https://doi.org/10.3389/fcell.2020.00073. PMID: 32117993; PMCID: PMC7033430.
[50] Cremer S, et al. Hematopoietic deficiency of the long non-coding RNA MALAT1 promotes atherosclerosis and plaque inflammation. Circulation 2018.
[51] Wang J, An FS, Zhang W, Gong L, Wei SJ, Qin WD, et al. Inhibition of c-Jun N-terminal kinase attenuates low shear stress-induced atherogenesis in apolipoprotein E-deficient mice. Mol Med 2011;17(9–10):990–9.
[52] Wu G, Cai J, Han Y, et al. LincRNA-p21 regulates neointima formation, vascular smooth muscle cell proliferation, apoptosis, and atherosclerosis by enhancing p53 activity. Circulation 2014;130:1452–65.
[53] Kishore S, Luber S, Zavolan M. Deciphering the role of RNA-binding proteins in the post-transcriptional control of gene expression. Brief Funct Genomics 2010;9:391–404.
[54] Alves LR, Goldenberg S. RNA-binding proteins related to stress response and differentiation in protozoa. World J Biol Chem 2016;7:78–87.
[55] Beckmann BM, Horos R, Fischer B, Castello A, Eichelbaum K, Alleaume AM, et al. The RNA binding proteomes from yeast to man harbour conserved enigmRBPs. Nat Commun 2015;6:10127.

[56] Ray D, Kazan H, Cook KB, et al. A compendium of RNA-binding motifs for decoding gene regulation. Nature 2013;499:172–7.

[57] Brooks SA, Blackshear PJ. Tristetraprolin (TTP): interactions with mRNA and proteins, and current thoughts on mechanisms of action. Biochim Biophys Acta 2013;1829:666–79.

[58] Blackshear PJ, Perera L. Phylogenetic distribution and evolution of the linked RNA-binding and NOT1-binding domains in the tristetraprolin family of tandem CCCH zinc finger proteins. J Interferon Cytokine Res 2014;34:297–306.

[59] Fabian MR, Frank F, Rouya C, Siddiqui N, Lai WS, Karetnikov A, et al. Structural basis for the recruitment of the human CCR4-NOT deadenylase complex by tristetraprolin. Nat Struct Mol Biol 2013;20:735–9.

[60] Sato T, Chang HC, Bayeva M, Shapiro JS, Ramos-Alonso L, Kouzu H, et al. mRNA-binding protein tristetraprolin is essential for cardiac response to iron deficiency by regulating mitochondrial function. Proc Natl Acad Sci USA July 3, 2018;115(27):E6291–300. https://doi.org/10.1073/pnas.1804701115. Epub 2018 Jun 18. PMID: 29915044; PMCID: PMC6142244.

[61] Zhang H, Taylor WR, Joseph G, Caracciolo V, Gonzales DM, Sidell N, et al. mRNA-binding protein ZFP36 is expressed in atherosclerotic lesions and reduces inflammation in aortic endothelial cells. Arterioscler Thromb Vasc Biol 2013;33:1212–20.

[62] Mukherjee N, et al. Integrative regulatory mapping indicates that the RNA-binding protein HuR couples pre-mRNA processing and mRNA stability. Mol Cell 2011;43:327–39.

[63] Lebedeva S, et al. Transcriptome-wide analysis of regulatory interactions of the RNA-binding protein HuR. Mol Cell 2011;43:340–52.

[64] Ghosh M, et al. Essential role of the RNA-binding protein HuR in progenitor cell survival in mice. J Clin Invest 2009;119:3530–43.

[65] Papadaki O, et al. Control of thymic T cell maturation, deletion and egress by the RNA-binding protein HuR. J Immunol 2009;182:6779–88.

[66] Misquitta CM, Iyer VR, Werstiuk ES, Grover AK. The role of 3′-untranslated region (3′-UTR) mediated mRNA stability in cardiovascular pathophysiology. Mol Cell Biochem 2001;224:53–67 [PubMed: 11693200].

[67] Osera C, Martindale JL, Amadio M, Kim J, Yang X, Moad CA, et al. Induction of VEGFA mRNA translation by $CoCl_2$ mediated by HuR. RNA Biol 2015;12:1121–30.

[68] Lin FY, Chen YH, Lin YW, Tsai JS, Chen JW, Wang HJ, et al. The role of human antigen R, an RNA-binding protein, in mediating the stabilization of toll-like receptor 4 mRNA induced by endotoxin: a novel mechanism involved in vascular inflammation. Arterioscler Thromb Vasc Biol 2006;26:2622–9.

[69] Cheng HS, Sivachandran N, Lau A, Boudreau E, Zhao JL, Baltimore D, et al. microRNA-146 represses endothelial activation by inhibiting pro-inflammatory pathways. EMBO Mol Med 2013;5:949–66.

[70] Misquitta CM, Chen T, Grover AK. Control of protein expression through mRNA stability in calcium signalling. Cell Calcium 2006;40:329–46.

[71] Yin H, Kanasty RL, Eltoukhy AA, Vegas AJ, Dorkin JR, Anderson DG. Non-viral vectors for gene-based therapy. Nat Rev Genet 2014;15:541–55.

[72] Aartsma-Rus A. New momentum for the field of oligonucleotide therapeutics. Mol Ther 2016;24:193–4. Darbelli L, Richard S. Emerging functions of the Quaking RNA-binding proteins and link to human diseases. Wiley Interdiscip Rev RNA. 2016 May;7(3):399-412. doi: 10.1002/wrna.1344. Epub 2016 Mar 14. PMID: 26991871.

[73] Thangaraj MP, Furber KL, Gan JK, et al. RNA-binding protein Quaking stabilizes Sirt2 mRNA during oligodendroglial differentiation. J Biol Chem 2017;292:5166–82.

[74] Hall MP, Nagel RJ, Fagg WS, et al. Quaking and PTB control overlapping splicing regulatory networks during muscle cell differentiation. RNA 2013;19:627–38.

[75] Wang Y, Vogel G, Yu Z, et al. The QKI-5 and QKI-6 RNA binding proteins regulate the expression of microRNA 7 in glial cells. Mol Cell Biol 2013;33:1233–43.
[76] Guo W, Shi X, Liu A, Yang G, Yu F, Zheng Q, et al. RNA binding protein QKI inhibits the ischemia/reperfusion-induced apoptosis in neonatal cardiomyocytes. Cell Physiol Biochem 2011;28:593–602.
[77] Yu F, Jin L, Yang G, Ji L, Wang F, Lu Z. Posttranscriptional repression of FOXO1 by QKI results in low levels of FOXO1 expression in breast cancer cells. Oncol Rep 2014;31:1459–65.
[78] De Bruin RG, van der Veer EP, Prins J, Lee DH, Dane MJ, Zhang H, et al. The RNA-binding protein quaking maintains endothelial barrier function and affects VE-cadherin and b-catenin protein expression. Sci Rep 2016;6.
[79] Van der Veer EP, de Bruin RG, Kraaijeveld AO, de Vries MR, Bot I, Pera T, et al. Quaking, an RNA-binding protein, is a critical regulator of vascular smooth muscle cell phenotype. Circ Res 2013;113:1065–75.

CHAPTER

microRNAs as critical regulators in heart development and diseases

11

Vibha Rani

Department of Biotechnology, Jaypee Institute of Information Technology, Noida, Uttar Pradesh, India

Introduction

Despite tremendous advancement in therapeutics for myocardial infarction the mortality rate due to cardiovascular diseases (CVDs) is inclining across the world [1]. Cardiovascular abnormalities including cardiac arrest, peripheral vascular diseases, and stroke are some of the prominent reasons for mortality and morbidity [2]. It is projected that approximately 422.7 million individuals are suffering from CVDs and 17.9 million people die worldwide from CVDs annually. India contributes almost one-fifth (18.6%) of the global CVD burden [3].

The heart is very sensitive to stress, which results in remodeling and pathological manifestation leading to pump failure and sudden death [4]. Cardiac myocytes are terminally differentiated cells with lost division potential once achieve a mature phenotype, however during certain stresses, these myocytes undergo hypertrophic conditions, defined as the enlarged size of terminally differentiated cardiomyocytes with no change in number [5]. Prolonger such hypertrophic enlargement leads to cardiac cell death and is one of the main causes of heart failure [6]. Left ventricular pathological hypertrophy has been a major self-governing prognosticator of CVD.

At a cellular level, various stresses lead to pathophysiological cardiac hypertrophy that causes hindrance in various interconnected intracellular signaling pathways resulting in cardiomyocytes stress-associated cardiac death [7]. The cardiac muscle cells try to compensate for the stress initially by enlarging the size of differentiated cardiomyoblasts through reprogramming of the fetal cardiac genes which are normally expressed during the development of the heart [8]. Many such genes include actin alpha1, alpha-actinin, titin, troponins, myosin-binding protein C, myosin-6 and 7, atrial natriuretic peptides, and brain natriuretic peptides are known to reexpress in cardiac hypertrophy [9]. Increased risk of cardiac failure occurs due to prolonged cardiac hypertrophy that transit cardiac cells from physiological to pathological hypertrophic conditions [4].

microRNAs: small RNA molecules with enormous potential

A unique class of small noncoding RNA regulators, microRNAs are 18–24 nucleotides long conserved small RNAs. MicroRNAs are known to regulate many biological functions by suppressing genes at the post-transcriptional level by mRNA inhibition or degradation. Most of the microRNAs are

Post-Transcriptional Gene Regulation in Human Disease, Volume 32. https://doi.org/10.1016/B978-0-323-91305-8.00005-3

originated from the intronic regions of protein-coding genes. These small RNAs are post-transcriptional regulators of gene expression. These single-stranded small nonmessenger RNAs can inhibit expression through partial or complete binding to the target mRNAs [10]. microRNAs are processed from their large hairpin precursors. microRNAs are known to deliver a specific layer of gene regulatory network with comprehensive series of numerous biological pathways via regulating mRNA expression levels and also by modifying the genes expression via transcription factors [11].

The single-stranded small nonmessenger RNAs are able to inhibit protein expression through partially or completely binding to the target mRNA. The microRNA genes occur in clusters and can be transcribed polycistronically and processed sequentially into pre-miRNA and then mature miRNA. In the human genome, there are more than 95% intronic regions; most of them are reported to encode the microRNAs [12]. To date, microRNAs are reported to regulate several cellular events such as cell cycle, proliferation, cell differentiation, cell/organs development, and apoptosis.

Journey of microRNAs research

MicroRNAs were discovered in the early 1980s during a study on loss of function mutation of heterochronic gene lin-4 that regulates the timing and sequence of larval development in *Caenorhabditis elegans* [13]. Later, lin-4 was found to regulate the expression of two heterochronic genes known as lin-14 and lin-28. Gain of function mutation within the 3′-untranslated region of lin-14 transcript resulted in high expression of the lin-14 protein, indicating that lin-4 regulates the lin-14 function not through RNA-protein interaction but by RNA–RNA interaction [14]. Since then, miRNA research has moved at a rapid pace due to advancements in miRNA identification techniques such as next-generation sequencing and the development of many robust, sensitive, and specific bioinformatics tools [15].

After 7 years of discovery of the first miRNA, lin-4 like noncoding small tiny regulatory gene let-7 was also found in the heterochronic pathway in *C. elegans* and is now accepted as critical development specific regulator from *C. elegans* to humans. Currently, there are 155 *C. elegans* miRNA genes in the miRBase database [16]. In 1998, small RNAs from humans, worms, and flies were cloned and found more than a hundred small noncoding regulatory genes approximately (60 in worms, 30 in humans, and 20 in Drosophila). The term microRNAs were used initially for these small regulatory genes but the function was unknown. With the development of miRBase, a database to provide complete information about the reported microRNAs by Griffith et al. in 2006 facilitated microRNAs annotation and research [17]. In 2005, the first study of microRNAs in the cardiac system was published where miR-1 was found to negatively regulate cardiac hypertrophy [18] Since then, many research studies have shown the importance of microRNAs in the cardiac system (Fig. 11.1). Fire and Mello received Nobel Prize in 2006 to discover RNAi, a silencing mechanism observed in humans, animals, and plants. At present, the human genome encodes approximately more than a thousand microRNAs, a few of them are characterized to regulate one-third of all genes, reported in the miRBase database [19]. The latest released miRBase contains ~38,589 mature miRNAs entries across all the species and submitted in the miRBase database, out of which ~2812 have been identified in humans. To date, the biological significance and functional characterization of a vast majority of annotated miRNAs remain unknown which need to be explored further for functional validation. MicroRNAs are highly conserved and predicted to be regulating the expression of >30% genes in humans thus involved in most of the biological processes including development and diseases [20].

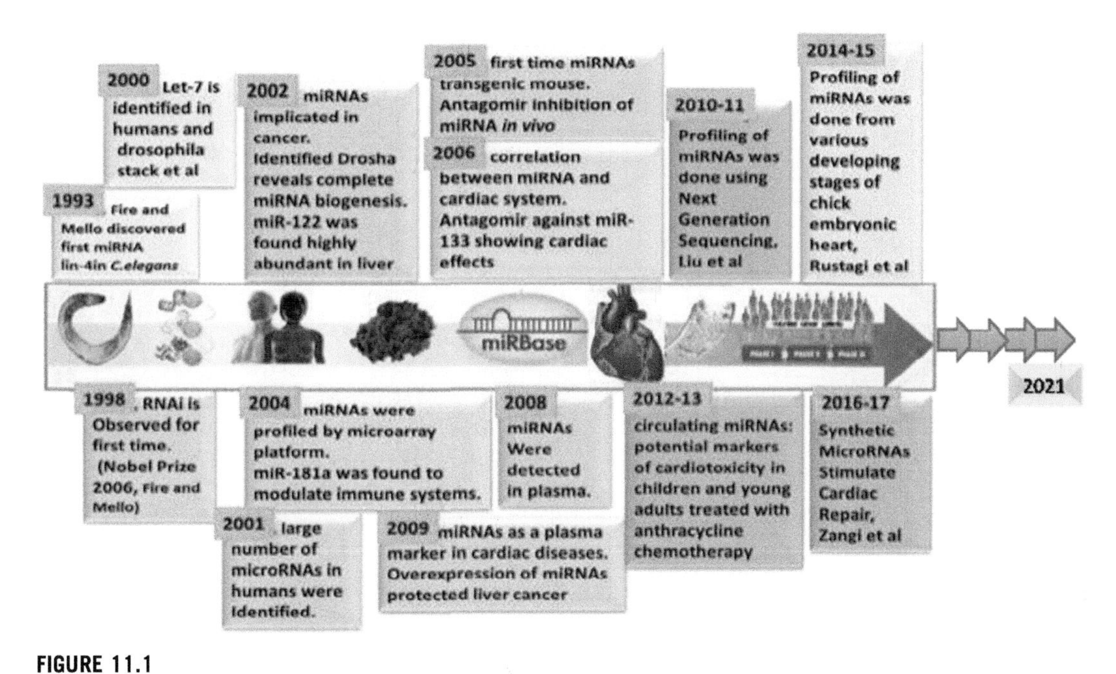

FIGURE 11.1

Brief history and milestones of miRNA research in heart.

MiRNAs get encoded from intronic, intergenic, and polycistronic regions of the genome that was considered as junk or nonprotein-coding portion. Nearly 80% of the genome undergoes transcription out of which only 1%–2% gets transcribed into coding RNA (mRNA) while the rest gives noncoding RNAs after transcription. Inside the nucleus, RNA polymerase II transcribes the genomic DNA into protein-coding (mRNA) as well as noncoding RNAs (primary miRNA, snRNA, snoRNAs, Piwi-interacting RNAs) (Fig. 11.2).

MicroRNA biogenesis

RNA polymerase II (Pol II) transcribed maximum microRNA genes and is capable of forming a characteristic hairpin structure by intramolecular pairing, which is called primary miRNA (pri-miRNA). microRNA. The size of the pri-miRNA ranges from 100 to 1000 nt. Primary RNA gets cleaved by the combination of enzymes called Drosha and Pasha into precursor miRNA (51–150 nucleotides). The precursor miRNAs form a hairpin loop structure by intramolecular pairing and get transported into the cytoplasm with the help of Ran/GTP/Exportin-5. Precursor miRNA gets sliced by another enzyme called Dicer (RNase III) into mature double-stranded miRNA (17–25 nucleotides) in the cytoplasm. Exportin-5 is a karyopherin protein that facilitates the nuclear export of double-stranded RNA binding proteins to the cytoplasm. RNase III Dicer and its cofactors (PACT and TRBP) recognize the pre-microRNAs and process the precursors into 18–24 nt long microRNA duplexes in the cytoplasm. RNase III Dicer and its cofactors (PACT and TRBP) recognize the pre-microRNAs and process the precursors into 18–24 nt long microRNA duplexes in the cytoplasm [21].

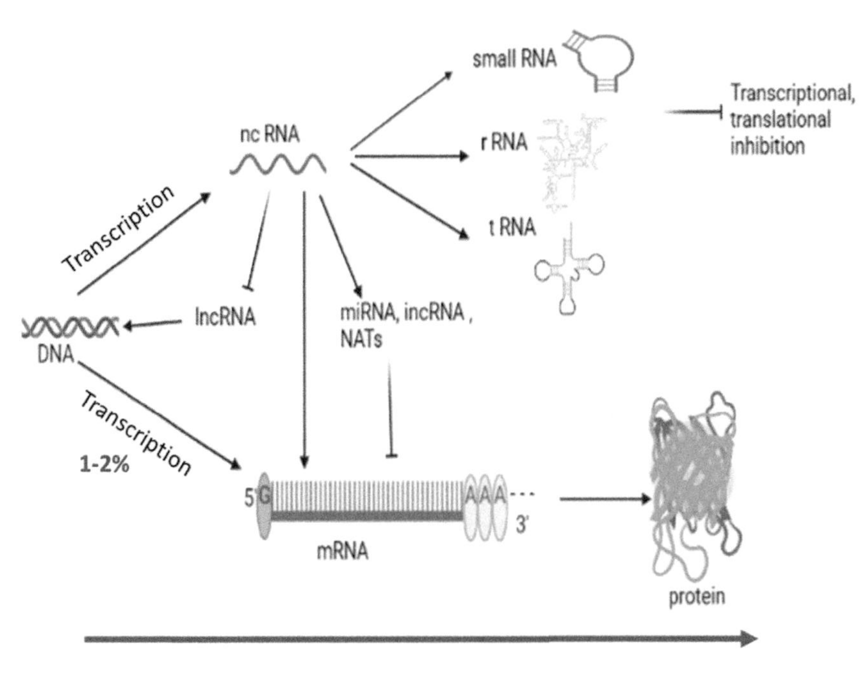

FIGURE 11.2

Diagrammatic representation depicting modified central dogma of life.

The double-stranded miRNA binds with RNA-induced silencing complex or (RISC) by attaching through Argonaute proteins where one strand called passenger gets degraded whereas the second strand called as guide remains intact. mature microRNA strand incorporates into RISC (the RNA-induced silencing complex) molecule. The seed region of mature microRNA-loaded RISC recognizes and binds to 3′UTR of mRNAs and inhibits mRNA expression and block translation (Fig. 11.3) [22]. The number of mismatches between microRNAs and target mRNA affects transcription repression. The most powerful feature of microRNA regulatory genes is the ability of a single microRNA to regulate more than hundreds of microRNAs. For example, miR-210 regulates multiple metabolic genes such as protein-tyrosine phosphatase1B (PTP1B), sirtuin3 (sirt3), and hepcidin (hamp). [23].

Dicer, a microRNA biogenesis component in cardiac development and disease

Any interruption in key factors involved in microRNA biogenesis described the status of microRNAs in fetal development. Abscission of dicer leads to deformities during the developmental process before gastrulation which leads to early embryonic death. For example, abnormalities in cardiac and neural development are seen in Ago2 null animals [24]. In various transgenic studies, the role of Ago2 and

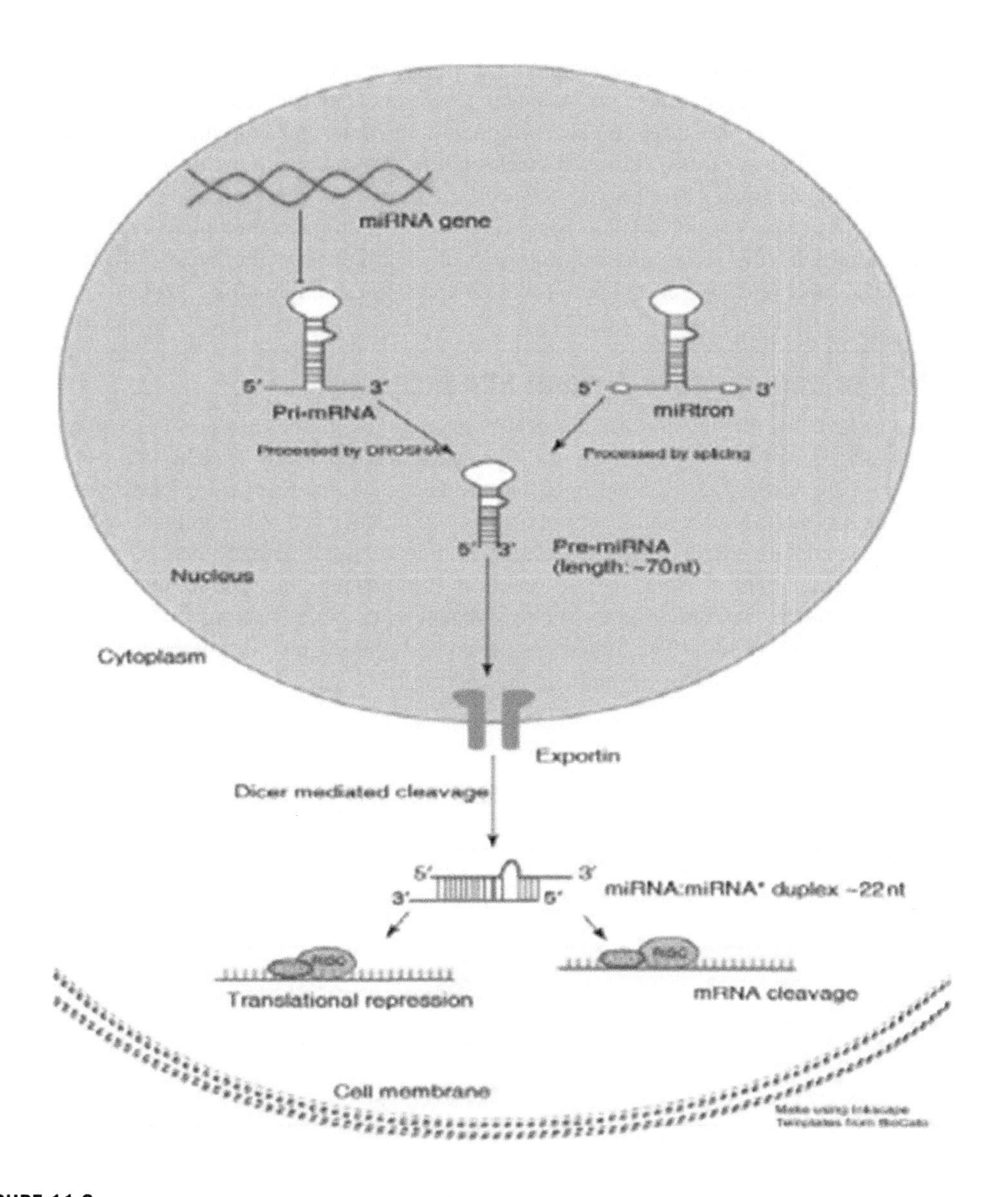

FIGURE 11.3

Diagrammatic representation of miRNA biogenesis as well as therapeutics.

Adopted from March 2011. Biochemical and Biophysical Research Communications 407(3):445-9.

dicer was found to be changed at different developmental levels. Floxed dicer transgenic mice and Nkx2.5-Cre were used to assess the involvement of Dicer in primary heart development and cause deformities in both myocardium formation and of heart outflow region along with missed chamber 11 septation. Due to the deletion of dicer, a decrease in contractility of the heart was observed along with, enlarged cardiomyopathy and fast progression toward cardiac death within 4 days after the birth [25]. To illuminate the microRNA function in postneonatal development and adult heart, α-HMC Cre/floxed dicer transgenic system was used after tamoxifen-induction in young and adult mice and observed that selective deletion of dicer in cardiac cells of 3-week-old mice caused death within 7 days. In 8-week-old mice, reduced expression of cardiomyocyte-specific microRNAs was seen along with changes in functions and morphology of heart cells including cell disarray, cardiac pathological hypertrophy, and reduced cardiac muscle contraction due to removal of myocardium specific Dicer [26]. In adults, a reduction in heart function was reported due to the reexpression of fetal cardiac genes.

MicroRNAs in cardiac development and diseases

Previous studies revealed the function of microRNAs in context with the regulation of sets of cardiac hypertrophic genes involved in by carrying out microarray analysis. miR-1, -23a, -195, -499, miR-199a, -133, -21, -29, -210, -212/132, and miR-208 are few well-characterized microRNAs reported in the literature in context with cardiac hypertrophy [27]. However, the complete cataloging of microRNAs involved in heart development and regulation of fetal cardiac gene programs is still missing and requires further research in this direction. Interestingly, the inconsistent expression of microRNAs is related to various diseases such as diabetes, diabetic cardiomyopathy, cancer, and cardiac failure (Fig. 11.4) [28]. The following section of the chapter discusses the microRNAs expressed in various stages of heart development and diseases.

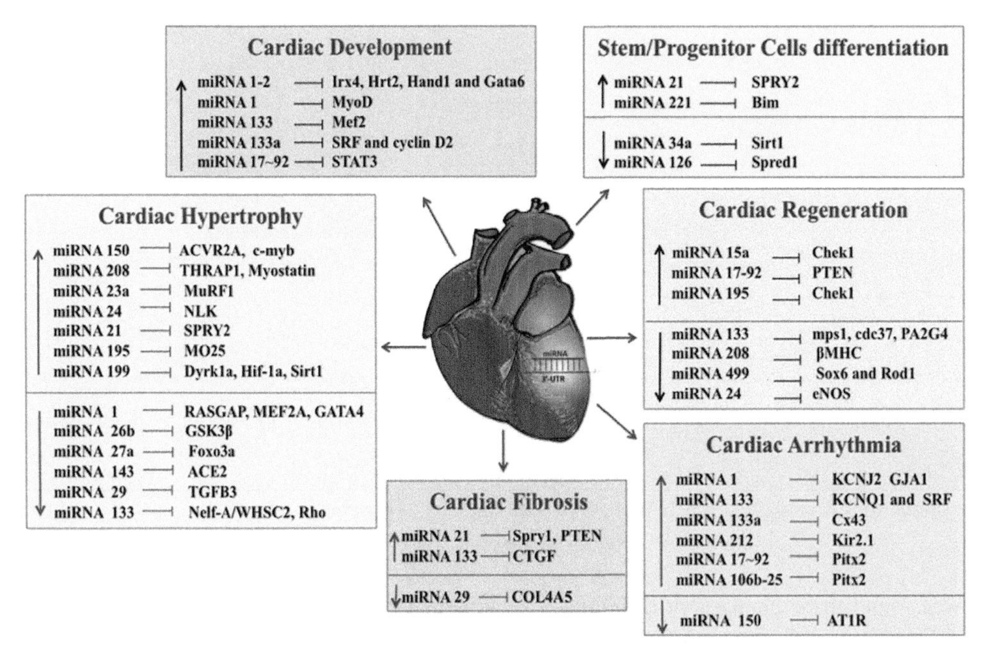

FIGURE 11.4

Schematic representation of specific microRNAs involved in cardiac development and diseases.

microRNA-1 and microRNA-133a family

The functions of the miR-1 and miR-133a microRNA family have been explored and characterized at various stages of development in animal models. These two miRNAs perform the function of promoting mesodermal differentiation and decreasing ectodermal and endodermal differentiation in embryonic stem cells [29,30]. Cardiomyocyte growth is affected by miR-1 by negative regulation of calcium-binding protein, calmodulin, and calmodulin-dependent nuclear import in activated-T cells. It also targets 3′-UTR of several cardiac-specific transcription factors such as GATA-binding protein 4 and myocyte enhancer factor 2A, Iroquois-class homeodomain protein-5, notch Delta ligand, the human homolog of the Notch Delta, Dll-1, and HAND2, which are crucial for cardiomyocyte growth and development. miR-1 also regulates important cardiac genes and potassium voltage-gated channel subfamily D member 2 (KCND2) [31–33]. There was an increase in the division of the cardiac cells and induced expression of genes related to smooth muscle due to decreased function of mutated miR-133a. It also regulates the propagation of cardiomyocytes by decreasing the expression of serum response factors (SRF) and cyclin-D2. Along with miR-1 and miR-133 also expressed during the initial stages of heart development and modulate Mef2c expression further controlling vascular differentiation and endothelial cell fate and plasticity.

microRNA-208 and microRNA-499 family

The intronic region of muscle-specific genes MYH-6, -7, and -7b, encode microRNA-miR-499, -208a and b respectively. These were expressed in a spatial-temporal manner in the embryonic and adult hearts. Overexpression of miR-208a in mice reported having cardiac-specific hypertrophic cell growth and abnormal heartbeats. The reduction of miR-208a in the knockout mouse model leads to a defective cardiac conduction system due to the defective expression of connexin 40 and homeodomain only protein [34]. Overexpression of miR-208a and miR-208b leads to post-transcriptional repression of myostatin, GATA4, and Trap-1, further indicating that this superfamily member is intricated in the synthesis of heart-specific myosin and activation of expression of cardiac genes. Excessive expression of microRNA-499 promoted early synthesis of clusters of precursors beating embryoid bodies that showed higher cardiac cell differentiation and also ablation of microRNA- 499 in mouse and human embryonic stem cell blocks cardiomyocyte differentiation [35].

microRNA-15 family

It consists of a group of microRNAs, that is, miR-15a, -15b, -16-1, -16-2, -195, and -497 sharing a common AGCAGC sequence in 5′ ends of cognate mature miRNAs. During postnatal development, upregulation of miR-15 was observed and miR-195 was found to be most abundant among all six members. Also, microRNA-195 shown to regulate key cell cycle regulators such as cell cycle checkpoint kinase 1 (Chek1), Baculoviral IAP repeat-containing protein 5 (Birc5), sperm-associated antigen 5 (Spag5), cell division cycle 2 homolog A (Cdc2a) nucleolar and spindle-associated protein 1 (Nusapl) [36]. It also reduces the cardiomyocytes proliferation and causes modification of regeneration of the heart in neonatal models.

microRNA-23 family

miR-23a was reported as prohypertrophic microRNA. Foxo3a gene was found to involve in the regulation of cardiac remodeling during various cell signaling cascades in human and other animals' hearts. The study was conducted to determine the regulation of the Foxo3a gene by miR23a. miR-23a-induced hypertrophic signal in models treated with phenylephrine and endothelin (ET-1). miR-23a antagomir reduced heart weight ratios and expression levels of Atrial natriuretic peptide (ANP), Brain natriuretic peptide (BNP), and Myosin heavy chain beta (β-MHC) hypertrophic markers, therefore miR-23a knockdown can attenuate the cardiac hypertrophic abnormalities. the expression of miR-23 is modulated by the transcription factor, Nuclear factor of activated T cells (NFATc3) [37]. Loss of function of upregulated miR-23a could attenuate hypertrophy because of upregulated expression of a direct antihypertensive target of miR-23a namely muscle-specific ring finger protein 1.

microRNA-212/132 families

Cardiomyocyte autophagy and cardiac hypertrophy are regulated by involving antihypertrophic transcription factors FoxO, which suppress prohypertrophic calcineurin/NFAT signaling. miR212/132 mimic was found to exert a maximum effect in cardiac hypertrophy in mice, resulting in cardiomyocytes size enlargement. Knockdown of miR-212/132 with their chemically modified AntagomiR in primary cardiomyocytes and h9c2 cell lines minimized the cell size, confirming prohypertrophic effects of microRNAs [38]. Mutation in both the miRNAs binding sites within the FoxO3 opposed the effects of miR212/132 by blocking the binding which indicates that FoxO3 is a direct target of both the miRNAs. inhibition of the endogenous miR-132 has shown promising results in a large animal model of nonischemic heart failure with cardiac fibrosis and hypertrophy via an antisense microRNA in preventing pathological hypertrophic cardiac conditions.

microRNA-29 family

The two human gene clusters encode microRNA-29 family members 29a, -29b, and -29c. All family members have a common seed region sequence and are predicted to target largely overlapping sets of genes. All three miR-29 isoforms showed reduced expression in infarcted myocardium in mice and miR-29b in human samples. Additionally, TGF-β-treated mouse cardiac fibroblasts showed less expression of all three members. In vivo and in vitro studies showed that loss of function of miR-29 family members modulates the expression of MMP genes involved with fibrogenesis during cardiac pressure overloading [39]. Matrix metalloproteinase 2 is a direct target of miR-29b, and it is downregulated in end-stage dilated cardiomyopathy. All such studies represent that miR-29 family members are important for regulating the synthesis and breakdown of extracellular matrix components during pathological cardiac hypertrophy. In vivo suppression of miR-29a and miR-29c is reported to have the protection of the heart from ischemia-reperfusion injury.

microRNA-30 family

miR-30 family contains microRNA-30a/b/c/d/e mature microRNA members, and all members share the same seed sequence regions and are projected to control mainly transcripts having overlapping sets. microRNA-30 family sequence is conserved among humans, drosophila, rat, mouse, nematodes, and zebrafish. In in vivo studies, members of miR-30 were found to modulate the activities of muscle-specific miR-206 and the expression of proteins. The gain of function studies was conducted for the miR-30 family in various models including primary cardiomyocyte hypoxic sets and myocardial infarction-induced mice. Higher expression of the miR-30 family was reported to suppress H2S synthesis and cystathionine-γlyase expression and then increased hypoxic cardiomyocyte injury. In the heart, CTGF expression is negatively regulated by microRNA-30c [40]. microRNA-30 and -133 family members are mostly expressed microRNAs in cardiomyocytes and their levels significantly decrease in the course of pathological hypertrophy, leading to overt heart failure. miR30a was found to regulate the myocardial fibrosis induced by atrial fibrillation (AF) via regulating snail 1, a direct miR-30a target, and is involved in AF-induced myocardial fibrosis [41]. Snail 1 protein is significantly present in myocardial tissues while miR-30a could inhibit the expression of snail 1.

microRNAs in cardiac hypertrophy

miR-1, which is considered as one of the most abundant cardiac microRNAs, plays an important role in hypertrophy as it has an opposite association between its expression and progression of hypertrophic conditions in a heart rodent model. Whereas a negative regulation was observed in neonatal myocyte culture. Various hypertrophic target genes of miR-1 include cyclin-dependent kinase, Ras GTPase activating protein (RasGAP), and Ras homolog present in fibronectin and brain [42]. It also suppresses hypertrophic conditions by inhibiting Hand2, Igf1, and twinfilin 1 and thereby control the cardiac events. The expression of ras-related protein-1A is reported to be controlled by miR-1 in angiotensin II and TAC-induced cardiac hypertrophic rodent models. miR-133 was also reported to target various antihypertrophic genes including signal transduction kinases Guanosine di phosphate (GDP)-Guanosine tri phosphate (GTP) exchange proteins, nuclear factors, negative elongation factor complexes, and cell division regulatory proteins [43]. Another microRNA, miR-145 is reported to reduce isoproterenol-induced hypertrophy by regulating the expression of the cardiac transcription factor, GATA 6 [44]. High glucose-induced cardiac hypertrophy is modulated by microRNA-150 controlling the expression of transcriptional coactivator protein, p300 [2]. In the TAC-induced hypertrophy rodent model, downregulation of miR-185 leads to antihypertrophic events by targeting crucial genes calcium/calmodulin-dependent protein kinase II (Camk2d), cytoplasmic and calcineurin-dependent 3 protein, Na^{+}/Ca^{+} exchangers, and NFAT expression thereby regulating calcium signaling. In neonatal rat myocytes, microRNA-185 is also reported to decrease in cellular size and reexpression of hypertrophic fetal gene markers thereby preventing endothelial 1-induced hypertrophy. miR-223 was also reported to reverse hypertrophy by targeting cardiac troponin I -interacting kinase (TNNI3k) whereas miR-26 targets GATA4 protein for hypertrophic reversal [45]. Downregulation of miR-34a is reported to upregulate the expression of the autophagy-related gene, Agt9a [46]. Many of the miRNAs are discussed for their prohypertrophic roles. miR-155 targets tumor suppressor protein p53-induced function of the nuclear protein (Tp53inp1) [47]. The miR-19a/b family targets the antihypertrophic genes atrogin 1 and E3 ubiquitin ligase, Murf1 which further activates NFAT signaling. Overexpression of miR-19a/b was reported to induce hypertrophic phenotypes in in vitro rat neonatal cardiomyocyte cells [48]. miR-199a is known to induce cardiac hypertrophy by impairing autophagy through activation of mTOR signaling [49]. miR was found to activate calcineurin/NFAT signaling by targeting NFAT kinase dual-specificity tyrosine-(Y)-phosphorylation-regulated kinase 1a (Dyrk1a).

miR-199b inhibitors have been found to suppress the development of cardiac hypertrophy and fibrosis in the heart failure model. miR-208a and miR-208b exerted prohypertrophic effects by targeting myostatin 2 and thyroid hormone-associated protein 1 [50]. Furthermore, targetting phosphatase and tensin homolog by mir-21 was reported to modulate AKT/mTOR pathway, in turn, modulated hypertrophy and cardiac fibrotic conditions. Interestingly, miR-21-3p employs cellular hypertrophy by targeting another set of proteins including Hdac8, Foxo3, Sorbs2, and Pdlim5. Another class of microRNAs including miR-212/132 and miR-23a activates calcineurin/NFAT signaling pathways and exerts prohypertrophy. miR-23a is reported to induce cardiac hypertrophy by regulating the expression of lysophosphatidic acid receptors [51]. Several evidences have been published regarding the involvement of miR-22 in controlling the expression of targeting Sirt1, Pten, and HDAC4 in the prevention of cardiac hypertrophic conditions [52]. miR-221 is another important microRNA having a significant role in the cellular differentiation and progression of hypertrophy in various cardiac pathological conditions. miR-27b was also reported to suppress the activity of prohypertrophic transcription factors. miR-30a is found to act on cardiac hypertrophy via downregulation of beclin 1 in angiotensin induced in vitro and in vivo experimental models. miR-30a-3p, the passenger strand is also reported to control the expression of Xbp1, a crucial protein involved in myocardial hypertrophy [53].

Involvement of microRNAs in various pathways leading to cardiac hypertrophy

A cardiac hypertrophy is regulated by multiple signaling pathways. A common feature of cardiac hypertrophic signaling is that there are multiple pathways operative with great cross-talk with one another to exert their effects. Cardiac hypertrophic responses are associated with multiple signal transduction pathways and microRNAs are reported to regulate multi-operative cardiac hypertrophy by targeting key signaling molecules. Table 11.1 summarizes the microRNAs and their target associated with signaling pathways.

MicroRNAs in ECM remodeling

Extracellular matrix (ECM) remodeling is one of the critical responses against various stress responses. Treatments for myocardium ischemia or myocardial infarction such as catheterization and revascularization fail to compensate for the loss of contractile myocardium resulting in permanent injury to the heart [54]. The damaged cells induce a number of cellular and ECM responses through inflammatory and immune system activation [55]. Induction of cytokines, interleukins, and reactive oxygen species further triggers the release/activation of matrix metalloproteinases (MMPs) responsible for ECM degradation [56]. At the onset, this process is considered as adaptive cardiac remodeling; however, the transformed cardiac fibroblast and myoblast alters the expression of MMPs enhancing ECM degradation leading to heart failure [57].

MMPs have emerged as critical therapeutic targets for a variety of diseases such as CVDs and cancer due to their involvement in ECM remodeling. MMPs constitute a large family of proteins (23 in humans) possessing high structural homology, unique functionality as well as bilateral role. The literature discusses many synthetic MMPs inhibitors including Batimastat, Marimastat, and PG-116800 that failed during the clinical trial to suppress left ventricle remodeling [58,59]. Since then,

Table 11.1 microRNA involved in cardiac hypertrophic pathways.

microRNAs	Targeted microRNAs	Signaling pathways	Upregulated/ downregulated
microRNA-1	FBLN2 TWF1 IGF1 FABP3	ECM, PI3K-Akt PPAR	Upregulated
microRNA-19a/b	ATROGIN1 MURF1	Calcineurin/NFAT PKC	Upregulated
microRNA21	PTEN	PI3K-AKT	Downregulated
microRNA-3p	HDAC8	AKT/GSK3β	Upregulated
microRNA-22	SIRT1 HDAC4 PTEN	AMPK AMPK PI3K-AKT	Upregulated
microRNA-23a	FOXO3A and LPA1	PI3K-AKT	Downregulated/ upregulated
microRNA-26b	GATA4	cGMP-PKG	Upregulated
microRNA-27b	PPARΓ	PPAR	Upregulated
microRNA-30a	BECLIN 1	Autophagy	Upregulated
microRNA-101	RAB1A	MAPKK	Upregulated
microRNA-133	RHOA CDC42 NELFA/WHSC2	cGMP-PKG/MAPK	Upregulated
microRNA-145	GATA6	cGMP-PKG	Upregulated
microRNA-150	P300	FoxO	Upregulated
microRNA-155	TP53INP1	p53	Upregulated
microRNA-185	CAMK D/NCX1 NFATC3	Calcium/cGMP-PKG	Upregulated
microRNA-199 a &b	GSK3B and DYRK1A	PI3K-AKT and calcineurin/NFAT	Upregulated
microRNA-212/132/221	FOXO3 and P27	PI3K-AKT	Upregulated
microRNA-223	TNI3K	Serine/threonine-protein kinase	Upregulated
microRNA-328	SERCA2A	cGMP-PKG	Upregulated
microRNA-350	MAPK-8/911/14	MAPK	Upregulated

several MMPs inhibitors based on small peptides, antibodies, and organic compounds were reported but none of them have successfully passed clinical trials [60]. Few reports suggested that single stimuli/compound such as cardiotoxic anticancer drugs or cardiomyopathy associated with the myocardial remodeling in different patients elicits differential expression of MMP-2 and -9 [61,62]. Therefore, there is a need to understand the gene expression profile of MMPs and their regulators under different pathological states. This will facilitate innovative strategies for the development of next-generation endogenous inhibitors with high specificity and selectivity. microRNAs have developed as crucial regulators of target expression due to their high specificity toward their target genes. Hence, miRNAs can be an ideal molecule for targeting, however; a question is raised that which tissue or pathway should be searched for identifying miRNAs that can regulate MMPs expression?

Previous reports suggest that developing fetal heart and an ischemic or hypertrophic heart share a common program known as fetal gene program. Reexpression of fetal genes such as β-MHC, ANF, TGF-β, and early responsive genes including c-myc and c-fos is a hallmark of adaptive cardiac remodeling [63], differentially expressed known/conserved miRNAs using in silico analysis and stem-loop RT-PCR. From chicken fetal cardiac miRNA libraries, miR-20 was identified as a potential 4 miRNA targeting the MMP-2 gene. Post-Myocardial infarction (MI), the cardiac cell undergoes hypertrophy and initiates an adaptive process through the remodeling of the extracellular matrix called cardiac remodeling. Several miRNAs including microRNA-21, -29, -199, and -208 have been described to be involved in hypertrophy and cardiac remodeling. Previous studies have suggested that the expression of both miR-21 and miR-29 elevates during hypertrophy or post-MI. Antagomir-mediated inhibition of miR-21 and miR-29 on rodent models has resulted in stabilization of the ECM turnover and prevention of collagen accumulation [64]. Although several miRNAs individually or in clusters have been found to regulate various signaling cascades implicated in hyperlipidemia, diabetes, cardiac contractile 20 functions, and cardiac hypertrophy, very few studies have shown the association between miRNAs and MMPs. In one of the reports, miR-21 has been found to regulate MMP-2 mRNA levels in cardiac fibroblasts through the PTEN pathway [65]. Likewise, miR-29b has been found to regulate the elevated expression of MMP-2/MMP-9 during atherosclerosis indicating that miRNAs possess the potential as MMPs inhibitors [66]. These reports indicate that post-transcriptional regulation of MMPs through miRNA is critical during atherosclerosis and myocardial ischemia. A thorough understanding of miRNA-mediated MMP regulation might provide novel tools to target imbalanced MMPs during adverse cardiac remodeling.

microRNAs as a therapeutic target

microRNAs have abundant perspectives to develop a novel class of therapeutics against human diseases such as diabetes, cardiovascular, cancer, and neurodegenerative disorders. For altering microRNA expression levels, two methods can be employed such as gene therapy and RNAi technologies. siRNAs, antimicroRNA oligonucleotides known as "AntagomiRs" are commonly used for microRNA inhibition [67]. Various types of AntagomiRs that are complementary to mature microRNA sequences have been utilized to suppress the expression of endogenous microRNAs in cell lines as well as in animal models [68]. There are three types of AntagomiRs with modified 2-OH residues of the ribose by (i) locked nucleic acid (LNA), (ii) 2′-O-methyl (2′-OMe), and (iii) 2′-O-methoxyethyl (2′-MOE). Most of the cardiovascular microRNAs mentioned in the previous section have been targeted by using specific AntagomiRs by using various in vivo delivery methods [69]. Another approach is microRNA replacement where a viral or plasmid vector (adenoviral vectors) comprising promoter region of polymerase II or III upstream to a small hairpin RNA expresses specific microRNAs, which is further cleaved into mature microRNA by dicer machinery before being processed by RISC [31]. This strategy is adopted to deliver miRNA mimics or inhibitors for normalizing the expression of up- and down-regulated microRNAs in the prevention of cardiovascular and other related diseases [32]. To manipulate the microRNAs-based regulatory mechanisms of several target sites, microRNA mimics are being designed to interact with homologous transcript sequences in many mRNAs. They can potentially target an uncontrolled set of messenger RNAs in an RNAi-like manner. Zhang and colleagues proposed the first time the concept of microRNAs mimics for regulating a specific group of target mRNAs with partially complementary with targets [33]. "microRNA decoy or Sponge" is another inventive approach, containing an RNAs segment that consists of 20 multiple tandems repeats binding sites that are complementary to the 7–20 nucleotide long seed region of mature microRNAs, leading to the reduction of active mRNAs [70]. The pathophysiological roles of adenoviral and lentiviral constructs of miR-21 and miR-133 sponges/decoy were constructed and tested against cardiac hypertrophy [71]. microRNAs are also able to circulate in the blood vesicles when they secrete from the cellular organ in the form of apoptotic bodies, microvesicles, or exosomes. Therefore, for delivering the microRNAs, circulating vessels can also be used [72]. miR-126 through apoptotic bodies derived from endothelial cells and miR-150 through microvesicles derived from THP-1 cells were successfully administered in an in vivo and confirmed miR-126 induced angioprotective effects in mice [73].

Limitation in microRNA-based therapy in the cardiovascular system

Although enough research has been conducted to explore the role of microRNAs in the regulation of progression and development of various diseases and also various delivery methods have been established still we do not see successive development at the translational front [74]. Clinical trials toward microRNA-based treatments to control viral replication in liver and tumor development have been initiated in the past. However, their results and outcomes are still awaited [75]. Any positive results will surely be energizing and motivating for improving the diagnostics and therapeutics against human pathologies [76]. For the delivery of microRNA mimics or antagomiRs, recent advancements and standards are generally embraced from MIR gene treatment. Feasible microRNA treatments depend on chemical alterations of properly-outlined microRNAs and antagomiRs particles to enhance stability and improve binding specificity [77]. Expression vector-based systems might be utilized as powerful transporters to carry small microRNAs to bind with the target in the specified destinations. Different liposomal and cationic polymeric conveyance vehicles are the most commonly used techniques for improved delivery of microRNAs [78]. High viability tissue-specific delivery strategy is desired to treat hypertrophic and cardiac failure conditions induced by various stress conditions. Our knowledge of microRNAs in signaling pathways is also emerging. However, more investigations for the identification of unique microRNAs in particular tissue and disease are still required for better understanding and benefits.

Conclusion

miRNAs have come up as a new therapeutic target during cardiac development and diseases. To attain a thorough understanding of molecular events occurring during adaptive and adverse cardiac remodeling, it is necessary identifying the genes and their regulators (miRNAs) differentially expressed during fetal cardiac development. Also, miRNAs, due to their specificity toward their target genes possess high therapeutic potential as endogenous MMP inhibitors. The genetic and epigenetic profile of developing fetal heart has provided crucial information regarding switching off cardiac metabolism, gene and miRNAs expression between developing heart, healthy adult and stressed heart. Studies should be designed in future to profile and characterize differentially conserved miRNAs from developing and diseased heart and the impact of modulated miRNAs on cardiac reprogramming.

References

[1] Chamnan P, Aekplakorn W. Cardiovascular risk assessment in developing world. In: Recent trends in cardiovascular risks, vol. 4; July 2017. p. 57–70.

[2] Duan Y, Zhou B, Su H, Liu Y, Du C. miR-150 regulates high glucose induced cardiomyocyte hypertrophy by targeting the transcriptional co-activator p300. Exp Cell Res 2013;319(3):173–84.

[3] Prabhakaran D, Singh K, Roth GA, Banerjee A, Pagidipati NJ, Huffman MD. Cardiovascular diseases in India compared with the United States. J Am Coll Cardiol July 2018;72:79–95.

[4] Juan EC, Jorge EJ. Cardiac hypertrophy: molecular and cellular events. Revi Esp Cardiol 2006;59(5): 473–86.

[5] Gaziano TA. Cardiovascular disease in the developing world and its costeffective management. Circulation 2005;112(23):3547–53.

[6] Van Rooij E, Sutherland LB, Liu N, Williams AH, McAnally J, Gerard RD, Richardson JA, Olson EN. A signature pattern of stress-responsive microRNAs that can evoke cardiac hypertrophy and heart failure. Proc Natl Acad Sci Unit States Am 2006;103(48):18255–60.

[7] Heineke J, Molkentin JD. Regulation of cardiac hypertrophy by intracellular signalling pathways. Nat Rev Mol Cell Biol 2006;7(8):589–600.

[8] Rosca MG, Hoppel CL. Mitochondria in cardiac hypertrophy and heart failure. J Mol Cell Cardiol 2013;55: 31–41.

[9] Kohli S, Rani V. Transcription factors in heart: promising therapeutic targets in cardiac hypertrophy. Curr Cardiol Rev 2011;7(4):262–71.

[10] Bartel DP. MicroRNAs: genomics, biogenesis, mechanism, and function. Cell 2004;116(2):281–97.

[11] Lee Y, Jeon K, Lee JT, Kim S, Kim VN. MicroRNA maturation: stepwise processing and subcellular localization. EMBO J 2002;21(17):4663–70.

[12] Hwang HW, Mendell JT. MicroRNAs in cell proliferation, cell death, and tumorigenesis. Br J Cancer 2006; 94(6):776–80.

[13] Lee RC, Feinbaum RL, Ambros V. The *C. elegans* heterochronic gene lin-4 encodes small RNAs with antisense complementarity to lin-14. Cell 1993;75(5):843–54.

[14] Ambros V. A hierarchy of regulatory genes controls a larva-to-adult developmental switch in *C. elegans*. Cell Apr. 1989;57:49–57.

[15] Saxena S, Gupta A, Shukla V, Rani V. Functional annotation of differentially expressed fetal cardiac microRNA targets: implication for microRNA-based cardiovascular therapeutics. 3 Biotech Dec. 2018;8: 494.

[16] Slack FJ, Basson M, Liu Z, Ambros V, Horvitz HR, Ruvkun G. The lin41 RBCC gene acts in the *C. elegans* heterochronic pathway between the let-7 regulatory RNA and the LIN-29 transcription factor. Mol Cell 2000;5(4):659–69.

[17] Griffiths-Jones S, Grocock RJ, Van Dongen S, Bateman A, Enright AJ. miRBase: microRNA sequences, targets and gene nomenclature. Nucleic Acids Res 2006;34(Suppl. 1_1):D140–4.

[18] Maatouk D, Harfe BD. MicroRNAs in development. Sci World J 2006;6:1828–40.

[19] Shivdasani RA. MicroRNAs: regulators of gene expression and cell differentiation. Blood 2006;108(12): 3646–53.

[20] Carthew RW, Sontheimer EJ. Origins and mechanisms of miRNAs and siRNAs. Cell 2009;136:642–55.

[21] Lau PW, MacRae IJ. The molecular machines that mediate microRNA maturation. J Cell Mol Med 2009; 13(1):54–60.

[22] He L, Hannon GJ. MicroRNAs: small RNAs with a big role in gene regulation. Nat Rev Genet 2004;5(7): 522–31.

[23] Noman MZ, Buart S, Romero P, Ketari S, Janji B, Mari B, Mami-Chouaib F, Chouaib S. Hypoxia-inducible miR-210 regulates the susceptibility of tumor cells to lysis by cytotoxic T cells. Cancer Res 2012;72(18): 4629−41.

[24] Liu J, Carmell MA, Rivas FV, Marsden CG, Thomson JM, Song JJ, Hammond SM, Joshua-Tor L, Hannon GJ. Argonaute2 is the catalytic engine of mammalian RNAi. Science 2004;305(5689):1437−41.

[25] Chen JF, Murchison EP, Tang R, Callis TE, Tatsuguchi M, Deng Z, Rojas M, Hammond SM, Schneider MD, Selzman CH, Meissner G. Targeted deletion of Dicer in the heart leads to dilated cardiomyopathy and heart failure. Proc Natl Acad Sci Unit States Am 2008;105(6):2111−6.

[26] Da Costa Martins PA, Bourajjaj M, Gladka M, Kortland M, van Oort RJ, Pinto YM, Molkentin JD, de Windt LJ. Conditional dicer gene deletion in the postnatal myocardium provokes spontaneous cardiac remodelling. Circulation 2008;118(15):1567−76.

[27] Da Costa Martins PA, De Windt LJ. MicroRNAs in control of cardiac hypertrophy. Cardiovasc Res 2012; 93(4):563−72.

[28] Reid G, Kirschner MB, van Zandwijk N. Circulating microRNAs: association with disease and potential use as biomarkers. Crit Rev Oncol Hematol 2011;80(2):193−208.

[29] Ikeda S, He A, Kong SW, Lu J, Bejar R, Bodyak N, Lee KH, Ma Q, Kang PM, Golub TR, Pu WT. MicroRNA-1 negatively regulates expression of the hypertrophy-associated calmodulin and Mef2a genes. Mol Cell Biol 2009;29(8):2193−204.

[30] Liu N, Bezprozvannaya S, Williams AH, Qi X, Richardson JA, Bassel-Duby R, Olson EN. MicroRNA-133a regulates cardiomyocyte proliferation and suppresses smooth muscle gene expression in the heart. Genes Dev 2008;22(23):3242−54.

[31] Nandi SS, Mishra PK. Harnessing fetal and adult genetic reprograming for therapy of heart disease. J Nat Sci 2015;1:4−6.

[32] Zhou J, Dong X, Zhou Q, Wang H, Qian Y, Tian W, Ma D, Li X. microRNA expression profiling of heart tissue during fetal development. Int J Mol Med 2014;33:1250−60.

[33] Glazov EA, Cottee PA, Barris WC, Moore RJ, Dalrymple BP, Tizard ML. A microRNA catalog of the developing chicken embryo identified by a deep sequencing approach. Genome Res 2008;18:957−64.

[34] Callis TE, Pandya K, Seok HY, Tang RH, Tatsuguchi M, Huang ZP, Chen JF, Deng Z, Gunn B, Shumate J, Willis MS. MicroRNA-208a is a regulator of cardiac hypertrophy and conduction in mice. J Clin Invest 2009;119(9):2772.

[35] Sluijter JP, van Mil A, van Vliet P, Metz CH, Liu J, Doevendans PA, Goumans MJ. MicroRNA-1 and -499 regulate differentiation and proliferation in human-derived cardiomyocyte progenitor cells. Arterioscler Thromb Vasc Biol 2010;30:859−68.

[36] Porrello ER, Johnson BA, Aurora AB, Simpson E, Nam YJ, Matkovich SJ, Dorn GW, van Rooij E, Olson EN. miR-15 family regulates postnatal mitotic arrest of cardiomyocytes. Circ Res 2011;109: 670−9.

[37] Köck J, Kreher S, Lehmann K, Riedel R, Bardua M, Lischke T, Jargosch M, Haftmann C, Bendfeldt H, Hatam F, Mashreghi MF. Nuclear factor of activated T cells regulates the expression of interleukin-4 in Th2 cells in an all-ornone fashion. J Biol Chem 2014;289(39):26752−61.

[38] Ucar A, Gupta SK, Fiedler J, Erikci E, Kardasinski M, Batkai S, Dangwal S, Kumarswamy R, Bang C, Holzmann A, Remke J. The miRNA-212/132 family regulates both cardiac hypertrophy and cardiomyocyte autophagy. Nat Commun 2012;3:1078.

[39] Kriegel AJ, Liu Y, Fang Y, Ding X, Liang M. The miR-29 family: genomics, cell biology, and relevance to renal and cardiovascular injury. Physiol Genom 2012;44(4):237−44.

[40] Guess MG, Barthel KKB, Harrison BC, Leinwand L. miR-30 family microRNAs regulate myogenic differentiation and provide negative feedback on the microRNA pathway. PLoS One 2015;10(2): e0118229.

[41] Shen Y, Shen Z, Miao L, Xin X, Lin S, Zhu Y, Guo W, Zhu YZ. miRNA-30 family inhibition protects against cardiac ischemic injury by regulating cystathionine-γ-lyase expression. Antioxid Redox Signal 2015;22(3): 224–40.
[42] Care A, Catalucci D, Felicetti F, Bonci D, Addario A, Gallo P, Bang ML, Segnalini P, Gu Y, Dalton ND, Elia L. MicroRNA-133 controls cardiac hypertrophy. Nat Med 2007;13(5):613–8.
[43] Uisters RF, Tijsen AJ, Schroen B, Leenders JJ, Lentink V, van der Made I, Herias V, van Leeuwen RE, Schellings MW, Barenbrug P, Maessen JG. miR-133 and miR-30 regulate connective tissue growth factor. Circ Res 2009;104(2):170–8.
[44] Li R, Yan G, Zhang Q, Jiang Y, Sun H, Hu Y, Sun J, Xu B. miR-145 inhibits isoproterenol-induced cardiomyocyte hypertrophy by targeting the expression and localization of GATA6. FEBS Lett 2013;587(12): 1754–61.
[45] Wang YS, Zhou J, Hong K, Cheng XS, Li YG. MicroRNA-223 displays a protective role against cardiomyocyte hypertrophy by targeting cardiac troponin I-interacting kinase. Cell Physiol Biochem 2015; 35(4):1546–56.
[46] Huang J, Sun W, Huang H, Ye J, Pan W, Zhong Y, Cheng C, You X, Liu B, Xiong L, Liu S. miR-34a modulates angiotensin II-induced myocardial hypertrophy by direct inhibition of ATG9A expression and autophagic activity. PLoS One 2014;9(4):e94382.
[47] He W, Huang H, Xie Q, Wang Z, Fan Y, Kong B, Huang D, Xiao Y. MiR-155 knockout in fibroblasts improves cardiac remodeling by targeting tumor protein p53-inducible nuclear protein 1. J Cardiovasc Pharmacol Therapeut 2016;21(4):423–35.
[48] Song DW, Ryu JY, Kim JO, Kwon EJ, Kim do H. The miR-19a/b family positively regulates cardiomyocyte hypertrophy by targeting atrogin-1 and MuRF1. Biochem J 2014;457(1):151–62.
[49] Li z, song y, liu l, hou n, an x, zhan d, li y, zhou l, li p, Yu L, Xia J. miR-199a impairs autophagy and induces cardiac hypertrophy through mTOR activation. Cell Death Differ 2015;1.
[50] Yan M, Chen C, Gong W, Yin Z, Zhou L, Chaugai S, Wang DW. miR-21-3p regulates cardiac hypertrophic response by targeting histone deacetylase-8. Cardiovasc Res 2015;105(3):340–52.
[51] Yang J, Nie Y, Wang F, Hou J, Cong X, Hu S, Chen X. Reciprocal regulation of miR-23a and lysophosphatidic acid receptor signaling in cardiomyocyte hypertrophy. Biochim Biophys Acta 2013;1831(8): 1386–94.
[52] Huang ZP, Chen J, Seok HY, Zhang Z, Kataoka M, Hu X, Wang DZ. MicroRNA-22 regulates cardiac hypertrophy and remodeling in response to stress. Circ Res 2013;112:1234–43.
[53] Wang J, Liew OW, Richards AM, Chen YT. Overview of microRNAs in cardiac hypertrophy, fibrosis, and apoptosis. Int J Mol Sci 2016;17(5):749–70.
[54] Gabisonia K, Prosdocimo G, Aquaro GD, Carlucci L, Zentilin L, Secco I, Braga H, Ali L, Gorgodze N, Bernini F, Burchielli S. MicroRNA therapy stimulates uncontrolled cardiac repair after myocardial infarction in pigs. Nature 2019;569:418–22.
[55] Zamilpa R, Lindsey ML. Extracellular matrix turnover and signaling during cardiac remodeling following MI: causes and consequences. J Mol Cell Cardiol 2010;48:558–63.
[56] Kondapalli SR, Galimudi MS, Srilatha RK, Sahu G, Hanumanth SK. Matrix metalloproteinases in coronary artery disease: a review. J Life Sci 2012;4:55–8.
[57] Kehat I, Molkentin JD. Molecular pathways underlying cardiac remodeling during pathophysiological stimulation. Circulation 2010;122(25):2727–35.
[58] Hudson MP, Armstrong PW, Ruzyllo W, Brum J, Cusmano L. Effects of selective matrix metalloproteinase inhibitor (PG-116800) to prevent ventricular remodeling after myocardial infarction: results of the PREMIER (Prevention of Myocardial Infarction Early Remodeling) trial. J Am Coll Cardiol 2006;48:15–20.
[59] Spinale FG, Villarreal F. Targeting matrix metalloproteinases in heart disease: lessons from endogenous inhibitors. Biochem Pharmacol 2014;90:7–15.

[60] Gaffney J, Solomonov I, Zehorai E, Sagi I. Multilevel regulation of matrix metalloproteinases in tissue homeostasis indicates their molecular specificity in vivo. Matrix Biol 2015;44–46:191–9.
[61] Sagi I, Talmi-Frank D, Arkadash V, Papo N, Mohan V. Matrix metalloproteinase protein inhibitors: highlighting a new beginning for metalloproteinases in medicine. Met Med 2016;3:31–47.
[62] Spallarossa P, Altieri P, Garibaldi S, Ghigliotti G, Barisione C, Manca V, Fabbi P, Ballestrero A, Brunelli C, Barsotti A. Matrix metalloproteinase-2 and -9 are induced differently by doxorubicin in H9c2 cells: the role of MAP kinases and NAD(P)H oxidase. Cardiovasc Res 2006;69:736–45.
[63] Nandi SS, Mishra PK. Harnessing fetal and adult genetic reprograming for therapy of heart disease. J Nat Sci 2015;1:e71.
[64] Maegdefessel L, Azuma J, Toh R, Deng A, Merk DR, Raiesdana A, Leeper NJ, Raaz U, Schoelmerich AM, McConnell MV, Dalman RL, Spin JM, Tsao PS. MicroRNA-21 blocks abdominal aortic aneurysm development and nicotine-augmented expansion. Sci Transl Med February 22, 2012;4(122):122ra22.
[65] Tu Y, Wan L, Bu L, Zhao D, Dong D, Huang T, Cheng Z, Shen B. MicroRNA-22 downregulation by atorvastatin in a mouse model of cardiac 147 hypertrophy: a new mechanism for antihypertrophic intervention. Cell Physiol Biochem 2013;31(6):997–1008.
[66] Abonnenc M, Nabeebaccus AA, Mayr U, Barallobre-Barreiro J, Dong X, Cuello F, Sur S, Drozdov I, Langley SR, Lu R, Stathopoulou K. Extracellular matrix secretion by cardiac fibroblasts: role of microRNA-29b and microRNA30c. Circ Res 2013;113:1138–47.
[67] Currie S, Smith GL. Enhanced phosphorylation of phospholamban and downregulation of sarco/endoplasmic reticulum Ca2+ ATPase type 2 (SERCA 2) in cardiac sarcoplasmic reticulum from rabbits with heart failure. Cardiovasc Res 1999;41:135–46.
[68] Hang CT, Yang J, Han P, Cheng HL, Shang C, Ashley E, Zhou B, Chang CP. Chromatin regulation by Brg1 underlies heart muscle development and disease. Nature 2010;466:62–7.
[69] Dietz JR. Mechanisms of atrial natriuretic peptide secretion from the atrium. Cardiovasc Res 2005;68:8–17.
[70] Lagana A, Veneziano D, Spata T, Tang R, Zhu H, Mohler PJ, Kilic A. Identification of general and heart-specific miRNAs in sheep (*Ovis aries*). PLoS One 2015;10:e0143313.
[71] Varkonyi-Gasic E, Wu R, Wood M, Walton EF, Hellens RP. Protocol: a highly sensitive RT-PCR method for detection and quantification of microRNAs. Plant Methods 2017;3:1–12.
[72] Yang LH, Wang SL, Tang LL, Liu B, Wang LL, Wang ZY, Zhou MT, Chen BC. Universal stem-loop primer method for screening and quantification of microRNA. PLoS One 2014;9:e115293.
[73] Kozomara A, Griffiths-Jones S. miRBase: annotating high confidence microRNAs using deep sequencing data. Nucleic Acids Res 2013;42:68–73.
[74] Gao D, Middleton R, Rasko JE, Ritchie W. miREval 2.0: a web tool for simple microRNA prediction in genome sequences. Bioinformatics 2013;29:3225–6.
[75] Jiang P, Wu H, Wang W, Ma W, Sun X, Lu Z. MiPred: classification of real and pseudo microRNA precursors using random forest prediction model with combined features. Nucleic Acids Res 2007;35:339–44.
[76] Gruber AR, Lorenz R, Bernhart SH, Neuböck R, Hofacker IL. The vienna RNA websuite. Nucleic Acids Res 2008;36:70–4.
[77] Wong N, Wang X. miRDB: an online resource for microRNA target prediction and functional annotations. Nucleic Acids Res 2014;43:146–52.
[78] Agarwal V, Bell GW, Nam JW, Bartel DP. Predicting effective microRNA target sites in mammalian mRNAs. Elife 2015;4:e05005.

CHAPTER

MicroRNA regulation in autoimmune diseases

12

Ishani Dasgupta

Horae Gene Therapy Center, Department of Pediatrics, University of Massachusetts Chan Medical School, Worcester, MA, United States

Introduction

MicroRNAs (miRNAs) are conserved 20–22 nucleotide long noncoding RNA molecules that regulate gene expression at the post-transcriptional level by targeting the 3′-untranslated region of specific messenger RNAs [1]. Robust sequencing technologies have led to the discovery of more than 2000 human miRNAs, and the field is constantly evolving [2]. miRNAs have emerged as crucial regulators of several biological processes, which include cell proliferation, differentiation, maturation, and apoptosis. Increasing evidence has validated their role in the pathogenesis of many autoimmune diseases [3–5]. Endoribonucleases, like dicer, that control miRNA biogenesis have been associated with inflammatory responses and autoimmunity, thereby affirming the contribution of miRNA dysregulation in the etiology and progression of autoimmune diseases. Here, we discuss how miRNAs regulate autoimmunity by affecting the activation and differentiation of immune cells and provide a comprehensive understanding of their differential expression in various autoimmune diseases. Elucidation of the mechanism of miRNA regulation in the disease pathogenesis and associated inflammatory responses will be insightful for their potential use as molecular diagnostic and novel therapeutic markers for treating autoimmune disorders.

miRNA in autoimmunity

Autoimmune diseases are multifactorial, chronic disorders resulting in sustained abnormal immunological response against the body's healthy cells and tissues and loss of immunological tolerance to self-antigens that lead to overproduction of autoreactive immune cells and autoantibodies [6]. The innate and adaptive immune responses are associated with the pathogenesis of autoimmune disorders [7]. Both environmental factors and genetic anomalies contribute to autoimmunity [8–10]. Epigenetic regulation, which includes histone modification, DNA methylation, and miRNA expression, can act as a connecting link between environmental and genetic aberrancies [11]. Several studies have shown the involvement of miRNAs in a multitude of autoimmune diseases, affecting their onset, diagnosis, and progression [12–14]. miRNAs are associated with the cytogenesis and function of dendritic cells, B

Post-Transcriptional Gene Regulation in Human Disease, Volume 32. https://doi.org/10.1016/B978-0-323-91305-8.00002-8

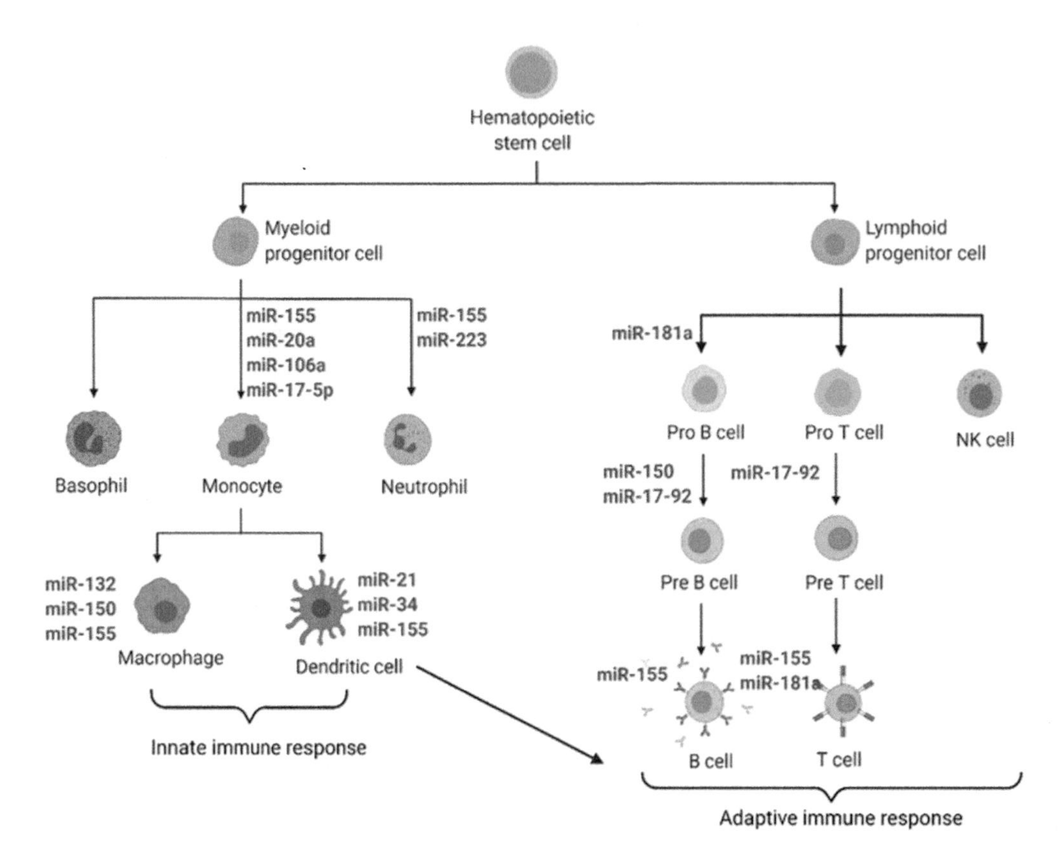

FIGURE 12.1 miRNAs involved in immune cell development.

Schematic illustration showing some of the key miRNAs implicated in immune cell differentiation and development. Immune cells of both the innate and adaptive immune components show high levels of miR-155, miR-21, miR-34, etc. that lead to the secretion of inflammatory immune response. Dysregulation of these miRNAs results in aberrant inflammatory cytokine release, which is one of the hallmarks of autoimmune diseases.

Created with Biorender.com.

cells, and T cells, thus regulating both innate and adaptive immunity. Moreover, miRNA dysregulation perturbs the immune system homeostasis by triggering the secretion of inflammatory cytokines and overproduction of autoantibodies [12,15]. In this section, we will briefly discuss the involvement of miRNAs in immune system regulation (Fig. 12.1).

miRNAs in the innate immune system

The initial defense against pathogens is accomplished by the innate immune system, comprising of a repertoire of immune cells, including monocytes, natural killer (NK) cells, and granulocytes. miRNAs play a pivotal role in the development of these immune cells via Toll-like receptor (TLR) and pattern recognition receptor signaling cascades.

miRNA in monocytes, macrophages, and dendritic cells

Monocytes and macrophages that function to phagocytose pathogens and secretion of inflammatory cytokines are the most common innate immune cells. miRNA expression profile in monocytes identified upregulation of miR-132, miR-146a, and miR-155. miR-146a is regulated by the NF-κB pathway, which drives the expression of TNF receptor-associated factor 6 (TRAF6) and IL-receptor-associated kinase 1 (IRAK1). TLR stimulation results in the activation of these proteins. miR-146a acts as a negative regulator by targeting TRAF6 and IRAK1 in macrophages, thereby disrupting the balanced immune responses [16]. Another miRNA regulated by the NF-κB pathway that plays a key role in the activation of macrophages is miR-125b [17]. High levels of miRNA-150 have also been reported in an autoimmune encephalitis mouse model, correlating with the activation of macrophages [18]. miR-155, one of the most characterized miRNAs, is induced in bone marrow-derived macrophages [19,20]. Elevated levels of miR-155 target the negative regulators of SH2-containing inositol5′-phosphatase 1 (SHIP1) and suppressor of cytokine signaling 1 (SOCS1), leading to subsequent activation of IFN-γ and cytokine release [21]. Apart from monocytes, miR-155, miR-21, and miR-34 trigger differentiation of dendritic cells, enabling IL-6, IL-1, and IFN-γ secretion [22–24]. These miRNAs are crucial for innate immune response in dendritic cells.

miRNAs in granulocytes and NK cells

miRNAs play an essential role in the cytogenesis of granulocytes, modulating their proliferation, activation, and differentiation. miR-223 functions as a modulator of granulocyte activation by targeting the myocyte enhancer factor 2C [25]. Both miR-21 and miR-9 are implicated in the activation of granulocytes and neutrophils respectively [26,27]. miR-155 is essential for granulocyte proliferation by targeting SHIP1. In vivo studies demonstrated an upregulation of miR-155 is related to myeloproliferative disease, reinforcing its importance in maintaining innate immune system balance [21,28]. Published reports reveal that invariant natural killer T cells were abrogated in the absence of Dicer, indicating the role of miRNAs in NK cell development and immunoregulatory cytokine production [29,30]. Overall, these studies illustrate that miRNA dysregulation is responsible for autoimmune disease onset and progression.

miRNAs in the adaptive immune system

The primary components of the adaptive immune system include B and T cells, that mount a specific and targeted immune response upon antigenic stimulation. Defects in miRNA biogenesis in lymphocyte progenitors impede the development of functional B and T cells, thus emphasizing the role of miRNAs as essential regulators of the adaptive immune system. Depletion of Dicer in lymphocytes and early B-cell progenitors impeded normal T-cell development, the proliferation of T helper cells, and obstructed B-cell development [31–33].

miRNAs in T-cell development

CD4+ T cells differentiate into different subtypes that include T regulatory cells (Tregs), T helper cells, Th1, Th2, and Th17, capable of secreting different cytokines in response to pathogens. Specific miRNAs are associated with T-cell differentiation, development, and regulation. Immature T lymphocytes differentiate into mature T cells and exit into the peripheral lymph tissues from the thymus after rounds of positive and negative selection. miR-181a modulates the selection process by regulating T-cell receptor signaling during T-cell development [34]. miRNAs that belong to the miR-17-92 repress Bcl-2-interacting mediator of cell death and phosphatase and tension homolog (PTEN) to regulate T-cell survival. Unwanted T-cell clones are removed by apoptosis during T-cell development and these miRNAs enable the survival of the desired T-cell population [35]. After mature T cells recognize their respective antigens, they differentiate into particular subtypes and exhibit distinct cytokine responses. T-cell differentiation is regulated by distinct miRNAs that target specific transcription factors. For example, miR-29 enabled Th1 cell differentiation by downregulating the Th2 transcriptional T-bet [36] respectively. Th1 cell expansion is also promoted by miR-19b and miR-17 by enabling interaction between PTEN and TGF-β receptors [37]. miR-155 is related to Th2 activation and expression of respective cytokines [38]. Th17 development is also driven by miRNA regulation. miR-326 facilitated Th17 cell development by repressing ETS1, an inhibitor of Th17 differentiation. miR-326 upregulation that correlated with Th17 cell activation has been implicated in severe experimental autoimmune encephalomyelitis (EAE) disease and multiple sclerosis [39]. Th17 cell development miR-183c overexpression triggered the activation of Th17 cells and the concomitant production of inflammatory cytokines, such as IL-17, IL-23, and IL-1β [40]. miR-17-92 cluster aids the development of Th2 [41] and T follicular helper cells that function to promote affinity antibody maturation in B cells [42]. miRNAs also act as negative modulators of T-cell differentiation. miR-23 targets SOCS1 to attenuate Th1 differentiation. Similarly, miR146a and miR-155 also suppress Th1 differentiation and production of inflammatory cytokines, such as IFN-γ and IL-2 [43,44]. miR-27 and mir-24 prohibit Th2 cell activation and differentiation by targeting STAT5 and inhibiting IL4 production [45]. miR-210 overexpression promotes Th1 and Th17 but suppresses Th2 differentiation, triggering the production of IFN-γ, IL-17, and IL-2 [46]. Tregs that are involved in the maintenance of immune tolerance and neutralization of inflammatory immune response have key roles in autoimmune diseases. miRNA regulation is associated with the development and differentiation of Tregs. miR-155, miR-142, and miR-17-92 cluster augments Treg expansion and maintains their inhibitory function [47–49].

miRNAs in B-cell development

B cells secrete antibodies to combat pathogens and are vital players of the adaptive immune system. They differentiate into plasma cells, responsible for antibody production, and memory cells, that enable the reactivation of the immune system when the body encounters a similar antigen. Over the years, there is increasing evidence that supports the involvement of miRNA in B-cell activation and development [50]. miR-17-92 cluster is associated with B-cell development [51]. miR-155, miR-148a, and miR-150 promote B-cell differentiation into antibody-secreting plasma cells [52–54]. miR-223 and miR-217 stimulate memory B-cell proliferation and differentiation [55]. Besides positive regulation, miRNAs act as negative regulators of B-cell development. miR-34a attenuates B cell and plasma cell development by blocking the Fox P1 pathway [56]. miR-125b impedes plasma cell differentiation by targeting the B lymphocyte-induced maturation protein 1 and miR-181b negatively affects memory B-cell differentiation [57,58].

Overall, these studies reflect the contribution of miRNA regulation in the activation of these immune cells implicated in autoimmune diseases. Further understanding of how these miRNAs orchestrate the development of immune cells and how their dysregulation results in autoimmune disorders are needed. The role of miRNAs associated with different autoimmune diseases is discussed in the following section.

miRNA in autoimmune diseases

miRNA in systemic lupus erythematosus

Systemic lupus erythematosus (SLE) is a chronic inflammatory autoimmune disease, with diverse clinical manifestations ranging from joint pains, glomerulonephritis, skin rashes and lesions, photosensitivity, and neurological disorders [59,60]. SLE symptoms vary across individuals and affect multiple organs such as the skin, joints, kidneys, central nervous system, lungs, and hematopoietic system. The reported prevalence of SLE in the United States is 20–150 cases per 100,000 individuals [61–64]. This disease is characterized by the loss of immune tolerance, aberrant activation of lymphocytes and dendritic cells, and the presence of autoantibodies against nuclear antigens, ribonucleoproteins, and phospholipids [65]. A combination of several factors, such as environmental, genetic, epigenetic regulation, hormonal is implicated in SLE. miRNA dysregulation is associated with disease progression in SLE [66], and ongoing research indicates their potential as biomarkers and therapeutic targets for treating this autoimmune disorder.

The involvement of miRNAs in SLE pathogenesis garnered attention after researchers reported anti-Su autoantibodies, found in lupus patients that recognize enzymes in miRNA pathways [67,68]. Evaluation of miRNA expression in peripheral blood mononuclear cells from SLE patients using microarray revealed 16 differentially expressed miRNAs. Of these 16, seven miRNAs (miR-196a, miR-17-5p, miR-409-3p, miR-141, miR-383, miR-112, and miR-184) were downregulated and the remaining 9 (miR-189, miR-61, miR-78, miR-21, miR-142-3p, miR-342, miR-299-3p, miR-198, and miR-298) were upregulated [69]. miRNA expression profile in kidney biopsies from lupus nephritis patients also uncovered 66 miRNAs that were differentially expressed [70]. A recent high throughput sequencing study identified a group of novel miRNA candidates, namely miR-221-5p, miR-380-3p, miR-556-5p, miR-758-3p, and miR-3074-3 in the plasma of lupus nephritis patients that act as potential diagnostic markers [71]. Besides nephritis, miRNA dysregulation is implicated in other SLE-related organ damage, such as cardiovascular disorders. For example, miR-125b, miR-101, and miR-375 were all differentially expressed in SLE patients with atherosclerosis [72]. These findings highlight the role of miRNAs in SLE pathogenesis and further investigation of the underlying mechanisms of miRNA dysregulation that contribute to SLE is critical.

miR-146a has been reported to be downregulated in lupus patients, this downregulation has been attributed to genetic mutations in SLE as well persistent activation of Type 1 Interferon (IFN) [73–75]. Type 1 IFN is among the most prominent proinflammatory cytokines in SLE [76]. miR-146a inhibited Type I IFN and STAT1, a key player in the IFN signaling pathway [75,77]. SLE is characterized by elevated levels of Type 1 IFN secretion in plasmacytoid dendritic cells (pDCs). miR-146a was shown to suppress inflammatory responses and impair the ability of pDCs to trigger the proliferation of CD4+ T cells and T helper cell accumulation [78,79]. miRNA-146a was able to reduce autoantibody production and conferred increased resistance to hemorrhagic pulmonary capillaritis in an SLE mouse model [80]. This immunomodulatory role of miRNA-146a renders it a promising therapeutic target. Other miRNAs such as miR-155 that mediate inflammatory responses, type 1 IFN signaling [81] and regulate adaptive immunity are associated with SLE. Inhibition of miR-155 reduces autoantibody levels, kidney inflammation, renal damage, and lupus-associated pulmonary hemorrhage in SLE mouse models [82–87]. miR-21 and miR-148a have been documented to be upregulated in SLE. DNA-methyltransferase-1 levels, one of the key epigenetic players attributed to DNA hypomethylation, in SLE are reduced by miR-21 and miR-148a [88]. Besides enhancing DNA hypomethylation, miR-21 also targets programmed cell death protein 4, implicated in T-cell proliferation, B-cell maturation, and IL10 production in SLE [89]. Other examples of miRs implicated in SLE pathogenesis include miR-23b and miR-125a, that is downregulated in SLE mouse models, and administering these miRs rescues autoimmune disease phenotypes [90,91]. Published reports identified miR-183 as a potential therapy for SLE since miR-183 successfully decreases immune responses and enhances survival in mouse models [92]. Other miRNAs implicated in lupus nephritis are miR-873 [93], miR148a-3p [94], and injection of miR-130b and miR-182-5p, which ameliorates renal impairments observed in lupus nephritis mouse models [95,96]. Most of the available treatment options rely heavily on steroids and immunosuppressive drugs that exert unwanted side effects. Since miRNAs can serve as biomarkers and therapeutic targets for SLE, miRNA-based therapeutics hold immense potential to treat this autoimmune disease.

miRNA in rheumatoid arthritis

Rheumatoid arthritis (RA) is a chronic systemic autoimmune disease that manifests as inflammation of joints, leading to impaired functionality and musculoskeletal defects [97]. Inhibiting inflammatory cytokines such as IL-6, IL-1β, and TNF-α implicated in RA pathogenesis can improve the disease phenotype in some patients [98,99]. A plethora of epigenetic modifications, primarily miRNA regulation is involved in RA pathogenesis [100–102]. Changes in miRNA expression resulted in the secretion of inflammatory cytokines, associated with RA [103].

Elevated levels of miR-146a [104,105], miR-155 [106,107], miR-16, miR-103a, miR-132 [108], miR-145 [109], miR221 [110], and miR-301a [111] and reduced levels of miR-21 [112], miR-125b [113], and miR-548a [114] were observed in the peripheral blood mononuclear cells (PBMCs) of RA patients. miR-146a was increased in peripheral Th17 and decreased in T regs, favoring the secretion of proinflammatory cytokines, resulting in disease aggravation [115,116]. Rheumatoid arthritis synovial fibroblasts (RASFs) exhibit increased production of matrix metalloproteases that erode the bone matrix, and proinflammatory cytokines [115]. Some of the prominent miRNAs that show enhanced expression, augmenting the inflammatory milieu in RA patients include miR-146a, miR-155, miR-203, miR-221 among several others [117,118]. RASF-derived exosomes also show elevations in these cytokines upon TNF-α stimulation [119]. Besides triggering inflammatory responses, these miRNAs play a role in reducing osteoclastogenesis. While miRNAs like miR-145, whose upregulation in PBMCs and synovial tissue in RA patients enhances osteoclastogenesis [109], elevated levels of miR-146a prevent the differentiation to osteoclasts, thereby preventing bone loss [117,120]. Besides upregulation in PBMCs and RASFs, miR-155 is overexpressed in macrophages and synovial tissues of RA patients, contributing to inflammatory cytokine production. Abrogation of miR-155 attenuates RA development [106,121], emphasizing its role as a promising therapeutic target. While most miRNAs, for example, miR-10 [122], miR-17 [123], miR-18 [124], miR-27a [125], miR-203 [126], miR-221 [127], miR-522 [128], and miR-663 [129] that are dysregulated in RASFs contribute to MMP production, resulting in bone erosion, increased expression of miR-155 in RASFs reduces MMP production, thus modulating downstream bone damage [107,118]. miR-223 is also upregulated in synovial macrophages [130] and T cells [131] derived from RA patients. Despite enhancing the inflammatory cytokine secretion, miR-223 negatively regulates osteoclast formation and subsequent bone erosion [132]. In addition to the elevated miRNAs, downregulation of miRNAs, such as miR-27a, miR30a, and miR-708 in synovial tissues of RA patients results in RASF invasion and disintegration of the cartilage matrix [125,133,134]. Further, reduced expression of miR-22 and miR10a-5p triggered inflammation and joint defects in individuals suffering from RA [135,136]. Alterations in the miRNA expression in the plasma and serum of RA patients reveal upregulation of several miRNAs, among which the prominent ones, such as miR-125a, miR-146a, miR-24, and miR-155 can be used as biomarkers for diagnosing RA. However, further studies on how miRNA levels in serum can differentiate between RA and other arthritis-related diseases need to be conducted.

miRNAs can be used as a potential therapeutic target for RA treatment. Intra-articular administration of miR-124 and miR-140 mimics reduces disease severity by inhibiting RASF proliferation [137,138]. Peritoneal injection of miR-26a, miR-150 mimics and anti-miR to silence miR-223 ameliorates the disease phenotype by suppressing inflammatory cytokine production, synovial hypertrophy, osteoclast formation, and bone erosion [139—141]. In vivo administration of miR-106b inhibitor, miR34a antagonist, miR-146a and miR-708 mimic in murine disease models reduce cytokine production, synovial hyperplasia, bone loss, thus alleviating the disease phenotype [120,133,142,143]. Some in vitro studies also report treating RASFs with miR-26b mimic, miR-124a, miR-451, and miR-573 to inhibit RASF proliferation and cytokine production [144—147]. Although these miRNAs have yielded encouraging results in disease models, investigating their effectiveness in clinical trials for RA needs to be determined.

miRNA in multiple sclerosis

Several studies have shown the involvement of miRNA dysregulation in multiple sclerosis (MS), thereby underscoring the importance of examining their role for early diagnosis and miRNA-based therapeutic interventions for MS. MS is a chronic autoimmune disease of the central nervous system (CNS), characterized by demyelination, inflammation, axonal damage, and neurological degeneration [148]. Improper functioning of the blood—brain barrier (BBB) and glial cells, and inflammatory immune responses that alter Th1/Th17 and Tregs are implicated in MS. miRNA dysregulation is associated with Th17 polarization and MS pathogenesis. Upregulation of miR-326, a Th17-related miRNA triggers IL-17 production and differentiation of Th17 cells by targeting Ets-1, a negative regulator of Th17 differentiation, thereby contributing to MS and EAE [39,149]. Silencing miR-326 reduces Th17 cells and causes mild EAE in vivo, rendering miR-326 as a potential therapeutic target for treating MS [39]. miR-155 has also been shown to be upregulated in MS and EAE, thereby inducing Th1 and Th17 differentiation, inflammatory responses, and impairing BBB function [150,151]. miR-155 knockdown results in significant amelioration of the inflammatory response in CNS and treatment with anti-miR-155 successfully inhibits EAE progression [152,153]. miR34a, mir-132, and mir-212 cluster, miR-128, miR-27 b, and miR-340 are similarly upregulated in MS patients and trigger Th1 and Th17 cell development [153—157]. Downregulating the expression of these miRNAs inhibits EAE progression. miR-223, a miRNA upregulated in MS patients is associated with macrophage differentiation. Thus, inhibiting the expression of miR-223 in combination with miR-15a and miR-16, known to regulate the NF-κB signaling pathway, prevents macrophage activation, thereby stalling disease progression [158]. Compared to healthy controls, 10 miRNAs are reported to be differentially expressed in B cells. miR-17-92 and miR-106b-25, involved in inflammatory responses are upregulated in B cells in MS patients [159]. Downregulation of miR-320a in B cells contributes to

MS by inducing reorganization of the extracellular matrix via matrix metalloproteinase (MMP-9) overexpression [160]. In addition to immune cells, miRNA dysregulation in glial cells such as astrocytes, microglia, and oligodendrocytes also contributes to inflammatory demyelination in MS. miR-873 overexpression in astrocytes triggers inflammatory cytokines, such as IL-6, MIP-1, and MCP-2 in EAE mice [161]. miRNA profiling of astrocytes has revealed the presence of all 10miRNAs that were highly upregulated in MS lesions including miR-155, miR-326, and miR-34a [162]. By promoting oligodendrocyte differentiation, miR-219 and miR-338 are involved in myelination [163]. miR-125a-3p, reported to inhibit oligodendrocyte cell maturation, is associated with MS pathogenesis [164].

The stability of miRNAs in serum and plasma renders them suitable diagnostic biomarkers for MS. Several studies have been conducted on test samples obtained from serum, plasma, cerebrospinal fluid (CSF), PBMCs, and T cells [165]. miR-145 levels, elevated in peripheral blood, are known to have the highest expression in plasma derived from MS patients [166,167]. miR-320a and miR-27a-3p levels are high in patient serum samples leading to disease progression. Plasma samples obtained from patients show upregulation of miR-22, miR-422-a, miR-572, miR-614, miR-648, and miR-1826 [168]. miRNA expression also correlates with the different stages of the disease. For example, evaluation of miR-326, miR-15b-5p, miR-23a-3p, miR-30b-5p, miR-223-3p, miR-374a-5p, miR-342-3p, miR-432-5p, miR-433-3p, and miR-485-3p expression can be used to distinguish between relapsing and remitting stages of MS [169,170]. RNA ChIP data from screening 297 miRNAs revealed 46 miRNAs that exhibited differential expression in PBMCs from MS patients. miR-26a and miR-145 are predominantly increased in patient-derived PBMCs [167,171]. Besides serum, plasma, and PBMCs, miRNA screening studies on CSF in MS patients indicate the following miRNAs to be specifically expressed in MS patient samples, namely miR-181-c, miR-633, and miR-922 [172]. These have the potential to serve as a biomarker for the relapsing-remitting disease condition. miR-150, miR-219, miR-21, and miR-146-a and b in CSF can all be putative biomarkers of MS [82,173–176]. The increasing repertoire of candidate biomarkers of MS paves the way for advancements in miRNA-based treatment.

The use of RNA interference technology based on modified oligonucleotide analogs that abrogate miRNAs implicated in disease development holds tremendous promise for treating MS. Use of extracellular vesicles to deliver miR-219a-5p triggers remyelination in MS patients, which is encouraging [177]. However, further improvements in the nanocarrier design for higher biocompatibility, stability, and specificity need to be done to achieve robust therapeutic efficiency.

miRNA in inflammatory bowel disease

Inflammatory bowel disease (IBD) is mainly categorized into two major chronic inflammatory conditions, ulcerative colitis (UC) and Crohn's disease (CD). Most of the available treatment options target the inflammatory pathways to abrogate the disease. A complete understanding of the underlying molecular pathogenesis of the disease is imperative for designing therapeutic approaches to treat IBD.

The first miRNA profiling studies of colon biopsies from patients suffering from active UC reported 11 differentially expressed miRNAs. miR-16, miR-21, miR-23a, miR-24, miR-29a, miR-126, miR-195 and let-7f were upregulated, and miR-192, miR-375, and miR-422b were downregulated [178]. Since then, additional miRNAs that shown abnormal expression, including overexpression of miR-21, miR-155, miR-31, miR-126, miR-7, miR-29a, miR-29b, miR-20b, miR-223, miR-150, and downregulation of miR-215, miR-320a, miR-346 have been identified in UC patients [179–184].

These studies suggest that miRNAs can be used as biomarkers to distinguish between IBD subtypes. Comparable to UC, microarray analysis has revealed differential miRNA expression profiles in individuals with active CD. miR-362-3p, miR-199a-5p, miR-340, miR-532-3p, miR-16, miR-23a, miR-29a, miR-106a, miR-107, miR-126, miR-191 are upregulated in peripheral blood from active CD patients [185,186]. miRNA expression in the peripheral blood can be used to differentiate between active UC and CD [186]. Published reports demonstrate some overlap between tissue and peripheral blood expression of miRNAs in both UC and CD. Here, we highlight some of the prominent miRNAs implicated in IBD. Patients suffering from active UC and ileal CD show elevated levels of miR-21 [178,187]. miR-21 is implicated in fibrosis, immune responses, and epithelial barrier function. In a colitis mouse model, miR-21 knockout mice were characterized by lower intestinal permeability, CD3 T cells, and epithelial cell death [125]. High levels of miR-21 are present in effector T cells, reinforcing their role in mounting adaptive immune responses [188]. Another widely studied miRNA implicated in inflammatory disorders is miR-155, which regulates both innate and adaptive immune system responses upon antigenic stimulation by targeting SOCS-1, a negative regulator of macrophage activation and antigen presentation [189]. AntimicroRNA oligonucleotides against miR-155 successfully repress acute inflammatory response [190] both in vitro and in vivo, thus underscoring the importance of miR-155 as a potential therapeutic target. miR-192 is a prominent miRNA known to be associated with UC. This miRNA targets macrophage inflammatory peptide (MIP)-2 alpha, a chemokine involved in IBD [178]. More studies in the future need to be undertaken to unravel the therapeutic potential of this miRNA. Other miRNAs that may act as predictive biomarkers include miR-126, whose upregulation has been linked to active UC [178]. However, this miRNA inhibits a tumor suppressor in acute myeloid leukemia and may result in unwanted off-target effects [191]. On the contrary, in gastric [192] and nonsmall cell lung cancers [193], miR-126 has a tumor-suppressive role. A detailed study for identifying any potential side effects is necessary before exploiting its therapeutic role in IBD. miR-106a inhibits IL-10, an antiinflammatory cytokine secreted by B cells, T cells, and antigen-presenting cells with an immunomodulatory role [194]. Reduction of IL-10 results in chronic colitis and its administration has achieved therapeutic efficiency in animal models, with limited progress in humans [195]. Identification of miRNA dysregulation in IBD has invigorated studies for delineating mechanistic and functional pathways of miRNA regulation and developing better IBD diagnostic and therapeutic approaches. Adenoviral and Adeno-associated viral vector-mediated delivery of IL-10 is known to help in ameliorating colitis in mouse models [196,197]. Additional approaches that include nanoparticles, antagomirs that include modifications of AMOs to enable targeted delivery are being studied for utilizing miRNAs as biomarkers and potential therapeutic targets in IBD.

miRNA in psoriasis

Psoriasis is a chronic autoimmune skin disorder, characterized by raised scaly patches on the skin caused by inflammation [198]. An assembly of environmental factors, keratinocytes, immune system, and genetic factors contribute to this condition. T helper cells, namely Th1 and Th17 cell population and their cytokines [199–201], dendritic cells [202], and antimicrobial peptides [203] are implicated in the immune-mediated inflammatory skin disorder. The first evidence showing miRNA involvement in psoriasis pathogenesis dated back to 2007 [204]. Since then, there is emerging evidence to support their role in disease development, with over 250 miRNAs identified to be dysregulated in this skin condition. Most of the aberrantly expressed miRNAs are present in the peripheral blood and psoriatic skin cells. miR-203 is significantly upregulated in psoriasis patients [204–206] and inhibits the SOCS3, resulting in prolonged activation of signal transducer and activator of transcription-3 (STAT3) and subsequent immune cells infiltration and development of psoriatic skin lesions [207]. Moreover, miR-203 abrogates proinflammatory immune response by inhibiting tumor necrosis factor-α (TNF-α), IL-8, and Il-24 and affects keratinocyte differentiation [208,209]. Another miRNA that is upregulated in psoriatic skin and PBMCs from patients is miR-146a [206,207,210]. Elevated miR-146a levels promote IL-17 production, an important cytokine in the regulation of the disease [210]. Besides, it negatively regulates TNF receptor-associated factor 6 (TRAF6) and IL-1 receptor-associated kinase 1 (IRAK1), which mediate proinflammatory cytokine production [211]. On the other hand, increased levels of miR-210 found in psoriatic patients' T cells regulate proinflammatory cytokine levels, mainly IFN-γ and Il-17 by directly targeting the master regulator of Tregs, FOXP3 [212]. Another oncomiR upregulated in psoriasis patients that promotes infiltration of inflammatory cells is miR-21 [213,214]. Upregulation of mir-31 in psoriasis impedes serine/threonine kinase 40, thereby regulating NF-κB signaling and subsequent keratinocyte proliferation [215]. It also exerts adverse effects on keratinocyte migration during wound healing and vascular differentiation [216]. A marked increase in miR-221 and miR-221 levels activates metalloproteinases, resulting in a high rate of epidermal proliferation and inflammation, a characteristic of psoriasis [206,217]. Apart from the upregulated miRNAs, downregulation of several miRNAs, such as miR-99a, miR-424, and miR-125b, is also involved in the immunopathogenesis of psoriasis [205,218,219]. In healthy individuals, mir-99a and miR-424 maintain epidermal homeostasis. However, in psoriatic skin, their levels are low, which leads to aberrant keratinocyte proliferation. Another study reported the increase in miR-242 levels in hair shafts of psoriasis patients [220]. Thus, further investigation on miR-424 dysregulation in the pathogenesis of psoriasis is required. Reduced levels of miR-12b in patients' keratinocytes led to defects in keratinocyte differentiation, aberrant increase in cell proliferation [219], and TNF-α signaling [221] causing psoriatic lesions. Although the existing studies confirm the involvement of miRNAs in the immunopathogenesis of psoriasis, these studies are limited to small patient cohorts. Further mechanistic studies that elucidate the function and expression patterns of specific miRNAs will be beneficial for identifying direct therapeutic targets in the future.

miRNA in type 1 diabetes

Type 1 diabetes (T1D) is a chronic autoimmune disease characterized by insulin deficiency and elevated blood glucose levels, caused by the destruction of insulin-producing pancreatic β cells by autoreactive T cells. Many studies have demonstrated the effects of miRNA-mediated modulation on T1D pathogenesis [222]. miR-21 overexpression has been reported to trigger β-cell death during disease development by targeting the antiapoptotic protein, Bcl-2 in T1D animal models [223]. Inflammatory cytokines impair pancreatic β-cell function and disease progression [224]. miR-30b and miR-101a play key roles in cytokine-mediated β-cell dysfunction, resulting in reduced insulin levels, defects in gene expression, and β-cell apoptosis [225]. Another prominent miRNA implicated in pancreatic β-cell dysfunction and apoptosis in diabetic children and adults is miR-181a, which targets SMAD7 [226]. Administration of miR-106b-5p and miR-222-3p mimics promotes pancreatic β-cell proliferation and regeneration in an insulin-deficient mouse model, thereby showing promise as therapeutic targets for ameliorating hyperglycemia [227]. miR-153 is another miR that prohibits insulin and dopamine secretion and miR-153 inhibitors can reverse this effect, rendering it a potential therapeutic target for insulin resistance [228]. Insulin-secreting β cells are destroyed by the autoreactive T cells, leading to the disease progression. T1D patients are characterized by an unrestricted expansion of diabetogenic T cells [229]. miRNAs such as miR-98, miR-23b, and miR-590-5p stimulate the uncontrolled proliferation of cytotoxic T cells [230]. miR-29b triggers the production of TNF-α, IL-6, and IL-10 in vitro and suppresses the antigen-specific T-cell response in a mouse diabetic model, making them potential targets for T1D therapy [231]. miRNAs in human plasma can be used as biomarkers for studying T1D pathogenesis. Serum and PBMCs derived from T1D patients have shown differential regulation of several miRNAs, including miR-21, miR-24, miR-25, miR-26a, miR-29a, miR-93, miR-146a, miR-148, miR-152, miR-181a, miR-200, etc. [232–234]. Overall, these studies reflect the importance of miRNAs as diagnostic markers that contribute to T1D treatment.

miRNA in Sjogren's syndrome

This chronic autoimmune disease primarily causes prolonged inflammation of the salivary and lachrymal glands, resulting in mouth and eye dryness [235]. Sjogren's syndrome (SjS) patients also encounter cognitive impairment, autoimmune hepatitis, neuropathy, vasculitis, fibromyalgia, arthritis, lung disease, and Raynaud's phenomenon [236–238]. Approximately 3% of worldwide adults are affected by this syndrome [239]. This multifactorial disease stems from genetic predisposition, environmental and hormonal factors [239–241]. The first study that reported miRNA dysregulation in SjS was published in 2011 [242,243]. Since then, several reports and ongoing research have examined miRNAs associated with SjS salivary gland inflammation. A variety of samples, ranging from PBMCs, monocytes, serum, lip biopsies, saliva have been used to detect miRNAs implicated in SjS [242–246]. Downregulation of the miR-17-92 family was reported in the salivary glands of SjS patients, mainly miR-17, miR-18a, miR-19a, miR-20a, and miR-92a-1. Deregulation of these miRNAs is linked to autoimmune and lymphoproliferative disorders [247]. Based on the available data to date, increased levels of miR-146a and miR-155 are observed in PBMCs and purified B and T cells from SjS patients [100,245,248,249]. miR-146a abrogates the expression of TRAF6 and IRAK1, thereby inhibiting NF-κB activity and expression of proinflammatory cytokines, IL-6, IL-8, IL-1β, and TNFα [250]. However, upregulation of miR-146a in PBMCs from SjS patients is correlated to IRAK1 inhibition but overexpression of TRAF6 [251]. Although miR-146a should negatively regulate the inflammatory response, it could have tissue- and disease-specific functions as well, which needs further exploration. High levels of miR-155 in epithelial cells from the salivary gland of SjS patients [252] may be activated by the FoxP3 transcription factor [253]. Increased miR-155 levels also resulted in activation of B and T cells and are correlated with eye dryness in SjS patients [248,254]. miRNA expression profiling in monocytes, macrophages, and dendritic cells from SjS patients revealed aberrant expression of miR-300, miR-609, miR-3162-3p, and miR-4701-5p by affecting the TGF-β signaling cascade [246]. Labial salivary gland tissues from SjS patients demonstrate downregulation of miR-181a and miR-16 expression [255]. However, miR-181a was elevated in PBMCS isolated from SjS patients, primarily in B cells [256]. Further studies on larger patient cohorts will be essential to determine the mechanisms underlying this contradictory miRNA-181a expression, thus improving our understanding of SjS pathogenesis. The standard diagnostic procedure for SjS is labial salivary gland biopsy, which is invasive. Published reports discussed here or otherwise illustrate that miRNA dysregulation in PBMCs, serum, and saliva obtained from SjS patients might serve as an important readout for SjS pathogenesis and will be useful for developing lesser invasive diagnostic approaches.

miRNAs as therapeutic targets

Autoimmune diseases have distinct miRNA expression profiles. Various studies, including some discussed here, have shown patient-derived serum, plasma, urine, saliva, PBMCs, etc. to have been used as specimens for evaluating miRNA expression. These studies have yielded promising results, suggesting miRNAs can be used as potential biomarkers for autoimmune diseases as well as predictive markers for studying drug response (Fig. 12.2). Additionally, miRNAs can be used as therapeutic targets. Small molecules that target these miRNAs either inhibit the miRNA expression directly or the corresponding signaling pathway. Several challenges including stability and delivery hinder the advancement of miRNA therapeutic agents [257]. Modified miRNAs, with a 2′-O-methyl instead of the unstable −OH group are more stable and thus preferred. miR antagonists have been used in clinical trials with considerable progress. Examples include MesomiR-1 and RG-012 targeting miR-16 and miR-21 in lung cancer and Alport syndrome, respectively. Administration of ABX-464 that targets miR-124 in patients suffering from ulcerative colitis has exhibited encouraging results in a small cohort of patients and large-scale trials are currently ongoing [257]. Clinical applications of miRNA-based therapeutic agents for treating autoimmune diseases need to be developed in the future.

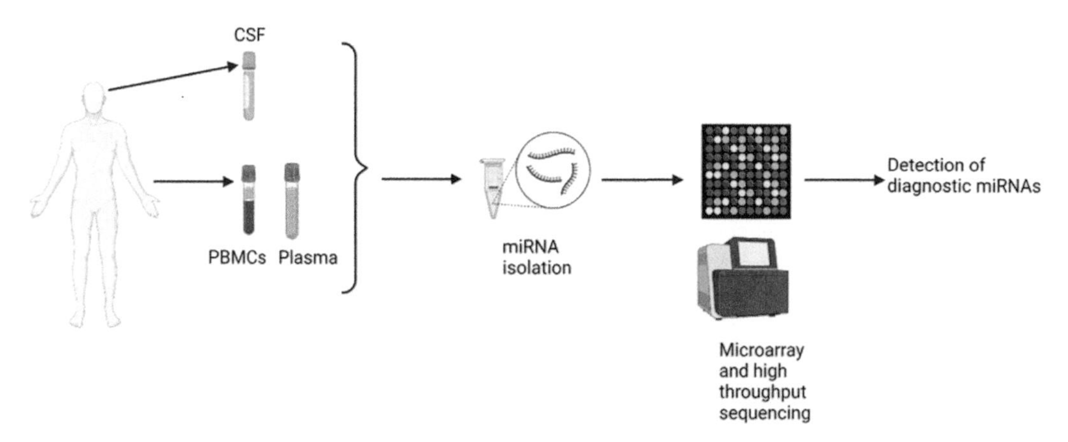

FIGURE 12.2 Schematic for using miRNAs as diagnostic markers for autoimmune diseases. miRNAs isolated from biofluids.

Cerebrospinal fluid (CSF), serum, plasma, peripheral blood mononuclear cells (PBMCs) derived from patient samples can be analyzed by microarray and high throughput sequencing to detect miRNAs that might serve as biomarkers for disease diagnosis.

Created with Biorender.com.

Summary

Aberrant miRNA expression that stimulates inflammatory cytokine production and autoreactive immune cells has been implicated in the pathogenesis of autoimmune disorders. Despite their prevalence, no miRNAs have been adopted as therapeutic targets in clinical trials for autoimmune diseases. Emerging high throughput technology and advancements in research warrant the need for future research on miRNA regulation and its role in the etiology of autoimmune diseases. Overcoming the challenges associated with miRNAs in clinical applications will expedite the development of novel therapeutic approaches in the future.

References

[1] Lee RC, Feinbaum RL, Ambros V. The *C. elegans* heterochronic gene lin-4 encodes small RNAs with antisense complementarity to lin-14. Cell 1993;75(5):843–54. https://doi.org/10.1016/0092-8674(93)90529-y. Epub 1993/12/03. PubMed PMID: 8252621.

[2] Griffiths-Jones S, Grocock RJ, van Dongen S, Bateman A, Enright AJ. miRBase: microRNA sequences, targets and gene nomenclature. Nucleic Acids Res 2006;34(Database issue):D140–4. https://doi.org/10.1093/nar/gkj112. Epub 2005/12/31. PubMed PMID: 16381832; PMCID: PMC1347474.

[3] Dai R, Ahmed SA. MicroRNA, a new paradigm for understanding immunoregulation, inflammation, and autoimmune diseases. Transl Res 2011;157(4):163–79. https://doi.org/10.1016/j.trsl.2011.01.007. Epub 2011/03/23. PubMed PMID: 21420027; PMCID: PMC3072681.

[4] Liu H, Lei C, He Q, Pan Z, Xiao D, Tao Y. Nuclear functions of mammalian MicroRNAs in gene regulation, immunity and cancer. Mol Cancer 2018;17(1):64. https://doi.org/10.1186/s12943-018-0765-5. Epub 2018/02/24. PubMed PMID: 29471827; PMCID: PMC5822656.

[5] Mehta A, Baltimore D. MicroRNAs as regulatory elements in immune system logic. Nat Rev Immunol 2016;16(5):279–94. https://doi.org/10.1038/nri.2016.40. Epub 2016/04/29. PubMed PMID: 27121651.

[6] Zhernakova A, Withoff S, Wijmenga C. Clinical implications of shared genetics and pathogenesis in autoimmune diseases. Nat Rev Endocrinol 2013;9(11):646–59. https://doi.org/10.1038/nrendo.2013.161. Epub 2013/08/21. PubMed PMID: 23959365.

[7] Alzabin S, Venables PJ. Etiology of autoimmune disease: past, present and future. Expet Rev Clin Immunol 2012;8(2):111–3. https://doi.org/10.1586/eci.11.88. Epub 2012/02/01. PubMed PMID: 22288447.

[8] Grolleau-Julius A, Ray D, Yung RL. The role of epigenetics in aging and autoimmunity. Clin Rev Allergy Immunol 2010;39(1):42–50. https://doi.org/10.1007/s12016-009-8169-3. Epub 2009/08/05. PubMed PMID: 19653133; PMCID: PMC2889224.

[9] Kreiner E, Waage J, Standl M, Brix S, Pers TH, Couto Alves A, Warrington NM, Tiesler CMT, Fuertes E, Franke L, Hirschhorn JN, James A, Simpson A, Tung JY, Koppelman GH, Postma DS, Pennell CE, Jarvelin MR, Custovic A, Timpson N, Ferreira MA, Strachan DP, Henderson J, Hinds D, Bisgaard H, Bonnelykke K. Shared genetic variants suggest common pathways in allergy and autoimmune diseases. J Allergy Clin Immunol 2017;140(3):771–81. https://doi.org/10.1016/j.jaci.2016.10.055. Epub 2017/02/12. PubMed PMID: 28188724.

[10] Li B, Selmi C, Tang R, Gershwin ME, Ma X. The microbiome and autoimmunity: a paradigm from the gut-liver axis. Cell Mol Immunol 2018;15(6):595–609. https://doi.org/10.1038/cmi.2018.7. Epub 2018/05/01. PubMed PMID: 29706647; PMCID: PMC6079090.

[11] Lu Q. The critical importance of epigenetics in autoimmunity. J Autoimmun 2013;41:1–5. https://doi.org/10.1016/j.jaut.2013.01.010. Epub 2013/02/05. PubMed PMID: 23375849.

[12] Garo LP, Murugaiyan G. Contribution of MicroRNAs to autoimmune diseases. Cell Mol Life Sci 2016; 73(10):2041–51. https://doi.org/10.1007/s00018-016-2167-4. Epub 2016/03/05. PubMed PMID: 26943802.

[13] Pauley KM, Cha S, Chan EK. MicroRNA in autoimmunity and autoimmune diseases. J Autoimmun 2009; 32(3–4):189–94. https://doi.org/10.1016/j.jaut.2009.02.012. Epub 2009/03/24. PubMed PMID: 19303254; PMCID: PMC2717629.

[14] Qu Z, Li W, Fu B. MicroRNAs in autoimmune diseases. BioMed Res Int 2014;2014:527895. https://doi.org/10.1155/2014/527895. Epub 2014/07/06. PubMed PMID: 24991561; PMCID: PMC4065654.

[15] Seeley JJ, Baker RG, Mohamed G, Bruns T, Hayden MS, Deshmukh SD, Freedberg DE, Ghosh S. Induction of innate immune memory via microRNA targeting of chromatin remodelling factors. Nature 2018; 559(7712):114–9. https://doi.org/10.1038/s41586-018-0253-5. Epub 2018/06/29. PubMed PMID: 29950719; PMCID: PMC6044474.

[16] Kamali K, Korjan ES, Eftekhar E, Malekzadeh K, Soufi FG. The role of miR-146a on NF-kappaB expression level in human umbilical vein endothelial cells under hyperglycemic condition. Bratisl Lek Listy 2016;117(7):376–80. https://doi.org/10.4149/bll_2016_074. Epub 2016/08/23. PubMed PMID: 27546538.

[17] Duroux-Richard I, Roubert C, Ammari M, Presumey J, Grun JR, Haupl T, Grutzkau A, Lecellier CH, Boitez V, Codogno P, Escoubet J, Pers YM, Jorgensen C, Apparailly F. miR-125b controls monocyte adaptation to inflammation through mitochondrial metabolism and dynamics. Blood 2016;128(26): 3125–36. https://doi.org/10.1182/blood-2016-02-697003. Epub 2016/10/21. PubMed PMID: 27702798; PMCID: PMC5335801.

[18] Shakerian L, Ghorbani S, Talebi F, Noorbakhsh F. MicroRNA-150 targets PU.1 and regulates macrophage differentiation and function in experimental autoimmune encephalomyelitis. J Neuroimmunol 2018;323: 167–74. https://doi.org/10.1016/j.jneuroim.2018.06.010. Epub 2018/09/11. PubMed PMID: 30196828.

[19] Gatto G, Rossi A, Rossi D, Kroening S, Bonatti S, Mallardo M. Epstein-Barr virus latent membrane protein 1 trans-activates miR-155 transcription through the NF-kappaB pathway. Nucleic Acids Res 2008;36(20): 6608–19. https://doi.org/10.1093/nar/gkn666. Epub 2008/10/23. PubMed PMID: 18940871; PMCID: PMC2582607.

[20] O'Connell RM, Taganov KD, Boldin MP, Cheng G, Baltimore D. MicroRNA-155 is induced during the macrophage inflammatory response. Proc Natl Acad Sci USA 2007;104(5):1604–9. https://doi.org/10.1073/pnas.0610731104. Epub 2007/01/24. PubMed PMID: 17242365; PMCID: PMC1780072.

[21] O'Connell RM, Chaudhuri AA, Rao DS, Baltimore D. Inositol phosphatase SHIP1 is a primary target of miR-155. Proc Natl Acad Sci USA 2009;106(17):7113–8. https://doi.org/10.1073/pnas.0902636106. Epub 2009/04/11. PubMed PMID: 19359473; PMCID: PMC2678424.

[22] Ceppi M, Pereira PM, Dunand-Sauthier I, Barras E, Reith W, Santos MA, Pierre P. MicroRNA-155 modulates the interleukin-1 signaling pathway in activated human monocyte-derived dendritic cells. Proc Natl Acad Sci USA 2009;106(8):2735–40. https://doi.org/10.1073/pnas.0811073106. Epub 2009/02/06. PubMed PMID: 19193853; PMCID: PMC2650335.

[23] Hashimi ST, Fulcher JA, Chang MH, Gov L, Wang S, Lee B. MicroRNA profiling identifies miR-34a and miR-21 and their target genes JAG1 and WNT1 in the coordinate regulation of dendritic cell differentiation. Blood 2009;114(2):404–14. https://doi.org/10.1182/blood-2008-09-179150. Epub 2009/04/29. PubMed PMID: 19398721; PMCID: PMC2927176.

[24] Lu Z, Liu M, Stribinskis V, Klinge CM, Ramos KS, Colburn NH, Li Y. MicroRNA-21 promotes cell transformation by targeting the programmed cell death 4 gene. Oncogene 2008;27(31):4373–9. https://doi.org/10.1038/onc.2008.72. Epub 2008/04/01. PubMed PMID: 18372920.

[25] Yuan X, Berg N, Lee JW, Le TT, Neudecker V, Jing N, Eltzschig H. MicroRNA miR-223 as regulator of innate immunity. J Leukoc Biol 2018;104(3):515–24. https://doi.org/10.1002/JLB.3MR0218-079R. Epub 2018/07/04. PubMed PMID: 29969525; PMCID: PMC6638550.

[26] Kim C, Hu B, Jadhav RR, Jin J, Zhang H, Cavanagh MM, Akondy RS, Ahmed R, Weyand CM, Goronzy JJ. Activation of miR-21-regulated pathways in immune aging selects against signatures characteristic of memory T cells. Cell Rep 2018;25(8):2148–62. https://doi.org/10.1016/j.celrep.2018.10.074. e5. Epub 2018/11/22. PubMed PMID: 30463012; PMCID: PMC6371971.

[27] Mussbacher M, Salzmann M, Brostjan C, Hoesel B, Schoergenhofer C, Datler H, Hohensinner P, Basilio J, Petzelbauer P, Assinger A, Schmid JA. Cell type-specific roles of NF-kappaB linking inflammation and thrombosis. Front Immunol 2019;10:85. https://doi.org/10.3389/fimmu.2019.00085. Epub 2019/02/20. PubMed PMID: 30778349; PMCID: PMC6369217.

[28] Seddiki N, Brezar V, Ruffin N, Levy Y, Swaminathan S. Role of miR-155 in the regulation of lymphocyte immune function and disease. Immunology 2014;142(1):32–8. https://doi.org/10.1111/imm.12227. Epub 2013/12/07. PubMed PMID: 24303979; PMCID: PMC3992045.

[29] Fedeli M, Napolitano A, Wong MP, Marcais A, de Lalla C, Colucci F, Merkenschlager M, Dellabona P, Casorati G. Dicer-dependent microRNA pathway controls invariant NKT cell development. J Immunol 2009;183(4):2506–12. https://doi.org/10.4049/jimmunol.0901361. Epub 2009/07/25. PubMed PMID: 19625646.

[30] Zhou L, Seo KH, He HZ, Pacholczyk R, Meng DM, Li CG, Xu J, She JX, Dong Z, Mi QS. Tie2cre-induced inactivation of the miRNA-processing enzyme Dicer disrupts invariant NKT cell development. Proc Natl Acad Sci USA 2009;106(25):10266–71. https://doi.org/10.1073/pnas.0811119106. Epub 2009/06/11. PubMed PMID: 19509335; PMCID: PMC2700920.

[31] Cobb BS, Nesterova TB, Thompson E, Hertweck A, O'Connor E, Godwin J, Wilson CB, Brockdorff N, Fisher AG, Smale ST, Merkenschlager M. T cell lineage choice and differentiation in the absence of the RNase III enzyme Dicer. J Exp Med 2005;201(9):1367–73. https://doi.org/10.1084/jem.20050572. Epub 2005/05/04. PubMed PMID: 15867090; PMCID: PMC2213187.

[32] Koralov SB, Muljo SA, Galler GR, Krek A, Chakraborty T, Kanellopoulou C, Jensen K, Cobb BS, Merkenschlager M, Rajewsky N, Rajewsky K. Dicer ablation affects antibody diversity and cell survival in the B lymphocyte lineage. Cell 2008;132(5):860–74. https://doi.org/10.1016/j.cell.2008.02.020. Epub 2008/03/11. PubMed PMID: 18329371.

[33] Muljo SA, Ansel KM, Kanellopoulou C, Livingston DM, Rao A, Rajewsky K. Aberrant T cell differentiation in the absence of Dicer. J Exp Med 2005;202(2):261–9. https://doi.org/10.1084/jem.20050678. Epub 2005/07/13. PubMed PMID: 16009718; PMCID: PMC2212998.

[34] Li QJ, Chau J, Ebert PJ, Sylvester G, Min H, Liu G, Braich R, Manoharan M, Soutschek J, Skare P, Klein LO, Davis MM, Chen CZ. miR-181a is an intrinsic modulator of T cell sensitivity and selection. Cell 2007;129(1):147–61. https://doi.org/10.1016/j.cell.2007.03.008. Epub 2007/03/27. PubMed PMID: 17382377.

[35] Xiao C, Srinivasan L, Calado DP, Patterson HC, Zhang B, Wang J, Henderson JM, Kutok JL, Rajewsky K. Lymphoproliferative disease and autoimmunity in mice with increased miR-17-92 expression in lymphocytes. Nat Immunol 2008;9(4):405–14. https://doi.org/10.1038/ni1575. Epub 2008/03/11. PubMed PMID: 18327259; PMCID: PMC2533767.

[36] Steiner DF, Thomas MF, Hu JK, Yang Z, Babiarz JE, Allen CD, Matloubian M, Blelloch R, Ansel KM. MicroRNA-29 regulates T-box transcription factors and interferon-gamma production in helper T cells. Immunity 2011;35(2):169–81. https://doi.org/10.1016/j.immuni.2011.07.009. Epub 2011/08/09. PubMed PMID: 21820330; PMCID: PMC3361370.

[37] Jeker LT, Bluestone JA. MicroRNA regulation of T-cell differentiation and function. Immunol Rev 2013; 253(1):65–81. https://doi.org/10.1111/imr.12061. Epub 2013/04/05. PubMed PMID: 23550639; PMCID: PMC3621017.
[38] Tahamtan A, Teymoori-Rad M, Nakstad B, Salimi V. Anti-inflammatory MicroRNAs and their potential for inflammatory diseases treatment. Front Immunol 2018;9:1377. https://doi.org/10.3389/fimmu.2018.01377. Epub 2018/07/11. PubMed PMID: 29988529; PMCID: PMC6026627.
[39] Du C, Liu C, Kang J, Zhao G, Ye Z, Huang S, Li Z, Wu Z, Pei G. MicroRNA miR-326 regulates TH-17 differentiation and is associated with the pathogenesis of multiple sclerosis. Nat Immunol 2009;10(12): 1252–9. https://doi.org/10.1038/ni.1798. Epub 2009/10/20. PubMed PMID: 19838199.
[40] Ichiyama K, Gonzalez-Martin A, Kim BS, Jin HY, Jin W, Xu W, Sabouri-Ghomi M, Xu S, Zheng P, Xiao C, Dong C. The MicroRNA-183-96-182 cluster promotes T helper 17 cell pathogenicity by negatively regulating transcription factor Foxo1 expression. Immunity 2016;44(6):1284–98. https://doi.org/10.1016/j.immuni.2016.05.015. Epub 2016/06/23. PubMed PMID: 27332731; PMCID: PMC4918454.
[41] Wu Y, Schutt S, Paz K, Zhang M, Flynn RP, Bastian D, Sofi MH, Nguyen H, Dai M, Liu C, Chang YJ, Blazar BR, Yu XZ. MicroRNA-17-92 is required for T-cell and B-cell pathogenicity in chronic graft-versus-host disease in mice. Blood 2018;131(17):1974–86. https://doi.org/10.1182/blood-2017-06-789321. Epub 2018/03/14. PubMed PMID: 29530952; PMCID: PMC5921962.
[42] Baumjohann D, Kageyama R, Clingan JM, Morar MM, Patel S, de Kouchkovsky D, Bannard O, Bluestone JA, Matloubian M, Ansel KM, Jeker LT. The microRNA cluster miR-17 approximately 92 promotes TFH cell differentiation and represses subset-inappropriate gene expression. Nat Immunol 2013; 14(8):840–8. https://doi.org/10.1038/ni.2642. Epub 2013/07/03. PubMed PMID: 23812098; PMCID: PMC3720769.
[43] Cho S, Lee HM, Yu IS, Choi YS, Huang HY, Hashemifar SS, Lin LL, Chen MC, Afanasiev ND, Khan AA, Lin SW, Rudensky AY, Crotty S, Lu LF. Differential cell-intrinsic regulations of germinal center B and T cells by miR-146a and miR-146b. Nat Commun 2018;9(1):2757. https://doi.org/10.1038/s41467-018-05196-3. Epub 2018/07/18. PubMed PMID: 30013024; PMCID: PMC6048122.
[44] Trotta R, Chen L, Ciarlariello D, Josyula S, Mao C, Costinean S, Yu L, Butchar JP, Tridandapani S, Croce CM, Caligiuri MA. miR-155 regulates IFN-gamma production in natural killer cells. Blood 2012; 119(15):3478–85. https://doi.org/10.1182/blood-2011-12-398099. Epub 2012/03/02. PubMed PMID: 22378844; PMCID: PMC3325038.
[45] Pua HH, Steiner DF, Patel S, Gonzalez JR, Ortiz-Carpena JF, Kageyama R, Chiou NT, Gallman A, de Kouchkovsky D, Jeker LT, McManus MT, Erle DJ, Ansel KM. MicroRNAs 24 and 27 suppress allergic inflammation and target a network of regulators of T helper 2 cell-associated cytokine production. Immunity 2016;44(4):821–32. https://doi.org/10.1016/j.immuni.2016.01.003. Epub 2016/02/07. PubMed PMID: 26850657; PMCID: PMC4838571.
[46] Wu R, Zeng J, Yuan J, Deng X, Huang Y, Chen L, Zhang P, Feng H, Liu Z, Wang Z, Gao X, Wu H, Wang H, Su Y, Zhao M, Lu Q. MicroRNA-210 overexpression promotes psoriasis-like inflammation by inducing Th1 and Th17 cell differentiation. J Clin Invest 2018;128(6):2551–68. https://doi.org/10.1172/JCI97426. Epub 2018/05/15. PubMed PMID: 29757188; PMCID: PMC5983326.
[47] Hippen KL, Loschi M, Nicholls J, MacDonald KPA, Blazar BR. Effects of MicroRNA on regulatory T cells and implications for adoptive cellular therapy to ameliorate graft-versus-host disease. Front Immunol 2018; 9:57. https://doi.org/10.3389/fimmu.2018.00057. Epub 2018/02/16. PubMed PMID: 29445371; PMCID: PMC5797736.
[48] Huang B, Zhao J, Lei Z, Shen S, Li D, Shen GX, Zhang GM, Feng ZH. miR-142-3p restricts cAMP production in CD4+CD25- T cells and CD4+CD25+ TREG cells by targeting AC9 mRNA. EMBO Rep 2009;10(2):180–5. https://doi.org/10.1038/embor.2008.224. Epub 2008/12/23. PubMed PMID: 19098714; PMCID: PMC2637310.

[49] Simpson LJ, Ansel KM. MicroRNA regulation of lymphocyte tolerance and autoimmunity. J Clin Invest 2015;125(6):2242–9. https://doi.org/10.1172/JCI78090. Epub 2015/06/02. PubMed PMID: 26030228; PMCID: PMC4497751.

[50] Leavy O. Immune memory: sequential evolution of B cell memory. Nat Rev Immunol 2016;16(2):72–3. https://doi.org/10.1038/nri.2016.15. Epub 2016/01/26. PubMed PMID: 26806486.

[51] Lai M, Gonzalez-Martin A, Cooper AB, Oda H, Jin HY, Shepherd J, He L, Zhu J, Nemazee D, Xiao C. Regulation of B-cell development and tolerance by different members of the miR-17 approximately 92 family microRNAs. Nat Commun 2016;7:12207. https://doi.org/10.1038/ncomms12207. Epub 2016/08/03. PubMed PMID: 27481093; PMCID: PMC4974641.

[52] Alivernini S, Gremese E, McSharry C, Tolusso B, Ferraccioli G, McInnes IB, Kurowska-Stolarska M. MicroRNA-155-at the critical interface of innate and adaptive immunity in arthritis. Front Immunol 2017;8: 1932. https://doi.org/10.3389/fimmu.2017.01932. Epub 2018/01/23. PubMed PMID: 29354135; PMCID: PMC5760508.

[53] Haftmann C, Stittrich AB, Zimmermann J, Fang Z, Hradilkova K, Bardua M, Westendorf K, Heinz GA, Riedel R, Siede J, Lehmann K, Weinberger EE, Zimmel D, Lauer U, Haupl T, Sieper J, Backhaus M, Neumann C, Hoffmann U, Porstner M, Chen W, Grun JR, Baumgrass R, Matz M, Lohning M, Scheffold A, Wittmann J, Chang HD, Rajewsky N, Jack HM, Radbruch A, Mashreghi MF. miR-148a is upregulated by Twist1 and T-bet and promotes Th1-cell survival by regulating the proapoptotic gene Bim. Eur J Immunol 2015;45(4):1192–205. https://doi.org/10.1002/eji.201444633. Epub 2014/12/10. PubMed PMID: 25486906; PMCID: PMC4406154.

[54] Xiao C, Calado DP, Galler G, Thai TH, Patterson HC, Wang J, Rajewsky N, Bender TP, Rajewsky K. MiR-150 controls B cell differentiation by targeting the transcription factor c-Myb. Cell 2007;131(1):146–59. https://doi.org/10.1016/j.cell.2007.07.021. Epub 2007/10/10. PubMed PMID: 17923094.

[55] Zheng B, Xi Z, Liu R, Yin W, Sui Z, Ren B, Miller H, Gong Q, Liu C. The function of MicroRNAs in B-cell development, lymphoma, and their potential in clinical practice. Front Immunol 2018;9:936. https://doi.org/10.3389/fimmu.2018.00936. Epub 2018/05/16. PubMed PMID: 29760712; PMCID: PMC5936759.

[56] Slabakova E, Culig Z, Remsik J, Soucek K. Alternative mechanisms of miR-34a regulation in cancer. Cell Death Dis 2017;8(10):e3100. https://doi.org/10.1038/cddis.2017.495. Epub 2017/10/13. PubMed PMID: 29022903; PMCID: PMC5682661.

[57] Barnes NA, Stephenson S, Cocco M, Tooze RM, Doody GM. BLIMP-1 and STAT3 counterregulate microRNA-21 during plasma cell differentiation. J Immunol 2012;189(1):253–60. https://doi.org/10.4049/jimmunol.1101563. Epub 2012/05/29. PubMed PMID: 22634616.

[58] Tsai DY, Hung KH, Chang CW, Lin KI. Regulatory mechanisms of B cell responses and the implication in B cell-related diseases. J Biomed Sci 2019;26(1):64. https://doi.org/10.1186/s12929-019-0558-1. Epub 2019/09/02. PubMed PMID: 31472685; PMCID: PMC6717636.

[59] Hochberg MC. Updating the American College of Rheumatology revised criteria for the classification of systemic lupus erythematosus. Arthritis Rheum 1997;40(9):1725. https://doi.org/10.1002/art.1780400928. Epub 1997/10/27. PubMed PMID: 9324032.

[60] Tan EM, Cohen AS, Fries JF, Masi AT, McShane DJ, Rothfield NF, Schaller JG, Talal N, Winchester RJ. The 1982 revised criteria for the classification of systemic lupus erythematosus. Arthritis Rheum 1982; 25(11):1271–7. https://doi.org/10.1002/art.1780251101. Epub 1982/11/01. PubMed PMID: 7138600.

[61] Chakravarty EF, Bush TM, Manzi S, Clarke AE, Ward MM. Prevalence of adult systemic lupus erythematosus in California and Pennsylvania in 2000: estimates obtained using hospitalization data. Arthritis Rheum 2007;56(6):2092–4. https://doi.org/10.1002/art.22641. Epub 2007/05/29. PubMed PMID: 17530651; PMCID: PMC2530907.

[62] Izmirly PM, Parton H, Wang L, McCune WJ, Lim SS, Drenkard C, Ferucci ED, Dall'Era M, Gordon C, Helmick CG, Somers EC. Prevalence of systemic lupus erythematosus in the United States: estimates from

a meta-analysis of the centers for disease control and prevention national lupus registries. Arthritis Rheumatol 2021;73(6):991–6. https://doi.org/10.1002/art.41632. Epub 2021/01/22. PubMed PMID: 33474834; PMCID: PMC8169527.

[63] Pons-Estel GJ, Alarcon GS, Scofield L, Reinlib L, Cooper GS. Understanding the epidemiology and progression of systemic lupus erythematosus. Semin Arthritis Rheum 2010;39(4):257–68. https://doi.org/10.1016/j.semarthrit.2008.10.007. Epub 2009/01/13. PubMed PMID: 19136143; PMCID: PMC2813992.

[64] Stojan G, Petri M. Epidemiology of systemic lupus erythematosus: an update. Curr Opin Rheumatol 2018;30(2):144–50. https://doi.org/10.1097/BOR.0000000000000480. Epub 2017/12/19. PubMed PMID: 29251660; PMCID: PMC6026543.

[65] Rekvig OP. Systemic lupus erythematosus: definitions, contexts, conflicts, enigmas. Front Immunol 2018;9:387. https://doi.org/10.3389/fimmu.2018.00387. Epub 2018/03/17. PubMed PMID: 29545801; PMCID: PMC5839091.

[66] Qu B, Shen N. miRNAs in the pathogenesis of systemic lupus erythematosus. Int J Mol Sci 2015;16(5):9557–72. https://doi.org/10.3390/ijms16059557. Epub 2015/05/01. PubMed PMID: 25927578; PMCID: PMC4463604.

[67] Bhanji RA, Eystathioy T, Chan EK, Bloch DB, Fritzler MJ. Clinical and serological features of patients with autoantibodies to GW/P bodies. Clin Immunol 2007;125(3):247–56. https://doi.org/10.1016/j.clim.2007.07.016. Epub 2007/09/18. PubMed PMID: 17870671; PMCID: PMC2147044.

[68] Jakymiw A, Ikeda K, Fritzler MJ, Reeves WH, Satoh M, Chan EK. Autoimmune targeting of key components of RNA interference. Arthritis Res Ther 2006;8(4):R87. https://doi.org/10.1186/ar1959. Epub 2006/05/11. PubMed PMID: 16684366; PMCID: PMC1779426.

[69] Dai Y, Huang YS, Tang M, Lv TY, Hu CX, Tan YH, Xu ZM, Yin YB. Microarray analysis of microRNA expression in peripheral blood cells of systemic lupus erythematosus patients. Lupus 2007;16(12):939–46. https://doi.org/10.1177/0961203307084158. Epub 2007/11/29. PubMed PMID: 18042587.

[70] Dai Y, Sui W, Lan H, Yan Q, Huang H, Huang Y. Comprehensive analysis of microRNA expression patterns in renal biopsies of lupus nephritis patients. Rheumatol Int 2009;29(7):749–54. https://doi.org/10.1007/s00296-008-0758-6. Epub 2008/11/11. PubMed PMID: 18998140.

[71] Navarro-Quiroz E, Pacheco-Lugo L, Lorenzi H, Diaz-Olmos Y, Almendrales L, Rico E, Navarro R, Espana-Puccini P, Iglesias A, Egea E, Aroca G. High-throughput sequencing reveals circulating miRNAs as potential biomarkers of kidney damage in patients with systemic lupus erythematosus. PLoS One 2016;11(11):e0166202. https://doi.org/10.1371/journal.pone.0166202. Epub 2016/11/12. PubMed PMID: 27835701; PMCID: PMC5106044.

[72] Kay SD, Carlsen AL, Voss A, Burton M, Diederichsen A, Poulsen MK, Heegaard N. Associations of circulating cell-free microRNA with vasculopathy and vascular events in systemic lupus erythematosus patients. Scand J Rheumatol 2019;48(1):32–41. https://doi.org/10.1080/03009742.2018.1450892. Epub 2018/07/10. PubMed PMID: 29985728.

[73] Luo X, Yang W, Ye DQ, Cui H, Zhang Y, Hirankarn N, Qian X, Tang Y, Lau YL, de Vries N, Tak PP, Tsao BP, Shen N. A functional variant in microRNA-146a promoter modulates its expression and confers disease risk for systemic lupus erythematosus. PLoS Genet 2011;7(6):e1002128. https://doi.org/10.1371/journal.pgen.1002128. Epub 2011/07/09. PubMed PMID: 21738483; PMCID: PMC3128113.

[74] Qu B, Cao J, Zhang F, Cui H, Teng J, Li J, Liu Z, Morehouse C, Jallal B, Tang Y, Guo Q, Yao Y, Shen N. Type I interferon inhibition of MicroRNA-146a maturation through up-regulation of monocyte chemotactic protein-induced protein 1 in systemic lupus erythematosus. Arthritis Rheumatol 2015;67(12):3209–18. https://doi.org/10.1002/art.39398. Epub 2015/09/01. PubMed PMID: 26315540.

[75] Tang Y, Luo X, Cui H, Ni X, Yuan M, Guo Y, Huang X, Zhou H, de Vries N, Tak PP, Chen S, Shen N. MicroRNA-146A contributes to abnormal activation of the type I interferon pathway in human lupus by targeting the key signaling proteins. Arthritis Rheum 2009;60(4):1065–75. https://doi.org/10.1002/art.24436. Epub 2009/04/01. PubMed PMID: 19333922.

[76] Banchereau J, Pascual V. Type I interferon in systemic lupus erythematosus and other autoimmune diseases. Immunity 2006;25(3):383–92. https://doi.org/10.1016/j.immuni.2006.08.010. Epub 2006/09/19. PubMed PMID: 16979570.

[77] Hou J, Wang P, Lin L, Liu X, Ma F, An H, Wang Z, Cao X. MicroRNA-146a feedback inhibits RIG-I-dependent Type I IFN production in macrophages by targeting TRAF6, IRAK1, and IRAK2. J Immunol 2009;183(3):2150–8. https://doi.org/10.4049/jimmunol.0900707. Epub 2009/07/15. PubMed PMID: 19596990.

[78] Chan VS, Nie YJ, Shen N, Yan S, Mok MY, Lau CS. Distinct roles of myeloid and plasmacytoid dendritic cells in systemic lupus erythematosus. Autoimmun Rev 2012;11(12):890–7. https://doi.org/10.1016/j.autrev.2012.03.004. Epub 2012/04/17. PubMed PMID: 22503660.

[79] Karrich JJ, Jachimowski LC, Libouban M, Iyer A, Brandwijk K, Taanman-Kueter EW, Nagasawa M, de Jong EC, Uittenbogaart CH, Blom B. MicroRNA-146a regulates survival and maturation of human plasmacytoid dendritic cells. Blood 2013;122(17):3001–9. https://doi.org/10.1182/blood-2012-12-475087. Epub 2013/09/10. PubMed PMID: 24014244; PMCID: PMC3811175.

[80] Pan Y, Jia T, Zhang Y, Zhang K, Zhang R, Li J, Wang L. MS2 VLP-based delivery of microRNA-146a inhibits autoantibody production in lupus-prone mice. Int J Nanomed 2012;7:5957–67. https://doi.org/10.2147/IJN.S37990. Epub 2012/12/13. PubMed PMID: 23233803; PMCID: PMC3518289.

[81] Wang P, Hou J, Lin L, Wang C, Liu X, Li D, Ma F, Wang Z, Cao X. Inducible microRNA-155 feedback promotes type I IFN signaling in antiviral innate immunity by targeting suppressor of cytokine signaling 1. J Immunol 2010;185(10):6226–33. https://doi.org/10.4049/jimmunol.1000491. Epub 2010/10/13. PubMed PMID: 20937844.

[82] Aboelenein HR, Hamza MT, Marzouk H, Youness RA, Rahmoon M, Salah S, Abdelaziz AI. Reduction of CD19 autoimmunity marker on B cells of paediatric SLE patients through repressing PU.1/TNF-alpha/BAFF axis pathway by miR-155. Growth Factors 2017;35(2–3):49–60. https://doi.org/10.1080/08977194.2017.1345900. Epub 2017/07/08. PubMed PMID: 28683581.

[83] Divekar AA, Dubey S, Gangalum PR, Singh RR. Dicer insufficiency and microRNA-155 overexpression in lupus regulatory T cells: an apparent paradox in the setting of an inflammatory milieu. J Immunol 2011; 186(2):924–30. https://doi.org/10.4049/jimmunol.1002218. Epub 2010/12/15. PubMed PMID: 21149603; PMCID: PMC3038632.

[84] Leiss H, Salzberger W, Jacobs B, Gessl I, Kozakowski N, Bluml S, Puchner A, Kiss A, Podesser BK, Smolen JS, Stummvoll GH. MicroRNA 155-deficiency leads to decreased autoantibody levels and reduced severity of nephritis and pneumonitis in pristane-induced lupus. PLoS One 2017;12(7):e0181015. https://doi.org/10.1371/journal.pone.0181015. Epub 2017/07/19. PubMed PMID: 28719617; PMCID: PMC5515414.

[85] Thai TH, Patterson HC, Pham DH, Kis-Toth K, Kaminski DA, Tsokos GC. Deletion of microRNA-155 reduces autoantibody responses and alleviates lupus-like disease in the Fas(lpr) mouse. Proc Natl Acad Sci USA 2013;110(50):20194–9. https://doi.org/10.1073/pnas.1317632110. Epub 2013/11/28. PubMed PMID: 24282294; PMCID: PMC3864325.

[86] Xin Q, Li J, Dang J, Bian X, Shan S, Yuan J, Qian Y, Liu Z, Liu G, Yuan Q, Liu N, Ma X, Gao F, Gong Y, Liu Q. miR-155 deficiency ameliorates autoimmune inflammation of systemic lupus erythematosus by targeting S1pr1 in Faslpr/lpr mice. J Immunol 2015;194(11):5437–45. https://doi.org/10.4049/jimmunol.1403028. Epub 2015/04/26. PubMed PMID: 25911753.

[87] Zhou S, Wang Y, Meng Y, Xiao C, Liu Z, Brohawn P, Higgs BW, Jallal B, Jia Q, Qu B, Huang X, Tang Y, Yao Y, Harley JB, Shen N. In vivo therapeutic success of MicroRNA-155 antagomir in a mouse model of lupus alveolar hemorrhage. Arthritis Rheumatol 2016;68(4):953–64. https://doi.org/10.1002/art.39485. Epub 2015/11/12. PubMed PMID: 26556607.

[88] Pan W, Zhu S, Yuan M, Cui H, Wang L, Luo X, Li J, Zhou H, Tang Y, Shen N. MicroRNA-21 and microRNA-148a contribute to DNA hypomethylation in lupus CD4+ T cells by directly and indirectly targeting DNA methyltransferase 1. J Immunol 2010;184(12):6773–81. https://doi.org/10.4049/jimmunol.0904060. Epub 2010/05/21. PubMed PMID: 20483747.

[89] Stagakis E, Bertsias G, Verginis P, Nakou M, Hatziapostolou M, Kritikos H, Iliopoulos D, Boumpas DT. Identification of novel microRNA signatures linked to human lupus disease activity and pathogenesis: miR-21 regulates aberrant T cell responses through regulation of PDCD4 expression. Ann Rheum Dis 2011; 70(8):1496–506. https://doi.org/10.1136/ard.2010.139857. Epub 2011/05/24. PubMed PMID: 21602271.

[90] Pan W, Zhu S, Dai D, Liu Z, Li D, Li B, Gagliani N, Zheng Y, Tang Y, Weirauch MT, Chen X, Zhu W, Wang Y, Chen B, Qian Y, Chen Y, Fang J, Herbst R, Richman L, Jallal B, Harley JB, Flavell RA, Yao Y, Shen N. MiR-125a targets effector programs to stabilize Treg-mediated immune homeostasis. Nat Commun 2015;6:7096. https://doi.org/10.1038/ncomms8096. Epub 2015/05/13. PubMed PMID: 25963922.

[91] Zhu S, Pan W, Song X, Liu Y, Shao X, Tang Y, Liang D, He D, Wang H, Liu W, Shi Y, Harley JB, Shen N, Qian Y. The microRNA miR-23b suppresses IL-17-associated autoimmune inflammation by targeting TAB2, TAB3 and IKK-alpha. Nat Med 2012;18(7):1077–86. https://doi.org/10.1038/nm.2815. Epub 2012/06/05. PubMed PMID: 22660635.

[92] Li X, Luo F, Li J, Luo C. MiR-183 delivery attenuates murine lupus nephritis-related injuries via targeting mTOR. Scand J Immunol 2019;90(5):e12810. https://doi.org/10.1111/sji.12810. Epub 2019/07/22. PubMed PMID: 31325389.

[93] Liu L, Liu Y, Yuan M, Xu L, Sun H. Elevated expression of microRNA-873 facilitates Th17 differentiation by targeting forkhead box O1 (Foxo1) in the pathogenesis of systemic lupus erythematosus. Biochem Biophys Res Commun 2017;492(3):453–60. https://doi.org/10.1016/j.bbrc.2017.08.075. Epub 2017/08/25. PubMed PMID: 28837808.

[94] Qingjuan L, Xiaojuan F, Wei Z, Chao W, Pengpeng K, Hongbo L, Sanbing Z, Jun H, Min Y, Shuxia L. miR-148a-3p overexpression contributes to glomerular cell proliferation by targeting PTEN in lupus nephritis. Am J Physiol Cell Physiol 2016;310(6):C470–8. https://doi.org/10.1152/ajpcell.00129.2015. Epub 2016/01/23. PubMed PMID: 26791485.

[95] Han X, Wang Y, Zhang X, Qin Y, Qu B, Wu L, Ma J, Zhou Z, Qian J, Dai M, Tang Y, Chan EK, Harley JB, Zhou S, Shen N. MicroRNA-130b ameliorates murine lupus nephritis through targeting the type I interferon pathway on renal mesangial cells. Arthritis Rheumatol 2016;68(9):2232–43. https://doi.org/10.1002/art.39725. Epub 2016/04/26. PubMed PMID: 27111096.

[96] Wang X, Wang G, Zhang X, Dou Y, Dong Y, Liu D, Xiao J, Zhao Z. Inhibition of microRNA-182-5p contributes to attenuation of lupus nephritis via Foxo1 signaling. Exp Cell Res 2018;373(1–2):91–8. https://doi.org/10.1016/j.yexcr.2018.09.026. Epub 2018/10/12. PubMed PMID: 30308195.

[97] Lee DM, Weinblatt ME. Rheumatoid arthritis. Lancet 2001;358(9285):903–11. https://doi.org/10.1016/S0140-6736(01)06075-5. Epub 2001/09/25. PubMed PMID: 11567728.

[98] Bresnihan B, Alvaro-Gracia JM, Cobby M, Doherty M, Domljan Z, Emery P, Nuki G, Pavelka K, Rau R, Rozman B, Watt I, Williams B, Aitchison R, McCabe D, Musikic P. Treatment of rheumatoid arthritis with recombinant human interleukin-1 receptor antagonist. Arthritis Rheum 1998;41(12):2196–204. https://doi.org/10.1002/1529-0131(199812)41:12<2196::AID-ART15>3.0.CO;2-2. Epub 1998/12/31. PubMed PMID: 9870876.

[99] Lipsky PE, van der Heijde DM, St Clair EW, Furst DE, Breedveld FC, Kalden JR, Smolen JS, Weisman M, Emery P, Feldmann M, Harriman GR, Maini RN. Anti-tumor necrosis factor trial in rheumatoid arthritis with concomitant therapy study G. Infliximab and methotrexate in the treatment of rheumatoid arthritis. Anti-tumor necrosis factor trial in rheumatoid arthritis with concomitant therapy study group. N Engl J Med 2000;343(22):1594–602. https://doi.org/10.1056/NEJM200011303432202. Epub 2000/11/30. PubMed PMID: 11096166.

[100] Chen JQ, Papp G, Szodoray P, Zeher M. The role of microRNAs in the pathogenesis of autoimmune diseases. Autoimmun Rev 2016;15(12):1171–80. https://doi.org/10.1016/j.autrev.2016.09.003. Epub 2016/09/18. PubMed PMID: 27639156.

[101] Ospelt C, Gay S, Klein K. Epigenetics in the pathogenesis of RA. Semin Immunopathol 2017;39(4): 409–19. https://doi.org/10.1007/s00281-017-0621-5. Epub 2017/03/23. PubMed PMID: 28324153.

[102] Singh RP, Massachi I, Manickavel S, Singh S, Rao NP, Hasan S, Mc Curdy DK, Sharma S, Wong D, Hahn BH, Rehimi H. The role of miRNA in inflammation and autoimmunity. Autoimmun Rev 2013; 12(12):1160–5. https://doi.org/10.1016/j.autrev.2013.07.003. Epub 2013/07/19. PubMed PMID: 23860189.

[103] Tavasolian F, Abdollahi E, Rezaei R, Momtazi-Borojeni AA, Henrotin Y, Sahebkar A. Altered expression of MicroRNAs in rheumatoid arthritis. J Cell Biochem 2018;119(1):478–87. https://doi.org/10.1002/jcb.26205. Epub 2017/06/10. PubMed PMID: 28598026.

[104] Abou-Zeid A, Saad M, Soliman E. MicroRNA 146a expression in rheumatoid arthritis: association with tumor necrosis factor-alpha and disease activity. Genet Test Mol Biomarkers 2011;15(11):807–12. https://doi.org/10.1089/gtmb.2011.0026. Epub 2011/08/04. PubMed PMID: 21810022.

[105] Pauley KM, Satoh M, Chan AL, Bubb MR, Reeves WH, Chan EK. Upregulated miR-146a expression in peripheral blood mononuclear cells from rheumatoid arthritis patients. Arthritis Res Ther 2008;10(4):R101. https://doi.org/10.1186/ar2493. Epub 2008/09/02. PubMed PMID: 18759964; PMCID: PMC2575615.

[106] Kurowska-Stolarska M, Alivernini S, Ballantine LE, Asquith DL, Millar NL, Gilchrist DS, Reilly J, Ierna M, Fraser AR, Stolarski B, McSharry C, Hueber AJ, Baxter D, Hunter J, Gay S, Liew FY, McInnes IB. MicroRNA-155 as a proinflammatory regulator in clinical and experimental arthritis. Proc Natl Acad Sci USA 2011;108(27):11193–8. https://doi.org/10.1073/pnas.1019536108. Epub 2011/06/22. PubMed PMID: 21690378; PMCID: PMC3131377.

[107] Long L, Yu P, Liu Y, Wang S, Li R, Shi J, Zhang X, Li Y, Sun X, Zhou B, Cui L, Li Z. Upregulated microRNA-155 expression in peripheral blood mononuclear cells and fibroblast-like synoviocytes in rheumatoid arthritis. Clin Dev Immunol 2013;2013:296139. https://doi.org/10.1155/2013/296139. Epub 2013/10/24. PubMed PMID: 24151514; PMCID: PMC3789322.

[108] Anaparti V, Smolik I, Meng X, Spicer V, Mookherjee N, El-Gabalawy H. Whole blood microRNA expression pattern differentiates patients with rheumatoid arthritis, their seropositive first-degree relatives, and healthy unrelated control subjects. Arthritis Res Ther 2017;19(1):249. https://doi.org/10.1186/s13075-017-1459-x. Epub 2017/11/12. PubMed PMID: 29126434; PMCID: PMC5681796.

[109] Chen Y, Wang X, Yang M, Ruan W, Wei W, Gu D, Wang J, Guo X, Guo L, Yuan Y. miR-145-5p increases osteoclast numbers in vitro and aggravates bone erosion in collagen-induced arthritis by targeting osteoprotegerin. Med Sci Mon Int Med J Exp Clin Res 2018;24:5292–300. https://doi.org/10.12659/MSM.908219. Epub 2018/07/31. PubMed PMID: 30059491; PMCID: PMC6080580.

[110] Abo ElAtta AS, Ali YBM, Bassyouni IH, Talaat RM. Upregulation of miR-221/222 expression in rheumatoid arthritis (RA) patients: correlation with disease activity. Clin Exp Med 2019;19(1):47–53. https://doi.org/10.1007/s10238-018-0524-3. Epub 2018/08/23. PubMed PMID: 30132091.

[111] Tang X, Yin K, Zhu H, Tian J, Shen D, Yi L, Rui K, Ma J, Xu H, Wang S. Correlation between the expression of MicroRNA-301a-3p and the proportion of Th17 cells in patients with rheumatoid arthritis. Inflammation 2016;39(2):759–67. https://doi.org/10.1007/s10753-016-0304-8. Epub 2016/01/20. PubMed PMID: 26782362.

[112] Dong L, Wang X, Tan J, Li H, Qian W, Chen J, Chen Q, Wang J, Xu W, Tao C, Wang S. Decreased expression of microRNA-21 correlates with the imbalance of Th17 and Treg cells in patients with rheumatoid arthritis. J Cell Mol Med 2014;18(11):2213–24. https://doi.org/10.1111/jcmm.12353. Epub 2014/08/29. PubMed PMID: 25164131; PMCID: PMC4224555.

[113] Hruskova V, Jandova R, Vernerova L, Mann H, Pecha O, Prajzlerova K, Pavelka K, Vencovsky J, Filkova M, Senolt L. MicroRNA-125b: association with disease activity and the treatment response of patients with early rheumatoid arthritis. Arthritis Res Ther 2016;18(1):124. https://doi.org/10.1186/s13075-016-1023-0. Epub 2016/06/04. PubMed PMID: 27255643; PMCID: PMC4890522.
[114] Wang Y, Zheng F, Gao G, Yan S, Zhang L, Wang L, Cai X, Wang X, Xu D, Wang J. MiR-548a-3p regulates inflammatory response via TLR4/NF-kappaB signaling pathway in rheumatoid arthritis. J Cell Biochem 2018. https://doi.org/10.1002/jcb.26659. Epub 2018/01/10. PubMed PMID: 29315763.
[115] Niimoto T, Nakasa T, Ishikawa M, Okuhara A, Izumi B, Deie M, Suzuki O, Adachi N, Ochi M. MicroRNA-146a expresses in interleukin-17 producing T cells in rheumatoid arthritis patients. BMC Muscoskel Disord 2010;11:209. https://doi.org/10.1186/1471-2474-11-209. Epub 2010/09/16. PubMed PMID: 20840794; PMCID: PMC2950393.
[116] Zhou Q, Haupt S, Kreuzer JT, Hammitzsch A, Proft F, Neumann C, Leipe J, Witt M, Schulze-Koops H, Skapenko A. Decreased expression of miR-146a and miR-155 contributes to an abnormal Treg phenotype in patients with rheumatoid arthritis. Ann Rheum Dis 2015;74(6):1265–74. https://doi.org/10.1136/annrheumdis-2013-204377. Epub 2014/02/25. PubMed PMID: 24562503.
[117] Nakasa T, Miyaki S, Okubo A, Hashimoto M, Nishida K, Ochi M, Asahara H. Expression of microRNA-146 in rheumatoid arthritis synovial tissue. Arthritis Rheum 2008;58(5):1284–92. https://doi.org/10.1002/art.23429. Epub 2008/04/29. PubMed PMID: 18438844; PMCID: PMC2749927.
[118] Stanczyk J, Pedrioli DM, Brentano F, Sanchez-Pernaute O, Kolling C, Gay RE, Detmar M, Gay S, Kyburz D. Altered expression of MicroRNA in synovial fibroblasts and synovial tissue in rheumatoid arthritis. Arthritis Rheum 2008;58(4):1001–9. https://doi.org/10.1002/art.23386. Epub 2008/04/03. PubMed PMID: 18383392.
[119] Alsaleh G, Nehmar R, Bluml S, Schleiss C, Ostermann E, Dillenseger JP, Sayeh A, Choquet P, Dembele D, Francois A, Salmon JH, Paul N, Schabbauer G, Bierry G, Meyer A, Gottenberg JE, Haas G, Pfeffer S, Vallat L, Sibilia J, Bahram S, Georgel P. Reduced DICER1 expression bestows rheumatoid arthritis synoviocytes proinflammatory properties and resistance to apoptotic stimuli. Arthritis Rheumatol 2016;68(8):1839–48. https://doi.org/10.1002/art.39641. Epub 2016/02/18. PubMed PMID: 26882526.
[120] Nakasa T, Shibuya H, Nagata Y, Niimoto T, Ochi M. The inhibitory effect of microRNA-146a expression on bone destruction in collagen-induced arthritis. Arthritis Rheum 2011;63(6):1582–90. https://doi.org/10.1002/art.30321. Epub 2011/03/23. PubMed PMID: 21425254.
[121] Jing W, Zhang X, Sun W, Hou X, Yao Z, Zhu Y. CRISPR/CAS9-Mediated genome editing of miRNA-155 inhibits proinflammatory cytokine production by RAW264.7 cells. BioMed Res Int 2015;2015:326042. https://doi.org/10.1155/2015/326042. Epub 2015/12/24. PubMed PMID: 26697483; PMCID: PMC4677169.
[122] Mu N, Gu J, Huang T, Zhang C, Shu Z, Li M, Hao Q, Li W, Zhang W, Zhao J, Zhang Y, Huang L, Wang S, Jin X, Xue X, Zhang W, Zhang Y. A novel NF-kappaB/YY1/microRNA-10a regulatory circuit in fibroblast-like synoviocytes regulates inflammation in rheumatoid arthritis. Sci Rep 2016;6:20059. https://doi.org/10.1038/srep20059. Epub 2016/01/30. PubMed PMID: 26821827; PMCID: PMC4731824.
[123] Akhtar N, Singh AK, Ahmed S. MicroRNA-17 suppresses TNF-alpha signaling by interfering with TRAF2 and cIAP2 association in rheumatoid arthritis synovial fibroblasts. J Immunol 2016;197(6):2219–28. https://doi.org/10.4049/jimmunol.1600360. Epub 2016/08/19. PubMed PMID: 27534557; PMCID: PMC5010933.
[124] Trenkmann M, Brock M, Gay RE, Michel BA, Gay S, Huber LC. Tumor necrosis factor alpha-induced microRNA-18a activates rheumatoid arthritis synovial fibroblasts through a feedback loop in NF-kappaB signaling. Arthritis Rheum 2013;65(4):916–27. https://doi.org/10.1002/art.37834. Epub 2013/01/03. PubMed PMID: 23280137.

[125] Shi DL, Shi GR, Xie J, Du XZ, Yang H. MicroRNA-27a inhibits cell migration and invasion of fibroblast-like synoviocytes by targeting Follistatin-like protein 1 in rheumatoid arthritis. Mol Cell 2016;39(8):611−8. https://doi.org/10.14348/molcells.2016.0103. Epub 2016/08/09. PubMed PMID: 27498552; PMCID: PMC4990753.

[126] Stanczyk J, Ospelt C, Karouzakis E, Filer A, Raza K, Kolling C, Gay R, Buckley CD, Tak PP, Gay S, Kyburz D. Altered expression of microRNA-203 in rheumatoid arthritis synovial fibroblasts and its role in fibroblast activation. Arthritis Rheum 2011;63(2):373−81. https://doi.org/10.1002/art.30115. Epub 2011/02/01. PubMed PMID: 21279994; PMCID: PMC3116142.

[127] Yang S, Yang Y. Downregulation of microRNA221 decreases migration and invasion in fibroblastlike synoviocytes in rheumatoid arthritis. Mol Med Rep 2015;12(2):2395−401. https://doi.org/10.3892/mmr.2015.3642. Epub 2015/04/22. PubMed PMID: 25891943.

[128] Wang X, Si X, Sun J, Yue L, Wang J, Yu Z. miR-522 modulated the expression of proinflammatory cytokines and matrix metalloproteinases partly via targeting suppressor of cytokine signaling 3 in rheumatoid arthritis synovial fibroblasts. DNA Cell Biol 2018;37(4):405−15. https://doi.org/10.1089/dna.2017.4008. Epub 2018/02/03. PubMed PMID: 29394098.

[129] Miao CG, Shi WJ, Xiong YY, Yu H, Zhang XL, Qin MS, Du CL, Song TW, Zhang B, Li J. MicroRNA-663 activates the canonical Wnt signaling through the adenomatous polyposis coli suppression. Immunol Lett 2015;166(1):45−54. https://doi.org/10.1016/j.imlet.2015.05.011. Epub 2015/06/02. PubMed PMID: 26028359.

[130] Ogando J, Tardaguila M, Diaz-Alderete A, Usategui A, Miranda-Ramos V, Martinez-Herrera DJ, de la Fuente L, Garcia-Leon MJ, Moreno MC, Escudero S, Canete JD, Toribio ML, Cases I, Pascual-Montano A, Pablos JL, Manes S. Notch-regulated miR-223 targets the aryl hydrocarbon receptor pathway and increases cytokine production in macrophages from rheumatoid arthritis patients. Sci Rep 2016;6:20223. https://doi.org/10.1038/srep20223. Epub 2016/02/04. PubMed PMID: 26838552; PMCID: PMC4738320.

[131] Fulci V, Scappucci G, Sebastiani GD, Giannitti C, Franceschini D, Meloni F, Colombo T, Citarella F, Barnaba V, Minisola G, Galeazzi M, Macino G. miR-223 is overexpressed in T-lymphocytes of patients affected by rheumatoid arthritis. Hum Immunol 2010;71(2):206−11. https://doi.org/10.1016/j.humimm.2009.11.008. Epub 2009/11/26. PubMed PMID: 19931339.

[132] Shibuya H, Nakasa T, Adachi N, Nagata Y, Ishikawa M, Deie M, Suzuki O, Ochi M. Overexpression of microRNA-223 in rheumatoid arthritis synovium controls osteoclast differentiation. Mod Rheumatol 2013;23(4):674−85. https://doi.org/10.1007/s10165-012-0710-1. Epub 2012/08/21. PubMed PMID: 22903258.

[133] Wu J, Fan W, Ma L, Geng X. miR-708-5p promotes fibroblast-like synoviocytes' cell apoptosis and ameliorates rheumatoid arthritis by the inhibition of Wnt3a/beta-catenin pathway. Drug Des Dev Ther 2018;12:3439−47. https://doi.org/10.2147/DDDT.S177128. Epub 2018/10/24. PubMed PMID: 30349197; PMCID: PMC6186895.

[134] Xu K, Xu P, Yao JF, Zhang YG, Hou WK, Lu SM. Reduced apoptosis correlates with enhanced autophagy in synovial tissues of rheumatoid arthritis. Inflamm Res 2013;62(2):229−37. https://doi.org/10.1007/s00011-012-0572-1. Epub 2012/11/28. PubMed PMID: 23178792.

[135] Hussain N, Zhu W, Jiang C, Xu J, Wu X, Geng M, Hussain S, Cai Y, Xu K, Xu P, Han Y, Sun J, Meng L, Lu S. Down-regulation of miR-10a-5p in synoviocytes contributes to TBX5-controlled joint inflammation. J Cell Mol Med 2018;22(1):241−50. https://doi.org/10.1111/jcmm.13312. Epub 2017/08/07. PubMed PMID: 28782180; PMCID: PMC5742673.

[136] Lin J, Huo R, Xiao L, Zhu X, Xie J, Sun S, He Y, Zhang J, Sun Y, Zhou Z, Wu P, Shen B, Li D, Li N. A novel p53/microRNA-22/Cyr61 axis in synovial cells regulates inflammation in rheumatoid arthritis. Arthritis Rheumatol 2014;66(1):49−59. https://doi.org/10.1002/art.38142. Epub 2014/01/23. PubMed PMID: 24449575.

[137] Nakamachi Y, Kawano S, Takenokuchi M, Nishimura K, Sakai Y, Chin T, Saura R, Kurosaka M, Kumagai S. MicroRNA-124a is a key regulator of proliferation and monocyte chemoattractant protein 1 secretion in fibroblast-like synoviocytes from patients with rheumatoid arthritis. Arthritis Rheum 2009; 60(5):1294–304. https://doi.org/10.1002/art.24475. Epub 2009/05/01. PubMed PMID: 19404929.

[138] Peng JS, Chen SY, Wu CL, Chong HE, Ding YC, Shiau AL, Wang CR. Amelioration of experimental autoimmune arthritis through targeting of synovial fibroblasts by intraarticular delivery of MicroRNAs 140-3p and 140-5p. Arthritis Rheumatol 2016;68(2):370–81. https://doi.org/10.1002/art.39446. Epub 2015/10/17. PubMed PMID: 26473405.

[139] Chen Z, Wang H, Xia Y, Yan F, Lu Y. Therapeutic potential of mesenchymal cell-derived miRNA-150-5p-expressing exosomes in rheumatoid arthritis mediated by the modulation of MMP14 and VEGF. J Immunol 2018;201(8):2472–82. https://doi.org/10.4049/jimmunol.1800304. Epub 2018/09/19. PubMed PMID: 30224512; PMCID: PMC6176104.

[140] Jiang C, Zhu W, Xu J, Wang B, Hou W, Zhang R, Zhong N, Ning Q, Han Y, Yu H, Sun J, Meng L, Lu S. MicroRNA-26a negatively regulates toll-like receptor 3 expression of rat macrophages and ameliorates pristane induced arthritis in rats. Arthritis Res Ther 2014;16(1):R9. https://doi.org/10.1186/ar4435. Epub 2014/01/16. PubMed PMID: 24423102; PMCID: PMC3978458.

[141] Li YT, Chen SY, Wang CR, Liu MF, Lin CC, Jou IM, Shiau AL, Wu CL. Brief report: amelioration of collagen-induced arthritis in mice by lentivirus-mediated silencing of microRNA-223. Arthritis Rheum 2012;64(10):3240–5. https://doi.org/10.1002/art.34550. Epub 2012/06/08. PubMed PMID: 22674011.

[142] Dang Q, Yang F, Lei H, Liu X, Yan M, Huang H, Fan X, Li Y. Inhibition of microRNA-34a ameliorates murine collagen-induced arthritis. Exp Ther Med 2017;14(2):1633–9. https://doi.org/10.3892/etm.2017.4708. Epub 2017/08/16. PubMed PMID: 28810629; PMCID: PMC5525646.

[143] Tao Y, Wang Z, Wang L, Shi J, Guo X, Zhou W, Wu X, Liu Y, Zhang W, Yang H, Shi Q, Xu Y, Geng D. Downregulation of miR-106b attenuates inflammatory responses and joint damage in collagen-induced arthritis. Rheumatology 2017;56(10):1804–13. https://doi.org/10.1093/rheumatology/kex233. Epub 2017/09/29. PubMed PMID: 28957555.

[144] Kawano S, Nakamachi Y. miR-124a as a key regulator of proliferation and MCP-1 secretion in synoviocytes from patients with rheumatoid arthritis. Ann Rheum Dis 2011;70(Suppl. 1):i88–91. https://doi.org/10.1136/ard.2010.138669. Epub 2011/02/26. PubMed PMID: 21339227.

[145] Sun J, Yan P, Chen Y, Chen Y, Yang J, Xu G, Mao H, Qiu Y. MicroRNA-26b inhibits cell proliferation and cytokine secretion in human RASF cells via the Wnt/GSK-3beta/beta-catenin pathway. Diagn Pathol 2015; 10:72. https://doi.org/10.1186/s13000-015-0309-x. Epub 2015/06/20. PubMed PMID: 26088648; PMCID: PMC4472173.

[146] Wang L, Song G, Zheng Y, Wang D, Dong H, Pan J, Chang X. miR-573 is a negative regulator in the pathogenesis of rheumatoid arthritis. Cell Mol Immunol 2016;13(6):839–49. https://doi.org/10.1038/cmi.2015.63. Epub 2015/07/15. PubMed PMID: 26166764; PMCID: PMC5101444.

[147] Wang ZC, Lu H, Zhou Q, Yu SM, Mao YL, Zhang HJ, Zhang PC, Yan WJ. MiR-451 inhibits synovial fibroblasts proliferation and inflammatory cytokines secretion in rheumatoid arthritis through mediating p38MAPK signaling pathway. Int J Clin Exp Pathol 2015;8(11):14562–7. Epub 2016/01/30. PubMed PMID: 26823778; PMCID: PMC4713564.

[148] Compston A, Coles A. Multiple sclerosis. Lancet 2008;372(9648):1502–17. https://doi.org/10.1016/S0140-6736(08)61620-7. Epub 2008/10/31. PubMed PMID: 18970977.

[149] Chen C, Zhou Y, Wang J, Yan Y, Peng L, Qiu W. Dysregulated MicroRNA involvement in multiple sclerosis by induction of T helper 17 cell differentiation. Front Immunol 2018;9:1256. https://doi.org/10.3389/fimmu.2018.01256. Epub 2018/06/20. PubMed PMID: 29915595; PMCID: PMC5994557.

[150] Kucukali CI, Kurtuncu M, Coban A, Cebi M, Tuzun E. Epigenetics of multiple sclerosis: an updated review. NeuroMolecular Med 2015;17(2):83–96. https://doi.org/10.1007/s12017-014-8298-6. Epub 2014/03/22. PubMed PMID: 24652042.

[151] Lopez-Ramirez MA, Wu D, Pryce G, Simpson JE, Reijerkerk A, King-Robson J, Kay O, de Vries HE, Hirst MC, Sharrack B, Baker D, Male DK, Michael GJ, Romero IA. MicroRNA-155 negatively affects blood-brain barrier function during neuroinflammation. Faseb J 2014;28(6):2551–65. https://doi.org/10.1096/fj.13-248880. Epub 2014/03/08. PubMed PMID: 24604078.

[152] Ma X, Zhou J, Zhong Y, Jiang L, Mu P, Li Y, Singh N, Nagarkatti M, Nagarkatti P. Expression, regulation and function of microRNAs in multiple sclerosis. Int J Med Sci 2014;11(8):810–8. https://doi.org/10.7150/ijms.8647. Epub 2014/06/18. PubMed PMID: 24936144; PMCID: PMC4057480.

[153] Thamilarasan M, Koczan D, Hecker M, Paap B, Zettl UK. MicroRNAs in multiple sclerosis and experimental autoimmune encephalomyelitis. Autoimmun Rev 2012;11(3):174–9. https://doi.org/10.1016/j.autrev.2011.05.009. Epub 2011/05/31. PubMed PMID: 21621006.

[154] Chen F, Hu SJ. Effect of microRNA-34a in cell cycle, differentiation, and apoptosis: a review. J Biochem Mol Toxicol 2012;26(2):79–86. https://doi.org/10.1002/jbt.20412. Epub 2011/12/14. PubMed PMID: 22162084.

[155] Ghadiri N, Emamnia N, Ganjalikhani-Hakemi M, Ghaedi K, Etemadifar M, Salehi M, Shirzad H, Nasr-Esfahani MH. Analysis of the expression of mir-34a, mir-199a, mir-30c and mir-19a in peripheral blood CD4+T lymphocytes of relapsing-remitting multiple sclerosis patients. Gene 2018;659:109–17. https://doi.org/10.1016/j.gene.2018.03.035. Epub 2018/03/20. PubMed PMID: 29551498.

[156] Guerau-de-Arellano M, Smith KM, Godlewski J, Liu Y, Winger R, Lawler SE, Whitacre CC, Racke MK, Lovett-Racke AE. Micro-RNA dysregulation in multiple sclerosis favours pro-inflammatory T-cell-mediated autoimmunity. Brain 2011;134(Pt 12):3578–89. https://doi.org/10.1093/brain/awr262. Epub 2011/11/18. PubMed PMID: 22088562; PMCID: PMC3235556.

[157] Nakahama T, Hanieh H, Nguyen NT, Chinen I, Ripley B, Millrine D, Lee S, Nyati KK, Dubey PK, Chowdhury K, Kawahara Y, Kishimoto T. Aryl hydrocarbon receptor-mediated induction of the microRNA-132/212 cluster promotes interleukin-17-producing T-helper cell differentiation. Proc Natl Acad Sci USA 2013;110(29):11964–9. https://doi.org/10.1073/pnas.1311087110. Epub 2013/07/03. PubMed PMID: 23818645; PMCID: PMC3718186.

[158] Li T, Morgan MJ, Choksi S, Zhang Y, Kim YS, Liu ZG. MicroRNAs modulate the noncanonical transcription factor NF-kappaB pathway by regulating expression of the kinase IKKalpha during macrophage differentiation. Nat Immunol 2010;11(9):799–805. https://doi.org/10.1038/ni.1918. Epub 2010/08/17. PubMed PMID: 20711193; PMCID: PMC2926307.

[159] Sievers C, Meira M, Hoffmann F, Fontoura P, Kappos L, Lindberg RL. Altered microRNA expression in B lymphocytes in multiple sclerosis: towards a better understanding of treatment effects. Clin Immunol 2012;144(1):70–9. https://doi.org/10.1016/j.clim.2012.04.002. Epub 2012/06/05. PubMed PMID: 22659298.

[160] Aung LL, Mouradian MM, Dhib-Jalbut S, Balashov KE. MMP-9 expression is increased in B lymphocytes during multiple sclerosis exacerbation and is regulated by microRNA-320a. J Neuroimmunol 2015;278:185–9. https://doi.org/10.1016/j.jneuroim.2014.11.004. Epub 2014/12/04. PubMed PMID: 25468268; PMCID: PMC4297694.

[161] Liu X, He F, Pang R, Zhao D, Qiu W, Shan K, Zhang J, Lu Y, Li Y, Wang Y. Interleukin-17 (IL-17)-induced microRNA 873 (miR-873) contributes to the pathogenesis of experimental autoimmune encephalomyelitis by targeting A20 ubiquitin-editing enzyme. J Biol Chem 2014;289(42):28971–86. https://doi.org/10.1074/jbc.M114.577429. Epub 2014/09/04. PubMed PMID: 25183005; PMCID: PMC4200254.

[162] Junker A, Krumbholz M, Eisele S, Mohan H, Augstein F, Bittner R, Lassmann H, Wekerle H, Hohlfeld R, Meinl E. MicroRNA profiling of multiple sclerosis lesions identifies modulators of the regulatory protein

CD47. Brain 2009;132(Pt 12):3342–52. https://doi.org/10.1093/brain/awp300. Epub 2009/12/03. PubMed PMID: 19952055.

[163] Pusic AD, Kraig RP. Youth and environmental enrichment generate serum exosomes containing miR-219 that promote CNS myelination. Glia 2014;62(2):284–99. https://doi.org/10.1002/glia.22606. Epub 2013/12/18. PubMed PMID: 24339157; PMCID: PMC4096126.

[164] Marangon D, Boda E, Parolisi R, Negri C, Giorgi C, Montarolo F, Perga S, Bertolotto A, Buffo A, Abbracchio MP, Lecca D. In vivo silencing of miR-125a-3p promotes myelin repair in models of white matter demyelination. Glia 2020;68(10):2001–14. https://doi.org/10.1002/glia.23819. Epub 2020/03/13. PubMed PMID: 32163190.

[165] Teuber-Hanselmann S, Meinl E, Junker A. MicroRNAs in gray and white matter multiple sclerosis lesions: impact on pathophysiology. J Pathol 2020;250(5):496–509. https://doi.org/10.1002/path.5399. Epub 2020/02/20. PubMed PMID: 32073139.

[166] Keller A, Leidinger P, Lange J, Borries A, Schroers H, Scheffler M, Lenhof HP, Ruprecht K, Meese E. Multiple sclerosis: microRNA expression profiles accurately differentiate patients with relapsing-remitting disease from healthy controls. PLoS One 2009;4(10):e7440. https://doi.org/10.1371/journal.pone.0007440. Epub 2009/10/14. PubMed PMID: 19823682; PMCID: PMC2757919.

[167] Sondergaard HB, Hesse D, Krakauer M, Sorensen PS, Sellebjerg F. Differential microRNA expression in blood in multiple sclerosis. Mult Scler 2013;19(14):1849–57. https://doi.org/10.1177/1352458513490542. Epub 2013/06/19. PubMed PMID: 23773985.

[168] Siegel SR, Mackenzie J, Chaplin G, Jablonski NG, Griffiths L. Circulating microRNAs involved in multiple sclerosis. Mol Biol Rep 2012;39(5):6219–25. https://doi.org/10.1007/s11033-011-1441-7. Epub 2012/01/11. PubMed PMID: 22231906.

[169] Ebrahimkhani S, Vafaee F, Young PE, Hur SSJ, Hawke S, Devenney E, Beadnall H, Barnett MH, Suter CM, Buckland ME. Exosomal microRNA signatures in multiple sclerosis reflect disease status. Sci Rep 2017;7(1):14293. https://doi.org/10.1038/s41598-017-14301-3. Epub 2017/11/01. PubMed PMID: 29084979; PMCID: PMC5662562.

[170] Honardoost MA, Kiani-Esfahani A, Ghaedi K, Etemadifar M, Salehi M. miR-326 and miR-26a, two potential markers for diagnosis of relapse and remission phases in patient with relapsing-remitting multiple sclerosis. Gene 2014;544(2):128–33. https://doi.org/10.1016/j.gene.2014.04.069. Epub 2014/05/06. PubMed PMID: 24792898.

[171] Mahmoud FM, ElSheshtawy NM, Zaki WK, Zamzam DM, Fahim NM. MicroRNA 26a expression in peripheral blood mononuclear cells and correlation with serum interleukin-17 in relapsing-remitting multiple sclerosis patients. Egypt J Immunol 2017;24(2):71–82. Epub 2018/03/13. PubMed PMID: 29528581.

[172] Haghikia A, Haghikia A, Hellwig K, Baraniskin A, Holzmann A, Decard BF, Thum T, Gold R. Regulated microRNAs in the CSF of patients with multiple sclerosis: a case-control study. Neurology 2012;79(22):2166–70. https://doi.org/10.1212/WNL.0b013e3182759621. Epub 2012/10/19. PubMed PMID: 23077021.

[173] Bergman P, Piket E, Khademi M, James T, Brundin L, Olsson T, Piehl F, Jagodic M. Circulating miR-150 in CSF is a novel candidate biomarker for multiple sclerosis. Neurol Neuroimmunol Neuroinflamm 2016;3(3):e219. https://doi.org/10.1212/NXI.0000000000000219. Epub 2016/05/05. PubMed PMID: 27144214; PMCID: PMC4841644.

[174] Bruinsma IB, van Dijk M, Bridel C, van de Lisdonk T, Haverkort SQ, Runia TF, Steinman L, Hintzen RQ, Killestein J, Verbeek MM, Teunissen CE, de Jong BA. Regulator of oligodendrocyte maturation, miR-219, a potential biomarker for MS. J Neuroinflammation 2017;14(1):235. https://doi.org/10.1186/s12974-017-1006-3. Epub 2017/12/06. PubMed PMID: 29202778; PMCID: PMC5716023.

[175] Munoz-San Martin M, Reverter G, Robles-Cedeno R, Buxo M, Ortega FJ, Gomez I, Tomas-Roig J, Celarain N, Villar LM, Perkal H, Fernandez-Real JM, Quintana E, Ramio-Torrenta L. Analysis of miRNA signatures in CSF identifies upregulation of miR-21 and miR-146a/b in patients with multiple sclerosis and active lesions. J Neuroinflammation 2019;16(1):220. https://doi.org/10.1186/s12974-019-1590-5. Epub 2019/11/16. PubMed PMID: 31727077; PMCID: PMC6857276.

[176] Ahlbrecht J, Martino F, Pul R, Skripuletz T, Suhs KW, Schauerte C, Yildiz O, Trebst C, Tasto L, Thum S, Pfanne A, Roesler R, Lauda F, Hecker M, Zettl UK, Tumani H, Thum T, Stangel M. Deregulation of microRNA-181c in cerebrospinal fluid of patients with clinically isolated syndrome is associated with early conversion to relapsing-remitting multiple sclerosis. Mult Scler 2016;22(9):1202−14. https://doi.org/10.1177/1352458515613641. Epub 2015/10/24. PubMed PMID: 26493127.

[177] Osorio-Querejeta I, Carregal-Romero S, Ayerdi-Izquierdo A, Mager I, N.L A, Wood M, Egimendia A, Betanzos M, Alberro A, Iparraguirre L, Moles L, Llarena I, Moller M, Goni-de-Cerio F, Bijelic G, Ramos-Cabrer P, Munoz-Culla M, Otaegui D. MiR-219a-5p enriched extracellular vesicles induce OPC differentiation and EAE improvement more efficiently than liposomes and polymeric nanoparticles. Pharmaceutics 2020;12(2). https://doi.org/10.3390/pharmaceutics12020186. Epub 2020/02/27. PubMed PMID: 32098213; PMCID: PMC7076664.

[178] Wu F, Zikusoka M, Trindade A, Dassopoulos T, Harris ML, Bayless TM, Brant SR, Chakravarti S, Kwon JH. MicroRNAs are differentially expressed in ulcerative colitis and alter expression of macrophage inflammatory peptide-2 alpha. Gastroenterology 2008;135(5):1624−35. https://doi.org/10.1053/j.gastro.2008.07.068. e24. Epub 2008/10/07. PubMed PMID: 18835392.

[179] Bian Z, Li L, Cui J, Zhang H, Liu Y, Zhang CY, Zen K. Role of miR-150-targeting c-Myb in colonic epithelial disruption during dextran sulphate sodium-induced murine experimental colitis and human ulcerative colitis. J Pathol 2011;225(4):544−53. https://doi.org/10.1002/path.2907. Epub 2011/05/19. PubMed PMID: 21590770.

[180] Fasseu M, Treton X, Guichard C, Pedruzzi E, Cazals-Hatem D, Richard C, Aparicio T, Daniel F, Soule JC, Moreau R, Bouhnik Y, Laburthe M, Groyer A, Ogier-Denis E. Identification of restricted subsets of mature microRNA abnormally expressed in inactive colonic mucosa of patients with inflammatory bowel disease. PLoS One 2010;5(10). https://doi.org/10.1371/journal.pone.0013160. Epub 2010/10/20. PubMed PMID: 20957151; PMCID: PMC2950152.

[181] Feng X, Wang H, Ye S, Guan J, Tan W, Cheng S, Wei G, Wu W, Wu F, Zhou Y. Up-regulation of microRNA-126 may contribute to pathogenesis of ulcerative colitis via regulating NF-kappaB inhibitor IkappaBalpha. PLoS One 2012;7(12):e52782. https://doi.org/10.1371/journal.pone.0052782. Epub 2013/01/04. PubMed PMID: 23285182; PMCID: PMC3532399.

[182] Lin J, Welker NC, Zhao Z, Li Y, Zhang J, Reuss SA, Zhang X, Lee H, Liu Y, Bronner MP. Novel specific microRNA biomarkers in idiopathic inflammatory bowel disease unrelated to disease activity. Mod Pathol 2014;27(4):602−8. https://doi.org/10.1038/modpathol.2013.152. Epub 2013/09/21. PubMed PMID: 24051693.

[183] Takagi T, Naito Y, Mizushima K, Hirata I, Yagi N, Tomatsuri N, Ando T, Oyamada Y, Isozaki Y, Hongo H, Uchiyama K, Handa O, Kokura S, Ichikawa H, Yoshikawa T. Increased expression of microRNA in the inflamed colonic mucosa of patients with active ulcerative colitis. J Gastroenterol Hepatol 2010;25(Suppl. 1):S129−33. https://doi.org/10.1111/j.1440-1746.2009.06216.x. Epub 2010/07/14. PubMed PMID: 20586854.

[184] Yang Y, Ma Y, Shi C, Chen H, Zhang H, Chen N, Zhang P, Wang F, Yang J, Yang J, Zhu Q, Liang Y, Wu W, Gao R, Yang Z, Zou Y, Qin H. Overexpression of miR-21 in patients with ulcerative colitis impairs intestinal epithelial barrier function through targeting the Rho GTPase RhoB. Biochem Biophys Res Commun 2013;434(4):746–52. https://doi.org/10.1016/j.bbrc.2013.03.122. Epub 2013/04/16. PubMed PMID: 23583411.

[185] Paraskevi A, Theodoropoulos G, Papaconstantinou I, Mantzaris G, Nikiteas N, Gazouli M. Circulating MicroRNA in inflammatory bowel disease. J Crohns Colitis 2012;6(9):900–4. https://doi.org/10.1016/j.crohns.2012.02.006. Epub 2012/03/06. PubMed PMID: 22386737.

[186] Wu F, Guo NJ, Tian H, Marohn M, Gearhart S, Bayless TM, Brant SR, Kwon JH. Peripheral blood microRNAs distinguish active ulcerative colitis and Crohn's disease. Inflamm Bowel Dis 2011;17(1): 241–50. https://doi.org/10.1002/ibd.21450. Epub 2010/09/03. PubMed PMID: 20812331; PMCID: PMC2998576.

[187] Wu F, Zhang S, Dassopoulos T, Harris ML, Bayless TM, Meltzer SJ, Brant SR, Kwon JH. Identification of microRNAs associated with ileal and colonic Crohn's disease. Inflamm Bowel Dis 2010;16(10):1729–38. https://doi.org/10.1002/ibd.21267. Epub 2010/09/18. PubMed PMID: 20848482; PMCID: PMC2946509.

[188] Wu H, Neilson JR, Kumar P, Manocha M, Shankar P, Sharp PA, Manjunath N. miRNA profiling of naive, effector and memory CD8 T cells. PLoS One 2007;2(10):e1020. https://doi.org/10.1371/journal.pone.0001020. Epub 2007/10/11. PubMed PMID: 17925868; PMCID: PMC2000354.

[189] Evel-Kabler K, Song XT, Aldrich M, Huang XF, Chen SY. SOCS1 restricts dendritic cells' ability to break self tolerance and induce antitumor immunity by regulating IL-12 production and signaling. J Clin Invest 2006;116(1):90–100. https://doi.org/10.1172/JCI26169. Epub 2005/12/17. PubMed PMID: 16357940; PMCID: PMC1312019.

[190] Worm J, Stenvang J, Petri A, Frederiksen KS, Obad S, Elmen J, Hedtjarn M, Straarup EM, Hansen JB, Kauppinen S. Silencing of microRNA-155 in mice during acute inflammatory response leads to derepression of c/ebp Beta and down-regulation of G-CSF. Nucleic Acids Res 2009;37(17):5784–92. https://doi.org/10.1093/nar/gkp577. Epub 2009/07/15. PubMed PMID: 19596814; PMCID: PMC2761263.

[191] Li Z, Lu J, Sun M, Mi S, Zhang H, Luo RT, Chen P, Wang Y, Yan M, Qian Z, Neilly MB, Jin J, Zhang Y, Bohlander SK, Zhang DE, Larson RA, Le Beau MM, Thirman MJ, Golub TR, Rowley JD, Chen J. Distinct microRNA expression profiles in acute myeloid leukemia with common translocations. Proc Natl Acad Sci USA 2008;105(40):15535–40. https://doi.org/10.1073/pnas.0808266105. Epub 2008/10/04. PubMed PMID: 18832181; PMCID: PMC2563085.

[192] Feng R, Chen X, Yu Y, Su L, Yu B, Li J, Cai Q, Yan M, Liu B, Zhu Z. miR-126 functions as a tumour suppressor in human gastric cancer. Cancer Lett 2010;298(1):50–63. https://doi.org/10.1016/j.canlet.2010.06.004. Epub 2010/07/14. PubMed PMID: 20619534.

[193] Crawford M, Brawner E, Batte K, Yu L, Hunter MG, Otterson GA, Nuovo G, Marsh CB, Nana-Sinkam SP. MicroRNA-126 inhibits invasion in non-small cell lung carcinoma cell lines. Biochem Biophys Res Commun 2008;373(4):607–12. https://doi.org/10.1016/j.bbrc.2008.06.090. Epub 2008/07/08. PubMed PMID: 18602365.

[194] Sharma A, Kumar M, Aich J, Hariharan M, Brahmachari SK, Agrawal A, Ghosh B. Posttranscriptional regulation of interleukin-10 expression by hsa-miR-106a. Proc Natl Acad Sci USA 2009;106(14):5761–6. https://doi.org/10.1073/pnas.0808743106. Epub 2009/03/25. PubMed PMID: 19307576; PMCID: PMC2659714.

[195] Marlow GJ, van Gent D, Ferguson LR. Why interleukin-10 supplementation does not work in Crohn's disease patients. World J Gastroenterol 2013;19(25):3931–41. https://doi.org/10.3748/wjg.v19.i25.3931. Epub 2013/07/11. PubMed PMID: 23840137; PMCID: PMC3703179.

[196] Lindsay J, Van Montfrans C, Brennan F, Van Deventer S, Drillenburg P, Hodgson H, Te Velde A, Sol Rodriguez Pena M. IL-10 gene therapy prevents TNBS-induced colitis. Gene Ther 2002;9(24):1715–21. https://doi.org/10.1038/sj.gt.3301841. Epub 2002/11/29. PubMed PMID: 12457286.

[197] Polyak S, Mach A, Porvasnik S, Dixon L, Conlon T, Erger KE, Acosta A, Wright AJ, Campbell-Thompson M, Zolotukhin I, Wasserfall C, Mah C. Identification of adeno-associated viral vectors suitable for intestinal gene delivery and modulation of experimental colitis. Am J Physiol Gastrointest Liver Physiol 2012;302(3):G296–308. https://doi.org/10.1152/ajpgi.00562.2010. Epub 2011/11/25. PubMed PMID: 22114116.

[198] Griffiths CE, Barker JN. Pathogenesis and clinical features of psoriasis. Lancet 2007;370(9583):263–71. https://doi.org/10.1016/S0140-6736(07)61128-3. Epub 2007/07/31. PubMed PMID: 17658397.

[199] Kryczek I, Bruce AT, Gudjonsson JE, Johnston A, Aphale A, Vatan L, Szeliga W, Wang Y, Liu Y, Welling TH, Elder JT, Zou W. Induction of IL-17+ T cell trafficking and development by IFN-gamma: mechanism and pathological relevance in psoriasis. J Immunol 2008;181(7):4733–41. https://doi.org/10.4049/jimmunol.181.7.4733. Epub 2008/09/20. PubMed PMID: 18802076; PMCID: PMC2677162.

[200] Langrish CL, Chen Y, Blumenschein WM, Mattson J, Basham B, Sedgwick JD, McClanahan T, Kastelein RA, Cua DJ. IL-23 drives a pathogenic T cell population that induces autoimmune inflammation. J Exp Med 2005;201(2):233–40. https://doi.org/10.1084/jem.20041257. Epub 2005/01/20. PubMed PMID: 15657292; PMCID: PMC2212798.

[201] Lowes MA, Kikuchi T, Fuentes-Duculan J, Cardinale I, Zaba LC, Haider AS, Bowman EP, Krueger JG. Psoriasis vulgaris lesions contain discrete populations of Th1 and Th17 T cells. J Invest Dermatol 2008; 128(5):1207–11. https://doi.org/10.1038/sj.jid.5701213. Epub 2008/01/18. PubMed PMID: 18200064.

[202] Glitzner E, Korosec A, Brunner PM, Drobits B, Amberg N, Schonthaler HB, Kopp T, Wagner EF, Stingl G, Holcmann M, Sibilia M. Specific roles for dendritic cell subsets during initiation and progression of psoriasis. EMBO Mol Med 2014;6(10):1312–27. https://doi.org/10.15252/emmm.201404114. Epub 2014/09/14. PubMed PMID: 25216727; PMCID: PMC4287934.

[203] Lande R, Botti E, Jandus C, Dojcinovic D, Fanelli G, Conrad C, Chamilos G, Feldmeyer L, Marinari B, Chon S, Vence L, Riccieri V, Guillaume P, Navarini AA, Romero P, Costanzo A, Piccolella E, Gilliet M, Frasca L. The antimicrobial peptide LL37 is a T-cell autoantigen in psoriasis. Nat Commun 2014;5:5621. https://doi.org/10.1038/ncomms6621. Epub 2014/12/04. PubMed PMID: 25470744.

[204] Sonkoly E, Wei T, Janson PC, Saaf A, Lundeberg L, Tengvall-Linder M, Norstedt G, Alenius H, Homey B, Scheynius A, Stahle M, Pivarcsi A. MicroRNAs: novel regulators involved in the pathogenesis of psoriasis? PLoS One 2007;2(7):e610. https://doi.org/10.1371/journal.pone.0000610. Epub 2007/07/12. PubMed PMID: 17622355; PMCID: PMC1905940.

[205] Lerman G, Avivi C, Mardoukh C, Barzilai A, Tessone A, Gradus B, Pavlotsky F, Barshack I, Polak-Charcon S, Orenstein A, Hornstein E, Sidi Y, Avni D. MiRNA expression in psoriatic skin: reciprocal regulation of hsa-miR-99a and IGF-1R. PLoS One 2011;6(6):e20916. https://doi.org/10.1371/journal.-pone.0020916. Epub 2011/06/21. PubMed PMID: 21687694; PMCID: PMC3110257.

[206] Zibert JR, Lovendorf MB, Litman T, Olsen J, Kaczkowski B, Skov L. MicroRNAs and potential target interactions in psoriasis. J Dermatol Sci 2010;58(3):177–85. https://doi.org/10.1016/j.jdermsci.2010.03.004. Epub 2010/04/27. PubMed PMID: 20417062.

[207] Sonkoly E, Wei T, Pavez Lorie E, Suzuki H, Kato M, Torma H, Stahle M, Pivarcsi A. Protein kinase C-dependent upregulation of miR-203 induces the differentiation of human keratinocytes. J Invest Dermatol 2010;130(1):124–34. https://doi.org/10.1038/jid.2009.294. Epub 2009/09/18. PubMed PMID: 19759552.

[208] Wei T, Xu N, Meisgen F, Stahle M, Sonkoly E, Pivarcsi A. Interleukin-8 is regulated by miR-203 at the post-transcriptional level in primary human keratinocytes. Eur J Dermatol 2013. https://doi.org/10.1684/ejd.2013.1997. Epub 2013/04/24. PubMed PMID: 23608026.

[209] Yi R, Poy MN, Stoffel M, Fuchs E. A skin microRNA promotes differentiation by repressing 'stemness. Nature 2008;452(7184):225–9. https://doi.org/10.1038/nature06642. Epub 2008/03/04. PubMed PMID: 18311128; PMCID: PMC4346711.

[210] Xia P, Fang X, Zhang ZH, Huang Q, Yan KX, Kang KF, Han L, Zheng ZZ. Dysregulation of miRNA146a versus IRAK1 induces IL-17 persistence in the psoriatic skin lesions. Immunol Lett 2012;148(2):151–62. https://doi.org/10.1016/j.imlet.2012.09.004. Epub 2012/09/29. PubMed PMID: 23018031.

[211] O'Connell RM, Rao DS, Baltimore D. microRNA regulation of inflammatory responses. Annu Rev Immunol 2012;30:295–312. https://doi.org/10.1146/annurev-immunol-020711-075013. Epub 2012/01/10. PubMed PMID: 22224773.

[212] Zhao M, Wang LT, Liang GP, Zhang P, Deng XJ, Tang Q, Zhai HY, Chang CC, Su YW, Lu QJ. Up-regulation of microRNA-210 induces immune dysfunction via targeting FOXP3 in CD4(+) T cells of psoriasis vulgaris. Clin Immunol 2014;150(1):22–30. https://doi.org/10.1016/j.clim.2013.10.009. Epub 2013/12/10. PubMed PMID: 24316592.

[213] Guinea-Viniegra J, Jimenez M, Schonthaler HB, Navarro R, Delgado Y, Concha-Garzon MJ, Tschachler E, Obad S, Dauden E, Wagner EF. Targeting miR-21 to treat psoriasis. Sci Transl Med 2014;6(225):225re1. https://doi.org/10.1126/scitranslmed.3008089. Epub 2014/02/28. PubMed PMID: 24574341.

[214] Meisgen F, Xu N, Wei T, Janson PC, Obad S, Broom O, Nagy N, Kauppinen S, Kemeny L, Stahle M, Pivarcsi A, Sonkoly E. MiR-21 is up-regulated in psoriasis and suppresses T cell apoptosis. Exp Dermatol 2012;21(4):312–4. https://doi.org/10.1111/j.1600-0625.2012.01462.x. Epub 2012/03/16. PubMed PMID: 22417311.

[215] Xu N, Meisgen F, Butler LM, Han G, Wang XJ, Soderberg-Naucler C, Stahle M, Pivarcsi A, Sonkoly E. MicroRNA-31 is overexpressed in psoriasis and modulates inflammatory cytokine and chemokine production in keratinocytes via targeting serine/threonine kinase 40. J Immunol 2013;190(2):678–88. https://doi.org/10.4049/jimmunol.1202695. Epub 2012/12/13. PubMed PMID: 23233723.

[216] Peng H, Kaplan N, Hamanaka RB, Katsnelson J, Blatt H, Yang W, Hao L, Bryar PJ, Johnson RS, Getsios S, Chandel NS, Lavker RM. microRNA-31/factor-inhibiting hypoxia-inducible factor 1 nexus regulates keratinocyte differentiation. Proc Natl Acad Sci USA 2012;109(35):14030–4. https://doi.org/10.1073/pnas.1111292109. Epub 2012/08/15. PubMed PMID: 22891326; PMCID: PMC3435188.

[217] Joyce CE, Zhou X, Xia J, Ryan C, Thrash B, Menter A, Zhang W, Bowcock AM. Deep sequencing of small RNAs from human skin reveals major alterations in the psoriasis miRNAome. Hum Mol Genet 2011;20(20):4025–40. https://doi.org/10.1093/hmg/ddr331. Epub 2011/08/03. PubMed PMID: 21807764; PMCID: PMC3177648.

[218] Ichihara A, Jinnin M, Yamane K, Fujisawa A, Sakai K, Masuguchi S, Fukushima S, Maruo K, Ihn H. microRNA-mediated keratinocyte hyperproliferation in psoriasis vulgaris. Br J Dermatol 2011;165(5):1003–10. https://doi.org/10.1111/j.1365-2133.2011.10497.x. Epub 2011/06/30. PubMed PMID: 21711342.

[219] Xu N, Brodin P, Wei T, Meisgen F, Eidsmo L, Nagy N, Kemeny L, Stahle M, Sonkoly E, Pivarcsi A. MiR-125b, a microRNA downregulated in psoriasis, modulates keratinocyte proliferation by targeting FGFR2. J Invest Dermatol 2011;131(7):1521–9. https://doi.org/10.1038/jid.2011.55. Epub 2011/03/18. PubMed PMID: 21412257.

[220] Tsuru Y, Jinnin M, Ichihara A, Fujisawa A, Moriya C, Sakai K, Fukushima S, Ihn H. miR-424 levels in hair shaft are increased in psoriatic patients. J Dermatol 2014;41(5):382–5. https://doi.org/10.1111/1346-8138.12460. Epub 2014/03/19. PubMed PMID: 24628460.

[221] Tili E, Michaille JJ, Cimino A, Costinean S, Dumitru CD, Adair B, Fabbri M, Alder H, Liu CG, Calin GA, Croce CM. Modulation of miR-155 and miR-125b levels following lipopolysaccharide/TNF-alpha stimulation and their possible roles in regulating the response to endotoxin shock. J Immunol 2007;179(8):5082–9. https://doi.org/10.4049/jimmunol.179.8.5082. Epub 2007/10/04. PubMed PMID: 17911593.

[222] Ventriglia G, Nigi L, Sebastiani G, Dotta F. MicroRNAs: novel players in the dialogue between pancreatic islets and immune system in autoimmune diabetes. BioMed Res Int 2015;2015:749734. https://doi.org/10.1155/2015/749734. Epub 2015/09/05. PubMed PMID: 26339637; PMCID: PMC4538424.

[223] Sims EK, Lakhter AJ, Anderson-Baucum E, Kono T, Tong X, Evans-Molina C. MicroRNA 21 targets BCL2 mRNA to increase apoptosis in rat and human beta cells. Diabetologia 2017;60(6):1057–65. https://doi.org/10.1007/s00125-017-4237-z. Epub 2017/03/11. PubMed PMID: 28280903; PMCID: PMC5425307.

[224] Tomita T. Apoptosis of pancreatic beta-cells in Type 1 diabetes. Bosn J Basic Med Sci 2017;17(3):183–93. https://doi.org/10.17305/bjbms.2017.1961. Epub 2017/04/04. PubMed PMID: 28368239; PMCID: PMC5581966.

[225] Zheng Y, Wang Z, Tu Y, Shen H, Dai Z, Lin J, Zhou Z. miR-101a and miR-30b contribute to inflammatory cytokine-mediated beta-cell dysfunction. Lab Invest 2015;95(12):1387–97. https://doi.org/10.1038/labinvest.2015.112. Epub 2015/09/15. PubMed PMID: 26367486.

[226] Nabih ES, Andrawes NG. The association between circulating levels of miRNA-181a and pancreatic beta cells dysfunction via SMAD7 in type 1 diabetic children and adolescents. J Clin Lab Anal 2016;30(5):727–31. https://doi.org/10.1002/jcla.21928. Epub 2016/02/20. PubMed PMID: 26892629; PMCID: PMC6807027.

[227] Tsukita S, Yamada T, Takahashi K, Munakata Y, Hosaka S, Takahashi H, Gao J, Shirai Y, Kodama S, Asai Y, Sugisawa T, Chiba Y, Kaneko K, Uno K, Sawada S, Imai J, Katagiri H. MicroRNAs 106b and 222 improve hyperglycemia in a mouse model of insulin-deficient diabetes via pancreatic beta-cell proliferation. EBioMedicine 2017;15:163–72. https://doi.org/10.1016/j.ebiom.2016.12.002. Epub 2016/12/16. PubMed PMID: 27974246; PMCID: PMC5233820.

[228] Xu H, Abuhatzira L, Carmona GN, Vadrevu S, Satin LS, Notkins AL. The Ia-2beta intronic miRNA, miR-153, is a negative regulator of insulin and dopamine secretion through its effect on the Cacna1c gene in mice. Diabetologia 2015;58(10):2298–306. https://doi.org/10.1007/s00125-015-3683-8. Epub 2015/07/05. PubMed PMID: 26141787; PMCID: PMC6754265.

[229] Roep BO, Kracht MJ, van Lummel M, Zaldumbide A. A roadmap of the generation of neoantigens as targets of the immune system in type 1 diabetes. Curr Opin Immunol 2016;43:67–73. https://doi.org/10.1016/j.coi.2016.09.007. Epub 2016/10/11. PubMed PMID: 27723537.

[230] de Jong VM, van der Slik AR, Laban S, van 't Slot R, Koeleman BP, Zaldumbide A, Roep BO. Survival of autoreactive T lymphocytes by microRNA-mediated regulation of apoptosis through TRAIL and Fas in type 1 diabetes. Gene Immun 2016;17(6):342–8. https://doi.org/10.1038/gene.2016.29. Epub 2016/07/29. PubMed PMID: 27467285.

[231] Salama A, Fichou N, Allard M, Dubreil L, De Beaurepaire L, Viel A, Jegou D, Bosch S, Bach JM. MicroRNA-29b modulates innate and antigen-specific immune responses in mouse models of autoimmunity. PLoS One 2014;9(9):e106153. https://doi.org/10.1371/journal.pone.0106153. Epub 2014/09/10. PubMed PMID: 25203514; PMCID: PMC4159199.

[232] Nielsen LB, Wang C, Sorensen K, Bang-Berthelsen CH, Hansen L, Andersen ML, Hougaard P, Juul A, Zhang CY, Pociot F, Mortensen HB. Circulating levels of microRNA from children with newly diagnosed type 1 diabetes and healthy controls: evidence that miR-25 associates to residual beta-cell function and glycaemic control during disease progression. Exp Diabetes Res 2012;2012:896362. https://doi.org/10.1155/2012/896362. Epub 2012/07/26. PubMed PMID: 22829805; PMCID: PMC3398606.

[233] Salas-Perez F, Codner E, Valencia E, Pizarro C, Carrasco E, Perez-Bravo F. MicroRNAs miR-21a and miR-93 are down regulated in peripheral blood mononuclear cells (PBMCs) from patients with type 1 diabetes. Immunobiology 2013;218(5):733–7. https://doi.org/10.1016/j.imbio.2012.08.276. Epub 2012/09/25. PubMed PMID: 22999472.

[234] Yang M, Ye L, Wang B, Gao J, Liu R, Hong J, Wang W, Gu W, Ning G. Decreased miR-146 expression in peripheral blood mononuclear cells is correlated with ongoing islet autoimmunity in type 1 diabetes patients 1miR-146. J Diabetes 2015;7(2):158–65. https://doi.org/10.1111/1753-0407.12163. Epub 2014/05/07. PubMed PMID: 24796653.
[235] Hu S, Vissink A, Arellano M, Roozendaal C, Zhou H, Kallenberg CG, Wong DT. Identification of autoantibody biomarkers for primary Sjogren's syndrome using protein microarrays. Proteomics 2011;11(8):1499–507. https://doi.org/10.1002/pmic.201000206. Epub 2011/03/18. PubMed PMID: 21413148; PMCID: PMC3209962.
[236] Karageorgas T, Fragioudaki S, Nezos A, Karaiskos D, Moutsopoulos HM, Mavragani CP. Fatigue in primary sjogren's syndrome: clinical, laboratory, psychometric, and biologic associations. Arthritis Care Res 2016;68(1):123–31. https://doi.org/10.1002/acr.22720. Epub 2015/09/01. PubMed PMID: 26315379.
[237] Kocer B, Tezcan ME, Batur HZ, Haznedaroglu S, Goker B, Irkec C, Cetinkaya R. Cognition, depression, fatigue, and quality of life in primary Sjogren's syndrome: correlations. Brain Behav 2016;6(12):e00586. https://doi.org/10.1002/brb3.586. Epub 2016/12/30. PubMed PMID: 28032007; PMCID: PMC5167008.
[238] Ramos-Casals M, Brito-Zeron P, Seror R, Bootsma H, Bowman SJ, Dorner T, Gottenberg JE, Mariette X, Theander E, Bombardieri S, De Vita S, Mandl T, Ng WF, Kruize A, Tzioufas A, Vitali C, Force ESST. Characterization of systemic disease in primary Sjogren's syndrome: EULAR-SS Task Force recommendations for articular, cutaneous, pulmonary and renal involvements. Rheumatology 2015;54(12):2230–8. https://doi.org/10.1093/rheumatology/kev200. Epub 2015/08/02. PubMed PMID: 26231345; PMCID: PMC6281074.
[239] Nair JJ, Singh TP. Sjogren's syndrome: review of the aetiology, Pathophysiology & Potential therapeutic interventions. J Clin Exp Dent 2017;9(4):e584–9. https://doi.org/10.4317/jced.53605. Epub 2017/05/05. PubMed PMID: 28469828; PMCID: PMC5410683 with regards to this work.
[240] Carsons SE, Patel BC. Sjogren syndrome. StatPearls. Treasure Island (FL). 2021.
[241] Garcia-Carrasco M, Fuentes-Alexandro S, Escarcega RO, Salgado G, Riebeling C, Cervera R. Pathophysiology of Sjogren's syndrome. Arch Med Res 2006;37(8):921–32. https://doi.org/10.1016/j.arcmed.2006.08.002. Epub 2006/10/19. PubMed PMID: 17045106.
[242] Alevizos I, Alexander S, Turner RJ, Illei GG. MicroRNA expression profiles as biomarkers of minor salivary gland inflammation and dysfunction in Sjogren's syndrome. Arthritis Rheum 2011;63(2):535–44. https://doi.org/10.1002/art.30131. Epub 2011/02/01. PubMed PMID: 21280008; PMCID: PMC3653295.
[243] Kapsogeorgou EK, Gourzi VC, Manoussakis MN, Moutsopoulos HM, Tzioufas AG. Cellular microRNAs (miRNAs) and Sjogren's syndrome: candidate regulators of autoimmune response and autoantigen expression. J Autoimmun 2011;37(2):129–35. https://doi.org/10.1016/j.jaut.2011.05.003. Epub 2011/06/03. PubMed PMID: 21632209.
[244] Gallo A, Tandon M, Alevizos I, Illei GG. The majority of microRNAs detectable in serum and saliva is concentrated in exosomes. PLoS One 2012;7(3):e30679. https://doi.org/10.1371/journal.pone.0030679. Epub 2012/03/20. PubMed PMID: 22427800; PMCID: PMC3302865.
[245] Pauley KM, Stewart CM, Gauna AE, Dupre LC, Kuklani R, Chan AL, Pauley BA, Reeves WH, Chan EK, Cha S. Altered miR-146a expression in Sjogren's syndrome and its functional role in innate immunity. Eur J Immunol 2011;41(7):2029–39. https://doi.org/10.1002/eji.201040757. Epub 2011/04/07. PubMed PMID: 21469088; PMCID: PMC3760391.
[246] Williams AE, Choi K, Chan AL, Lee YJ, Reeves WH, Bubb MR, Stewart CM, Cha S. Sjogren's syndrome-associated microRNAs in CD14(+) monocytes unveils targeted TGFbeta signaling. Arthritis Res Ther 2016;18(1):95. https://doi.org/10.1186/s13075-016-0987-0. Epub 2016/05/05. PubMed PMID: 27142093; PMCID: PMC4855899.

[247] Alevizos I, Illei GG. MicroRNAs in Sjogren's syndrome as a prototypic autoimmune disease. Autoimmun Rev 2010;9(9):618–21. https://doi.org/10.1016/j.autrev.2010.05.009. Epub 2010/05/12. PubMed PMID: 20457282; PMCID: PMC3408312.

[248] Shi H, Zheng LY, Zhang P, Yu CQ. miR-146a and miR-155 expression in PBMCs from patients with Sjogren's syndrome. J Oral Pathol Med 2014;43(10):792–7. https://doi.org/10.1111/jop.12187. Epub 2014/06/17. PubMed PMID: 24931100.

[249] Wang-Renault SF, Boudaoud S, Nocturne G, Roche E, Sigrist N, Daviaud C, Bugge Tinggaard A, Renault V, Deleuze JF, Mariette X, Tost J. Deregulation of microRNA expression in purified T and B lymphocytes from patients with primary Sjögren's syndrome. Ann Rheum Dis 2018;77(1):133–40. https://doi.org/10.1136/annrheumdis-2017-211417. Epub 2017/09/17. PubMed PMID: 28916716; PMCID: PMC5754740.

[250] Taganov KD, Boldin MP, Chang KJ, Baltimore D. NF-kappaB-dependent induction of microRNA miR-146, an inhibitor targeted to signaling proteins of innate immune responses. Proc Natl Acad Sci USA 2006;103(33):12481–6. https://doi.org/10.1073/pnas.0605298103. Epub 2006/08/04. PubMed PMID: 16885212; PMCID: PMC1567904.

[251] Zilahi E, Tarr T, Papp G, Griger Z, Sipka S, Zeher M. Increased microRNA-146a/b, TRAF6 gene and decreased IRAK1 gene expressions in the peripheral mononuclear cells of patients with Sjogren's syndrome. Immunol Lett 2012;141(2):165–8. https://doi.org/10.1016/j.imlet.2011.09.006. Epub 2011/10/29. PubMed PMID: 22033216.

[252] Le Dantec C, Varin MM, Brooks WH, Pers JO, Youinou P, Renaudineau Y. Epigenetics and Sjogren's syndrome. Curr Pharmaceut Biotechnol 2012;13(10):2046–53. https://doi.org/10.2174/138920112802273326. Epub 2012/01/03. PubMed PMID: 22208659.

[253] Lu LF, Thai TH, Calado DP, Chaudhry A, Kubo M, Tanaka K, Loeb GB, Lee H, Yoshimura A, Rajewsky K, Rudensky AY. Foxp3-dependent microRNA155 confers competitive fitness to regulatory T cells by targeting SOCS1 protein. Immunity 2009;30(1):80–91. https://doi.org/10.1016/j.immuni.2008.11.010. Epub 2009/01/16. PubMed PMID: 19144316; PMCID: PMC2654249.

[254] Fulci V, Chiaretti S, Goldoni M, Azzalin G, Carucci N, Tavolaro S, Castellano L, Magrelli A, Citarella F, Messina M, Maggio R, Peragine N, Santangelo S, Mauro FR, Landgraf P, Tuschl T, Weir DB, Chien M, Russo JJ, Ju J, Sheridan R, Sander C, Zavolan M, Guarini A, Foa R, Macino G. Quantitative technologies establish a novel microRNA profile of chronic lymphocytic leukemia. Blood 2007;109(11):4944–51. https://doi.org/10.1182/blood-2006-12-062398. Epub 2007/03/01. PubMed PMID: 17327404.

[255] Wang Y, Zhang G, Zhang L, Zhao M, Huang H. Decreased microRNA-181a and -16 expression levels in the labial salivary glands of Sjogren syndrome patients. Exp Ther Med 2018;15(1):426–32. https://doi.org/10.3892/etm.2017.5407. Epub 2018/02/02. PubMed PMID: 29387196; PMCID: PMC5769212.

[256] Peng L, Ma W, Yi F, Yang YJ, Lin W, Chen H, Zhang X, Zhang LH, Zhang F, Du Q. MicroRNA profiling in Chinese patients with primary Sjogren syndrome reveals elevated miRNA-181a in peripheral blood mononuclear cells. J Rheumatol 2014;41(11):2208–13. https://doi.org/10.3899/jrheum.131154. Epub 2014/08/17. PubMed PMID: 25128511.

[257] Bonneau E, Neveu B, Kostantin E, Tsongalis GJ, De Guire V. How close are miRNAs from clinical practice? A perspective on the diagnostic and therapeutic market. EJIFCC 2019;30(2):114–27. Epub 2019/07/03. PubMed PMID: 31263388; PMCID: PMC6599191.

CHAPTER

Post-transcriptional regulation of inflammatory disorder

13

Savita Devi

Department of Pathology and Laboratory Medicine, Cedars Sinai Medical Center, Los Angeles, CA, United States

Introduction

Rheumatoid arthritis (RA) affects approximately 0.5%−1% of the population worldwide. RA is a chronic autoimmune disease, demonstrated by the hyperactivation of the immune system and unrestricted secretion of cytokines and other inflammatory mediators in synovial joints. One of its features is synovitis, wherein the inflammation in the synovial membrane majorly affects both, large and small peripheral joints of the hand and feet. Systemic inflammation increases the comorbidities such as inflammation in lungs, blood vessels together with an enhanced likelihood of cardiovascular disease [1].

The inflammatory milieu in the joints is controlled by a network of cytokines and chemokines and clinical studies have confirmed the essential role of Tumor necrosis factor (TNF), interleukin-6 (IL-6), IL-1b, and granulocyte-monocyte colony-stimulating factor [2] in the process. Production of chemokines and cytokines aggravates the inflammatory response by infiltrating the synovial joints with macrophages and monocytes that further secretes the cytokines such as TNF and IL-1b and activates the fibroblast-like synoviocytes that eventually causes bone and cartilage degeneration [3]. Many mouse model studies also confirmed the importance of TNF-α in RA. TNF expression level is regulated at many different levels but in this chapter, we focus on the role of adenosine-uridine-rich element (ARE)-binding proteins, micro-RNA, and MAP kinases in post-transcriptional regulation of genes in RA.

Activation of the innate immune system in RA

Many different innate immune cells are present in the synovial membrane, whereas synovial fluid majorly holds neutrophils, which plays an imperative function during the inflammatory process inside the synovial fluid. The role of the macrophage colony-stimulating factor, granulocyte colony-stimulating factor, and granulocyte-macrophage colony-stimulating factor is to increase the maturation of these immune cells, effluxion from the bone marrow, and transfer to the synovium [4]. Particularly, macrophages play a pivotal function as a central effector of synovitis [5]. Macrophages signal mainly via the production of cytokines such as IL-23, IL-12, IL-1, IL-15, IL-18, IL-6, TNF-α, superoxide anion radicals, hydrogen peroxide, hydroxyl radical, nitrogen intermediates, prostaglandins, and antigen presentation.

Post-Transcriptional Gene Regulation in Human Disease, Volume 32. https://doi.org/10.1016/B978-0-323-91305-8.00009-0

This specific way of induction of nitric oxide synthase and proinflammatory cytokines propose an M1 macrophage phenotype predominantly. Macrophages are stimulated by toll-like receptors (TLRs) and nucleotide-binding oligomerization domain-like receptors (NLRs) that detect a wide variety of damage-associated molecular patterns and pathogen-associated molecular patterns that mostly comprises putative viral, bacterial ligands, and endogenous molecules [6]. Furthermore, microRNAs also play a crucial function in the pathogenesis of RA for example microRNA-155 regulates synovial cytokine expression [7,8].

Different clinical trials have provided the importance of the intracellular signaling pathways that are active in the synovium. In particular, kinases that control cytokine-receptor-mediated activities may aid in the generation of small-molecule inhibitors. For example, tofacitinib, a JAK1 and JAK3 inhibitor that intervene in the production of several cytokines and growth factors are useful in the treatment of RA.

Regulation: post-transcriptionally

Regulation of gene expression post-transcriptionally is an important adaptive mechanism executed by eukaryotic cells in response to different types of cellular modulations, for example, proliferation, cellular stress, or immune modulation [9]. This adaptive mechanism is very pivotal in regulating the induction of gene expression of early response genes, through the association with gene transcription and under the universal management of phosphorylation-mediated signaling events [10]. The crucial function of post-transcriptional gene regulation called attention in a gene array study, in which around 50% of the transcript that is upregulated by stress was found to be regulated mainly at the level of mRNA stability [11]. The half-life of unstable mRNAs is mostly in the range of 2-4-fold fluctuations [12]. Although it appears fair, however, perturbations of this range are shown to induce more than 1000-fold differences in steady-state transcript levels, which could eventually affect the rate of translation heavily [13].

The different stages of mRNA from transport, localization, stability, and translation are subject to the presence of the *cis*-regulatory elements that are found throughout the mRNA [13,14]. one of such sequences is the AREs that are conserved RNA motifs found within the 3′-untranslated region (UTR) of an mRNA and regulates a different subset of transcripts [15—17].

Regulation through ARE

Eukaryotic cells can adjust gene expression in response to an external environment by regulating the rates of mRNA degradation [18]. mRNA turnover is a very crucial step in the regulation of gene expression. TNF gene expression is also regulated at the level of mRNA turnover that is moderated through 5′- or 3′-UTRs of the mature mRNA, located adjacent to the protein-coding sequence and contain seven to eight copies of AREs with a sequence of AUUUA in their 3′-UTR region.

ARE of TNF-α is pivotal for post-transcriptional regulation. They mainly control the TNF-α mRNA transport from the nucleus to the cytoplasm [19], destabilize the encoded message, and also inhibit the TNF-α translation. ARE effects the translation of TNF mRNA via interactions between ARE-specific proteins and miRNA [20,21].

ARE-binding proteins

Various RNA-binding proteins have been discovered both in vitro and in vivo [22,23]. These proteins regulate the translation, transport, and stability of RNA. TNF-α translation is regulated RNA binding proteins such as heterogeneous nuclear ribonucleoprotein (hnRNP-A1), Hu antigen R (HuR, ELAVL1), T-cell intracellular Ag (TIA-1), TIA-1-related protein (TIAR), and tristetraprolin (TTP). These RNA binding proteins are recruited to the 3′-untranslated region and can inhibit the translation of TNF-α by various means. hnRNP-A1 inhibits the TNF translation by binding to ARE; furthermore, TIA-1 and TTP act downstream of Toll-like receptor signaling pathway to prevent the pathological consequences. TIA-1 impedes the translation of TNF-α transcripts and TTP promotes the degradation of TNF-α transcripts [24–26].

Different numbers of ARE-binding proteins that promote or suppress the degradation have been identified. One of them is TTP: a zinc finger protein that destabilizes TNF-α mRNA by binding to TNF-a AREs and promotes deadenylation by exonucleases [17,26,27]. This has also been confirmed in mice studies wherein mice lacking TTP showed an upregulated TNF-α mRNA level [20,28,29]. TTP is associated with cytoplasmic adenylasases and activates shortening of poly-A tail by promoting deadenylation of target mRNA. This suggests that TTP regulates the TNF-α expression by a negative feedback loop by destabilizing mRNA in an ARE-dependent fashion [20,30,31].

HuR is another RNA binding protein, which uniquely stabilizes the TNF-α RNA transcripts [21,32]. HuR is has a specialized shuttling domain that assists in its shuttling between nuclei and cytoplasm, though it mainly localizes in the nucleus. HuR has a dual regulatory role as it regulates the mRNA transport as well as stabilizes mRNA by modulating mRNA translation. During Inflammation HuR binds to the cytokine mRNA thus stabilizing the transcript levels [22,33].

hnRNP-A1 also localizes both in the nucleus and cytoplasm and shuttles between the two; it has a role in mRNA export and translation. hnRNP-A1 downregulates the TNF-α mRNA levels by binding to ARE whereas MnK a kinase decreases the interaction between ARE and hnRNP-A1 thereby reactivating TNF-a translation [24,34,35].

Multiple cells signaling pathways regulate mRNA decay however, more studies need to be done to study how different signals are communicated to the decay machinery. Various external stresses activate the p38 mitogen-activated protein kinase (MAPK) pathway that impairs the deadenylation of ARE mRNA in vivo [20,28,30].

p38 MAP kinase pathways

Behavior of cells following an external stimulus is via intracellular signaling cascades; for example, the mitogen-activated protein (MAP) kinase pathways [36]. They are members of individually separate and distinct signaling pathways that act as a central point in response to multiple external stimuli. Majorly, four different subgroups have been described for the MAP kinase family: c-jun N- terminal or stress-activated protein kinases (JNK/SAPK), extracellular signal-regulated kinases (ERKs), ERK/big MAP kinase1 (BMK1), and p38 group of protein kinases.

The regulatory proteins mRNA that controls the inflammatory responses are mostly unstable and can be stabilized by the p38 MAP kinase pathway. Well balance control of mRNA stability also supports an increased expression of a gene, allows fast changes in transcript levels and termination of protein synthesis.

RNA-binding proteins control the cytokine mRNA expression via post-translational modification. These modifications are mainly mediated via phosphatases and kinases that might control the interaction between RNA binding proteins and ARE sites. It has been demonstrated earlier that the deletion of the TNF-α ARE sequence in the mouse genome generates a mutant that is unresponsive to p38 kinases and jun N-terminal kinase [20]. p38/MK-2 signaling controls the TTP phosphorylation and encourages TTP sequestration that is further regulated by 14-3-3 regulatory proteins. These regulatory proteins prevent binding to the stress granules where mRNA is stored and thus lead to upregulated cytokine secretion since it cannot bind to ARE and destabilize mRNA [37]. MAP kinase enhances the localization of HUR in the cytoplasm and therefore assists in the translation of TNF-α, interleukin-8, and COX2 [38]. Interestingly, HuR also promotes translation of MAPK phosphatase-1 (MKP-1), which dephosphorylates thereby inactivating extracellular MAP kinase, JNK, and p38 [39]. MK2 kinase assists in the direct phosphorylation of RNA binding proteins and thus regulates signaling. MK2 is mainly activated through p38 MAPKα/β and is required for Lipopolysaccharide (LPS)-induced TNF-1 production in mouse macrophages [40]. Studies have shown that MK-2 kinase phosphorylates three RNA-binding proteins TTP, HuR, and hnRNP-A1 that are associated with TNF-α mRNA [40–43].

Mechanism of miRNA-mediated cytokine regulation

miRNAs are noncoding RNA guide molecules, which are about 20–22 nt long and mostly found in different species that regulate gene expression through interaction with "microribonucleoprotein" or "miRNP" and "cognate mRNAs" [44–47]. Nearly half of the miRNAs are found within clusters and have the same common promoter that suggests a possibility of gene duplications event [48,49]. miRNAs inhibit gene expression through various manners such as deadenylation and mRNA cleavage [50]. The majority of observations have pointed out that miRNA functions as effector complexes instead of as naked RNAs. These complexes act as effectors of post-transcriptional gene silencing that are known as miRNP, miRgonaute, or miRISC. In addition to effectors, they are also capable of perturbing gene expression either by activating or suppressing it. Argonaute, one of the principal constituents of all miRNPs [51], which is the determinant for miRNA specific target recognition is based on Watson-Crick pairing of 5′-proximal "seed" region (nucleotide 2–8) to the match site in 3′-UTR of the target mRNA [52]. Moreover, some studies hint toward the possibility that few miRNAs perturb the expression by targeting the 5′-UTR or coding region of some mRNAs [53,54]. Several factors can influence the outcome of miRNA-mRNA binding that either contributes to the binding strength or repression effect of a target site [55]. One of the factors is base pairing between the miRNA seed region and target site [56], others are AU-rich nucleotide composition near the site or other measures of site accessibility, proximity to sites for coexpressed miRNAs, positioning away from the center of long UTRs [57–59].

Contrary to the conventional belief that miRNA-mediated downregulation is irreversible and decreases mRNA stability or translational inhibition; it could be reversible in some cases [60]. In some conditions, few miRNAs can increase gene expression in some particular cells and conditions. One single miRNA can modulate the gene expression based on some peculiar factors. For example, miR-145 upregulates myocardin gene expression in muscle differentiation and, however, decreases ROCK1 expression in osteosarcoma [61,62]. miR-206 upregulates KLF-4 expression in nontumor cells and miR344 downregulates KLF-4 expression in normal proliferating cells [63]. These examples confirm the fact that miRNA-mediated gene expression upregulation is dependent on the type, condition, and local environment of a cell. Table 13.1 summarizes the genes upregulated by different miRNAs.

There are different ways through which miRNA controls or regulates gene expression; for example, miRNA can function as decoys and also act as a competitive binding partner for RNA-binding protein. For an instance, miR-29 functions as a decoy for HuR (RNA-binding protein).

Translational repression caused by miRNA could be relaxed by the binding of a positive-regulator RNA-binding protein; cellular stress induces the relaxation of CAT-1 mRNA repression by miR-122 (Table 13.2).

One study has also shown that miR-579 miR-221 and miR-125b are expressed in an LPS tolerance state wherein TNF-a mRNA is degraded. miR-221 induces TNF-a degradation and miR-579 and miR-125b inhibits or stop TNF translation, via TIAR [30,84]. Furthermore, a few of these effects might be through upregulation of miR-125b expression that further destabilizes TNF-α.

MiRNA as a regulator of NLRP3 inflammasome in RA

Inflammasomes are the central pillars of innate immunity and their discovery in 2002 was a breakthrough in inflammation and cell death pathway [85]. Inflammasomes are protein complexes, which comprise a cytoplasmic sensor, adaptor, and downstream effector proteins; all these are assembled upon detection of pathogen-associated molecular patterns or damage-associated molecular patterns by an innate immune receptor [86,87]. Various innate immune receptors act as sensors and the best studied so far is NLRP3. NLRP3 regulates cell stress and its expression is firmly regulated in resting cells. The assembly of the NLRP3 sensor-ASC-caspase-1 complex leads to the secretion of IL-18 and IL-1β [88]. With recent advancements in the field, we have understood that the two-tiered regulation is imperative to avoid any unwanted IL-18 IL-1β and secretion, as deregulated inflammasome activation has been associated with autoinflammatory diseases, cancer, and neurological and cardiovascular disorders. NLRP3 transcript expression levels are regulated at the post-transcriptional level and numerous miRNA have been known for regulating inflammasome post-transcriptionally [89]. Hence, miRNA-based transcriptional control has become a candidate for a potential therapeutic approach.

Inflammasome activation is regulated by various factors. A two-step regulation is vital for the full activation of NLRP3 inflammasome. The first step of priming is mediated by TLR ligands [90,91] that have been studied extensively [92]. It has been confirmed by multiple studies that the TLR ligand

Table 13.1 List of miRNA-mediated upregulation of different mRNAs.

miRNA	Target mRNA	References
miR-328	C/EBP alpha	[51,63,64]
miR-125b	B-Ras2	[65,66]
miR-34a/b-5	Beta-actin	[67]
miR-346	RIP140	[68]
miR-10a	TOP RNA	[69]
miR122	HCV	[70,71]
miR-360-3p	TNF-a	[60]

Table 13.2 List of miRNA-mediated downregulation of different mRNAs.

miRNA	Target mRNA	References
miR-122	CAT-1	[72,73]
miR-134	LIMK1	[74,75]
miR-1	RhoB	[76]
miR-548c-3p	TOP2A	[77]
miR-430	Nanos1 and TDRD7	[78]
miR-221 and miR-222	P27KIP1	[78]
miR-19b	PTEN	[79]
miR-166/165	Homeodomain leucine zipper transcript factor	[80]
miR-184 and let-7	LRRK2	[81,82]
miR-26a/b	IL-6	[83]

binding upregulates the expression of different miRNAs, such as *miR-155*, *miR-146a*, *miR-21*, and *miR-132*, which were associated with TLR4/MyD88/NF-κB pathways inhibition [93–98]. This suggests that an increment of miRNAs copies is a part of a negative feedback loop evolved to lower the inflammation post microbial stimuli [99]. In addition to miRNA, TLRs driven priming is also NF-KB dependent, which is moderated by Fas-associated death domain and caspase-8. These signaling cascades eventually regulate the transcripts of NLRP3 and pro-form of IL-1β, thereby preparing the cell for the activators.

Chromatin-mediated inflammatory gene regulation

Chromatin plays an important part in the regulation of inflammatory gene expression [100]. Many different histone modifications are known for regulating subsets of genes, which are mainly induced by LPS. One of the studies demonstrated that H3S10 phosphorylation might induce NF-κB activation [101]. Post LPS challenge, IL-12p40, IL-6, and CC-chemokine ligand 2 (CCL2) transcripts, except CCL3 and TNF a, phosphorylates of H3S10 that depends on the p38, and impediment of p38 activation specifically discourage inflammatory gene signaling through H3S10 phosphorylation, NF-κB activation [107]. However, it is unclear how the H3S10 phosphorylation and NF-κB are linked; however, it is congruous with the results of other research, demonstrating that the H3S10 phosphorylation is also linked with the transcription. Furthermore, the control of inflammation triggered by a few microbes is accompanied by the interference of H3S10 phosphorylation [102].

Coregulators

Transcriptional regulators that do not bind to DNA unlike transcription factors are known as coregulators. These can be divided into two major groups based on their mode of action: coactivators and corepressors. As multiple coactivators (coregulators that enhance) and corepressors (downregulates gene expression) function in an LPS-induced transcriptional manner, here we are discussing a few of them [103].

Coactivators

Activation of transcription is regulated via the coordination between the transcription factors and coactivators. Coactivator proteins diversely induce transcription and one of them is indirect binding to transcription factors. Coactivators further enhanced transcriptional rate either through DNA binding or by adding chromatin-modifying enzymes. Coactivators possess a histone-modifying enzymatic function that remodels chromatin at target genes to induce gene expression; whereas, other coactivators are deficient in an internal enzymatic activity that may stimulate the coordination of a transactivating complex. One such example is the transcriptional activation of IκBζ through TLR signaling, which assists in the exchange of inhibitory p50 units for p50–p65 heterodimers on some target gene promoters [103,104]. Inflammatory gene expression regulation by transcription factor NF-κB depends also upon the coactivator proteins.

Corepressors

Silencing mediator of retinoid and thyroid receptors (SMRT) and Nuclear corepressor protein (NCoR) are pivotal regulators of inflammatory transcript levels. These corepressors contain histone deacetylases, which regulate gene transcription through modifying the DNA-histone interactions. Therefore, they might have other roles for hindering gene expression, together with the stimulus inducible derepression at the promoters, which is indispensable for the induction-dependent gene expression.

Transcription factor JUN or AP-1 direct NCOR to different inflammatory genes and its clearance is induced by signal-induced exchange of JUN homodimers for transcriptionally active JUN-FOS heterodimers [105,106].On the other hand, SMRT is directed toward promoters through translocation–E26 transformation-specific (ETS)–leukemia repressor [105]. Moreover, few of the inflammatory genes require both NCOR and SMRT corepressor complexes to inhibit transcription, suggesting these genes can be regulated by more diverse signals [105,107].

Inflammatory transcript regulation via negative regulators

The commencement of antiinflammatory cascades is required for the host defense mechanism in an infection; however, timely management of inflammatory storms is also very imperative to inhibit the damaging effects of inflammation [108].

This stage of inflammatory surge evokes the induction of negative regulators. There are mainly two types of negative regulators: signal-specific and gene-specific regulators. Signal-specific regulators comprise regulators that function through TLRs and other inflammatory pathways and include A20, ST2, IL-1R-associated kinase M, and suppressors of cytokine signaling proteins [108]. Although these regulatory proteins reduce inflammatory signaling by multiple mechanisms, they all function in a close association with the receptor, therefore, expected to inhibit gene induction globally. The second types are transcriptional repressors or other negative regulators that function to perturb transcript levels.

Summary

The severity of the inflammatory disease is also determined by the post-transcriptional mechanisms that eventually modulate the levels of different key genes such as chemokine. The presence of various regulatory mechanisms suggests tight control at different checkpoints. A much-detailed understanding of multiple layers of the regulatory system is pivotal and could be harnessed in designing the binding partners and regulatory sequences that influence the perturbation in mRNA turnover and translation. One of the important factors in the regulation of cytokinesis is the three prime untranslated (3′ -UTR) regions present in their mRNA. Many ARE-binding proteins are known to be involved in the inflammatory responses [109–111].

Additionally, comprehension of regulatory proteins and signaling pathways is crucial for the design of new anti-TNF-based therapies for RA patients. The capacity of NLRP3 inflammasome in accelerating or controlling the pathophysiology of RA and in diverse other diseases was illustrated, including cancers, CAPS, and autoimmune disorders [112–116]. An enhanced IL-1β, majorly detected in such diseases, indicates NLRP3 inflammasome activation. Additionally, major evidence denoted that IL-1β has a pivotal function in disease pathogenesis. Hence, aiming for IL-1β could be a promising therapeutic approach. This was properly applied to combat NLRP3 inflammasome-mediated cancer and autoimmune diseases [117,118]. These data lay out a compelling verification for the NLRP3 as a probable prospective therapeutic target for the cure of the disorders linked with an inflated amount of IL-1β. Therefore, targeting miRNAs to inhibit NLRP3 inflammasome-mediated cytokine secretion has therapeutic potential as they could prevent the transcript levels and, subsequently, avoid IL-1β production. Several clinical and preclinical studies have been done for miRNA-based replacement and silencing therapeutic approaches [119]. As compared to the small molecule drugs, oligonucleotide approaches (anti-miRNA or mimic) are shown to have higher efficacy and effectivity by virtue of their potential to affect multiple gene targets concurrently. Oligonucleotide that targets miRNA is designed to interact and bind to the targeted miRNA [119]. In general, miRNA targets more than one gene in any signaling pathways that enhance their value as therapeutic candidates. However, there are some limitations to overcome, such as toxicity, specificity of miRNA-targeting oligonucleotides [120]. Furthermore, administration of miRNA without a carrier, modifications of anti-miRNAs have been shown to enhance the specificity [121,122]. In addition to the above-mentioned limitations, one major concern about anti-miRNA is that they can be detected by the innate immune receptors and hence eliminated by the immune system [122,123]. To avoid this, antibodies coated with nanoparticles and carrier proteins have been utilized for the advancement of specificity and efficacy of delivery, which further demonstrates immense caliber for miRNA-based treatments of disorders linked to NLRP3 inflammasome abnormality. A limited understanding of NLRP3 inflammasome in various disease pathogenesis thus hinders the expansion of miRNA-based therapeutics. Moreover, novel clinical discoveries have deciphered immense caliber for the treatment of diseases linked to NLRP3 inflammasome hyperactivation.

References

[1] Guo Q, Wang Y, Xu D, Nossent J, Pavlos NJ, Xu J. Rheumatoid arthritis: pathological mechanisms and modern pharmacologic therapies. Bone Res April 27, 2018;6(1):1–4.

[2] Feldmann M, Maini SR. Role of cytokines in rheumatoid arthritis: an education in pathophysiology and therapeutics. Immunol Rev June 2008;223(1):7–19.

[3] Smolen JS, Aletaha D, Mcinnes IB. Rheumatoid arthritis. Lancet 2016;388:2023–38. https://doi.org/10.1016/S0140-6736(16)30173-8.

[4] Cornish AL, Campbell IK, McKenzie BS, Chatfield S, Wicks IP. G-CSF and GM-CSF as therapeutic targets in rheumatoid arthritis. Nat Rev Rheumatol October 2009;5(10):554–9.

[5] Haringman JJ, Gerlag DM, Zwinderman AH, Smeets TJ, Kraan MC, Baeten D, et al. Synovial tissue macrophages: a sensitive biomarker for response to treatment in patients with rheumatoid arthritis. Ann Rheum Dis June 1, 2005;64(6):834–8.

[6] Seibl R, Birchler T, Loeliger S, Hossle JP, Gay RE, Saurenmann T, et al. Expression and regulation of Toll-like receptor 2 in rheumatoid arthritis synovium. Am J Pathol April 1, 2003;162(4):1221–7.

[7] Blüml S, Bonelli M, Niederreiter B, Puchner A, Mayr G, Hayer S, et al. Essential role of microRNA-155 in the pathogenesis of autoimmune arthritis in mice. Arthritis Rheum May 2011;63(5):1281–8.

[8] Kurowska-Stolarska M, Alivernini S, Ballantine LE, Asquith DL, Millar NL, Gilchrist DS, et al. MicroRNA-155 as a proinflammatory regulator in clinical and experimental arthritis. Proc Natl Acad Sci USA July 5, 2011;108(27):11193–8.

[9] Wilusz CJ, Wormington M, Peltz SW. The cap-to-tail guide to mRNA turnover. Nat Rev Mol Cell Biol April 2001;2(4):237–46.

[10] Kracht M, Saklatvala J. Transcriptional and post-transcriptional control of gene expression in inflammation. Cytokine November 1, 2002;20(3):91–106.

[11] Fan J, Yang X, Wang W, Wood WH, Becker KG, Gorospe M. Global analysis of stress-regulated mRNA turnover by using cDNA arrays. Proc Natl Acad Sci USA August 6, 2002;99(16):10611–6.

[12] Ross J. mRNA stability in mammalian cells. Microbiol Rev September 1995;59(3):423–50.

[13] Peltz SW, Brewer G, Bernstein P, Hart PA, Ross J. Regulation of mRNA turnover in eukaryotic cells. Crit Rev Eukaryot Gene Expr 1991;1(2):99–126.

[14] Guhaniyogi J, Brewer G. Regulation of mRNA stability in mammalian cells. Gene March 7, 2001; 265(1–2):11–23.

[15] Keene JD. Ribonucleoprotein infrastructure regulating the flow of genetic information between the genome and the proteome. Proc Natl Acad Sci USA June 19, 2001;98(13):7018–24.

[16] Keene JD, Tenenbaum SA. Eukaryotic mRNPs may represent post-transcriptional operons. Mol Cell June 1, 2002;9(6):1161–7.

[17] Chen CY, Shyu AB. AU-rich elements: characterization and importance in mRNA degradation. Trends Biochem Sci November 1, 1995;20(11):465–70.

[18] Garneau NL, Wilusz J, Wilusz CJ. The highways and byways of mRNA decay. Nat Rev Mol Cell Biol February 2007;8(2):113–26.

[19] Dumitru CD, Ceci JD, Tsatsanis C, Kontoyiannis D, Stamatakis K, Lin JH, et al. TNF-α induction by LPS is regulated post-transcriptionally via a Tpl2/ERK-dependent pathway. Cell December 22, 2000;103(7): 1071–83.

[20] Kontoyiannis D, Pasparakis M, Pizarro TT, Cominelli F, Kollias G. Impaired on/off regulation of TNF biosynthesis in mice lacking TNF AU-rich elements: implications for joint and gut-associated immunopathologies. Immunity March 1, 1999;10(3):387–98.

[21] Han J, Brown T, Beutler B. Endotoxin-responsive sequences control cachectin/tumor necrosis factor biosynthesis at the translational level. J Exp Med February 1, 1990;171(2):465–75.

[22] Blackshear PJ. Tristetraprolin and other CCCH tandem zinc-finger proteins in the regulation of mRNA turnover. Biochem Soc Trans November 2002;30(6):945–52.

[23] Dean JL, Sully G, Clark AR, Saklatvala J. The involvement of AU-rich element-binding proteins in p38 mitogen-activated protein kinase pathway-mediated mRNA stabilisation. Cell Signal October 1, 2004; 16(10):1113–21.
[24] Buxadé M, Parra JL, Rousseau S, Shpiro N, Marquez R, Morrice N, et al. The Mnks are novel components in the control of TNF alpha biosynthesis and phosphorylate and regulate hnRNP A1. Immunity August 2005;23(2):177–89.
[25] Piecyk M, Wax S, Beck AR, Kedersha N, Gupta M, Maritim B, et al. TIA-1 is a translational silencer that selectively regulates the expression of TNF-alpha. EMBO J August 2000;19(15):4154–63.
[26] Gueydan C, Droogmans L, Chalon P, Huez G, Caput D, Kruys V. Identification of TIAR as a protein binding to the translational regulatory AU-rich element of tumor necrosis factor α mRNA. J Biol Chem January 22, 1999;274(4):2322–6.
[27] Sakai K, Kitagawa Y, Hirose G. Binding of neuronal ELAV-like proteins to the uridine-rich sequence in the 3′-untranslated region of tumor necrosis factor-α messenger RNA. FEBS Lett March 5, 1999;446(1): 157–62.
[28] Carballo E, Lai WS, Blackshear PJ. Feedback inhibition of macrophage tumor necrosis factor-α production by tristetraprolin. Science August 14, 1998;281(5379):1001–5.
[29] Sneezum L, Eislmayr K, Dworak H, Sedlyarov V, Le Heron A, Ebner F, et al. Context-dependent IL-1 mRNA-destabilization by TTP prevents dysregulation of immune homeostasis under steady state conditions. Front Immunol July 7, 2020;11:1398.
[30] Lai WS, Carballo E, Strum JR, Kennington EA, Phillips RS, Blackshear PJ. Evidence that tristetraprolin binds to AU-rich elements and promotes the deadenylation and destabilization of tumor necrosis factor alpha mRNA. Mol Cell Biol June 1, 1999;19(6):4311–23.
[31] Clark A. Post-transcriptional regulation of pro-inflammatory gene expression. Arthritis Res Ther April 2000;2(3):1–3.
[32] Dean JL, Wait R, Mahtani KR, Sully G, Clark AR, Saklatvala J. The 3′ untranslated region of tumor necrosis factor alpha mRNA is a target of the mRNA-stabilizing factor HuR. Mol Cell Biol February 1, 2001; 21(3):721–30.
[33] Cok SJ, Acton SJ, Sexton AE, Morrison AR. Identification of RNA-binding proteins in RAW 264.7 cells that recognize a lipopolysaccharide-responsive element in the 3-untranslated region of the murine cyclooxygenase-2 mRNA. J Biol Chem February 27, 2004;279(9):8196–205.
[34] Piñol-Roma S, Dreyfuss G. Shuttling of pre-mRNA binding proteins between nucleus and cytoplasm. Nature February 1992;355(6362):730–2.
[35] Svitkin YV, Ovchinnikov LP, Dreyfuss G, Sonenberg N. General RNA binding proteins render translation cap dependent. EMBO J December 1996;15(24):7147–55.
[36] Rouse J, Cohen P, Trigon S, Morange M, Alonso-Llamazares A, Zamanillo D, et al. A novel kinase cascade triggered by stress and heat shock that stimulates MAPKAP kinase-2 and phosphorylation of the small heat shock proteins. Cell September 23, 1994;78(6):1027–37.
[37] Mahtani KR, Brook M, Dean JL, Sully G, Saklatvala J, Clark AR. Mitogen-activated protein kinase p38 controls the expression and posttranslational modification of tristetraprolin, a regulator of tumor necrosis factor alpha mRNA stability. Mol Cell Biol October 1, 2001;21(19):6461–9.
[38] Doller A, Pfeilschifter J, Eberhardt W. Signalling pathways regulating nucleo-cytoplasmic shuttling of the mRNA-binding protein HuR. Cell Signal December 1, 2008;20(12):2165–73.
[39] Kuwano Y, Kim HH, Abdelmohsen K, Pullmann Jr R, Martindale JL, Yang X, et al. MKP-1 mRNA stabilization and translational control by RNA-binding proteins HuR and NF90. Mol Cell Biol July 15, 2008; 28(14):4562–75.

[40] Neininger A, Kontoyiannis D, Kotlyarov A, Winzen R, Eckert R, Volk HD, et al. MK2 targets AU-rich elements and regulates biosynthesis of tumor necrosis factor and interleukin-6 independently at different post-transcriptional levels. J Biol Chem February 1, 2002;277(5):3065–8.
[41] Chrestensen CA, Schroeder MJ, Shabanowitz J, Hunt DF, Pelo JW, Worthington MT, et al. MAPKAP kinase 2 phosphorylates tristetraprolin on in vivo sites including Ser178, a site required for 14-3-3 binding. J Biol Chem March 12, 2004;279(11):10176–84.
[42] Tran H, Maurer F, Nagamine Y. Stabilization of urokinase and urokinase receptor mRNAs by HuR is linked to its cytoplasmic accumulation induced by activated mitogen-activated protein kinase-activated protein kinase 2. Mol Cell Biol October 15, 2003;23(20):7177–88.
[43] Rousseau S, Morrice N, Peggie M, Campbell DG, Gaestel M, Cohen P. Inhibition of SAPK2a/p38 prevents hnRNP A0 phosphorylation by MAPKAP-K2 and its interaction with cytokine mRNAs. EMBO J December 1, 2002;21(23):6505–14.
[44] Carthew RW, Sontheimer EJ. Origins and mechanisms of miRNAs and siRNAs. Cell February 20, 2009; 136(4):642–55.
[45] Ghildiyal M, Zamore PD. Small silencing RNAs: an expanding universe. Nat Rev Genet February 2009; 10(2):94–108.
[46] Kim VN, Han J, Siomi MC. Biogenesis of small RNAs in animals. Nat Rev Mol Cell Biol February 2009; 10(2):126–39.
[47] Siomi H, Siomi MC. On the road to reading the RNA-interference code. Nature January 2009;457(7228): 396–404.
[48] Bartel DP. MicroRNAs: genomics, biogenesis, mechanism, and function. Cell January 23, 2004;116(2): 281–97.
[49] Seitz H, Youngson N, Lin SP, Dalbert S, Paulsen M, Bachellerie JP, et al. Imprinted microRNA genes transcribed antisense to a reciprocally imprinted retrotransposon-like gene. Nat Genet July 2003;34(3): 261–2.
[50] Cai Y, Yu X, Hu S, Yu J. A brief review on the mechanisms of miRNA regulation. Dev Reprod Biol December 1, 2009;7(4):147–54.
[51] Steitz JA, Vasudevan S. miRNPs: versatile regulators of gene expression in vertebrate cells. Biochem Soc Trans October 1, 2009;37(5):931–5.
[52] Lewis BP, Burge CB, Bartel DP. Conserved seed pairing, often flanked by adenosines, indicates that thousands of human genes are microRNA targets. Cell January 14, 2005;120(1):15–20.
[53] Lee I, Ajay SS, Yook JI, Kim HS, Hong SH, Kim NH, et al. New class of microRNA targets containing simultaneous 5′-UTR and 3′-UTR interaction sites. Genome Res July 1, 2009;19(7):1175–83.
[54] Brümmer A, Hausser J. MicroRNA binding sites in the coding region of mRNAs: extending the repertoire of post-transcriptional gene regulation. Bioessays June 2014;36(6):617–26.
[55] Carroll AP, Goodall GJ, Liu B. Understanding principles of miRNA target recognition and function through integrated biological and bioinformatics approaches. Wiley Inter Rev RNA May 2014;5(3):361–79.
[56] Wang X. Composition of seed sequence is a major determinant of microRNA targeting patterns. Bioinformatics May 15, 2014;30(10):1377–83.
[57] Ohler UW, Yekta S, Lim LP, Bartel DP, Burge CB. Patterns of flanking sequence conservation and a characteristic upstream motif for microRNA gene identification. RNA September 1, 2004;10(9):1309–22.
[58] Majoros WH, Ohler U. Spatial preferences of microRNA targets in 3′untranslated regions. BMC Genom December 2007;8(1):1–9.
[59] Brodersen P, Voinnet O. Revisiting the principles of microRNA target recognition and mode of action. Nat Rev Mol Cell Biol February 2009;10(2):141–8.
[60] Vasudevan S, Steitz JA. AU-rich-element-mediated upregulation of translation by FXR1 and Argonaute 2. Cell March 23, 2007;128(6):1105–18.

[61] Cordes KR, Sheehy NT, White MP, Berry EC, Morton SU, Muth AN, et al. miR-145 and miR-143 regulate smooth muscle cell fate and plasticity. Nature August 2009;460(7256):705–10.
[62] Li E, Zhang J, Yuan T, Ma B. MiR-145 inhibits osteosarcoma cells proliferation and invasion by targeting ROCK1. Tumor Biol August 2014;35(8):7645–50.
[63] Lin CC, Liu LZ, Addison JB, Wonderlin WF, Ivanov AV, Ruppert JM. A KLF4–miRNA-206 autoregulatory feedback loop can promote or inhibit protein translation depending upon cell context. Mol Cell Biol June 15, 2011;31(12):2513–27.
[64] Vasudevan S, Tong Y, Steitz JA. Switching from repression to activation: microRNAs can up-regulate translation. Science December 21, 2007;318(5858):1931–4.
[65] Eiring AM, Harb JG, Neviani P, Garton C, Oaks JJ, Spizzo R, et al. miR-328 functions as an RNA decoy to modulate hnRNP E2 regulation of mRNA translation in leukemic blasts. Cell March 5, 2010;140(5): 652–65.
[66] Murphy AJ, Guyre PM, Pioli PA. Estradiol suppresses NF-κB activation through coordinated regulation of let-7a and miR-125b in primary human macrophages. J Immunol May 1, 2010;184(9):5029–37.
[67] Ghosh T, Soni K, Scaria V, Halimani M, Bhattacharjee C, Pillai B. MicroRNA-mediated up-regulation of an alternatively polyadenylated variant of the mouse cytoplasmic β-actin gene. Nucleic Acids Res November 1, 2008;36(19):6318–32.
[68] Tsai NP, Lin YL, Wei LN. MicroRNA mir-346 targets the 5′-untranslated region of receptor-interacting protein 140 (RIP140) mRNA and up-regulates its protein expression. Biochem J December 15, 2009; 424(3):411–8.
[69] Ørom UA, Nielsen FC, Lund AH. MicroRNA-10a binds the 5′ UTR of ribosomal protein mRNAs and enhances their translation. Mol Cell May 23, 2008;30(4):460–71.
[70] Niepmann M. Activation of hepatitis C virus translation by a liver-specific microRNA. Cell Cycle May 15, 2009;8(10):1473–7.
[71] Henke JI, Goergen D, Zheng J, Song Y, Schüttler CG, Fehr C, et al. microRNA-122 stimulates translation of hepatitis C virus RNA. EMBO J December 17, 2008;27(24):3300–10.
[72] Jopling CL, Schütz S, Sarnow P. Position-dependent function for a tandem microRNA miR-122-binding site located in the hepatitis C virus RNA genome. Cell Host & Microbe July 17, 2008;4(1):77–85.
[73] Li Y, Masaki T, Yamane D, McGivern DR, Lemon SM. Competing and noncompeting activities of miR-122 and the 5′ exonuclease Xrn1 in regulation of hepatitis C virus replication. Proc Natl Acad Sci USA January 29, 2013;110(5):1881–6.
[74] Schratt GM, Tuebing F, Nigh EA, Kane CG, Sabatini ME, Kiebler M, et al. A brain-specific microRNA regulates dendritic spine development. Nature January 2006;439(7074):283–9.
[75] Khudayberdiev S, Fiore R, Schratt G. MicroRNA as modulators of neuronal responses. Commun Integr Biol September 1, 2009;2(5):411–3.
[76] Glorian V, Maillot G, Polès S, Iacovoni JS, Favre G, Vagner S. HuR-dependent loading of miRNA RISC to the mRNA encoding the Ras-related small GTPase RhoB controls its translation during UV-induced apoptosis. Cell Death Differ November 2011;18(11):1692–701.
[77] Srikantan S, Abdelmohsen K, Lee EK, Tominaga K, Subaran SS, Kuwano Y, et al. Translational control of TOP2A influences doxorubicin efficacy. Mol Cell Biol September 15, 2011;31(18):3790–801.
[78] Kedde M, Strasser MJ, Boldajipour B, Vrielink JA, Slanchev K, le Sage C, et al. RNA-binding protein Dnd1 inhibits microRNA access to target mRNA. Cell December 28, 2007;131(7):1273–86.
[79] Poliseno L, Salmena L, Zhang J, Carver B, Haveman WJ, Pandolfi PP. A coding-independent function of gene and pseudogene mRNAs regulates tumour biology. Nature June 2010;465(7301):1033–8.
[80] Zhu H, Hu F, Wang R, Zhou X, Sze SH, Liou LW, et al. Arabidopsis Argonaute10 specifically sequesters miR166/165 to regulate shoot apical meristem development. Cell April 15, 2011;145(2):242–56.

[81] Gehrke S, Imai Y, Sokol N, Lu B. Pathogenic LRRK2 negatively regulates microRNA-mediated translational repression. Nature July 2010;466(7306):637−41.
[82] Chan SP, Ramaswamy G, Choi EY, Slack FJ. Identification of specific let-7 microRNA binding complexes in *Caenorhabditis elegans*. RNA October 1, 2008;14(10):2104−14.
[83] Jones MR, Quinton LJ, Blahna MT, Neilson JR, Fu S, Ivanov AR, et al. Zcchc11-dependent uridylation of microRNA directs cytokine expression. Nat Cell Biol September 2009;11(9):1157−63.
[84] O'neill LA, Sheedy FJ, McCoy CE. MicroRNAs: the fine-tuners of Toll-like receptor signalling. Nat Rev Immunol March 2011;11(3):163−75.
[85] Martinon F, Burns K, Tschopp J. The inflammasome: a molecular platform triggering activation of inflammatory caspases and processing of proIL-β. Mol Cell August 1, 2002;10(2):417−26.
[86] Lamkanfi M. Emerging inflammasome effector mechanisms. Nat Rev Immunol March 2011;11(3):213−20.
[87] Evavold CL, Kagan JC. Inflammasomes: threat-assessment organelles of the innate immune system. Immunity October 15, 2019;51(4):609−24.
[88] Aglietti RA, Estevez A, Gupta A, Ramirez MG, Liu PS, Kayagaki N, et al. GsdmD p30 elicited by caspase-11 during pyroptosis forms pores in membranes. Proc Natl Acad Sci USA July 12, 2016;113(28):7858−63.
[89] Tezcan G, Martynova EV, Gilazieva ZE, McIntyre A, Rizvanov AA, Khaiboullina SF. MicroRNA post-transcriptional regulation of the NLRP3 inflammasome in immunopathologies. Front Pharmacol May 1, 2019;10:451.
[90] Bauernfeind FG, Horvath G, Stutz A, Alnemri ES, MacDonald K, Speert D, et al. (n.d.). Cutting edge: NF-kappaB activating pattern recognition and cytokine.
[91] Franchi L, Eigenbrod T, Muñoz-Planillo R, Nuñez G. The inflammasome: a caspase-1-activation platform that regulates immune responses and disease pathogenesis. Nat Immunol March 2009;10(3):241−7.
[92] Groslambert M, Py BF. Spotlight on the NLRP3 inflammasome pathway. J Inflamm Res 2018;11:359.
[93] Coll RC, O'Neill LA. New insights into the regulation of signalling by toll-like receptors and nod-like receptors. J Innat Imm 2010;2(5):406−21.
[94] Sheedy FJ, Palsson-McDermott E, Hennessy EJ, Martin C, O'leary JJ, Ruan Q, et al. Negative regulation of TLR4 via targeting of the proinflammatory tumor suppressor PDCD4 by the microRNA miR-21. Nat Immunol February 2010;11(2):141−7.
[95] Nahid P, Jarlsberg LG, Rudoy I, de Jong BC, Unger A, Kawamura LM, et al. Factors associated with mortality in patients with drug-susceptible pulmonary tuberculosis. BMC Infect Dis December 2011;11(1):1−7.
[96] Anzola A, González R, Gámez-Belmonte R, Ocón B, Aranda CJ, Martínez-Moya P, et al. miR-146a regulates the crosstalk between intestinal epithelial cells, microbial components and inflammatory stimuli. Sci Rep November 26, 2018;8(1):1−2.
[97] Tan Y, Yu L, Zhang C, Chen K, Lu J, Tan L. miRNA-146a attenuates inflammation in an in vitro spinal cord injury model via inhibition of TLR4 signaling. Exp Ther Med October 1, 2018;16(4):3703−9.
[98] Zhi H, Yuan N, Wu JP, Lu LM, Chen XY, Wu SK, et al. MicroRNA−21 attenuates BDE-209-induced lipid accumulation in THP-1 macrophages by downregulating Toll-like receptor 4 expression. Food Chem Toxicol March 1, 2019;125:71−7.
[99] Ceppi M, Pereira PM, Dunand-Sauthier I, Barras E, Reith W, Santos MA, et al. MicroRNA-155 modulates the interleukin-1 signaling pathway in activated human monocyte-derived dendritic cells. Proc Natl Acad Sci USA February 24, 2009;106(8):2735−40.
[100] Bernstein BE, Meissner A, Lander ES. The mammalian epigenome. Cell February 23, 2007;128(4):669−81.
[101] Saccani S, Pantano S, Natoli G. p38-dependent marking of inflammatory genes for increased NF-κB recruitment. Nat Immunol January 2002;3(1):69−75.

[102] Komar D, Juszczynski P. Rebelled epigenome: histone H3S10 phosphorylation and H3S10 kinases in cancer biology and therapy. Clin Epigenet December 2020;12(1):1–4.
[103] Medzhitov R, Horng T. Transcriptional control of the inflammatory response. Nat Rev Immunol October 2009;9(10):692–703.
[104] Yamamoto M, Yamazaki S, Uematsu S, Sato S, Hemmi H, Hoshino K, et al. Regulation of Toll/IL-1-receptor-mediated gene expression by the inducible nuclear protein IκBζ. Nature July 2004;430(6996): 218–22.
[105] Ghisletti S, et al. Cooperative NCoR/SMRT interactions establish a corepressor-based strategy for integration of inflammatory and anti-inflammatory signaling pathways. Genes Dev 2009;23:681–93.
[106] Huang W, Ghisletti S, Perissi V, Rosenfeld MG, Glass CK. Transcriptional integration of TLR2 and TLR4 signaling at the NCoR derepression checkpoint. Mol Cell July 10, 2009;35(1):48–57.
[107] Ogawa S, Lozach J, Jepsen K, Sawka-Verhelle D, Perissi V, Sasik R, et al. A nuclear receptor corepressor transcriptional checkpoint controlling activator protein 1-dependent gene networks required for macrophage activation. Proc Natl Acad Sci USA October 5, 2004;101(40):14461–6.
[108] Liew FY, Xu D, Brint EK, O'Neill LA. Negative regulation of toll-like receptor-mediated immune responses. Nat Rev Immunol June 2005;5(6):446–58.
[109] Sadri N, Schneider RJ. Auf1/Hnrnpd-deficient mice develop pruritic inflammatory skin disease. J Invest Dermatol March 1, 2009;129(3):657–70.
[110] Katsanou V, Papadaki O, Milatos S, Blackshear PJ, Anderson P, Kollias G, et al. HuR as a negative post-transcriptional modulator in inflammation. Mol Cell September 16, 2005;19(6):777–89.
[111] Phillips K, Kedersha N, Shen L, Blackshear PJ, Anderson P. Arthritis suppressor genes TIA-1 and TTP dampen the expression of tumor necrosis factor α, cyclooxygenase 2, and inflammatory arthritis. Proc Natl Acad Sci USA February 17, 2004;101(7):2011–6.
[112] Aganna E, Martinon F, Hawkins PN, Ross JB, Swan DC, Booth DR, et al. Association of mutations in the NALP3/CIAS1/PYPAF1 gene with a broad phenotype including recurrent fever, cold sensitivity, sensorineural deafness, and AA amyloidosis. Arthritis Rheum September 2002;46(9):2445–52.
[113] Martinon F, Pétrilli V, Mayor A, Tardivel A, Tschopp J. Gout-associated uric acid crystals activate the NALP3 inflammasome. Nature March 2006;440(7081):237–41.
[114] Masters SL, Simon A, Aksentijevich I, Kastner DL. Horror autoinflammaticus: the molecular pathophysiology of autoinflammatory disease. Annu Rev Immunol April 23, 2009;27:621–68.
[115] Bauer C, Duewell P, Mayer C, Lehr HA, Fitzgerald KA, Dauer M, et al. Colitis induced in mice with dextran sulfate sodium (DSS) is mediated by the NLRP3 inflammasome. Gut September 1, 2010;59(9): 1192–9.
[116] Wen H, Gris D, Lei Y, Jha S, Zhang L, Huang MT, et al. Fatty acid–induced NLRP3-ASC inflammasome activation interferes with insulin signaling. Nat Immunol May 2011;12(5):408–15.
[117] Larsen CM, Faulenbach M, Vaag A, Vølund A, Ehses JA, Seifert B, et al. Interleukin-1–receptor antagonist in type 2 diabetes mellitus. N Engl J Med April 12, 2007;356(15):1517–26.
[118] Lust JA, Lacy MQ, Zeldenrust SR, Dispenzieri A, Gertz MA, Witzig TE, et al. Induction of a chronic disease state in patients with smoldering or indolent multiple myeloma by targeting interleukin 1β-induced interleukin 6 production and the myeloma proliferative component. In: Mayo clinic proceedings, vol. 84. Elsevier; February 1, 2009. p. 114–22 (2).
[119] Li Z, Rana TM. Therapeutic targeting of microRNAs: current status and future challenges. Nat Rev Drug Discov August 2014;13(8):622–38.
[120] Merhautova J, Demlova R, Slaby O. MicroRNA-based therapy in animal models of selected gastrointestinal cancers. Front Pharmacol September 27, 2016;7:329.
[121] Hogan DJ, Vincent TM, Fish S, Marcusson EG, Bhat B, Chau BN, et al. Anti-miRs competitively inhibit microRNAs in Argonaute complexes. PLoS One July 3, 2014;9(7):e100951.

[122] Bennett CF, Swayze EE. RNA targeting therapeutics: molecular mechanisms of antisense oligonucleotides as a therapeutic platform. Annu Rev Pharmacol Toxicol February 10, 2010;50:259–93.
[123] Heil F, Hemmi H, Hochrein H, Ampenberger F, Kirschning C, Akira S, et al. Species-specific recognition of single-stranded RNA via toll-like receptor 7 and 8. Science March 5, 2004;303(5663):1526–9.

CHAPTER

Post-transcriptional gene regulation in metabolic syndrome

14

Rashmi Pathak[1] and Avinash Kumar[2,3]

[1]*Comparative Biomedical Sciences, School of Veterinary Medicine, Louisiana State University, Baton Rouge, LA, United States;* [2]*Department of Biological Sciences, Louisiana State University, Baton Rouge, LA, United States;* [3]*Pennington Biomedical Research Center, Baton Rouge, LA, United States*

Abbreviations

CVD-	cardiovascular disease
NAFLD-	nonalcoholic fatty liver disease
ncRNA-	noncoding ribonucleic acid
RBP-	RNA-binding protein
T2D-	Type 2 diabetes

Introduction to metabolic syndrome

Metabolic syndrome is a serious health condition affecting about one-third of the US population and is currently on the rise in the United States and other countries [1]. It is characterized by a cluster of risk factors which include high body mass index (BMI), blood sugar, blood pressure, cholesterol and triglycerides [2]. One of the major factors responsible for metabolic syndrome is excess energy intake coupled with a sedentary lifestyle that can eventually lead to the accumulation of adipose tissue in the body [3]. Obesity, insulin resistance, abnormal levels of lipids (dyslipidemia), and hypertension can further lead to the development of other comorbidities such as cardiovascular disease (CVD) and nonalcoholic fatty liver disease (NAFLD) (Fig. 14.1) [4,5].

Metabolic syndrome is associated with metabolism, which is an indispensable process vital to all forms of life on earth. Metabolism comprises all the physical and biochemical processes that take place within the live organisms including synthesis of macromolecules, transport of matter, excretion of waste materials, and chemical breakdown of molecules to generate energy. Homeostasis in intermediary metabolic pathways such as glucose, lipid, and cholesterol is highly regulated through multiple processes and is important for the normal functioning of the body. Any perturbation in these pathways can lead to the development of metabolic syndrome. Studies with intermediary metabolism have mostly focused on post-translational protein modification as a major regulatory mechanism. Recently, it has become more and more evident that several post-transcriptional gene regulation mechanisms are also involved in metabolism and cellular homeostasis (Fig. 14.2).

Post-Transcriptional Gene Regulation in Human Disease, Volume 32. https://doi.org/10.1016/B978-0-323-91305-8.00003-X

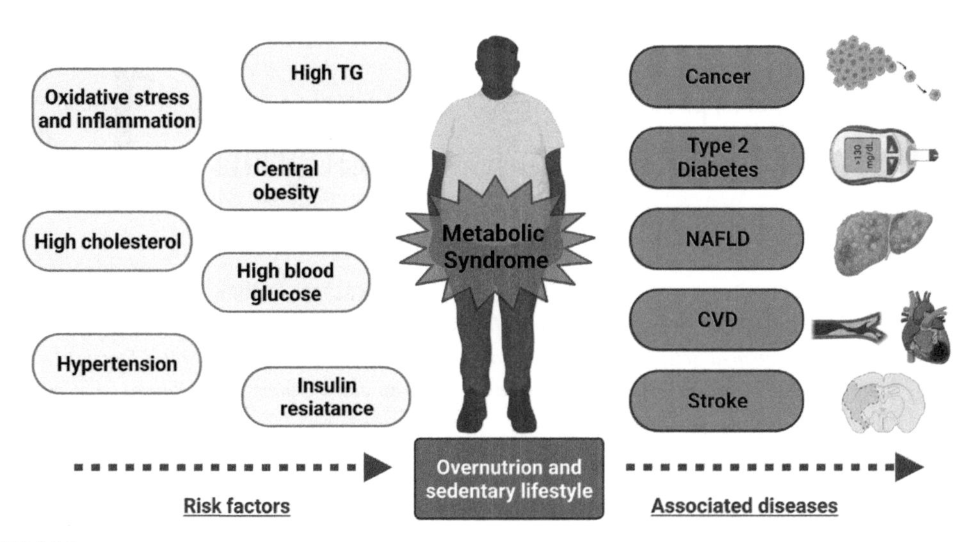

FIGURE 14.1

Metabolic syndromes, risk factors for metabolic syndrome, and associated disorders.

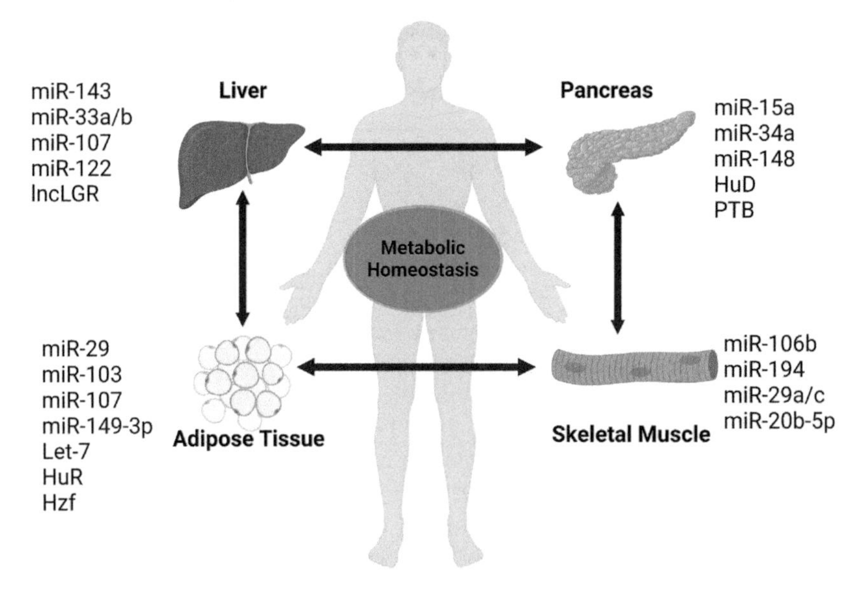

FIGURE 14.2

Schematic representation of different miRNAs, RNA-binding proteins involved in maintaining metabolic homeostasis in different organs.

Post-transcriptional gene regulation in metabolic pathways

Regulation of intermediary metabolic pathways has mainly focused on post-translational enzyme modification, enzyme synthesis, allosteric enzyme regulation, and substrate availability. However, recent studies have made it clear that many post-transcriptional gene regulatory mechanisms play an important role in enzyme synthesis and breakdown, thus maintaining homeostasis. Eukaryotic transcription and translation are compartmentalized, and they take place in the nucleus and cytoplasm respectively. Due to this, the mRNA that is synthesized in the nucleus must go to the cytoplasm for protein synthesis. During the journey from the nucleus to the cytoplasm, the mRNA transcript interacts with numerous factors and goes through extensive remodeling [6]. This highly regulated mRNA remodeling process includes splicing, polyadenylation, and capping [6]. These modifications are important in regulating the gene expression by changing the nucleotide sequence, translation efficiency, stability, and localization of the RNA. Taken together, all these processes are known as post-transcriptional gene regulatory mechanisms. The post-translational and post-transcriptional mechanisms provide flexibility to organisms to adapt to varying physiological situations since they operate at different levels. This gives the organisms the ability to regulate gene expression at all these steps. Although the cells may need all the mechanisms at the same time, they act toward a common goal of maintaining homeostasis by responding to different stimuli.

Role of noncoding RNAs in metabolic regulation

Metabolism in the mammalian system is a very intricate process starting from the cellular level then going up to tissue and organism level. Depending on the cell, tissue, or organ type, metabolic processes differ significantly. However, the regulated metabolic process from each of the tissues or organs acts synergistically to provide vigorous homeostasis. Dysregulation of the metabolic processes in a single type of cell or tissue can trigger physiological imbalance which can further lead to the onset of metabolic syndrome. To maintain cellular homeostasis, acute and quick changes in mRNA expression levels are essential for accordingly modifying the cellular protein content in response to a variety of metabolic stimuli. Completion of the human genome project has made it clear that not all transcripts are translated into proteins [7]. The RNAs that are not translated into proteins are known as noncoding RNAs (ncRNAs). This group of ncRNAs includes, but is not limited to, microRNA, siRNAs, and long noncoding RNAs. Recent studies have illustrated that many ncRNAs play important roles in regulating the expression of genes in a highly dynamic manner [8]. Additionally, numerous metabolic disorders and pathological states such as obesity and diabetes are known to be caused by the aberrant expression of ncRNAs.

Post-transcriptional regulation by microRNA

Role of microRNA in glucose homeostasis

Glucose plays an essential role in energy generation, and macromolecules such as carbohydrates, proteins, and lipids are ultimately broken down to form glucose. After a meal, the glucose level can rise to 140 mg/dL, which is sensed by the body's feedback mechanism, and the level is brought back to the normal (80–90 mg/dL) within 2 h. This glucose homeostasis is maintained by two opposing hormones, insulin and glucagon, both produced by the pancreas. Post food intake, blood glucose level rises which initiates insulin signaling resulting in the conversion of excess glucose to glycogen in the liver. During fasting conditions, the blood glucose level falls, which triggers glucagon secretion from the pancreas. Glucagon then acts on the liver to convert liver glycogen to glucose, which is released into the blood circulation, thus maintaining normal glucose levels.

Insulin receptor belongs to the receptor tyrosine kinase family and after binding to insulin, the receptor activates complex intracellular signaling pathways through insulin receptor substrate (IRS) proteins and canonical PI3K and extracellular signal-regulated kinase (ERK) signaling. Lack of insulin production by the pancreas or signaling at target organs can result in the development of diabetes. Recent studies have suggested that misregulated miRNA can contribute to the development of type 2 diabetes (T2D). It was demonstrated by Kornfeld et al. [9] that miR-802 is overexpressed in the liver of obese mice as well as in the obese human volunteers. They further confirmed its role by overexpressing miR-802 in mice, which resulted in reduced insulin sensitivity and impaired glucose tolerance. In contrast, suppression of miR-802 resulted in improved insulin action and glucose tolerance. Subsequent stringent bioinformatic analysis revealed 26 target genes for miR-802, including *hepatocyte nuclear factor 1 beta (Hnf1b)* [9]. *Hnf1b* is known to be associated with the maturity-onset diabetes of the young type 5 [10] and also SNPs in this gene are an important contributor to the development of T2D [11]. Further by using shRNA-mediated silencing of *the Hnf1b* gene, it was validated to be the target of miR-802 [9].

In addition to the liver, several miRNAs play an important role in other tissues to maintain glucose homeostasis. The lethal-7 (let-7) miRNA is a well-known tumor suppressor microRNA that plays an important role in the regulation of the oncogenes. Zhu et al. [12] uncovered a novel role of the Lin28/let-7 pathway in metabolism. They found that whole-body overexpression of let-7 in mice resulted in dysregulated glucose metabolism [12]. Although, tissue-specific overexpression of let-7 in adipocytes, liver, neurons, and muscle had no effect on glucose tolerance as suggested by another study [13]. Expression of let-7 was independent of obesity in mice with insulin resistance, however, treatment with let-7 anti-miRNA was enough to improve glucose tolerance. Another study with human skeletal muscle cells found that miR-1 and miR-133a were significantly decreased after stimulation with insulin [14]. The authors further demonstrated that myocyte enhancer factor 2C and sterol regulatory element-binding protein 1c (SREBP1c) were involved in mediating the effect of insulin on miR-1 and miR-133a. In addition to that, altered SREBP1c activity in the skeletal muscle of type 2 diabetic patients in response to insulin caused impaired regulation of miRNA [14]. Recent studies have found several miRNAs playing important role in T2D and altering insulin sensitivity such as miR-26a and miR-143 in the liver, miR-16 in skeletal muscle, and miR-103/107 in both liver and adipose tissue to mention some [15–18].

Role of microRNA in lipid homeostasis

Lipids such as triglycerides and cholesterol are absorbed from the small intestine and transported all over the body via lipoproteins for bile acid formation, steroid production, and, most importantly, energy production. The most important molecules of these pathways are low-density lipoprotein (LDL), high-density lipoprotein (HDL) cholesterol, triglycerides, and cholesterol. Any kind of imbalance in the levels of cholesterol, triglycerides, LDL, or HDL cholesterol can lead to dyslipidemia [19]. Several factors affect the level of lipids such as nutrition, obesity, physical inactivity, and tobacco use. In addition to nutrition factors, dyslipidemia can arise due to genetic factors also. Familial hypercholesterolemia is a genetic disorder that is most commonly caused by an autosomal dominant mutation in the *LDLR* gene and can lead to high levels of LDL cholesterol [20]. Dyslipidemia is very common in patients with T2D and those having insulin resistance [21]. Importantly, T2D and dyslipidemia both are the major risk factors responsible for the development of atherosclerosis and other cardiovascular diseases [21,22]. Dyslipidemia and abnormal glucose metabolism can also lead to the development of NAFLD [23]. Due to their role in several pathological states, homeostasis in lipid metabolic pathways is very important. Recent studies have demonstrated that in addition to glucose metabolism, miRNAs are also known to regulate many aspects of lipid metabolism. There are several tissue-specific miRNAs that play important regulatory functions at different stages of lipid metabolism. Also, there are several miRNAs such as miR-33a and miR-33b that are found in many organs such as the pancreas, adipose tissue, brain, and liver and are involved in cholesterol and lipid metabolism [24]. Both of these miRNAs are located in the intronic region of the *SREBP* (sterol regulatory element-binding protein) coding genes. miR-33a is encoded by the intronic region of *SREBF2* (sterol regulatory element-binding transcription factor 2) and miR-33b is encoded by the *SREBF1* intronic region [24]. The SREBPs are a family of membrane-bound transcription factors in animal cells that act as the master controller of lipid synthesis [25,26]. miR-33a/b are also known to regulate insulin signaling in the liver by targeting insulin receptor substrate 2 [27]. It was also demonstrated that inhibition of miR-33a/b increased insulin signaling and fatty acid oxidation while overexpressing both miRNAs decreased these metabolic pathways [27]. Activated SREBPs positively regulate the transcription of cholesterolemic and lipogenic genes during the scarcity of fatty acids or sterols. In addition to that, fatty acid degradation and cholesterol export pathways are shut down concurrently by miR-33a/b to enhance the lipid levels [28,29]. miR-33a/b particularly targets and inhibits the expression of genes involved in β-oxidation of fatty acids such as *carnitine palmitoyltransferase 1A (CPT1A), trifunctional enzyme subunit beta (HADHB),* and *carnitine O-octanoyl transferase (CROT)* [28,29]. miR-33a/b also inhibits cholesterol efflux by inhibiting the expression of genes *ATP-binding cassette transporter A1 (ABCA1)* and *ATP-binding cassette transporter G1 (ABCG1)*, which are members of the ATP-binding cassette superfamily [30].

In addition to directly regulating the lipid homeostasis like miR-33a/b, several miRNAs are known to influence the level of triglycerides and cholesterol in circulation. Using the data available in genome-wide association studies from more than 188,000 individuals, Wagschal et al. found several miRNAs near SNPs were associated with dyslipidemia [31]. Many of these, such as miR-148a, miR-130b, miR-301b, and miR-128-1, regulate the expression levels of ABCA1 cholesterol transporter, one of the proteins crucial in the process of cholesterol-lipoprotein trafficking. Indeed, silencing or overexpressing miR-148a and miR-128-1 in a mouse model resulted in altered lipid metabolism and trafficking [31]. Similarly, Goedeke et al. developed a high throughput genome-wide screening assay and identified several miRNAs that regulate LDL receptor (LDLR) levels in the liver. They further identified and characterized miRNA-148a and found that it acts as a negative regulator of LDLR and regulated the uptake of LDL cholesterol [32]. Collectively, these findings provide a clear indication that miRNAs are involved in regulating the circulating lipid levels and misregulation at any step can lead to dyslipidemia.

Role of RNA-binding proteins in metabolic regulation

Role of RBPs in glucose homeostasis

Adenylate-uridylate-rich elements (ARE)/poly(U)-binding factor 1 (AUF1) proteins are known to regulate mRNA translation and stability by binding to uridine and adenosine-rich regions of target mRNA. In a study by Roggli et al. [33], it was demonstrated that AUF1 mediates the cytotoxic effects of proinflammatory cytokines on β cells of the pancreas, indicating its role in the initial phases of type 1 diabetes. AUF1 proteins are mostly localized in the nucleus, however, treatment with cytokines can cause its translocation to the cytoplasm where it can activate the ERK signaling. Overexpression of AUF1 protein leads to a decrease in anti-apoptotic proteins causing β-cell death in the pancreas. Contrary to this, silencing of AUF1 restores the levels of anti-apoptotic proteins and protects the β cells from cytokine-induced cytotoxicity and cell death [33].

In another study, researchers identified RNA-binding protein (RBP) human antigen D (HuD), which is expressed in pancreatic β cells and controls insulin mRNA stability and translation [34]. They also demonstrated that insulin production was more in HuD knockout mice and decreased in HuD overexpressing animals. HuD transgenic mice displayed impaired glucose homeostasis indicating the vital role of HuD in glucose metabolism [34]. Findings by Tillmar et al. have shown that glucose-induced binding of heterogeneous nuclear ribonucleoprotein I (hnRNP) or polypyrimidine tract-binding (PTB) protein to the 3′-UTR of preproinsulin is a necessary step in stabilizing the insulin mRNA [35]. Similarly, hypoxia was shown to promote PTB's binding to 3′-UTR of insulin mRNA resulting in its increased levels and stability [36].

Role of RBPs in lipid homeostasis

Hematopoietic zinc-finger (Hzf), which is a well-known p53 transcriptional target, can also regulate C/EBPα mRNA expression during adipogenesis [37]. Its suppression by shRNA interrupts adipogenesis as Hzf is required during the differentiation of adipocytes. As Hzf directly interacts with the 3′-UTR regions of C/EBPα mRNA to increase its stability, the deficiency of Hzf led to reduced C/EBPα mRNA expression.

Similar to Hzf, human antigen R (HuR) is a universally expressed RBP that regulates translational efficiency and stability of mRNA by interacting with ARE in 3′-UTR regions [38]. HuR is constitutively expressed and mostly localized in the nucleus of the preadipocytes. During the differentiation process from preadipocyte to adipocytes, HuR translocates to the cytosol where it forms a complex with CCAAT enhancer-binding protein β (C/EBPβ). Inhibition of HuR results in a decline in C/EBPβ expression which further prevents adipogenesis signifying its important role in the establishment of adipocyte phenotype [38].

Adverse health conditions associated with metabolic syndromes

Obesity

Overweight and obesity are defined as the excessive or abnormal accumulation of fat that represents a health risk. A BMI greater than 25 is considered overweight and more than 30 is considered obese. It is a metabolic disease influenced by several factors such as biological [39], socioeconomic [40], and environmental [41], which can lead to adverse health outcomes [42]. The Global Burden of Disease obesity investigator estimated that there were more than 600 million obese adults in 2015. The menace of obesity has doubled in 73 countries between 1980 and 2015 and has increased steadily in most other countries [43]. A different study has estimated that between 39% and 49% of the global population (between 2.8 and 3.5 billion people) is overweight or obese [44]. Obesity and associated health effects have a significant direct and indirect economic impact including associated medical costs in the United States ($147 billion in 2008) [45,46]. Additionally, excess adiposity can be accompanied by structural and functional abnormalities such as left ventricular hypertrophy, liver steatosis/fibrosis, gastrointestinal reflux, and disability/immobility [42]. Most importantly, it can increase the comorbidity risk by providing favorable conditions for the development of more than 200 chronic diseases including CVD, some types of cancers, cerebrovascular diseases, type 2 diabetes mellitus, hypertension, asthma, and nonalcoholic fatty liver disease [42]. Collectively, the disease of obesity leads indirectly or directly to a reduction in the quality of life of the affected people and is associated with considerable social and economic costs.

Cardiovascular diseases

According to World Health Organization (WHO), CVDs are the leading cause of death with an estimated 17.9 million deaths annually the world over. CVDs encompass a group of disorders of heart and blood vessels such as coronary artery disease, rheumatic heart disease, cerebrovascular disease, and other heart conditions [47]. In the United States, approximately 1 in 4 deaths are due to heart disease, of which coronary artery disease is the most common, which results in poor blood flow to the heart consequently leading to heart attack [48,49]. It is reported that one person dies every 36 s in the United States from CVDs and the economic burden associated with CVDs is more than $350 billion each year, which includes health care costs and lost productivity on the job [49,50]. The key risk factors for CVDs are metabolic syndromes such as hypertension, hypercholesterolemia, and smoking [51]. Additionally, excessive alcohol consumption, diabetes, obesity, an unhealthy diet, and a sedentary lifestyle pose a higher risk for CVDs [4,51–53]. Among these major modifiable risk factors, hypertension, which is one of the components of metabolic syndrome, is strongly related to the increased incidence of CVDs and its associated mortality risks [54].

Nonalcoholic fatty liver disease

NAFLD is a chronic liver disease and is an important public health problem affecting 25%–30% of the general population worldwide [55]. Approximately 15%–46% of the adult population in the United States is affected by NAFLD, and it is also estimated as one of the most prevalent causes of liver cirrhosis and hepatocellular carcinoma (HCC) [56–59]. NAFLD is strongly associated with the hepatic presentation of metabolic syndrome [59]. A wide spectrum of histopathological changes is observed in NAFLD ranging from asymptomatic hepatic lipid accumulation (steatosis) to nonalcoholic steatohepatitis (NASH) [60,61]. The NASH is characterized by varying degrees of hepatic necrosis, inflammation, and fibrosis, in addition to steatosis [62]. The pathogenesis of NAFLD is influenced by multiple parallel factors including the inactive lifestyle and unhealthy dietary habits [63,64]. It is reported that approximately 10%–30% of patients with NAFLD eventually develop NASH [65]. While simple steatosis is relatively benign with a low degree of developing cirrhosis, NASH can progress to cirrhosis and HCC in a variable proportion of patients [66]. Patients with NAFLD often have dyslipidemia, which is one of the features of metabolic syndrome [67]. Dyslipidemia in NAFLD is marked by elevated serum triglycerides, increased LDL particles, and decreased HDL cholesterol [67,68]. Other components of metabolic syndrome such as obesity and its associated complications of insulin resistance and diabetes are recognized as the important risk factors for the development of NAFLD [69,70].

Type 2 diabetes

The prevalence of diabetes has been rising rapidly affecting more than 400 million adults globally according to WHO's data. An estimated 1.5 million deaths were directly caused by diabetes in the year 2019. Moreover, diabetes can lead to serious health complications and is a major cause of blindness, kidney failure, heart attacks, stroke, and lower limb amputation [71–75]. According to the CDC, T2D is the most common type of diabetes accounting for 90%–95% of all diagnosed cases of diabetes. T2D is a complex chronic disease characterized by elevated blood glucose levels resulting from the body's inefficiency to utilize insulin [76,77]. Numerous studies have established that the prevalence of metabolic syndrome increases the risk for the incidence of T2D [78–80]. Independent of the other components of metabolic syndrome, impaired fasting glucose and/or impaired glucose tolerance confer a higher risk for developing T2D, and current estimates suggest that about 70% of individuals with these prediabetic conditions eventually develop diabetes [81,82]. Apart from genetics, age, and family history of diabetes, certain behavioral risk factors such as excess body weight and physical inactivity increase the likelihood of developing T2D [83]. Moreover, patients with NAFLD are also at higher risk for type 2 diabetes [70]. The onset of T2D can be delayed by improved lifestyle management such as a healthy diet, regular exercise, and maintaining a normal body weight.

Summary

The diseases that are associated with metabolic syndrome can be chronic and lethal, thus, presenting a serious public health concern. Because they cause health, social, and economic loss, it is imperative to find a common mechanism or pathway for new therapeutic options. Understanding the pathways that lead to the origin of metabolic syndromes is crucial in the prevention and treatment of the affected population. Due to the changing lifestyles and dietary habits, a person can develop metabolic syndrome starting from childhood to a later stage of his/her life. In this present chapter, we have discussed several post-transcriptional gene regulators that have been identified as playing important roles in the metabolic processes. Owing to advancements in sequencing methods and recent research, there is a consensus that miRNAs, lncRNAs, and RBPs play significant roles in the pathogenesis of metabolic syndrome and disorders such as type 2 diabetes. In fact, studies point toward an aberrant plasma miRNA profiling in cancer and diabetic patients [84,85]. A loss of miRNA-126 in addition to a distinctive plasma miRNA signature in diabetic patients was found [84]. Likewise, studies have suggested that numerous miRNAs are altered in various pathological conditions and many of them are uniquely related to pathology, making them an ideal candidate to serve as a biomarker [86–88]. This can further help in the prevention, diagnosis, and treatment of metabolic disorders. Furthermore, as miRNAs are implicated in disease pathology, changing their expression pattern via inhibitors or mimetics can potentially be used as a therapeutic strategy. In a study by Krützfeldt et al. [89] the authors have demonstrated the efficacy of chemically engineered and synthesized oligonucleotides known as oligomers in silencing specific miRNAs. They intravenously administered the antagomirs against miR-194, miR-192, miR-122, and miR-16 and found a significant decrease in the corresponding miRNA levels in ovaries, muscles, bone marrow, skin, fat, intestine, heart, kidney, lung, and liver. They further studied the biological importance of suppressing miRNA by using antagomir against miR-122, which is highly expressed in the liver. Bioinformatics coupled with gene expression analysis revealed several genes which were up or downregulated due to miR-122 silencing. Functional annotation analysis of the downregulated genes revealed that the genes associated with cholesterol metabolism were specifically affected by miR-122. Indeed, mice treated with antagomir-122 had significantly low levels of plasma cholesterol demonstrating its effectiveness in treating metabolic diseases [89]. There are several other antagomirs designed to target different miRNAs such as miR-33 [28,90,91], miR-103/107 [15] in different diseases. Besides that, increasing the efficacy of the miRNA delivery system is also very important since silencing efficiency is often limited due to the loss of miRNA. Several delivery systems have been developed for efficient miRNA therapeutic such as viral vectors [92], poly(lactide-co-glycolide) particles [93,94], neutral lipid emulsions [95], EnGeneIC Delivery Vehicle nanocells [96,97], and synthetic polyethyleneimine [98] to name a few. One of the biggest hurdles in developing a miRNA-based therapeutic strategy is the requirement of in-depth knowledge of the post-transcriptional influence of miRNAs and RBPs so that we can identify the most suitable miRNA or miRNA target for disease. Another task is identifying the best suitable miRNA delivery system that can provide higher stability in conjunction with tissue-specific targeting avoiding off-target effects and toxicities [99,100]. The increased research coupled with preclinical analysis should enable miRNA therapeutic safety and effectiveness.

Acknowledgments

All the figures in this chapter were created using BioRender.com.

References

[1] Saklayen MG. The global epidemic of the metabolic syndrome. Curr Hypertens Rep 2018;20(2):1–8.

[2] Rochlani Y, Pothineni NV, Kovelamudi S, Mehta JL. Metabolic syndrome: pathophysiology, management, and modulation by natural compounds. Ther Adv Cardiovasc Dis 2017;11(8):215–25.

[3] Grundy SM. Obesity, metabolic syndrome, and cardiovascular disease. J Clin Endocrinol Metabol 2004; 89(6):2595–600.

[4] Powell-Wiley TM, Poirier P, Burke LE, Despres JP, Gordon-Larsen P, Lavie CJ, et al. Obesity and cardiovascular disease: a scientific statement from the American heart association. Circulation 2021;143(21): e984–1010.

[5] Fabbrini E, Sullivan S, Klein S. Obesity and nonalcoholic fatty liver disease: biochemical, metabolic, and clinical implications. Hepatology 2010;51(2):679–89.

[6] Hocine S, Singer RH, Grünwald D. RNA processing and export. Cold Spring Harb Perspect Biol 2010; 2(12):a000752.

[7] Sequencing HG. Finishing the euchromatic sequence of the human genome. Nature 2004;431(7011): 931–45.

[8] Rinn JL, Chang HY. Genome regulation by long noncoding RNAs. Annu Rev Biochem 2012;81:145–66.

[9] Kornfeld J-W, Baitzel C, Könner AC, Nicholls HT, Vogt MC, Herrmanns K, et al. Obesity-induced overexpression of miR-802 impairs glucose metabolism through silencing of Hnf1b. Nature 2013;494(7435):111–5.
[10] Horikawa Y, Iwasaki N, Hara M, Furuta H, Hinokio Y, Cockburn BN, et al. Mutation in hepatocyte nuclear factor–1β gene (TCF2) associated with MODY. Nat Genet 1997;17(4):384–5.
[11] Han X, Luo Y, Ren Q, Zhang X, Wang F, Sun X, et al. Implication of genetic variants near SLC30A8, HHEX, CDKAL1, CDKN2A/B, IGF2BP2, FTO, TCF2, KCNQ1, and WFS1 in type 2 diabetes in a Chinese population. BMC Med Genet 2010;11(1):1–9.
[12] Zhu H, Shyh-Chang N, Segrè AV, Shinoda G, Shah SP, Einhorn WS, et al. The Lin28/let-7 axis regulates glucose metabolism. Cell 2011;147(1):81–94.
[13] Frost RJ, Olson EN. Control of glucose homeostasis and insulin sensitivity by the Let-7 family of microRNAs. Proc Natl Acad Sci USA 2011;108(52):21075–80.
[14] Granjon A, Gustin M-P, Rieusset J, Lefai E, Meugnier E, Güller I, et al. The microRNA signature in response to insulin reveals its implication in the transcriptional action of insulin in human skeletal muscle and the role of a sterol regulatory element-binding protein-1c/myocyte enhancer factor 2C pathway. Diabetes 2009;58(11):2555–64.
[15] Trajkovski M, Hausser J, Soutschek J, Bhat B, Akin A, Zavolan M, et al. MicroRNAs 103 and 107 regulate insulin sensitivity. Nature 2011;474(7353):649–53.
[16] Lee DE, Brown JL, Rosa ME, Brown LA, Perry Jr RA, Wiggs MP, et al. microRNA-16 is downregulated during insulin resistance and controls skeletal muscle protein accretion. J Cell Biochem 2016;117(8):1775–87.
[17] Jordan SD, Krüger M, Willmes DM, Redemann N, Wunderlich FT, Brönneke HS, et al. Obesity-induced overexpression of miRNA-143 inhibits insulin-stimulated AKT activation and impairs glucose metabolism. Nat Cell Biol 2011;13(4):434–46.
[18] Fu X, Dong B, Tian Y, Lefebvre P, Meng Z, Wang X, et al. MicroRNA-26a regulates insulin sensitivity and metabolism of glucose and lipids. J Clin Invest 2015;125(6):2497–509.
[19] Rader DJ, Hoeg JM, Brewer HB. Quantitation of plasma apolipoproteins in the primary and secondary prevention of coronary artery disease. Ann Intern Med 1994;120(12):1012–25.
[20] Defesche JC, Gidding SS, Harada-Shiba M, Hegele RA, Santos RD, Wierzbicki AS. Familial hypercholesterolaemia. Nat Rev Dis Prim 2017;3(1):1–20.
[21] Adiels M, Olofsson S-O, Taskinen M-R, Borén J. Overproduction of very low–density lipoproteins is the hallmark of the dyslipidemia in the metabolic syndrome. Arterioscler Thromb Vasc Biol 2008;28(7):1225–36.
[22] Koba S, Hirano T. Dyslipidemia and atherosclerosis. Nihon Rinsho 2011;69(1):138–43. Japanese Journal of Clinical Medicine.
[23] Parekh S, Anania FA. Abnormal lipid and glucose metabolism in obesity: implications for nonalcoholic fatty liver disease. Gastroenterology 2007;132(6):2191–207.
[24] Najafi-Shoushtari SH, Kristo F, Li Y, Shioda T, Cohen DE, Gerszten RE, et al. MicroRNA-33 and the SREBP host genes cooperate to control cholesterol homeostasis. Science 2010;328(5985):1566–9.
[25] Brown MS, Goldstein JL. The SREBP pathway: regulation of cholesterol metabolism by proteolysis of a membrane-bound transcription factor. Cell 1997;89(3):331–40.
[26] Rawson RB. The SREBP pathway—insights from Insigs and insects. Nat Rev Mol Cell Biol 2003;4(8):631–40.
[27] Dávalos A, Goedeke L, Smibert P, Ramírez CM, Warrier NP, Andreo U, et al. miR-33a/b contribute to the regulation of fatty acid metabolism and insulin signaling. Proc Natl Acad Sci USA 2011;108(22):9232–7.
[28] Rayner KJ, Suárez Y, Dávalos A, Parathath S, Fitzgerald ML, Tamehiro N, et al. MiR-33 contributes to the regulation of cholesterol homeostasis. Science 2010;328(5985):1570–3.

[29] Gerin I, Clerbaux L-A, Haumont O, Lanthier N, Das AK, Burant CF, et al. Expression of miR-33 from an SREBP2 intron inhibits cholesterol export and fatty acid oxidation. J Biol Chem 2010;285(44): 33652−61.

[30] Horie T, Ono K, Horiguchi M, Nishi H, Nakamura T, Nagao K, et al. MicroRNA-33 encoded by an intron of sterol regulatory element-binding protein 2 (Srebp2) regulates HDL in vivo. Proc Natl Acad Sci USA 2010; 107(40):17321−6.

[31] Wagschal A, Najafi-Shoushtari SH, Wang L, Goedeke L, Sinha S, Andrew S, et al. Genome-wide identification of microRNAs regulating cholesterol and triglyceride homeostasis. Nat Med 2015;21(11):1290−7.

[32] Goedeke L, Rotllan N, Canfrán-Duque A, Aranda JF, Ramírez CM, Araldi E, et al. MicroRNA-148a regulates LDL receptor and ABCA1 expression to control circulating lipoprotein levels. Nat Med 2015; 21(11):1280−9.

[33] Roggli E, Gattesco S, Pautz A, Regazzi R. Involvement of the RNA-binding protein ARE/poly (U)-binding factor 1 (AUF1) in the cytotoxic effects of proinflammatory cytokines on pancreatic beta cells. Diabetologia 2012;55(6):1699−708.

[34] Lee EK, Kim W, Tominaga K, Martindale JL, Yang X, Subaran SS, et al. RNA-binding protein HuD controls insulin translation. Mol Cell 2012;45(6):826−35.

[35] Tillmar L, Carlsson C, Welsh N. Control of insulin mRNA stability in rat pancreatic islets: regulatory role of a 3′-untranslated region pyrimidine-rich sequence. J Biol Chem 2002;277(2):1099−106.

[36] Tillmar L, Welsh N. Hypoxia may increase rat insulin mRNA levels by promoting binding of the poly-pyrimidine tract-binding protein (PTB) to the pyrimidine-rich insulin mRNA 3′-untranslated region. Mol Med 2002;8(5):263−72.

[37] Kawagishi H, Wakoh T, Uno H, Maruyama M, Moriya A, Morikawa S, et al. Hzf regulates adipogenesis through translational control of C/EBPα. EMBO J 2008;27(10):1481−90.

[38] Gantt K, Cherry J, Tenney R, Karschner V, Pekala PH. An early event in adipogenesis, the nuclear selection of the CCAAT enhancer-binding protein β (C/EBPβ) mRNA by HuR and its translocation to the cytosol. J Biol Chem 2005;280(26):24768−74.

[39] Loos RJ. Genetic determinants of common obesity and their value in prediction. Best Pract Res Clin Endocrinol Metab 2012;26(2):211−26.

[40] Sommer I, Griebler U, Mahlknecht P, Thaler K, Bouskill K, Gartlehner G, et al. Socioeconomic inequalities in non-communicable diseases and their risk factors: an overview of systematic reviews. BMC Public Health 2015;15(1):1−12.

[41] Franks PW, McCarthy MI. Exposing the exposures responsible for type 2 diabetes and obesity. Science 2016;354(6308):69−73.

[42] Jastreboff AM, Kotz CM, Kahan S, Kelly AS, Heymsfield SB. Obesity as a disease: the obesity society 2018 position statement. Obesity 2019;27(1):7−9.

[43] Afshin A, Forouzanfar M, Reitsma M, Sur P, Estep K, Lee A, et al. Health effects of overweight and obesity in 195 countries over 25 years. GBD 2015 Obesity Collaborators. N Engl J Med 2017;377(1):13−27.

[44] Maffetone PB, Rivera-Dominguez I, Laursen PB. Overfat and underfat: new terms and definitions long overdue. Front Public Health 2017;4:279.

[45] Finkelstein EA, Trogdon JG, Cohen JW, Dietz W. Annual medical spending attributable to obesity: payer- and service-specific estimates: amid calls for health reform, real cost savings are more likely to be achieved through reducing obesity and related risk factors. Health Aff 2009;28(Suppl. 1):w822−31.

[46] Trogdon JG, Finkelstein EA, Hylands T, Dellea PS, Kamal-Bahl S. Indirect costs of obesity: a review of the current literature. Obes Rev 2008;9(5):489−500.

[47] Flora GD, Nayak MK. A brief review of cardiovascular diseases, associated risk factors and current treatment regimes. Curr Pharmaceut Des 2019;25(38):4063−84.

[48] Brown JC, Gerhardt TE, Kwon E. Risk factors for coronary artery disease. 2020.

[49] Virani SS, Alonso A, Aparicio HJ, Benjamin EJ, Bittencourt MS, Callaway CW, et al. Heart disease and stroke statistics—2021 update: a report from the American Heart Association. Circulation 2021;143(8): e254–743.

[50] Control CfD, Prevention. Underlying cause of death 1999–2018. CDC WONDER Online Database; 2020.

[51] Hajar R. Risk factors for coronary artery disease: historical perspectives. Heart Views 2017;18(3):109. The Official Journal of the Gulf Heart Association.

[52] Piano MR. Alcohol's effects on the cardiovascular system. Alcohol Res Curr Rev 2017;38(2):219.

[53] Lavie CJ, Ozemek C, Carbone S, Katzmarzyk PT, Blair SN. Sedentary behavior, exercise, and cardiovascular health. Circ Res 2019;124(5):799–815.

[54] Fuchs FD, Whelton PK. High blood pressure and cardiovascular disease. Hypertension 2020;75(2): 285–92.

[55] Rinella ME, Sanyal AJ. Management of NAFLD: a stage-based approach. Nat Rev Gastroenterol Hepatol 2016;13(4):196–205.

[56] Wong RJ, Singal AK. Trends in liver disease etiology among adults awaiting liver transplantation in the United States, 2014–2019. JAMA Netw Open 2020;3(2). e1920294-e.

[57] Sheth SG, Chopra S. Epidemiology, clinical features, and diagnosis of nonalcoholic fatty liver disease in adults. Waltham, MA: UpToDate; 2017.

[58] Dhamija E, Paul SB, Kedia S. Non-alcoholic fatty liver disease associated with hepatocellular carcinoma: an increasing concern. Indian J Med Res 2019;149(1):9.

[59] Paschos P, Paletas K. Non alcoholic fatty liver disease and metabolic syndrome. Hippokratia 2009;13(1):9.

[60] Kopec KL, Burns D. Nonalcoholic fatty liver disease: a review of the spectrum of disease, diagnosis, and therapy. Nutr Clin Pract 2011;26(5):565–76.

[61] Hardy T, Oakley F, Anstee QM, Day CP. Nonalcoholic fatty liver disease: pathogenesis and disease spectrum. Annu Rev Pathol 2016;11:451–96.

[62] Hashimoto E, Taniai M, Tokushige K. Characteristics and diagnosis of NAFLD/NASH. J Gastroenterol Hepatol 2013;28:64–70.

[63] Arab JP, Arrese M, Trauner M. Recent insights into the pathogenesis of nonalcoholic fatty liver disease. Annu Rev Pathol 2018;13:321–50.

[64] Smati S, Polizzi A, Fougerat A, Ellero-Simatos S, Blum Y, Lippi Y, et al. Integrative study of diet-induced mouse models of NAFLD identifies PPARα as a sexually dimorphic drug target. Gut 2022;71(4):807–21.

[65] Dyson JK, Anstee QM, McPherson S. Non-alcoholic fatty liver disease: a practical approach to diagnosis and staging. Frontline Gastroenterol 2014;5(3):211–8.

[66] Cholankeril G, Patel R, Khurana S, Satapathy SK. Hepatocellular carcinoma in non-alcoholic steatohepatitis: current knowledge and implications for management. World J Hepatol 2017;9(11):533.

[67] Dyslipidemia in patients with nonalcoholic fatty liver disease. In: Chatrath H, Vuppalanchi R, Chalasani N, editors. Seminars in liver disease. Thieme Medical Publishers; 2012.

[68] Deprince A, Haas JT, Staels B. Dysregulated lipid metabolism links NAFLD to cardiovascular disease. Mol Metabol 2020:101092.

[69] Sarwar R, Pierce N, Koppe S. Obesity and nonalcoholic fatty liver disease: current perspectives. Diabetes Metab Syndr Obes Targets Ther 2018;11:533.

[70] Gastaldelli A, Cusi K. From NASH to diabetes and from diabetes to NASH: mechanisms and treatment options. JHEP Rep 2019;1(4):312–28.

[71] Skarbez K, Priestley Y, Hoepf M, Koevary SB. Comprehensive review of the effects of diabetes on ocular health. Expet Rev Ophthalmol 2010;5(4):557–77.

[72] Thomas MC, Brownlee M, Susztak K, Sharma K, Jandeleit-Dahm KA, Zoungas S, et al. Diabetic kidney disease. Nat Rev Dis Prim 2015;1(1):1–20.

[73] Leon BM, Maddox TM. Diabetes and cardiovascular disease: epidemiology, biological mechanisms, treatment recommendations and future research. World J Diabetes 2015;6(13):1246.

[74] Chen R, Ovbiagele B, Feng W. Diabetes and stroke: epidemiology, pathophysiology, pharmaceuticals and outcomes. Am J Med Sci 2016;351(4):380–6.

[75] Moxey P, Gogalniceanu P, Hinchliffe R, Loftus I, Jones K, Thompson M, et al. Lower extremity amputations—a review of global variability in incidence. Diabet Med 2011;28(10):1144–53.

[76] Fonseca VA. Defining and characterizing the progression of type 2 diabetes. Diabetes Care 2009;32(Suppl. 2):S151–6.

[77] Henninger J, Rawshani A, Hammarstedt A, Eliasson B. Metabolic characteristics of individuals at a high risk of type 2 diabetes–a comparative cross-sectional study. BMC Endocr Disord 2017;17(1):1–9.

[78] Shin JA, Lee JH, Lim SY, Ha HS, Kwon HS, Park YM, et al. Metabolic syndrome as a predictor of type 2 diabetes, and its clinical interpretations and usefulness. J Diabetes Investig 2013;4(4):334–43.

[79] Lorenzo C, Okoloise M, Williams K, Stern MP, Haffner SM. The metabolic syndrome as predictor of type 2 diabetes: the San Antonio heart study. Diabetes Care 2003;26(11):3153–9.

[80] Najarian RM, Sullivan LM, Kannel WB, Wilson PW, D'Agostino RB, Wolf PA. Metabolic syndrome compared with type 2 diabetes mellitus as a risk factor for stroke: the Framingham Offspring Study. Arch Intern Med 2006;166(1):106–11.

[81] Di Bonito P, Pacifico L, Chiesa C, Valerio G, Del Giudice EM, Maffeis C, et al. Impaired fasting glucose and impaired glucose tolerance in children and adolescents with overweight/obesity. J Endocrinol Invest 2017;40(4):409–16.

[82] Tabák AG, Herder C, Rathmann W, Brunner EJ, Kivimäki M. Prediabetes: a high-risk state for developing diabetes. Lancet 2012;379(9833):2279.

[83] Bellou V, Belbasis L, Tzoulaki I, Evangelou E. Risk factors for type 2 diabetes mellitus: an exposure-wide umbrella review of meta-analyses. PLoS One 2018;13(3):e0194127.

[84] Zampetaki A, Kiechl S, Drozdov I, Willeit P, Mayr U, Prokopi M, et al. Plasma microRNA profiling reveals loss of endothelial miR-126 and other microRNAs in type 2 diabetes. Circ Res 2010;107(6): 810–7.

[85] Chen X, Ba Y, Ma L, Cai X, Yin Y, Wang K, et al. Characterization of microRNAs in serum: a novel class of biomarkers for diagnosis of cancer and other diseases. Cell Res 2008;18(10):997–1006.

[86] Condrat CE, Thompson DC, Barbu MG, Bugnar OL, Boboc A, Cretoiu D, et al. miRNAs as biomarkers in disease: latest findings regarding their role in diagnosis and prognosis. Cells 2020;9(2):276.

[87] Han SA, Jhun BW, Kim S-Y, Moon SM, Yang B, Kwon OJ, et al. miRNA expression profiles and potential as biomarkers in nontuberculous mycobacterial pulmonary disease. Sci Rep 2020;10(1):1–13.

[88] Ramzan F, D'Souza R, Durainayagam B, Milan A, Markworth J, Miranda-Soberanis V, et al. Circulatory miRNA biomarkers of metabolic syndrome. Acta Diabetol 2020;57(2):203–14.

[89] Krützfeldt J, Rajewsky N, Braich R, Rajeev KG, Tuschl T, Manoharan M, et al. Silencing of microRNAs in vivo with 'antagomirs. Nature 2005;438(7068):685–9.

[90] Marquart TJ, Allen RM, Ory DS, Baldán Á. miR-33 links SREBP-2 induction to repression of sterol transporters. Proc Natl Acad Sci USA 2010;107(27):12228–32.

[91] Price NL, Zhang X, Fernández-Tussy P, Singh AK, Burnap SA, Rotllan N, et al. Loss of hepatic miR-33 improves metabolic homeostasis and liver function without altering body weight or atherosclerosis. Proc Natl Acad Sci USA 2021;118(5).

[92] Van Rooij E, Kauppinen S. Development of micro RNA therapeutics is coming of age. EMBO Mol Med 2014;6(7):851–64.

[93] Kulkarni RK, Moore E, Hegyeli A, Leonard F. Biodegradable poly (lactic acid) polymers. J Biomed Mater Res 1971;5(3):169–81.

[94] Blum JS, Saltzman WM. High loading efficiency and tunable release of plasmid DNA encapsulated in submicron particles fabricated from PLGA conjugated with poly-L-lysine. J Control Release 2008;129(1): 66–72.
[95] Trang P, Wiggins JF, Daige CL, Cho C, Omotola M, Brown D, et al. Systemic delivery of tumor suppressor microRNA mimics using a neutral lipid emulsion inhibits lung tumors in mice. Mol Ther 2011;19(6): 1116–22.
[96] MacDiarmid JA, Mugridge NB, Weiss JC, Phillips L, Burn AL, Paulin RP, et al. Bacterially derived 400 nm particles for encapsulation and cancer cell targeting of chemotherapeutics. Cancer Cell 2007;11(5): 431–45.
[97] Taylor K, Howard CB, Jones ML, Sedliarou I, MacDiarmid J, Brahmbhatt H, et al. Nanocell targeting using engineered bispecific antibodies. mAbs 2015;7(1):53–65 [Taylor & Francis].
[98] Akhtar S, Benter IF. Nonviral delivery of synthetic siRNAs in vivo. J Clin Invest 2007;117(12):3623–32.
[99] Li Z, Rana TM. Therapeutic targeting of microRNAs: current status and future challenges. Nat Rev Drug Discov 2014;13(8):622–38.
[100] Rupaimoole R, Han H-D, Lopez-Berestein G, Sood AK. MicroRNA therapeutics: principles, expectations, and challenges. Chin J Cancer 2011;30(6):368.

CHAPTER 15

Post-transcriptional regulation of HIV-1 gene expression: role of viral and host factors

Anjali Tripathi, Alapani Mitra, Anindita Dasgupta and Debashis Mitra
National Centre for Cell Science, S P Pune University Campus, Pune, Maharashtra, India

Introduction

According to the last UNAIDS report, approximately 38 million people in the world are currently living with HIV/AIDS, out of which 1.7 million people were infected in 2019 [1]. Although the rate of infection has slowed down significantly in recent years, the numbers are still alarming and have been a matter of concern as the virus is mutating continuously due to its extremely fast rate of replication, homologous recombination as well as lack of proofreading activity. There are two different types of HIV: HIV-1 (the predominant type throughout the world) and HIV-2 (less infectious and mainly found in western Africa). HIV-1 primarily infects and reduces the number of vital immune cells that include CD4+ T cells, monocytes, and macrophages and establishes latent reservoirs in these cells [2]. Infection of these immune cells alters the cellular microenvironment that ultimately results in the death of these cells over a period of time [3].

HIV-1 is an enveloped virus, the genome of which consists of two copies of positive-sense, single-stranded RNA of about 9.7 kb length. It has nine open reading frames that code for 15 different viral proteins and has long terminal repeats (LTR) at both ends. The 15 different viral proteins belong to four different categories-structural (Gag and Env), catalytic (Pol), regulatory proteins (Tat and Rev), and accessory (Nef, Vif, Vpu, and Vpr/Vpx) proteins. Apart from the LTRs and coding sequences, the HIV-1 genome also consists of various cis-acting elements. These elements are responsible for various functions, some of which are as follows: trans-activation response (TAR) element for the enhancement of viral transcription, Rev-response element (RRE) for the mRNA transport, and packaging of the viral genome into virions mediated by the packaging signal [4].

The HIV-1 life cycle starts with the attachment of the virus envelope glycoprotein gp120 to two different host cell membrane proteins: primary receptor CD4 and a chemokine coreceptor, which could be either CCR5 or CXCR4 depending on the tropism of the virus. This binding enables the fusion of the viral envelope and the host plasma membrane leading to the delivery of the viral core into the cytoplasm [5]. This is followed by uncoating, which is coupled with the reverse transcription of the viral RNA into double-stranded DNA [6]. After the completion of reverse transcription, the dsDNA genome of the virus interacts with various viral as well as host factors to form a structure that is known as the preintegration complex (PIC), and this complex is transported into the nucleus. However,

Post-Transcriptional Gene Regulation in Human Disease, Volume 32. https://doi.org/10.1016/B978-0-323-91305-8.00004-1

recently there is strong evidence that suggests that the uncoating, as well as the reverse transcription, is completed in the nucleus [7]. With the help of the viral integrase enzyme, the viral dsDNA is integrated into the host genome and it is then referred to as the provirus [8]. Once the provirus is formed, it utilizes the host transcription machinery to produce viral genome length transcripts as well as spliced variants to produce viral proteins. The polyproteins, the genomic RNA (gRNA), and various other necessary factors assemble at the plasma membrane of the host cells and bud out of the cell as an immature virus. As soon as the viral particle buds out of the cell, the Protease enzyme becomes catalytically active and cleaves the Gag and Gag-Pol to create mature virions [9].

As described earlier, in the life cycle of HIV-1, post-transcriptional events comprise a major part of the virus life span and have a diverse range of roles in the regulation of viral propagation and pathogenicity. These post-transcriptional events begin with the co- and post-transcriptional modifications of viral pre-mRNAs including the processing of the 5′- and 3′-ends of the mRNA, alternative splicing, and epitranscriptomic modifications. Following viral pre-mRNA synthesis, HIV-1 uses both viral and host post-transcriptional machinery such as viral RNA instability elements, Rev-CRM1-dependent RNA nuclear export pathway, and noncoding RNAs to regulate the stability and translation of viral RNAs. Apart from regulation at the post-transcription level, HIV-1 along with various host factors tends to regulate the viral protein synthesis. In addition to utilizing the host machinery for classical translation, some specialized translation methods such as ribosomal shunting as well as ribosomal frameshifting provide uniqueness to the translation of the viral mRNAs. The viral proteins also undergo various post-translational modifications that play a crucial role in their functioning, such as downregulation of the CD4 receptors, counteraction of the host restriction factors, and viral infectivity. HIV-1 proteins are also post-translationally regulated through various post-translational events such as proteasomal degradation, lysosomal degradation, and ER-associated protein degradation (ERAD) as well as virion maturation.

In the following sections, we have tried to briefly consolidate the present knowledge regarding the post-transcriptional, translational, and post-translational regulatory events of the HIV-1 life cycle and their respective roles in viral propagation and infectivity. Understanding these crucial steps might help us in designing novel strategies to eradicate the HIV-1 infection and overcome the deficiencies of the current therapeutic strategies.

HIV-1 transcription

Following integration, HIV-1 uses the host transcription machinery to initially generate only short fully spliced transcripts encoding Nef and viral regulatory proteins Tat and Rev. With the progression of infection, viral transcription rate is sharply increased, and larger partially spliced and unspliced transcripts (that also acts a genomic RNA material) encoding viral structural proteins (Gag and Env), viral enzymes, as well as accessory proteins (Vif, Vpr, and Vpu) are produced.

Molecular mechanism of HIV-1 transcription

HIV-1 transcription is driven by duplicate transcriptional control sequence referred to as LTRs that are regulated at distinct phases mainly by Tat and Rev. In the case of acute HIV-1 infection in activated T cells, several transcription activators including NFκB (p65-p50 heterodimer), NFAT, and Sp1 start viral transcription by assembling and recruiting the preinitiation complex at the 5′-LTR promoter [10]. Initially, the transcription initiation factor TFIIH activates RNAPII to clear the HIV-1 LTR promoter by phosphorylating serine-5 at the C-terminal domain (CTD) of RNAPII. Following transcription initiation, RNAPII is soon stalled after transcribing the TAR region by transcription elongation factors DSIF (DRB sensitivity inducing factor) and negative elongation factor (NELF). Viral Tat binds to TAR and recruits the P-TEFb complex comprising CDK9 and Cyclin T1 (CycT1). The RNAPII pausing is relieved by phosphorylation of serine-2 at RNAPII-CTD by Tat-recruited P-TEFb complex that further stimulates HIV-1 transcription elongation and cotranscriptional processing. Conversely, HIV-1 latency is established in resting T cells where the P-TEFb becomes enzymatically inactive through interacting with cellular 7SK snRNA and HEXIM1. This in turn prevents the release of RNAPII arrest, thereby blocking the transition from viral transcription initiation to elongation as depicted in Fig. 15.1.

While the Tat-TAR interaction activates the elongation of viral transcription through 5′-LTR, the Rev-RRE interaction transports the otherwise unstable partially spliced and unspliced HIV-1 RNAs from nucleus to cytoplasm for their translation into proteins. Apart from these two viral regulatory proteins, HIV-1 infection is regulated at the post-transcriptional level by a plethora of host factors as well as other viral machinery, which will be discussed in the following sections [10,11].

Epigenetic regulation of HIV-1 transcription

Transcription from HIV-1 LTR of integrated provirus is subject to epigenetic regulation encompassing DNA methylation, histone modifications including acetylation and methylation, as well as chromosomal remodeling, which are discussed in the following based on modifications.

The CpG islands in the LTR region are hypermethylated by DNA methyltransferase (DNMT) and have been reported to promote transcriptional silencing during HIV-1 latency [12,13].

Regardless of the integration site, 5′-LTR of HIV-1 proviruses have precisely five nucleosomes, that is, nuc-0 to nuc-4. During latency, nuc-1 is maintained in the hypoacetylated state by the Histone deacetylase (HDACs) recruited to the LTR through multiple transcription factors including YY1, LSF, AP4, NFκB (p50-p50 homodimer), c-Myc, Sp1, CTIP-2, and CBF-1. Upon activation of the latent proviruses, p65 gets accumulated in the nucleus with subsequent recruitment of HATs including CBP, p300, GCN5, and p300/CBP-associated factor (PCAF) to the HIV-1 LTR by several LTR-binding transcription factors such as NFkB (p65-p50 heterodimer), AP-1, C/EBP, LEF-1, NFAT, Sp1, IRF, and viral Tat. The p300/CBP also acetylates Tat to modulate its function [14–18].

In the case of histone methylation mediated by histone methyltransferase (HMT), the Tat-recruited P-TEFb kinase induces H3K4 and H3K36 methylation, which leads to the transcriptional activation of HIV-1 genes. On the contrary, among the histone methylation marks associated with repressive chromatin, H3K9me2 (mediated by methyltransferase G9a), H3K9me3 (mediated by methyltransferases SUV39H1 and HP1γ), and H3K27me3 (mediated by methyltransferase EZH2, a component of polycomb repressive complex-2 [PRC-2]) induce HIV-1 latency [19,20].

Following modification of histones, the chromatin remodeling complex, polybromo-associated BAF (PBAF), a subclass of SWI/SNF complex gets recruited to the HIV-1 promoter and reactivates the HIV-1 genome from the latency. The nucleosomes nuc-0 and nuc-1 are precisely positioned around promoter blocking transcription elongation. Upon reactivation from latency, nuc-1 is specifically remodeled, which is necessary for the successful transcription of the HIV-1 genome. While PBAF is needed for remodeling of nuc-1 for transcription activation, BRG1/BRM-associated factor (BAF),

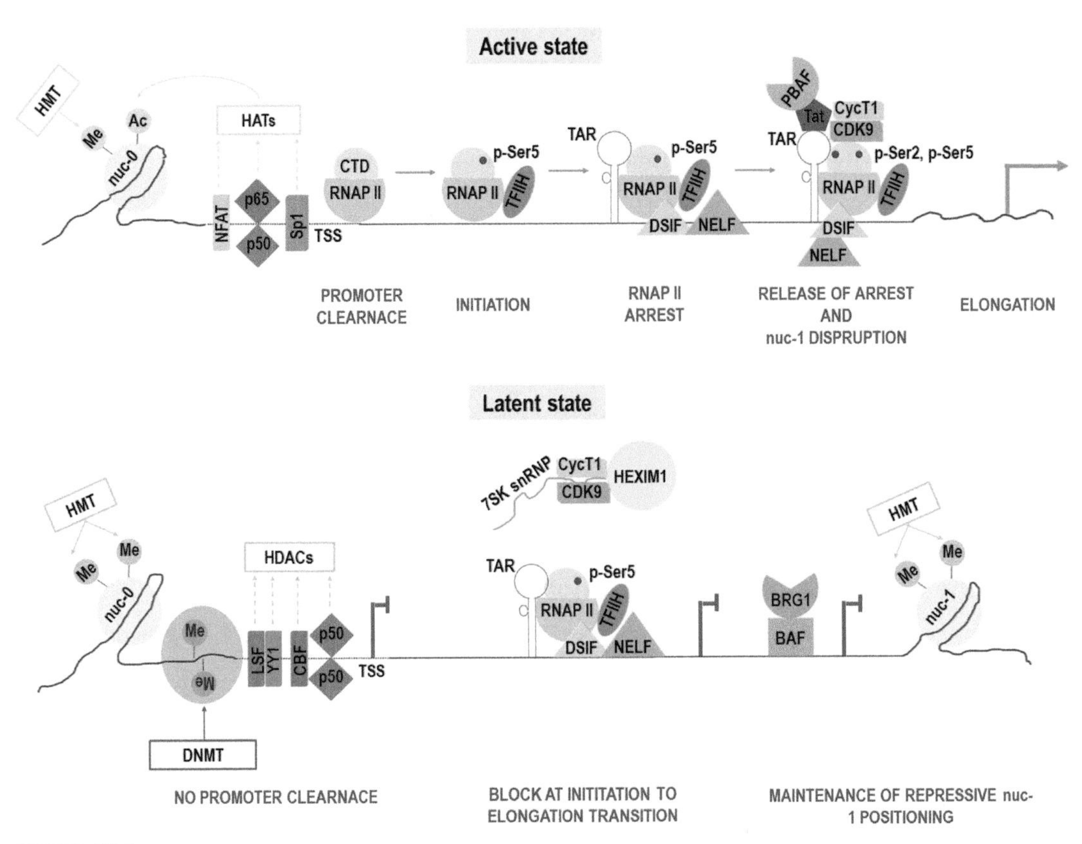

FIGURE 15.1

Regulation of HIV-1 transcription in productive and latent infection. In the case of productive infection, viral transcription starts with the help of transcription factors like NFkB (p65-p50 heterodimer), NFAT, and Sp1, which recruit HATs to HIV-1 5′-LTR. Following HAT-mediated histone acetylation and HMT-mediated histone methylation of the LTR promoter, the preinitiation complex containing RNAPII is recruited to the viral transcription start site. Phosphorylation of RNAPII CTD by initiation factor TFIIH activates viral transcription. After the TAR region is transcribed, RNAPII is paused by DSIF and NELF. This halt is relieved by hyper-phosphorylation of RNAPII CTD by the Tat-recruited P-TEFb complex (comprising CDK9 and CycT1), and nuc-1 is remodeled with the help of PBAF that in turn stimulates viral transcription elongation. In the case of HIV-1 latency, transcription is blocked at several steps. Viral LTR promoter activity is hindered by HDACs recruited by transcription factors (YY1, LSF, CBF, p50 homodimers), repressive histone methylation marks, and CpG island methylation by DNMTs. HIV-1 transcription is also inhibited at the stage of initiation to elongation transition by maintaining inactive P-TEFb through interaction with 7SK snRNA and HEXIM1. In latent infection, the repressive nuc-1 positioning is maintained by chromatin remodeling factor BAF that hinders the HIV-1 transcription elongation.

another subclass of SWI/SNF complex has been associated with HIV-1 transcriptional silencing by repressive positioning of nuc-1 (shown in Fig. 15.1). A nuclear matrix protein SMAR1 also acts as a repressor of HIV-1 LTR through chromatin remodeling. Acetylated Tat recruits two ATPase subunits of SWI/SNF, that is, Brg-1 and Brm to HIV-1 promoter for Tat-mediated LTR transactivation. INI1, a component of the SWI/SNF complex, has also been found to interact with HIV-specific integrase suggesting its involvement in viral post-transcriptional events [19,21].

Co- and post-transcriptional modifications of HIV-1 pre mRNAs

After Tat-induced HIV-1 LTR-driven transcription of viral pre-mRNAs, its 5′- and 3′-ends are co- and post-transcriptionally modified in a classical way that occurs for the host mRNAs. However, hypermethylated cap formation of some HIV-1 RNAs and differential use of 5′- and 3′-poly (A) sites present in duplicate viral LTRs render uniqueness to the post-transcriptional modification of HIV-1 transcripts. Alternative splicing of HIV-1 transcripts plays a huge role in optimal viral gene expression by balancing the levels of spliced and unspliced HIV-1 RNAs. RNA editing by epitranscriptomic modifications is also involved in the modulation of different phases of HIV-1 pathogenesis and in establishing latency.

5′-end-capping

The viral Tat protein with its potent nucleic acid-chaperoning activities promotes cap formation on HIV-1 RNA by two distinct mechanisms: (1) it stimulates capping of TAR RNA by the recruitment and activation of the capping enzyme through its direct interaction with the MceI subunit of the capping enzyme. (2) It also induces phosphorylation of RNAP II CTD by P-TEFb, stimulating the guanylyltransferase activity of the capping enzyme and thereby capping the HIV-1 pre-mRNA. Capping of HIV-1 pre-mRNAs happens less efficiently when RNAPII catalyzes the elongation without being halted initially at the promoter-proximal positions, which suggest the importance of transcription elongation checkpoint in HIV-1 gene expression. Some HIV-1 RNAs relying on Rev/RRE-dependent nuclear export pathways have hypermethylated 2,2,7 trimethylguanosine (TMG) caps. These hypermethylated HIV-1 caps may disrupt the biogenesis of U snRNA and/or their nuclear export [22,23].

3′-end polyadenylation

In the case of the integrated HIV-1 provirus, due to the presence of 5′- and 3′-LTRs where poly(A) sites are located at each R/U5 junction, the 3′-mRNA processing does not follow the default process. The promoter elements are utilized in the proviral 5′-LTR and 3′-end modification is carried out only in 3′-LTR to regulate proper levels of HIV-1 gene products (Fig. 15.2A). In the case of the 3′-poly(A) signal, HIV-1 U3 sequences 56–94 bases upstream of AAUAAA hexamer acts as the essential enhancer element for its polyadenylation [24,25]. The stem-loop RNA structure of TAR and of R region encompassing the poly(A) signals is called poly(A) hairpin, and it is required for the full activation of the 3′-poly(A) site of the transcript. HIV-1 leader sequences, exclusively located downstream of the 5′-poly(A) site inhibit the formation of poly(A) complex at the 5′-poly(A) site. The computational folding predictions also show that the metastable nature of poly(A) signals of the HIV-1 3′-poly(A) site

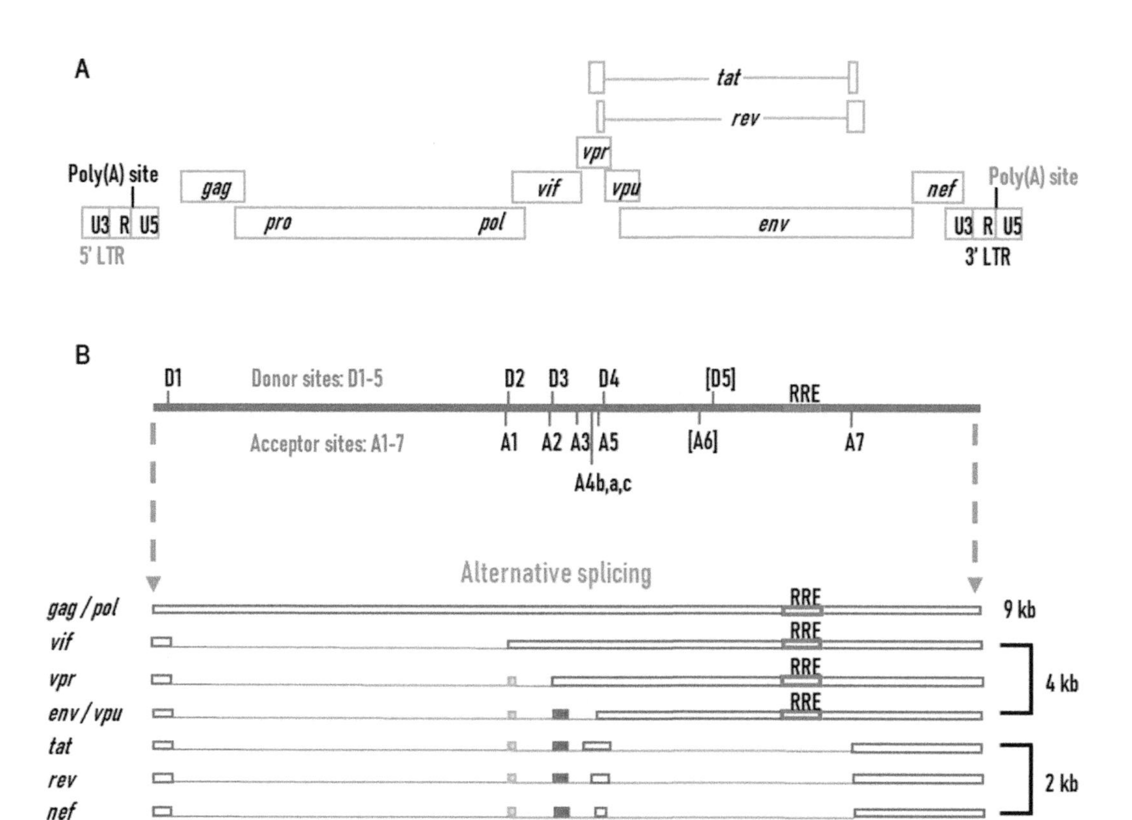

FIGURE 15.2

Schematic representation of HIV-1 transcription control sequences, genes, splice sites, and spliced elements in the HIV-1 genome. (A) Diagram of HIV-1 genome organization. Location of 5′- and 3′-LTR is also shown. The 5′-LTR (highlighted in green) is used as an HIV-1 promoter element and the 3′-poly (A) site (highlighted in green) is used for the polyadenylation step. (B) Location of major donor and acceptor splice sites in HIV-1 genome. The rarely used splice sites D5 and A6 are also shown. The location of RRE is shown in the red box. HIV-1 pre-mRNA undergoes alternative splicing to generate three size-classes of HIV-1 transcripts. The HIV-1 exons designated by boxes are spliced by differential use of 5′ and 3′ ss. Exon two is shown in light blue and exon three is shown in dark blue.

influences this viral mRNA processing step [26–28]. HIV-1 also promotes its 3′-mRNA processing through a novel pathway by Tat-mediated upregulation of CPSF3 expression [29].

Alternative splicing

The process of alternative splicing of HIV-1 transcripts is unique from that of cellular counterparts [30–32]. Splicing of the viral pre-mRNA by the host splicing machinery is inefficient and results in the accumulation of unspliced mRNA that indeed is required for the optimal expression of viral proteins and polyproteins Gag, Gag-Pro, and Gag-Pro-Pol. HIV-1 genome contains at least four donor splice sites (D1-D4), eight acceptor splice sites (A1, A2, A3, A4c, A4a, A4b, A5, A7), and several other rarely used splice sites (ss) that are used in various combinations (such as D5 and A6 as shown in Fig. 15.2B). The *cis*-acting enhancer and silencer sequences recognized by serine and arginine-rich (SR) proteins and heterogeneous nuclear ribonucleoproteins (hnRNPs), respectively, also regulate the usage of these splice sites.

The primary viral transcript goes through extensive alternative splicing generating more than 50 mRNA isoforms ensuring the optimal expression of all viral proteins [32]. Based on the sizes, HIV-1 mRNAs are divided into three classes: (1) the 2 kb intronless completely spliced multiple mRNAs viz., re *tat*, *rev*, and *nef* transcripts (2) the 4 kb intron-containing partially spliced multiple mRNAs viz., *vif*, *vpr*, *vpu*, and *env* transcripts and (3) full length 9 kb unspliced mRNAs producing Gag and Pol that also serves as gRNA (depicted in Fig. 15.2B). Recently, transcriptome analysis based on next-generation sequencing has identified an additional minor class of 1 kb mRNAs, but no functionality has been assigned to those yet [32,33].

The changes in the splicing of the HIV-1 transcript can alter the usage of the poly(A) site. The major splice donor site D1 positioned immediately downstream of the 5′-LTR has been identified as the main determinant of the differential use of HIV-1 poly(A) sites. The host factors Sam68 and eIF3f positively and negatively regulate HIV-1 mRNA 3′-end processing, respectively, through modulating the splicing reactions [10,30,31].

Regulation of splicing of HIV-1 transcripts at terminal 5′ ss D1 and 3′ ss A7

Activation of the 5′-terminal major splice site D1 is essential for the generation of fully and partially spliced HIV-1 mRNAs. Once first splicing is completed from D1 in the 5′-UTR to one of the downstream 3′-splice sites A1 through A5, the HIV-1 transcripts may or may not undergo further splicing from D4 to A7 to excise the RRE sequence. After initial splicing at D1, HIV-1 transcripts splicing out from D4 to A7 generate all the fully spliced intron less transcripts. Transcripts that no longer splice out D4 to A7 after first splicing at D1 are longer and give rise to the partially/incompletely spliced transcripts carrying the “env” intron. There can also be optional inclusion of either one or both small noncoding exons, that is, exon 2 (defined by 3′ ss A1 and 5′ ss D2) and exon 3 (defined by 3′ ss A2 and 5′ ss D3) in some 2 kb and 4 kb-size class HIV-1 transcripts [33–35].

Regulation of splicing of HIV-1 transcripts at other splice sites

HIV-1 *vif* transcript is of low abundance, constituting around 1% of the partially spliced mRNA, produced by the splicing from D1 to A1. The abundance of *vif* mRNA depends on both the usage of A1 and the repression of downstream 5′ ss D2.

HIV-1 *vpr* is also encoded in a low abundance of mRNAs contributing around 2% of the partially spliced mRNA pool. It is primarily produced by splicing from D1 to A2. Recognition of A2 and inactivation of 5′ ss D3 is necessary for the formation of the *vpr* transcript.

The transcript of 2 kb size-class encoding full length Tat is generated by splicing out both exon four and exon seven from D1 to A3 and from D4 to A7, respectively, which is around 9% of the fully spliced mRNAs. Moreover, a truncated isoform of Tat (1–72 aa) coding transcript carrying only first *tat* exon is encoded by a 4 kb size-class mRNAs representing approximately 5% of incompletely spliced viral transcripts.

The four 3′ ss A4b, A4a, A4c, and A5 are present near the middle of the HIV-1 genome within a 40-nt region (3′ ss central cluster) and give rise to completely spliced *rev* transcripts (A4c, A4a, A4b) and *nef* transcripts (A5), and incompletely spliced *env/vpu* bi-cistronic transcripts [33,35].

Epitranscriptomic modifications of the HIV-1 transcripts

Among the nuclear post-transcriptional covalent modifications of the HIV-1 transcripts, methylation of adenosines at N6 position (m^6A) is the most prevalent and it affects various aspects of RNA metabolism. The m^6A is added by the "writers," that is, RNA methyltransferase complex (MTase), it is recognized by the "readers," that is, m^6A-binding proteins YTHDF1-3, and can be removed by the "erasers," that is, RNA demethylases. HIV-1 infection increases the m^6A levels in both host and viral mRNAs. Methylation of two conserved adenosines present in the RRE enhances the Rev-RRE interaction and facilitates the nuclear export of HIV-1 RNAs. It has been reported that different HIV-1 isolates carry around four clusters of m^6A in the 3′-UTRs, promoting HIV-1 gene expression. Another common epitranscriptomic modification is the methylation of cytosine at the C5 position (m^5C) that is mainly mediated by writer NOP2/Sun RNA Methyltransferase 2 (NSUN2). The 3′-end of the HIV-1 genome shows the extensive level of NSUN2-mediated m^5C modification that promotes viral gene expression by regulating alternate splicing and supporting the translation of HIV-1 mRNAs. To escape innate immune recognition, HIV-1 uses 2′-*O*-methylation of viral RNA catalyzed by the FtsJ RNA 2′-*O*-methyltransferase 3 (FTSJ3) in complex with TAR RNA-binding protein (TRBP). Recently, it has been demonstrated that acetylation of cytidine at N4 position (ac^4C) is another RNA modification by cellular N-acetyltransferase 10 (NAT10) at multiple sites of the viral transcripts. It enhances HIV-1 replication by increasing the mRNA stability [36–40].

Post-transcriptional regulation of HIV-1 mRNAs

Following viral pre-mRNA synthesis and modifications, the stability, nucleocytoplasmic transport, and translation of HIV-1 RNAs are controlled by multiple host machinery/factors. Additionally, HIV-1 uses its machinery, which includes unique RNA instability elements present within the viral genome, nucleocytoplasmic shuttling protein Rev and virus-derived noncoding RNAs to regulate the stabilization and nuclear export of HIV-1 RNAs.

Regulation by HIV-1 RNA instability elements

In absence of Rev, the viral mRNAs undergo nuclear degradation that is partly due to the inhibitory sequences referred to as *cis*-acting repressor (CRS), or instability (INS) elements present within the HIV-1 genome. Such sequences have been identified in *pol* (CRS), *gag/pol* intersection (IN), $p17^{gag}$ (INS1), *env*, and RRE. Although the INSs are poorly characterized, they are considered as AU-rich elements (AREs). These INSs have been reported to affect mRNA stability, nuclear export, and translation. To overcome these barriers, HIV-1 uses Rev for the stabilization and nuclear export of partially spliced and unspliced HIV-1 RNAs with RRE sequence. Multiple host factors have been identified to associate with the INSs including hnRNP I or polypyrimidine tract-binding protein, poly(A)-binding protein 1 (PABP1), and heterodimeric p54nrb/PSF transcription/splicing factor

[39,41]. The host factor RVB2 has recently been shown to inhibit HIV-1 Gag protein expression by promoting mRNA degradation that is antagonized by Env protein [39].

Nuclear export of HIV-1 RNAs

For their nuclear export, the partially spliced and unspliced HIV RNAs follow a discrete pathway as compared to the one used by the fully spliced HIV-1 RNAs (Fig. 15.3). During the early phase of HIV-1 infection, like any other completely spliced cellular mRNAs, fully spliced HIV-1 RNAs are not subjected to nuclear retention as they exit the nucleus and get translated into regulatory proteins Rev, Tat, and Nef. The nuclear retention of partially spliced and unspliced viral RNAs during the late phase of HIV-1 infection has been attributed to either splicing commitment factors or the presence of inhibitory sequences within HIV-1 transcripts. To overcome these restricting factors, Rev specifically binds to the RNA structure called RRE present in the partially spliced and unspliced HIV-1 RNAs and helps their nuclear export and translation into viral structural proteins. Rev has an N-terminal arginine-

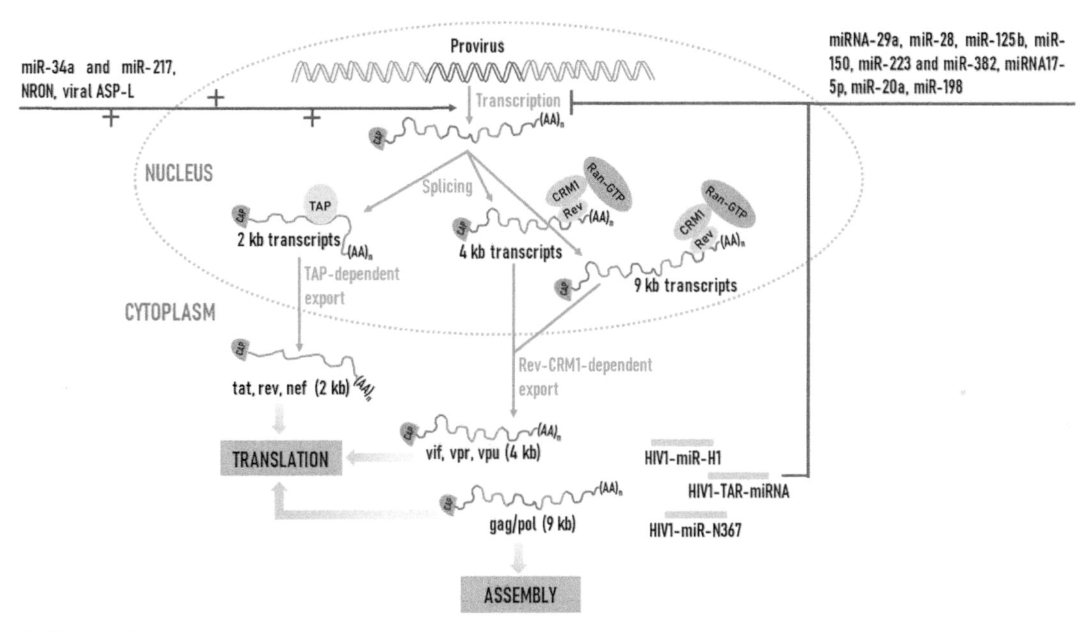

FIGURE 15.3

HIV-1 post-transcriptional modification and regulation. HIV-1 pre-mRNA undergoes co- and post-transcriptional processing including capping, polyadenylation, and alternative splicing. Alternative splicing generates 2 kb fully spliced intronless transcripts (encoding Tat, Rev, and Nef), 4 kb intron-containing partially spliced transcripts (encoding Vif, Vpr, and Vpu), and 9 kb unspliced transcripts (encoding Gag and Pol and acting as gRNA). The introns are shown in blue, and exons are shown in orange. The fully spliced transcripts are exported to the cytoplasm through the classical TAP/NFX1 dependent pathway. The intron-containing HIV-1 transcripts go through the Rev-CRM1-dependent nuclear export pathway for their successful translation into viral proteins. HIV-1 transcription is also positively and negatively regulated by both cellular and virus-derived noncoding RNAs, some of which are listed in the figure.

rich nuclear localization signal (NLS) that is responsible for its interaction with RRE, and a C-terminal leucine-rich nuclear export signal (NES), which is also called activation domain for its interaction with cellular factors involved in nucleocytoplasmic transport of viral RNAs. Thus Rev shuttles between the nucleus and cytoplasm enabling the nuclear export of RRE-containing RNAs. In the late phase of viral infection, Rev monomer recognizes RRE region followed by a cooperative and sequential assembly of Rev oligomers. The NES domains of oligomerized Rev are then associated with cellular nuclear export factor CRM1 with subsequent recruitment of RanGTP. The ternary complex Rev-CRM1-RanGTP complex then binds to FG-repeats of nucleoporins via Crm1p to move through the nuclear pore. Following the nuclear exit, hydrolysis of RanGTP to RanGDP leads to cytoplasmic cargo release that provides directionality to the CRM1-mediates nuclear export pathway. After this, nuclear import of Rev occurs by the recognition of its NLS by transport mediator importin-β [42–44].

Several host proteins, viz., ANP32A and ANP32B, hnRNPA1, hRIP, Martin-3, DEAD-box RNA helicases, Sam68, mRNA decay factors Staufen2, and UPF1 have been shown to positively regulate Rev activity in the Rev/RRE-dependent nuclear export. On the other hand, cellular proteins RREBP49 and Hax-1 have been identified as negative regulators of Rev function [39,44–46]. Moreover, an element termed constitutive transport element from the genome of simple retrovirus Mason-Pfizer monkey experimentally inserted into HIV-1 proviral plasmid has been reported to support Rev-independent expression of HIV-1 structural proteins through classical cellular mRNA nuclear export pathway involving TAP/NFX1 [46,47].

Regulation by host and HIV-1 noncoding RNAs

Several reports have identified noncoding RNAs (ncRNAs) as the crucial elements in the regulation of HIV-1 expression (Fig. 15.3). The ncRNAs are classified based on length into small (<200-nt) and long (>200-nt) ncRNAs (lncRNAs). The micro RNAs (miRNAs) are small ds RNA of ~21–24 nucleotides that control target gene expression at the post-transcriptional level. The first hint of miRNA-based regulation of HIV-1 came from the observation that HIV-1 gene expression increases following the knockdown of Drosha and Dicer, the two key RNases involved in miRNA biogenesis. Thereafter, many profiling studies have reported differential expression of cellular miRNAs as well as their function in HIV-1 infection. The miRNA-29a and miR-29b targeting the viral *nef* gene directly suppress viral replication [48]. The miRNAs, viz., miR-28, miR-125b, miR-150, miR-223, and miR-382 targeting HIV-1 3′-LTR, that is, also inhibit HIV-1 replication. These "anti-HIV" miRNAs might also participate in the differential susceptibility of differentiated macrophages and monocytes to HIV-1 infection. Some miRNAs target host factors important for HIV-1 replication. For example, polycistronic miR-17/92 cluster encoding miR17-5p and miR-20a suppress Tat cofactor PCAF, and miR-198 targets another cofactor Cyclin T1 resulting in a decrease in transcription of HIV-1 genome [49–51].

Deep sequencing of HIV-1 infected cells has revealed virus-encoded small ncRNAs comprising antisense transcripts, siRNA duplexes, and miRNAs that are likely to take part in viral replication. The viral TAR region is the most known source of HIV-derived miRNAs. The TAR-derived miRNAs are involved in chromatin remodeling of the LTR region and modulation of HIV-1 pathogenesis. Earlier studies have identified another virus-derived miRNA miR-N367, which is produced from the HIV-1 *nef* gene that post-transcriptionally regulates viral transcription. A novel miRNA named miR-H3 emanating from the coding sequence of HIV-1 RT has recently been reported. It enhances viral transcription upon its overexpression. Moreover, Tat has evolved to be a suppressor of RNA silencing machinery of the host [50,52,53].

Following HIV-1 infection, differential expression of cellular lncRNAs has also been reported. One such lncRNA, NEAT1 was found to be crucial for the integrity of the nuclear paraspeckle body, which is an important substructure for HIV-1 replication. Another lncRNA, NRON is reportedly involved in NFAT-mediated regulation of HIV-1 replication [54,55].

As in the case of small ncRNAs, HIV-encoded lncRNA named ASP-L also negatively regulates HIV-1 replication. Another virus-derived antisense lncRNA that originates from 3′-LTR and *nef* overlapping region epigenetically regulates HIV-1 transcription [50,56]. Apart from linear ncRNAs, a recent study has also reported differential expression of circular RNAs (circRNAs) and their potential contribution to HIV-1 replication [57].

Regulation of HIV-1 protein translation

Translation of HIV-1 mRNA is a complex process that utilizes the host machinery to synthesize the viral proteins. It has been hypothesized that similar to the cellular mRNAs, HIV-1 mRNA translation is also regulated at the initiation step. Several different mechanisms have been proposed for translational regulation that include the following: the cap-dependent, internal ribosome entry sites (IRES), ribosomal frameshifting, and ribosome shunting [58–60].

Cap-dependent translation

As described earlier, all HIV-1 mRNAs have 5′-caps and poly(A) tails. Some HIV-1 transcripts contain a trimethylguanosine cap that facilitates the binding of eIF4E. In the TAR element of the HIV-1 transcript, there is an inefficient unwinding step of the ribosomal scanning process. It is because the free energy at 5′-UTRs of HIV-I mRNA is less than −100 kcal/mol while a free energy value below −60 kcal/mol completely inhibits the scanning by the 43S preinitiation complex. Another reason for the inefficient unwinding is the formation of a stable stem-loop by the TAR directly next to the 5′-cap, inhibiting the binding of 43S preinitiation complex. Several proteins including La, TRBP, Staufen, DExH-box helicase 9 (DHX9), and DEAD-box RNA Helicase: (DDX3), bind to the HIV-1 5′-UTR and facilitate viral translation (Fig. 15.4). DDX3 binds to the 5′-UTR where it interacts with eIF4G and PABP, promoting the initiation of translation. To promote the translation of HIV-1 mRNAs, DDX3 is recruited to the TAR through interaction with Tat. Rev interacts with DDX3, stimulating the translation of HIV-I mRNAs at lower concentrations, whereas at higher concentrations it inhibits the translation in a nonspecific manner. Depletion of DDX3 results in decreased translation of Gag, Vif, and Nef in HIV-1-infected cells. DHX9 has been reported to promote polysome association of the HIV-1 gRNA via binding to the post-transcriptional control element located in the 5′-UTR. Depletion of DHX9 leads to decreased expression of Gag, Vif, Nef, and Rev [58,59].

Ribosome shunting

HIV-1 transcripts consist of a bicistronic *vpu-env* mRNA coding for both Vpu and Env. Translating different proteins from a common mRNA is one of the examples of how HIV-1 increases the coding

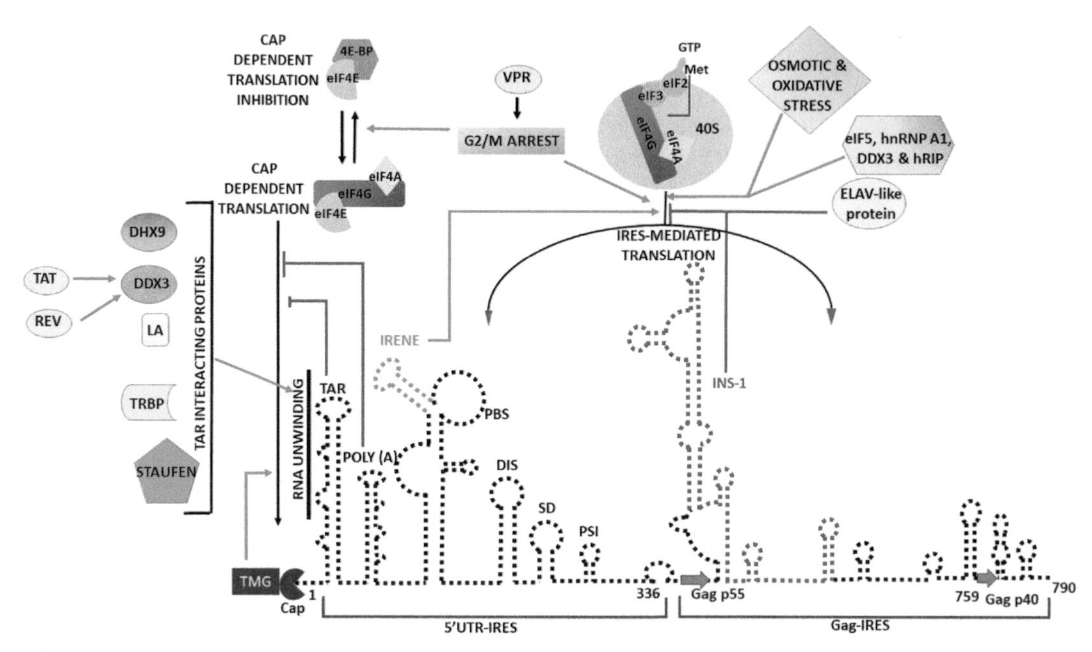

FIGURE 15.4

A schematic representation of translation initiation of the HIV-1 unspliced mRNA. HIV-1 RNA structure showing 5′-UTR and Gag IRES with different cellular conditions and cellular factors affecting the cap-dependent as well as IRES-mediated translation has been indicated. The IRENE and INS-1 regions of the RNA structure regulate the IRES-mediated translation positively (highlighted in green) and negatively (highlighted in red), respectively.

capacity of its genome. The Vpu and Env protein-coding sequences are arranged in such a way that the open reading frame (ORF) of Vpu precedes the coding sequence of Env. There are two models regarding the translation of these proteins: In the leaky scanning mechanism, the 43S preinitiation complex passes through the Vpu initiation codon and initiates the translation of Env. Such modulation of scanning by the 43S PIC is achieved due to the presence of weak Kozak context around the Vpu start codon [61]. Another one is the ribosome shunting model (Fig. 15.5) [62] in which Env translation is stimulated by an upstream ORF (uORF) present in the Vpu start codon region. The uORF is a six nucleotide sequence (AUGUAA) located in the start codon region of Vpu that acts as a ribosome pausing site [63].

IRES-dependent translation

There are two IRES in the HIV-1 genomic mRNA (Fig. 15.4). One of the IRES sites is in the 5′-UTR (HIV-1 IRES) [64] and the second one is present within the *gag* coding region and is referred to as HIV-1 Gag IRES [65]. The region spanning nucleotides 104 to 336 harbors the minimal activity of HIV-1 UTR IRES [64]. The IRES is present in the 5′-UTR of HIV-1 mRNAs that include the coding

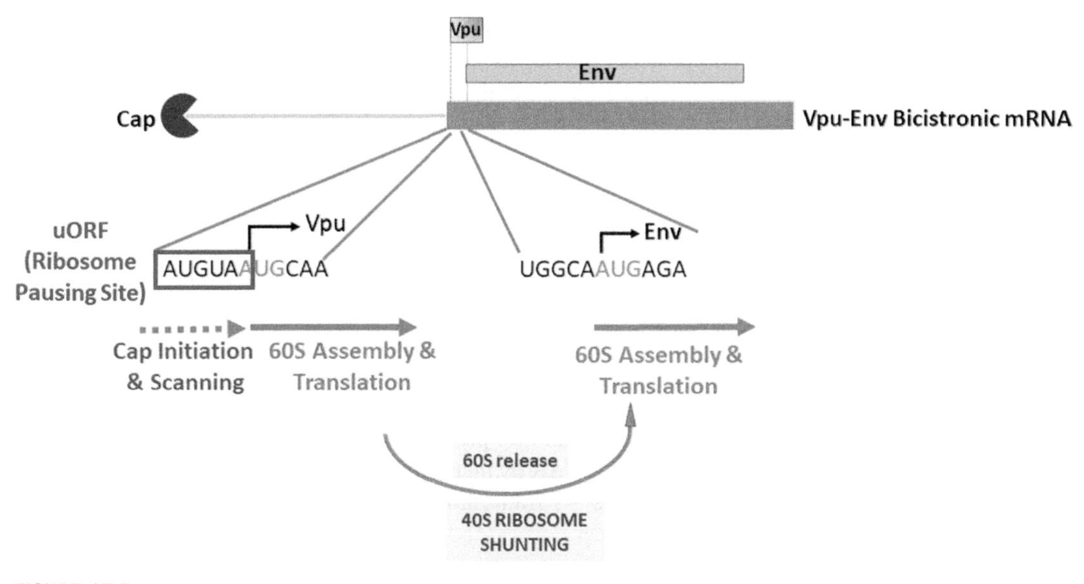

FIGURE 15.5

HIV-1 Vpu-Env bicistronic mRNA. A schematic representation of HIV-1 ribosomal shunting model for the translation of both Vpu and Env from a common bicistronic mRNA. The uORF has been indicated in the red box. The start sites of both Vpu and Env have also been marked in the figure.

sequences for the gRNA, Nef, Vif, Vpr, and Vpu. 5′-UTR IRES-mediated initiation can take place by two mechanisms either by direct recruitment of 40S ribosome to the start codon or by recruitment at upstream sites of the initiation codon followed by scanning to the start site by eIF4A [58,59]. It has been reported that HIV-1 5′-UTR IRES is more active under specific conditions as compared to others. For example, 5′-UTR IRES activity is enhanced during various stress conditions such as oxidative stress, osmotic stress, G2/M arrest; whereas, ER stress or hypoxia does not affect the IRES activity (Fig. 15.4) [58,59].

The HIV-1 IRES-mediated translation is functional in various cellular situations where the cap-mediated translation is disabled. Studies show that for the early (24—48 h) replication, the initiation of HIV-1 translation is largely cap-dependent; whereas, during later time points, to ensure virion production, the IRES-mediated translation is needed [66].

hnRNP A1 is one of the best characterized IRES trans-acting factors (ITAF) that enhances the IRES-mediated translation. As ITAFs are found only in CD4+T cells and under specific stress conditions, hnRNP A1-mediated induction of translation suggests that the reason for the HIV-1 5′-UTR IRES activity changes in these conditions. Several other cellular proteins such as eIF5, DDX3, and Rev-interacting protein (RIP) also positively regulate the HIV-1 IRES activity. On the other hand, the human embryonic lethal abnormal vision-like protein and the instability element 1 (INS-1) have been shown as the negative regulators of the HIV-1 IRES activity (Fig. 15.4) [58,59].

The Gag-IRES activity is within the first ~425 nucleotides of *gag* [67,68]. The Gag IRES mediates the translation of the full-length polyprotein Pr55Gag as well as a 40-kDa Gag (p40) isoform, which is N-truncated at methionine 142 and lacks the matrix domain, in an independent manner (Fig. 15.4).

Ribosomal frameshift

The Gag is translated from a bicistronic transcript. The Gag-Pol polyprotein is translated via a −1 nucleotide ribosomal frameshift [69,70] which is regulated by a slippery heptanucleotide sequence (UUUUUUA) where the frameshift takes place and is facilitated by a downstream RNA element known as the frameshift stimulatory signal (FSS) (Fig. 15.6) [71]. Due to this shift, the termination codon for Gag comes into an out-of-frame context, followed by continued translation toward the downstream *pol* sequence. The most widely accepted model suggests that the FSS forces pausing of the 80S complexes result in shifting one nucleotide backward into the *pol* sequence [72]. Recently, it has been proposed that four pseudoknots present in FSS promote −1 frameshift [73,74]. Tat acts as an inhibitor of frameshift [75] as it destabilizes the TAR by increasing the recruitment of helicases such as DDX3 [76] whereas helicase DDX17 induces −1 frameshift [77].

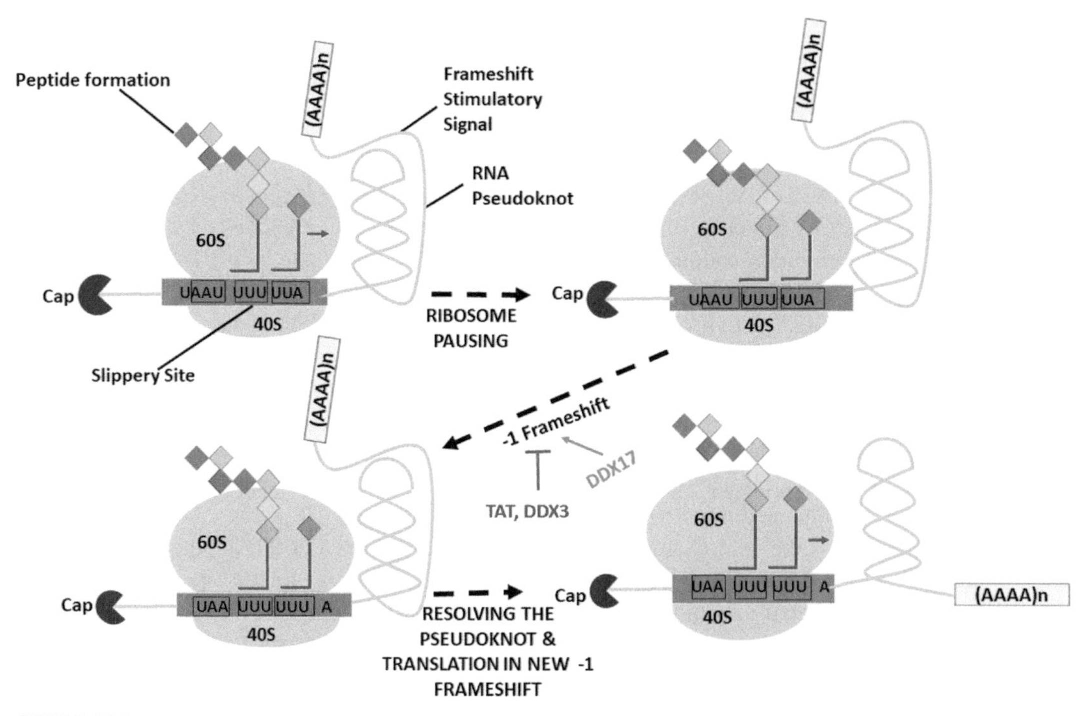

FIGURE 15.6

A schematic representation of HIV-1 ribosomal frameshift. −1 frameshift is mediated by a slippery sequence (UUUUUUA) and frameshift stimulatory signal (FSS). Tat inhibits the frameshift through the recruitment of DDX3; whereas, DDX17 is a positive regulator of the −1-nucleotide frameshift.

The fate of cellular translation factors

HIV-1 viral proteins try to inhibit the functions of various cellular translation factors [58]; whereas, host factors tend to restrict the function of viral proteins. HIV-1 protease cleaves the initiation factor eIF4G and eIF3D, leading to inhibition of cap-dependent translation. HIV-1 protease also inhibits PABP-dependent initiation of translation by degrading the PABP. During HIV-1 replication, the Protein Kinase R (PKR) is activated during the early phase of infection and is quickly inhibited by viral and cellular factors. In the early stages of HIV-1 replication, the TAR RNA element activates PKR; whereas, during late events of HIV-1 infection, a high concentration of TAR inhibits PKR activation. Activated PKR phosphorylates the alpha subunit of eIF2, blocking its recycling, resulting in global translational attenuation. HIV-1 Tat protein directly interacts with PKR that prevents autophosphorylation of PKR, an essential factor for its function. PKR activation is inhibited by dsDNA as well as by cellular factors: adenosine deaminase acting on RNA 1 (ADAR1), TRBP, and a functional change of PKR activator [58,59], which are involved in HIV-1-induced mechanisms.

Post-translational modifications of HIV-1 proteins

Immediately after HIV-1 infection, the viral and the cellular proteins undergo post-translational modifications (Fig. 15.7). HIV-1 targets the post-translational modification machinery of its host and turns the hostile host environment into a hospitable one. In recent years, post-translational modifications (PTMs) of viral and cellular proteins have gained a lot of focus as they regulate almost every step of the viral replication cycle [78]. These PTMs help in viral replication by enhancing viral

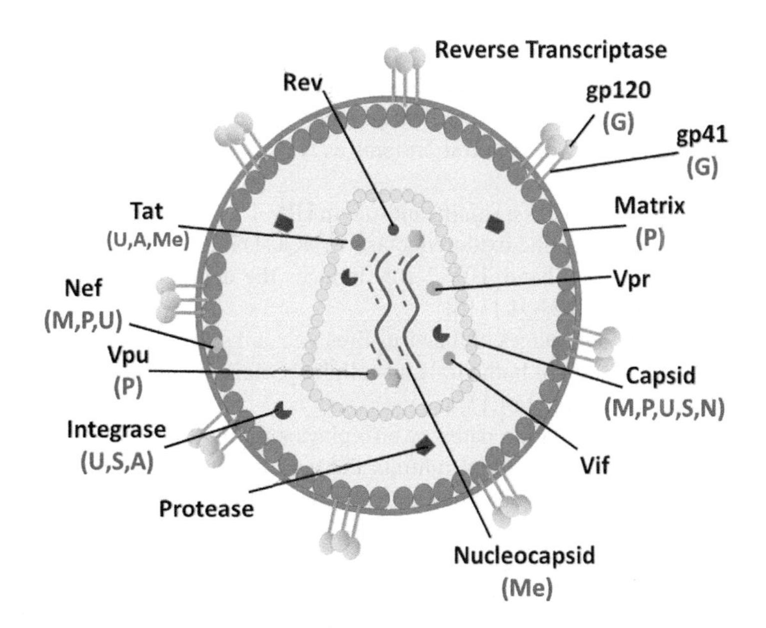

FIGURE 15.7

A schematic diagram of HIV-1 virion indicating various PTMs (highlighted in red) of the viral proteins. G: glycosylation; P: phosphorylation; M: myristoylation; U: ubiquitination; S: Sumoylation; Me: methylation; A: acetylation; N: NEDDylation.

assembly, release, and the inhibition of host interferon response. On the other hand, the host overcomes the viral infection by either modifying the enzymatic activity of viral proteins by removing the crucial PTMs or by the inactivation and/or proteasomal degradation of the viral proteins through the attachment of small molecules such as ubiquitin or ubiquitin-like proteins. A list of PTMs of HIV-1 proteins in the infected cells, the sites involved, and their functions are given in Table 15.1.

Post-translational regulation of HIV-1: protein degradation and maturation

The HIV-1 proteins after undergoing different PTMs, manipulate the host regulatory pathways such as proteasomal and lysosomal machinery to achieve various crucial functions. These regulatory pathways are also utilized by various host factors to decide the turnover and stability of viral proteins. One of the most important events of post-translational regulation is virion maturation that takes place after a fully assembled virus particle is released from the infected cell and plays an important role in the production of the infectious virion.

Protein degradation pathways

The interaction between viral proteins and their cellular counterparts sometimes requires ubiquitination and is important for the cell to restrict viral proliferation as well as counteraction by the virus to overcome this restriction. HIV-1 proteins manipulate the host 26S proteasomal machinery. For example, HIV-1 Tat, Nef, Vif, Vpr, and Vpu induces the proteasomal degradation of various host factors that serves a wide range of functions including viral replication, immune evasion, virion infectivity, inhibition of apoptosis, interference of DNA damage response, latency, cell cycle arrest, virion release as well as the downregulation of host surface receptors (Fig. 15.8) [114,115]. The viral proteins are themselves targeted by different host proteins for proteasomal degradation that thereby regulates the stability and activity of the viral proteins as well as their interactions with other cellular factors (Fig. 15.8) [114,115].

20S proteasomal pathway also has a significant role in HIV-1 replication as it is responsible for the degradation of Tat. NAD(P)H: quinine oxidoreductase 1 (NQO1) inhibits the degradation of Tat in the presence of NADH. On the other hand, HIV-1 Rev induces the degradation of Tat through the 20S proteasome by down-regulating NQO1 [116].

Thus, during the early and late stages of the virus life cycle, all structural and regulatory proteins of HIV-1 participate in the regulation of both viral and cellular proteins in some way or other, through the proteasomal degradation pathways [114,115].

The lysosome-mediated degradation pathway also plays an important role in HIV-1 pathogenesis. HIV-1 inhibits lysosomal acidification and inhibits the activity of lysosomal enzymes such as cathepsins. This phenomenon affects the digestion of viral particles, hence antigen processing and presentation [117]. Breast cancer-associated gene 2 (BCA2) is a RING-finger E3 ubiquitin ligase that targets HIV-1 Gag in a Tetherin-independent manner for lysosomal degradation, impairing the viral assembly and release [118]. Also, BCA2 targets the bone marrow stromal cell antigen 2 (BST2)-trapped virions for lysosomal degradation through its direct interaction with BST2 [119]. On the other hand, Vpu-mediated polyubiquitination of BST2 targets it to lysosomal degradation [120]. Another

Table 15.1 Post-translational modifications and their designated roles in HIV-1 replication.

Type of PTM	Viral and host proteins	Sites modified	Attributed functions	References
Glycosylation	gp120, gp41	N-linked	Immune evasion	[79]
Myristoylation	Gag and Gag-Pol	Glycine 2	Insertion into the plasma membrane, virion infectivity	[80]
	Nef	Glycine 2	Cell activation, MHC-I, CD4, and CD28 molecule downregulation and decreased infectivity through antagonizing SERINC3/5	[81]
Phosphorylation	Nef	Serine, threonine	Virion infectivity	[82,83]
	Vpu	Two serine residues in the DSGNES motif (AA 52–56) of vpu and SCFβTrCP recognition motif	Surface CD4 downregulation	[84–86]
		Cytoplasmic domain	Tetherin downregulation and virion release	[87–92]
	p6	Serine 40	Association of p6 with plasma membrane phospholipids	[93]
		Tryptophan 471	Virion release and maturation viral replication	[80]
	Matrix (MA)	Serine 9, 62, 72, and 77 Tyrosine 132	Localization of Pr55Gag at the plasma membrane	[80]
	Capsid (CA)	Serine 148	Incorporation of ERK2 in the virions	[80]
Ubiquitination	p6	Monoubiquitination Lysine 27and 33	Interaction with Tsg101, ALIX, and other ESCRT components, and virion release	[85,93,94]
	Integrase (IN)	Lysine 211, 215, 219, and 273	Proteasomal degradation of IN	[95]
	Tat	Polyubiquitination Lysine 71	Mdm2 stability by enhancing its phosphorylation	[96]
	Nef	Diubiquitinated Lys144	CD4 downregulation	[97]

Continued

Table 15.1 Post-translational modifications and their designated roles in HIV-1 replication.—cont'd

Type of PTM	Viral and host proteins	Sites modified	Attributed functions	References
Sumoylation	p6	Lysine 27	Defected budding activity	[98]
	IN	Lysine 46,136, and 244	Integration and binding to LEDGF/ p75 and p300	[95,99,100]
Acetylation	IN	Lysine 258, 264, 266, and 273	Affinity with DNA as well as DNA strand transfer activity, integration, and viral replication	[101–103].
	Tat	TAR RNA binding domain (Lysine 50)	Release of Tat from TAR RNA during early transcription elongation	[16,104,105]
		Lysine 28	Binding to Tat-associated kinase, CDK9/P-TEFb	[104,105]
		Lysine 50 and 51	Inhibition of HIV-1 mRNA splicing	[106]
Methylation	Tat	Arginine 52	Decreased interaction of Tat with TAR and cyclin T1 complex formation and LTR activity	[107]
		Lysine 50 and 51	Decreased LTR activity	[108]
		Lysine 71	Tat transactivation	[109]
	The nucleocapsid (NC)	Arginine 10 and 32	Defected RNA annealing and initiation of the reverse transcription	[110]
NEDDylation	Cullin-RING ligases (CLRs)	Lysine in the cullin C-terminal domain	Virion infectivity. Together with viral proteins vif and vpr facilitate the polyubiquitination and subsequent proteasomal degradation of the host restriction elements APOBEC3 and SAMHD1	[111,112]

Table 15.1 Post-translational modifications and their designated roles in HIV-1 replication.—cont'd

Type of PTM	Viral and host proteins	Sites modified	Attributed functions	References
ISGylation	Gag		Impairs HIV-1 replication and virion release	[113]

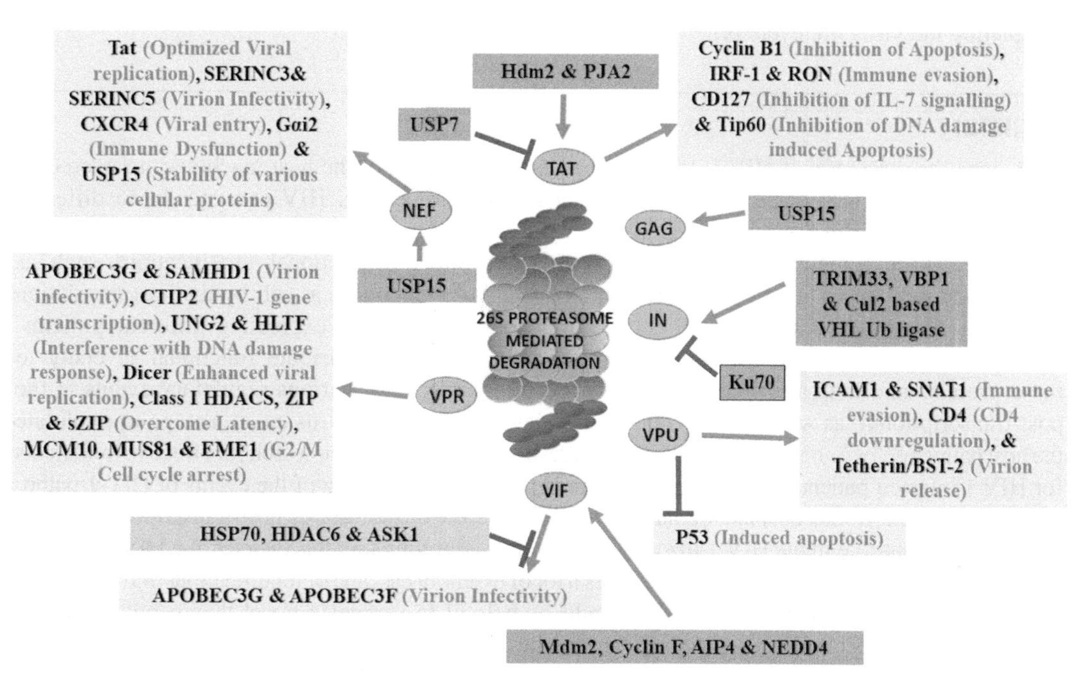

FIGURE 15.8

The interplay of host 26S proteasomal degradation machinery, HIV-1 proteins, and cellular factors. The targets of HIV-1 proteins for proteasomal degradation as well as their associated functions (written in blue color) are mentioned in shaded green boxes; whereas, the list of proteins that inhibit the proteasomal degradation of host factors by viral protein are indicated in a gray box. The purple boxes highlight the host factors that target HIV-1 proteins for proteasomal degradation.

antiviral protein membrane-associated RING-CH 2 (MARCH 2) suppresses HIV-1 infectivity by the ubiquitination of Env and its subsequent degradation in the lysosomes [121].

Recently another pathway is being studied that regulates the stability of HIV-1 proteins. ERAD has been shown to reduce HIV-1 infectivity by the depletion of the viral envelope [122]. Vpr has been reported to enhance the folding of Env in the ER and thereby prevent its degradation through the ERAD pathway [123].

Virion maturation

The Gag polypeptide is assembled in the immature virion in such a way that it projects radially inwards in the virion. After the virion buds out, the HIV-1 protease enzyme recognizes the specific cleavage sites in Gag and cleaves it into its various components that include MA, CA, NC, and p6 proteins as well as the SP1 and SP2 peptides. This cleavage converts the immature virion into a mature one through condensing and stabilizing the viral genome as well as assembling the capsid around the genome-NC complex that all together prepares the virion to enter, replicate, and uncoat in the next round of infection. Thus, virion maturation is the final switch that converts the immature virion budding out from an infected cell into a virion that can enter and replicate in a new host, hence completing the virus life cycle [9].

Summary

Ever since the discovery of HIV-1, the virus has evolved to control the host machinery to overcome different obstacles in the cell. Despite having a small genome size, HIV can encode 15 different proteins playing distinct roles during its life cycle. This is in part achieved via an orchestrated post-transcriptional regulation including a variety of post-transcriptional modifications such as capping and poly(A) tailing that provides the advantage of recognition as self by the host translation machinery as well as alternative splicing leading to splice variants expressing different proteins. Moreover, modifications of these proteins after being translated attribute functional diversity to these proteins and help at different stages of the virus life cycle. Various regulatory events at the post-transcriptional, as well as post-translational level, provide the virus an advantage to modulate the host pathways for its benefit and evade the host restriction factors. As the search for effective treatments for HIV-1 infected patients is still ongoing, a fundamental understanding of the events of HIV-1 pathogenesis is needed. In this chapter, we have given an overview of the various post-transcriptional events playing important role in the HIV-1 life cycle. As evidenced throughout this chapter, the HIV-1 genome after being transcribed in the host cell undergoes a lot of essential cascade of modifications and regulatory events to produce functional virions. This could be helpful to strategize novel therapeutic targets to overcome the drawbacks of current approaches to eliminate the virus from an infected individual.

Acknowledgements

The work in the laboratory is supported by the intramural funding from National Centre for Cell Science (NCCS), Pune, and the JC Bose Fellowship of SERB to DM. AT and AD were supported by NCCS research fellowship and AM is thankful to Department of Biotechnology (DBT) for her research fellowship.

References

1. UNAIDS data 2020 | UNAIDS. https://www.unaids.org/en/resources/documents/2020/unaids-data.
2. Carter CC, et al. HHS public access, vol. 16; 2010. p. 446–51.
3. Gaiha GD, Brass AL. The fiery side of HIV-induced T cell death. Science 2014;343:383–4.

4. Bushman FD, et al. HIV: From Biology to Prevention and Treatment. 1st Edition. NY: Cold Spring Harbor Laboratory Press, Cold Spring Harbor; 2012.
5. Wilen CB, Tilton JC, Doms RW. HIV: cell binding and entry. Cold Spring Harb Perspect Med 2012;2:1–14.
6. Hu WS, Hughes SH. HIV-1 reverse transcription. Cold Spring Harb Perspect Med 2012;2:1–22.
7. Dharan A, Bachmann N, Talley S, Zwikelmaier V, Campbell EM. Nuclear pore blockade reveals that HIV-1 completes reverse transcription and uncoating in the nucleus. Nat Microbiol 2020;5:1088–95.
8. Figueiredo A, Hope TJ. KAPs off for HIV-1 integration. Cell Host Microbe 2011;9:447–8.
9. Sundquist WI, Kra H. HIV-1 assembly, budding, and maturation. Cold Spring Harb Perspect Med 2012: 1–24.
10. Dandekar DH, Ganesh KN, Mitra D. HIV-1 Tat directly binds to NFκB enhancer sequence: role in viral and cellular gene expression. Nucleic Acids Res 2004;32:1270–8.
11. Karn J, Stoltzfus CM. Transcriptional and post-transcriptional regulation of HIV-1 gene expression. Cold Spring Harb Perspect Med 2012;2:a006916.
12. Blazkova J, et al. CpG methylation controls reactivation of HIV from latency. PLoS Pathog 2009;5: e1000554.
13. Kauder SE, Bosque A, Lindqvist A, Planelles V, Verdin E. Epigenetic regulation of HIV-1 latency by cytosine methylation. PLoS Pathog 2009;5:e1000495.
14. Van Lint C. Transcriptional activation and chromatin remodeling of the HIV-1 promoter in response to histone acetylation. EMBO J 1996;15:1112–20.
15. Coull JJ, et al. The human factors YY1 and LSF repress the human immunodeficiency virus type 1 long terminal repeat via recruitment of histone deacetylase 1. J Virol 2000;74:6790–9.
16. Ott M, et al. Acetylation of the HIV-1 tat protein by p300 is important for its transcriptional activity. Curr Biol 1999;9:1489–93.
17. Hakre S, Chavez L, Shirakawa K, Verdin E. Epigenetic regulation of HIV latency. Curr Opin HIV AIDS 2011;6:19–24.
18. Lusic M, Marcello A, Cereseto A, Giacca M. Regulation of HIV-1 gene expression by histone acetylation and factor recruitment at the LTR promoter. EMBO J 2003;22:6550–61.
19. Turner AMW, Margolis DM. Chromatin regulation and the histone code in HIV latency. Yale J Biol Med 2017;90:229–43.
20. Zhou M, et al. Coordination of transcription factor phosphorylation and histone methylation by the P-TEFb kinase during human immunodeficiency virus type 1 transcription. J Virol 2004;78:13522–33.
21. Tréand C, et al. Requirement for SWI/SNF chromatin-remodeling complex in Tat-mediated activation of the HIV-1 promoter. EMBO J 2006;25:1690–9.
22. Wilusz J. Putting an 'End' to HIV mRNAs: capping and polyadenylation as potential therapeutic targets. AIDS Res Ther 2013;10:31.
23. Chiu Y-L, et al. Tat stimulates cotranscriptional capping of HIV mRNA. Mol Cell 2002;10:585–97.
24. Bohnlein S, Hauber J, Cullen BR. Identification of a U5-specific sequence required for efficient polyadenylation within the human immunodeficiency virus long terminal repeat. J Virol 1989;63:421–4.
25. Valsamakis A, Zeichnert S, Carswell S, Alwine JC. The human immunodeficiency virus type 1 polyadenylylation signal: a 3' long terminal repeat element upstream of the AAUAAA necessary for efficient polyadenylylation (RNA processing/upstream element). Proc. Natl. Acad. Sci. USA 1991;88:2108–12.
26. Calzado MA, Sancho R, Muñoz E. Human immunodeficiency virus type 1 tat increases the expression of cleavage and polyadenylation specificity factor 73-kilodalton subunit modulating cellular and viral expression. J Virol 2004;78:6846–54.
27. Gilmartin GM, Fleming ES, Oetjen J, Keller W. Activation of HIV-1 pre-mRNA 3' processing in vitro requires both an upstream element and TAR. EMBO J 1992;11:4419–28.

28. Klasens BIF, Thiesen M, Virtanen A, Berkhout B. The ability of the HIV-1 AAUAAA signal to bind polyadenylation factors is controlled by local RNA structure. Nucleic Acids Res 1999;27:446–54.
29. Gee AH, Kasprzak W, Shapiro BA. Structural differentiation of the HIV-1 Poly(a) signals. J Biomol Struct Dyn 2006;23:417–28.
30. McLaren M, Asai K, Cochrane A. A novel function for Sam68: enhancement of HIV-1 RNA 3′ end processing. RNA 2004;10:1119–29.
31. Valente ST, Gilmartin GM, Venkatarama K, Arriagada G, Goff SP. HIV-1 mRNA 3′ end processing is distinctively regulated by eIF3f, CDK11, and splice factor 9G8. Mol Cell 2009;36:279–89.
32. Ocwieja KE, et al. Dynamic regulation of HIV-1 mRNA populations analyzed by single-molecule enrichment and long-read sequencing. Nucleic Acids Res 2012;40:10345–55.
33. Martin Stoltzfus C. Chapter 1 regulation of HIV-1 alternative RNA splicing and its role in virus replication. Adv Virus Res 2009;74:1–40.
34. Bohne J, Wodrich H, Kräusslich HG. Splicing of human immunodeficiency virus RNA is position-dependent suggesting sequential removal of introns from the 5′ end. Nucleic Acids Res 2005;33:825–37.
35. Purcell DFJ, Martin MA. Alternative splicing of human immunodeficiency virus type 1 mRNA modulates viral protein expression, replication, and infectivity. J Virol 1993;67:6365–78.
36. Tsai K, et al. Acetylation of cytidine residues boosts HIV-1 gene expression by increasing viral RNA stability. Cell Host Microbe 2020;28:306–12. e6.
37. Lichinchi G, et al. Dynamics of the human and viral m(6)A RNA methylomes during HIV-1 infection of T cells. Nat. Microbiol. 2016;1:16011.
38. Courtney DG, et al. Epitranscriptomic addition of m5C to HIV-1 transcripts regulates viral gene expression. Cell Host Microbe 2019;26:217–27. e6.
39. Toro-Ascuy D, Rojas-Araya B, Valiente-Echeverría F, Soto-Rifo R. Interactions between the HIV-1 unspliced mRNA and host mRNA decay machineries. Viruses 2016;8:320.
40. Ringeard M, Marchand V, Decroly E, Motorin Y, Bennasser Y. FTSJ3 is an RNA 2′-O-methyltransferase recruited by HIV to avoid innate immune sensing. Nature 2019;565:500–4.
41. Cochrane AW, et al. Identification and characterization of intragenic sequences which repress human immunodeficiency virus structural gene expression. J Virol 1991;65:5305–13.
42. Felber BK, Hadzopoulou-Cladaras M, Cladaras C, Copeland T, Pavlakis GN. Rev protein of human immunodeficiency virus type 1 affects the stability and transport of the viral mRNA. Proc Natl Acad Sci USA 1989;86:1495–9.
43. Fritz CC, Green MR. HIV Rev uses a conserved cellular protein export pathway for the nucleocytoplasmic transport of viral RNAs. Curr Biol 1996;6:848–54.
44. Kula A, Marcello A. Dynamic post-transcriptional regulation of HIV-1 gene expression. Biology 2012;1: 116–33.
45. Wang Y, et al. ANP32A and ANP32B are key factors in the Rev-dependent CRM1 pathway for nuclear export of HIV-1 unspliced mRNA. J Biol Chem 2019;294:15346–57.
46. Cochrane AW, McNally MT, Mouland AJ. The retrovirus RNA trafficking granule: from birth to maturity. Retrovirology 2006;3:1–17.
47. Bray M, et al. A small element from the Mason-Pfizer monkey virus genome makes human immunodeficiency virus type 1 expression and replication Rev-independent. Proc Natl Acad Sci USA 1994;91: 1256–60.
48. Ahluwalia JK, et al. Human cellular microRNA hsa-miR-29a interferes with viral nef protein expression and HIV-1 replication. Retrovirology 2008;5:117.
49. Triboulet R, et al. Suppression of micro-RNA-silencing pathway by HIV-1 during virus replication. Science 2007;315:1579–83.

50. Lazar DC, Morris KV, Saayman SM. The emerging role of long non-coding RNAs in HIV infection. Virus Res 2016;212:114–26.
51. Houzet L, Jeang KT. MicroRNAs and human retroviruses. Biochim. Biophys. Acta Gene Regul. Mech. 2011;1809:686–93.
52. Bennasser Y, Le SY, Yeung ML, Jeang KT. HIV-1 encoded candidate micro-RNAs and their cellular targets. Retrovirology 2004;1:1–5.
53. Bennasser Y, Le SY, Benkirane M, Jeang KT. Evidence that HIV-1 encodes an siRNA and a suppressor of RNA silencing. Immunity 2005;22:607–19.
54. Zhang Q, Chen CY, Yedavalli VSRK, Jeang KT. NEAT1 long noncoding RNA and paraspeckle bodies modulate HIV-1 post-transcriptional expression. mBio 2013;4:1–9.
55. Imam H, Bano AS, Patel P, Holla P, Jameel S. The lncRNA NRON modulates HIV-1 replication in a NFAT-dependent manner and is differentially regulated by early and late viral proteins. Sci Rep 2015;5:1–10.
56. Ludwig LB, et al. Human Immunodeficiency Virus-Type 1 LTR DNA contains an intrinsic gene producing antisense RNA and protein products. Retrovirology 2006;3:1–20.
57. Zhang Y., et al. Crosstalk in competing endogenous RNA networks reveals new circular RNAs involved in the pathogenesis of early HIV infection. J Transl Med 2018;16:1–11.
58. Guerrero S, et al. HIV-1 replication and the cellular eukaryotic translation apparatus. Viruses 2015;7: 199–218.
59. Hidalgo L, Swanson CM. Regulation of human immunodeficiency virus type 1 (HIV-1) mRNA translation. Biochem Soc Trans 2017;45:353–64.
60. de Breyne S, Ohlmann T. Focus on translation initiation of the HIV-1 mRNAs. Int J Mol Sci 2019;20:1–29.
61. Schwartz S, et al. Env and Vpu proteins of human immunodeficiency virus type 1 are produced from multiple bicistronic mRNAs. J Virol 1990;64:5448–56.
62. Anderson JL, Johnson AT, Howard JL, Purcell DFJ. Both linear and discontinuous ribosome scanning are used for translation initiation from bicistronic human immunodeficiency virus type 1 env mRNAs. J Virol 2007;81:4664–76.
63. Krummheuer J, et al. A minimal uORF within the HIV-1 vpu leader allows efficient translation initiation at the downstream env AUG. Virology 2007;363:261–71.
64. Brasey A, et al. The leader of human immunodeficiency virus type 1 genomic RNA harbors an internal ribosome entry segment that is active during the G 2/M phase of the cell cycle. J Virol 2003;77:3939–49.
65. Buck CB, et al. The human immunodeficiency virus type 1 gag gene encodes an internal ribosome entry site. J Virol 2001;75:181–91.
66. Amorim R, Costa SM, Cavaleiro NP, Da Silva EE, Da Costa LJ. HIV-1 transcripts use ires-initiation under conditions where cap-dependent translation is restricted by poliovirus 2A protease. PLoS One 2014;9:1–13.
67. Weill L, et al. A new type of IRES within gag coding region recruits three initiation complexes on HIV-2 genomic RNA. Nucleic Acids Res 2009;38:1367–81.
68. Locker N, Chamond N, Sargueil B. A conserved structure within the HIV gag open reading frame that controls translation initiation directly recruits the 40S subunit and eIF3. Nucleic Acids Res 2011;39: 2367–77.
69. Brierley I, Dos Ramos FJ. Programmed ribosomal frameshifting in HIV-1 and the SARS-CoV. Virus Res 2006;119:29–42.
70. Staple DW, Butcher SE. Solution structure and thermodynamic investigation of the HIV-1 frameshift inducing element. J Mol Biol 2005;349:1011–23.
71. Giedroc DP, Cornish PV. Frameshifting RNA pseudoknots: structure and mechanism. Virus Res 2009;139: 193–208.

72. Liao PY, Choi YS, Dinman JD, Lee KH. The many paths to frameshifting: kinetic modelling and analysis of the effects of different elongation steps on programmed−1 ribosomal frameshifting. Nucleic Acids Res 2011;39:300−12.
73. Huang X, Yang Y, Wang G, Cheng Q, Du Z. Highly conserved RNA pseudoknots at the gag-pol junction of HIV-1 suggest a novel mechanism of −1 ribosomal frameshifting. RNA 2014;20:587−93.
74. Wang G, Yang Y, Huang X, Du Z. Possible involvement of coaxially stacked double pseudoknots in the regulation of −1 programmed ribosomal frameshifting in RNA viruses. J Biomol Struct Dyn 2015;33: 1547−57.
75. Charbonneau J, Gendron K, Ferbeyre G, Brakier-Gingras L. The 5′ UTR of HIV-1 full-length mRNA and the Tat viral protein modulate the programmed −1 ribosomal frameshift that generates HIV-1 enzymes. RNA 2012;18:519−29.
76. Lai MC, et al. Human DDX3 interacts with the HIV-1 tat protein to facilitate viral mRNA translation. PLoS One 2013;8:1−14.
77. Lorgeoux RP, Pan Q, Le Duff Y, Liang C. DDX17 promotes the production of infectious HIV-1 particles through modulating viral RNA packaging and translation frameshift. Virology 2013;443:384−92.
78. Chen L, Keppler OT, Schölz C. Post-translational modification-based regulation of HIV replication. Front Microbiol 2018;9:1−22.
79. Lambert GS, Upadhyay C. HIV-1 envelope glycosylation and the signal peptide. Vaccines 2021;9:1−19.
80. Bussienne C, Marquet R, Paillart JC, Bernacchi S. Post-translational modifications of retroviral HIV-1 Gag precursors: an overview of their biological role. Int J Mol Sci 2021;22:1−28.
81. Wang B, et al. Protein N-myristoylation: functions and mechanisms in control of innate immunity. Cell Mol Immunol 2021;18:878−88.
82. Peter F. HIV Nef: the mother of all evil? Immunity 1998;9:433−7.
83. Li PL, et al. Phosphorylation of HIV Nef by cAMP-dependent protein kinase. Virology 2005;331:367−74.
84. Estrabaud E, et al. Regulated degradation of the HIV-1 Vpu protein through a βTrCP-independent pathway limits the release of viral particles. PLoS Pathog 2007;3:0995−1004.
85. Friedrich M, et al. Glutamic acid residues in HIV-1 p6 regulate virus budding and membrane association of gag. Viruses 2016;8:117.
86. Schubert U, Strebel K. Differential activities of the human immunodeficiency virus type 1-encoded Vpu protein are regulated by phosphorylation and occur in different cellular compartments. J Virol 1994;68: 2260−71.
87. Hinz A, et al. Structural basis of HIV-1 tethering to membranes by the BST-2/tetherin ectodomain. Cell Host Microbe 2010;7:314−23.
88. Neil SJD, Eastman SW, Jouvenet N, Bieniasz PD. HIV-1 Vpu promotes release and prevents endocytosis of nascent retrovirus particles from the plasma membrane. PLoS Pathog 2006;2:354−67.
89. Janvier K, et al. The ESCRT-0 component HRS is required for HIV-1 Vpu-mediated BST-2/tetherin down-regulation. PLoS Pathog 2011;7:e1001265.
90. Kueck T, et al. Serine phosphorylation of HIV-1 vpu and its binding to tetherin regulates interaction with clathrin adaptors. PLoS Pathog 2015;11:1−26.
91. Pujol FM, et al. Mediated exclusion from virus assembly sites. J. Virol 2016;90:6709−23.
92. Madjo U, et al. LC3C contributes to vpu-mediated antagonism of BST2/tetherin restriction on HIV-1 release through a non-canonical autophagy pathway. Cell Rep 2016;17:2221−33.
93. Solbak SMØ, et al. HIV-1 p6—a structured to flexible multifunctional membrane-interacting protein. Biochim Biophys Acta Biomembr 2013;1828:816−23.
94. Gottwein E, Kräusslich H-G. Analysis of human immunodeficiency virus type 1 gag ubiquitination. J Virol 2005;79:9134−44.

95. Zheng Y, Yao X. Posttranslational modifications of HIV-1 integrase by various cellular proteins during viral replication. Viruses 2013;5:1787–801.
96. Chen S, et al. Immune regulator ABIN1 suppresses HIV-1 transcription by negatively regulating the ubiquitination of Tat. Retrovirology 2017;14:12–4.
97. Jin Y-J, Cai CY, Zhang X, Burakoff SJ. Lysine 144, a ubiquitin attachment site in HIV-1 Nef, is required for Nef-mediated CD4 down-regulation. J Immunol 2008;180:7878–86.
98. Gurer C, Berthoux L, Luban J. Covalent modification of human immunodeficiency virus type 1 p6 by SUMO-1. J Virol 2005;79:910–7.
99. Zamborlini A, et al. Impairment of human immunodeficiency virus type-1 integrase SUMOylation correlates with an early replication defect. J Biol Chem 2011;286:21013–22.
100. Zheng Y, et al. Noncovalent SUMO-interaction motifs in HIV integrase play important roles in SUMOylation, cofactor binding, and virus replication. Virol J 2019;16:1–14.
101. Topper M, et al. Posttranslational acetylation of the human immunodeficiency virus type 1 integrase carboxyl-terminal domain is dispensable for viral replication. J Virol 2007;81:3012–7.
102. Terreni M, et al. GCN5-dependent acetylation of HIV-1 integrase enhances viral integration. Retrovirology 2010;7:1–16.
103. Cereseto A, et al. Acetylation of HIV-1 integrase by p300 regulates viral integration. EMBO J 2005;24: 3070–81.
104. Kiernan RE, et al. HIV-1 Tat transcriptional activity is regulated by acetylation. EMBO J 1999;18:6106–18.
105. Ott M, et al. Tat acetylation: a regulatory switch between early and late phases in HIV transcription elongation. Novartis Found Symp 2004;259:182–96.
106. Berro R, et al. Acetylated tat regulates human immunodeficiency virus type 1 splicing through its interaction with the splicing regulator p32. J Virol 2006;80:3189–204.
107. Boulanger M-C, et al. Methylation of tat by PRMT6 regulates human immunodeficiency virus type 1 gene expression. J Virol 2005;79:124–31.
108. Van Duyne R, et al. Lysine methylation of HIV-1 Tat regulates transcriptional activity of the viral LTR. Retrovirology 2008;5:1–13.
109. Ali I, et al. The HIV-1 Tat protein is monomethylated at lysine 71 by the lysine methyltransferase KMT7. J Biol Chem 2016;291:16240–8.
110. Invernizzi CF, et al. Arginine methylation of the HIV-1 nucleocapsid protein results in its diminished function. AIDS 2007;21:795–805.
111. Pan ZQ, Kentsis A, Dias DC, Yamoah K, Wu K. Nedd8 on cullin: building an expressway to protein destruction. Oncogene 2004;23:1985–97.
112. Nekorchuk MD, Sharifi HJ, Furuya AKM, Jellinger R, De Noronha CMC. HIV relies on neddylation for ubiquitin ligase-mediated functions. Retrovirology 2013;10:1–12.
113. Okumura A, Lu G, Pitha-Rowe I, Pitha PM. Innate antiviral response targets HIV-1 release by the induction of ubiquitin-like protein ISG15. Proc Natl Acad Sci USA 2006;103:1440–5.
114. Lata S, Mishra R, Banerjea AC. Proteasomal degradation machinery: favorite target of HIV-1 proteins. Front Microbiol 2018;9:1–17.
115. Proulx J, Borgmann K, Park IW. Post-translational modifications inducing proteasomal degradation to counter HIV-1 infection. Virus Res 2020;289:198142.
116. Lata S, Ali A, Sood V, Raja R, Banerjea AC. HIV-1 Rev downregulates Tat expression and viral replication via modulation of NAD(P)H:quinine oxidoreductase 1 (NQO1). Nat Commun 2015;6.
117. Zhou D, Kang KH, Spector SA. Production of interferon α by human immunodeficiency virus type 1 in human plasmacytoid dendritic cells is dependent on induction of autophagy. J Infect Dis 2012;205: 1258–67.

118. Nityanandam R, Serra-Moreno R. BCA2/Rabring7 targets HIV-1 gag for lysosomal degradation in a tetherin-independent manner. PLoS Pathog 2014;10:e1004151.
119. Miyakawa K, et al. BCA2/Rabring7 promotes tetherin-dependent HIV-1 restriction. PLoS Pathog 2009;5: 21–5.
120. Mangeat B, et al. HIV-1 Vpu neutralizes the antiviral factor Tetherin/BST-2 by binding it and directing its beta-TrCP2-dependent degradation. PLoS Pathog 2009;5:e1000574.
121. Zhang Y, Lu J, Liu X. MARCH2 is upregulated in HIV-1 infection and inhibits HIV-1 production through envelope protein translocation or degradation. Virology 2018;518:293–300.
122. Casini A, Olivieri M, Vecchi L, Burrone OR, Cereseto A. Reduction of HIV-1 infectivity through endoplasmic reticulum-associated degradation-mediated Env depletion. J Virol 2015;89:2966–71.
123. Zhang X, Zhou T, Frabutt DA, Zheng YH. HIV-1 Vpr increases Env expression by preventing Env from endoplasmic reticulum-associated protein degradation (ERAD). Virology 2016;496:194–202.

CHAPTER

Post-transcriptional regulation of gene expression in *Entamoeba histolytica*

16

Sandeep Ojha[1] and Sudha Bhattacharya[2]

[1]*Institute of Physical Chemistry, Westfälische Wilhelms-Universität, Münster, Germany;* [2]*Department of Biology, Ashoka University, Sonipat, Haryana, India*

Introduction

Entamoeba histolytica is a unicellular protistan parasite, which causes amebic dysentery and liver abscess in humans, that affects 90 million individuals globally [1,2]. Infection in humans occurs by the ingestion of quadrinucleated cysts through contaminated food or water [3]. Excystation occurs in the ileocecal region, with the emergence of trophozoites. These migrate to the colon where they feed on the microflora and actively multiply. In most infections, the trophozoites convert back into cysts, which are released into the feces, completing the life cycle. However, occasionally the trophozoites could invade the intestinal mucosa causing ulcers, and further invade other organs, mainly the liver resulting in liver abscess. Invasive disease, if not effectively treated, could be fatal.

E. histolytica belongs to the kingdom Amoebozoa and is one of the earliest branching eukaryotes. Its phylogenetic position and parasitic lifestyle make it an interesting system to reveal gene regulatory mechanisms that, perhaps, may not operate in free-living model organisms. The 20.7 Mb genome of *E. histolytica* is rich in AT content (76%) and contains several repetitive sequences. About 10% of the genome is composed of ribosomal RNA genes that are located exclusively on extrachromosomal circles [4]. Among other repeated sequences are the tandemly repeated arrays of tRNA genes [5,6], the long- and short-interspersed nuclear elements (LINEs and SINEs) that are retrotransposons and occupy 11% of the genome [7–10], and *Entamoeba*-specific repeats accounting for 8.4% of the genome [10]. Here, we review the current knowledge about various post-transcriptional gene regulatory mechanisms that operate in *E. histolytica.*

The new assembly and reannotation of *the E. histolytica* genome sequence revealed 8201 predicted genes [11]. The mean length of intergenic regions is estimated to be 708.7 bp and the mean gene length is 1260.9 bp [11]. Most genes lack introns, contributing to the compactness of the genome. Only 24.4% of *E. histolytica* genes contain introns, with only 6% containing more than one intron [6,12]. The mRNA 5′- and 3′-untranslated regions (UTRs) are typically short (5–21 nucleotides) [13]. Hence, alternative splicing and polyadenylation, which are important for generating protein diversity and for post-transcriptional regulation of developmental processes in eukaryotes [14], are expected to be less prevalent in *E. histolytica*. A detailed analysis of *E. histolytica* poly (A)+ transcriptome data was

Post-Transcriptional Gene Regulation in Human Disease, Volume 32. https://doi.org/10.1016/B978-0-323-91305-8.00017-X

carried out to determine the extent of alternative splicing and polyadenylation, and, importantly, to distinguish inherently stochastic events from physiologically relevant ones. This analysis showed that most events were likely to be solely stochastic, and alternative polyadenylation could be detected only for 1.9%–2.4% of genes. Thus, these processes being limited to only a small set of genes, would not significantly expand the proteome or generate transcriptome diversity in *E. histolytica*.

Small RNA-mediated gene regulation in *E. histolytica*

The first evidence of gene silencing in *E. histolytica* came from studies with the amoebapore a gene (*Ehap-a*). Trophozoites transfected with a plasmid containing the Ehap-a gene along with a 473 bp upstream segment showed silencing of the chromosomal copy of Ehap-a. Silencing was maintained even after the removal of the plasmid [15,16]. Further transfection of this strain (called G3) with a plasmid containing a second gene ligated to the upstream region of *Ehap-a* enabled the silencing, in-trans, of other genes of choice. Subsequent analysis linked the silencing mechanism in G3 with small RNAs.

Small silencing RNAs are a diverse set of regulatory noncoding RNAs found ubiquitously in plants, animals, and unicellular eukaryotes. These include microRNAs and small interfering RNAs. Their unifying features are their small size (20–30 nt), their association with the Argonaute (Ago) family of proteins, and their sequence-specific interaction with target genes, resulting in reduced expression [17–19]. This RNA silencing mechanism, referred to as RNA interference (RNAi), has emerged as one of the key gene regulatory pathways in most eukaryotes and is an important pathway in *E. histolytica* as well.

Endogenous small silencing RNAs (sRNAs) of *E. histolytica* display unique properties [20–23]. The major population of sRNAs identified initially was 27 nt with a 5′-polyphosphate structure [20,22]. Recently, a 31 nt population has also been identified that is longer as it contains three to four nontemplated adenosines at the 3′-end [21]. The 27-nt sRNAs have been shown to associate with EhAgo2-2. They have several properties in common with secondary sRNAs of nematodes, not reported in other organisms. These include the presence of 5′-polyphosphate termini, antisense (AS) orientation, and bias toward the 5′-ends of target genes [24].

E. histolytica sRNAs can perform RNAi

The functional involvement of *E. histolytica* sRNAs in RNAi has been demonstrated. The levels of AS sRNAs correlate inversely with mRNA levels of their cognate genes [22]. Direct demonstration of gene silencing was done by introducing a plasmid with a fusion construct in which the gene to be silenced was fused to a small portion of a coding region to which large numbers of AS sRNAs were known to map. The latter served as a "trigger" to silence a variety of genes fused to it, which included the highly expressed *E. histolytica* rhomboid protease, and a firefly luciferase reporter gene [23]. Silencing was associated with the appearance of 27-nt AS sRNAs corresponding to the silenced gene. These had 5′-polyphosphate termini and were derived from mature and nascent mRNA, as they mapped to introns. Silencing was independent of the orientation of the trigger with the fused gene (5′ or 3′) or to the gene segment included in the construction of the trigger. Further insights into the silencing mechanism came from the observation that transcription of the trigger-gene fusion construct was required for silencing and that silencing persisted after removal of the trigger plasmid, suggesting the involvement of an amplification pathway. In *C. elegans* such an amplification results in the generation of secondary sRNAs mediated by RNA-dependent RNA polymerase (RdRP) starting from dicer-derived primary siRNAs [25]. However, a canonical dicer has not been identified in *E. histolytica.*

Experiments to understand the mechanism of sRNA-mediated gene silencing in *E. histolytica* showed a prominent role in histone modification. In the presence of silencing Trigger, repression of gene transcription resulted from the deposition of H3K27Me2 and Ago proteins at the repressed loci. Subsequently, when the Trigger was absent, silencing was still maintained by increased deposits of repressive H3K27Me, in the absence of Ago [26]. Thus, methylated histones had an important role in initiating and maintaining gene silencing.

Proteins associated with *E. histolytica* sRNAs

sRNAs perform their RNAi function in association with proteins to form a ribonucleoprotein complex called the RNA-induced silencing complex (RISC) [27]. The Ago proteins are some of the most conserved, and *E. histolytica* possesses all three EhAgo proteins (EhAgo2-1, -2-2, -2-3) that associate with sRNA populations [22]. Of these Ago2-2 is the most abundant. It is nuclear-localized and associated with 27-nt sRNAs to mediate transcriptional gene silencing [20]. *E. histolytica* also contains two genes with RdRP domains. The components of RISC complexes vary depending on the nature of the underlying gene silencing mechanism. For example, RISC complexes involved in post-transcriptional gene silencing through siRNAs and miRNAs contain highly conserved dicer [28], while those involved in transcriptional gene silencing contain chromatin-interacting and remodeling proteins [29]. Sequence searches failed to detect a dicer enzyme in *E. histolytica* with all the canonical dicer-associated domains (helicase, PAZ, dsRBD, and two RNase III domains). However, a protein with a single RNase III domain has been detected in *E. histolytica* [30,31]., which is similar to the noncanonical dicer enzyme found in budding yeast. It is not known whether this protein interacts with EhAgo and can function in RNAi. Recently, the RISC components associated with each EhAgo have been characterized [32]. EhAgo2-2 RISC lacked protein factors associated with siRNA/miRNA pathways; rather it contained RecQ helicase and Tudor-like protein, found in Piwi RISC. This is interesting, especially since sequence analysis has shown that the three EhAgo proteins are related to the PIWI-like clade [33]. Further functional studies will reveal whether the Eh RNAi pathway indeed has common features with Piwi RISC. The presence of 5′-polyphosphate termini in Eh sRNAs suggests that they could be generated by RdRP. A number of the EhAgo RISC components were common with EhRdRP-associated proteins, suggesting a common role for these proteins in achieving RNAi in *E. histolytica* [32].

Possible role of *E. histolytica* sRNAs in retrotransposon silencing

The properties of *E. histolytica* sRNAs described earlier suggest that they are likely to function in a dicer-independent manner. EhRISC and EhAgo proteins share more similarities with the Piwi-mediated pathway. Piwi-interacting RNAs (piRNAs) are implicated in the silencing of retrotransposons and repetitive DNAs in flies and mammals, both transcriptionally and post-transcriptionally [34–38]. The piRNA pathway is considered as a host defense system against the potentially mutagenic activity of transposable elements and is especially active in germ cells where a majority of piRNAs are derived from retrotransposons and transposable elements.

The *E. histolytica* genome contains over 700 copies of a nonlong terminal repeat (LTR) element, EhLINE1. Recent transcriptomic analysis showed that only ~6% of EhLINE1 copies were transcriptionally active. Importantly, apart from sense strand transcripts, there was massive antisense transcription of a part of EhLINE1 corresponding to ORF2 that encodes reverse transcriptase (RT). A 1.5 kb RT-antisense long noncoding (lnc) RNA could be detected both by RNA-Seq and northern hybridization. Antisense transcripts were specific to the RT region as very few of these mapped to ORF1 [39]. That antisense transcript could have a regulatory role was suggested by the observation that ORF1p was constitutively expressed in *E. histolytica*, while ORF2p was undetectable [40]. As both polypeptides are required for active retrotransposition, the low levels of ORF2p could be an effective mechanism to keep retrotransposition in check. While direct evidence for a regulatory role of antisense RNAs in EhLINE1 expression is lacking, quantitative analysis of sRNA reads in *E. histolytica* showed that of all genomic loci, retrotransposons contributed the most to sRNAs, and these mapped in the antisense orientation. About 28% of 31 nt sRNAs and ~10% of 27 nt sRNAs mapped to EhLINEs, indicating a possible link between sRNAs and retrotransposon control in this parasite [21]. Further work on the mechanism of sRNA-mediated gene silencing in *E. histolytica* may reveal how these RNAs could contribute to EhLINE silencing, and whether a possible connection exists between the sRNAs and the 1.5 kb RT-antisense lncRNA.

LncRNAs in *E.histolytica*

LncRNAs, as the name implies, are transcripts longer than 200 nts that are structurally similar to mRNAs, but have little to no protein-coding potential and are expressed at low levels [41]. They are usually transcribed by RNA pol II and could be of sense or antisense orientation [42]. Transcriptomic analysis has shown that lncRNAs are present in a wide variety of organisms. They are versatile regulators of gene expression, that affect processes ranging from cellular differentiation to cell cycle [43,44], and human disease [45]. lncRNAs also play important roles in response to both biotic and abiotic stresses [46].

Stress-responsive lncRNA in *E. histolytica*

Studies on serum-starved *E. histolytica* showed for the first time a lncRNA that was upregulated in response to stress. Earlier studies had shown that a functional transmembrane kinase of the B1 family (EhTMKB1-9) was downregulated during serum starvation and was induced by serum replenishment. Conversely, another gene with homology to kinase domain sequence of TMK B1 family (EhTMKB1-18) was induced to high levels in response to serum starvation [47]. Serum-dependent selective expression of EhTMKB1-9, a member of *Entamoeba histolytica* B1 family of transmembrane kinases [47]. The latter gene encoded a transcript of 2.6 kb that lacked any ORFs >150bp and had the general features of a sense-strand long noncoding RNA (EhslncRNA). The transcript was polyadenylated and mainly associated with monosomes in the cytoplasm under serum starvation. It was nuclear-localized in normal proliferating cells but accumulated in the cytoplasm in starved cells [48]. The region between −437 and −346 acted as a negative repressor of serum response. In addition to serum starvation, the upregulation of EhslncRNA was observed in response to a variety of other stresses like oxygen stress (*E. histolytica* being microaerophilic), and high temperature, raising the possibility that this

RNA could act as a general stress regulator. The use of lncRNAs to regulate gene expression during growth stress has been reported in many other systems and could be an important role of these RNAs [49,50].

An interesting property of EhslncRNA was its association with monosomes. This has been observed in other systems as well. The ribosomal association could be linked to processes like lncRNA translation, or nonsense-mediated decay. Although lncRNAs lack long ORFs, their translation into peptides cannot be ruled out [51]. However, the data do not seem to generally support the translation of lncRNAs [52,53]. Rather, it seems possible that lncRNAs could regulate translation in a general, or sequence-specific manner by keeping the ribosomes in a poised state and inhibiting translation until positive signals are received [54]. As EhslncRNA associates with ribosomes and accumulates in the cytosol during growth stress, it is possible that it could be serving a translation attenuation function in stressed cells when growth-related pathways are shut down. Further functional analysis of this RNA is needed to understand its possible role.

Retrotransposon-encoded antisense lncRNA

In contrast to EhslncRNA, a sense-strand transcript, another lncRNA that is an antisense transcript has recently been described in *E. histolytica* [39]. This 1.5 kb transcript originates from the RT region of the non-LTR retrotransposon EhLINE1, as described briefly in the previous section. It is polyadenylated and can potentially code for four nonoverlapping peptide sequences of sizes 150, 47, 42, and 28 aa. Codon usage of these peptides was checked to determine the frequency of low-usage codons. There was a significantly greater frequency of low-usage codons in these peptides compared with sense sequences of EhLINE1 ORF1p and ORF2p. In addition, the putative peptides did not show any matches with known sequences in protein databases. Thus, the 1.5 kb transcript is likely to be a noncoding RNA derived from a LINE element. Antisense transcription of LINEs has been reported in other systems. The primate-specific LINE1 contains an antisense promoter (ASP) located in the 5′-UTR between nucleotides 400 and 600 [55]. It contains a short ORF termed ORF0 between nucleotides 452–236 in the antisense orientation, which is translated [56]. In addition, the ASP also drives read-through transcription of nearby genes, giving rise to chimeric transcripts [57]. Antisense transcription from ORF2 has also been observed in adult neurogenesis, a process in which mammalian LINE-1 activation is important. Both sense and antisense transcription originate from the mammalian LINE-1 ORF2 during neuronal differentiation. These transcripts possibly arise from the multiple, overlapping Sox/LEF binding sites in ORF2, which could promote transcription of nearby neuronal loci in both directions, mediated by Wnt/β catenin activation [58].

Compared with the antisense transcription reported for mammalian LINEs, the RT-antisense RNA of EhLINE1 is distinctly different. It is a discrete 1.5 kb transcript from the RT region, and the RNA-Seq data so far have not shown any evidence of read-through antisense transcription, or splicing leading to transcriptome diversity [39]. Further work is needed to understand the physiological functions of this antisense RNA. As EhLINE1-ORF1p is expressed constitutively whereas ORF2p is undetectable [40], a direct role of this RT-antisense transcript could be in the suppression of intracellular EhLINE1-RT expression to restrict active retrotransposition. Another intriguing possibility is that the EhLINE1 antisense transcript could have roles unrelated to retrotransposition, and could have been coopted by the host to serve as a lncRNA for other gene regulatory functions.

LncRNA from the intergenic spacer of rRNA genes of *Entamoeba invadens*

E. histolytica has a dimorphic life cycle comprising the dormant, nonmotile, quadrinucleate cyst and the actively dividing, motile, uninucleate trophozoite. The cyst is the infectious stage and is responsible for disease transmission, as the trophozoite does not survive outside the human body. Understanding the stage conversion process is key to breaking the disease transmission cycle. However, attempts to convert *E. histolytica* trophozoites to cysts under lab conditions have been unsuccessful so far, although recently some success in this direction has been reported [59]. On the other hand, *Entamoeba invadens*, a species that causes amebiasis in reptiles, can readily encyst in the lab and is used widely as a model system to study *Entamoeba* stage conversion. The rRNA genes in both *E. histolytica* and *E. invadens* are located on extrachromosomal circular molecules [4]. Transcriptional analysis of *E. invadens* rDNA showed that the rDNA circle contained at least two promoters. Pre-rRNA synthesis takes place from the rDNA promoter upstream of 5′-external transcribed spacer (ETS) and it appears to terminate downstream of 28S rRNA 3′-end. A second promoter exists downstream of 28S rRNA 3′-end that transcribes almost the entire intergenic spacer (IGS) into a lncRNA of >10 kb [60]. Studies with most model organisms show that under conditions of growth stress when the requirement for ribosomes goes down, rDNA transcription ceases [61,62]. Stage conversion is also a response to growth stress and is induced in *E. invadens* by shifting trophozoites to a low glucose medium. The process of cyst formation is generally completed by 72 h with the formation of mature cysts [59,63]. Transcript levels were measured by northern hybridization during stage conversion in *E. invadens*. In a time-course study, it was shown that immediately after shifting the trophozoites to a low glucose medium there was an initial drop in the levels of pre-rRNA. However, after 24 h, the levels of pre-rRNA and its processing intermediates began to rise and increased 4.8-fold by 72 h. Concomitantly the lncRNA from the IGS also showed an increase of 2.2-fold by 72 h [60]. Thus rDNA transcription did not shut down; rather all rDNA transcripts, including the lncRNA, accumulated coordinately during encystation. The functional importance of this lncRNA accumulating in mature cysts has not been explored yet. A large variety of lncRNAs originating from the rDNA IGS both in sense and antisense orientation with respect to rRNAs have been reported in yeast and mammals. Some of these IGSRNAs are induced during growth stress and have been associated with various functions, including repression of rDNA transcription, binding to proteins with nucleolar detention sequences, and sequestering them to the nucleolus [64,65]. These lncRNAs from the IGS are viewed as multifunctional regulators of nucleolar activities, responsive to various physiological conditions and growth stress. Whether the lncRNA from *E. invadens* IGS has similar functions, especially during cyst formation, needs to be investigated.

Post-transcriptional regulation of ribosome biogenesis in *Entamoeba*

Ribosome biogenesis, being a highly energy-consuming process, is tightly regulated, and rDNA transcription ceases during growth stress in most organisms [66]. Of the hundreds to thousands of rDNA copies in a cell, only a fraction are transcriptionally active while the others are silenced by epigenetic modifications [67,68]. Apart from the transcriptionally silent and active states, rRNA genes also exist in a poised state with hallmarks of active genes. The poised state is designed to respond to environmental cues and presumably serves to rapidly activate transcription upon resumption of normal growth. It appears that in *Entamoeba*, an early-branching eukaryote, the poised state may be less important. Rather the control of ribosome biogenesis appears to be shifted to a post-transcriptional level whereby the mature rRNAs are not available for ribosome assembly due to the accumulation of unprocessed precursors.

When *E. invadens* trophozoites were induced to form cysts under low glucose conditions, there was an initial drop in pre-rRNA levels after which the levels began to rise, and in mature cysts, there was an accumulation of unprocessed pre-rRNA [60]. Similarly, the transcripts of r-protein genes also increased in mature cysts [69]. Ultrastructural studies have shown that due to the reduced need for protein synthesis, *E. invadens* cysts contain large arrays of free ribosomes in their cytoplasm [70,71]. In addition to these ribosomes, mature cysts also store precursors like pre-rRNAs and r-protein mRNAs, which could presumably be rapidly processed or translated, respectively, and assembled to meet the immediate need for new ribosomes during excystation.

When *E. histolytica* cells were subjected to growth stress due to serum starvation, there was an accumulation of unprocessed pre-rRNAs and other intermediates, notably fragments of the 5′-ETS, pointing to a slowing down of pre-rRNA processing [72]. The mRNAs of r-protein (RP) genes also persisted in serum-starved cells [73]. The regulation of RP genes was studied in these cells using a luciferase reporter under the control of two RP genes (RPS19 and RPL30). Luciferase transcript levels remained unchanged during starvation; however, luciferase activity driven by RPS19 and RPL30 declined to 7.8% and 15% of control cells, respectively after 24 h of starvation. The activity reverted to normal levels upon serum replenishment, suggesting post-transcriptional control of gene expression. Further, it was shown that the stress-related repressive effect on translation was mediated by the 5′-UTR [73]. Mutations in the sequence −2 to −9 upstream of the AUG start codon in the RPL30 gene abolished the translational repression, and in this mutant, there was no decline in luciferase activity during serum starvation. Thus, during serum starvation, *E. histolytica* effectively blocked ribosome biogenesis post-transcriptional by inhibiting pre-rRNA processing on the one hand, and the translation of RP mRNAs on the other.

Post-transcriptional control of RP gene expression was also shown for RPL21 in *E. histolytica* and *E. dispar*, a sibling species that is nonpathogenic [74]. Both species contain two copies of the RPL21 gene that are transcribed to similar levels. However, only one of the genes is preferentially associated with polyribosomes, and the preferred copy is not the same in each species [74]. Here too, the translational control appears to be exerted by the 5′-UTR.

Regulation via mRNA stability in *E. histolytica*

mRNA stability plays a pivotal role in the regulation of gene expression. Expression of P-glycoprotein gene, *EhPgp5,* involved in drug resistance was studied. Half-life measurement of the mRNA of *EhPgp5* in trophozoites grown at different concentrations of emetine showed that the stability of this mRNA increased at high concentrations of the drug [75]. It was reported that *EhPgp5* mRNA contained a longer poly(A) tail in clones growing at higher concentrations of emetine [75]. It is well recognized that longer poly(A) tails give higher stability to mRNA and promote more efficient translation [76]. These results indicate that the higher levels of EhPGP5 protein in multidrug-resistant trophozoites could be influenced by increased mRNA stability [75].

Post-transcriptional regulation through RNA homeostasis mechanisms

The first indication that exoribonucleases of the RNA surveillance machinery might have regulatory roles in *E. histolytica* came from the observation that the rRNA 5′-ETS fragments, that are rapidly degraded after pre-rRNA processing, accumulate to very high levels in serum-starved *E. histolytica* cells [72,77]. These fragments are normally the targets of Rrp6p, a 3′−5′ exoribonuclease associated with the multisubunit protein complex, the exosome, which is highly conserved across eukaryotes [78,79]. The exosome is involved in RNA homeostasis and turnover [80], and in surveillance pathways [81], for a variety of nuclear and cytoplasmic RNAs [82−85]. The core exosome, composed of nine subunits (Exo9) lacks catalytic activity; rather through its barrel-shaped structure, it provides a central channel for ssRNA to pass through. The catalytic activity is provided by two exonucleases associated with the exosome-Rrp6, a distributive 3′−5′ exonuclease, and Rrp44, an enzyme with endoribonuclease and processive 3′−5′ exonuclease activities.

A study was undertaken to characterize EhRrp6 and determine whether it had a role in the accumulation of 5′-ETS fragments in stressed cells. Although EhRrp6 sequence differed from the *S. cerevisiae* and human homologs in having large deletions at both the N- and C-termini, it seemed to be functionally conserved as it could complement the growth defect of yeast Scrrp6ts mutant [77], and the recombinant EhRrp6p displayed 3′−5′ exonuclease activity with synthetic RNA substrates. Downregulation of EhRrp6 led to an increase in levels of 5′-ETS subfragments, suggesting that this role of Rrp6 was conserved in *E. histolytica* as well. Interestingly, subcellular localization studies showed that the enzyme was lost from the nuclei in starved cells, resulting in the accumulation of 5′-ETS fragments.

In model organisms, Rrp6 is a key enzyme in RNA homeostasis involved in the processing and degradation of many stable RNA precursors, aberrant transcripts, and noncoding RNAs. It, thus, has important regulatory roles. The properties of EhRrp6 also suggest its metabolic importance and that it could act as a stress sensor by regulating the levels of key RNA species. EhRrp6-depleted *E. histolytica* cells were severely growth restricted. Conversely, EhRrp6 overexpression protected the cells against stress. Importantly, EhRrp6 depleted cells showed a marked reduction in erythrophagocytosis, a process required for *E. histolytica* growth and virulence. Transcript levels of some of the phagocytosis-related genes, for example, EhCaBP3 and EhRho1 dropped in EhRrp6 depleted cells, while other genes (EhCaBP1, EhCaBP6, EhC2PK, and EhARP2/3) were unaffected [77].

EhRrp6 could have important gene-regulatory roles in *E. histolytica* by maintaining the levels of its target RNAs that could be specific mRNAs or small regulatory RNAs. For example, in budding yeast, noncoding RNAs such as MUTs, CUTs, and rsSUTs are direct targets of Rrp6 during vegetative growth. Rrp6 negatively regulates meiotic development by maintaining these ncRNAs at low levels. In response to nutrient-related stress budding yeast cells are induced to sporulate. During this developmental shift, the Rrp6p that was stable in mitotic growth begins to rapidly decline in sporulating cells [86] leading to increased levels of its target ncRNAs, and a resultant shift in the gene expression landscape. While such studies have not been performed in *E. histolytica* so far, one can speculate that similar to yeast sporulation, EhRrp6 could regulate the stage conversion process of a trophozoite to a cyst in *Entamoeba*. Further studies will show whether this important regulatory system mediates parasite response to the host environment, and in pathogenesis.

In conclusion, *E. histolytica* clearly utilizes post-transcriptional gene regulation through a variety of molecules and pathways shared with model organisms. The mechanisms involved are likely to have interesting unique aspects due to the evolutionary position and parasitic lifestyle of this organism. Further work will enrich our understanding of these processes.

Acknowledgments

This work was supported by a fellowship to SB from the Indian National Science Academy.

References

[1] Ximénez C, Morán P, Rojas L, Valadez A, Gómez A. Reassessment of the epidemiology of amebiasis: state of the art. Infect Genet Evol J Mol Epidemiol & Evol Genet Inf Dis 2009;9(6):1023–32. https://doi.org/10.1016/j.meegid.2009.06.008.

[2] WHO estimates of the global burden of foodborne diseases. Geneva, Switzerland: World Health Organization; 2015.

[3] Stanley SL. Amoebiasis. Lancet 2003;361(9362):1025–34. https://doi.org/10.1016/S0140-6736(03)12830-9.

[4] Bhattacharya S, Som I, Bhattacharya A. The ribosomal DNA plasmids of Entamoeba. Parasitol Today 1998; 14(5):181–5. https://doi.org/10.1016/S0169-4758(98)01222-8.

[5] Clark CG, Ali IKM, Zaki M, Loftus BJ, Hall N. Unique organisation of tRNA genes in *Entamoeba histolytica*. Mol Biochem Parasitol 2006;146(1):24–9. https://doi.org/10.1016/j.molbiopara.2005.10.013.

[6] Loftus B, Anderson I, Davies R, Alsmark UCM, Samuelson J, Amedeo P, et al. The genome of the protist parasite *Entamoeba histolytica*. Nature 2005;433(7028):865–8. https://doi.org/10.1038/nature03291.

[7] Sharma R, Bagchi A, Bhattacharya A, Bhattacharya S. Characterization of a retrotransposon-like element from *Entamoeba histolytica*. Mol Biochem Parasitol 2001;116(1):45–53. https://doi.org/10.1016/S0166-6851(01)00300-0.

[8] van Dellen K, Field J, Wang Z, Loftus B, Samuelson J. LINEs and SINE-like elements of the protist *Entamoeba histolytica*. Gene 2002;297(1–2):229–39. https://doi.org/10.1016/s0378-1119(02)00917-4.

[9] Bakre AA, Rawal K, Ramaswamy R, Bhattacharya A, Bhattacharya S. The LINEs and SINEs of *Entamoeba histolytica*: comparative analysis and genomic distribution. Exp Parasitol 2005;110(3):207–13. https://doi.org/10.1016/j.exppara.2005.02.009.

[10] Lorenzi H, Thiagarajan M, Haas B, Wortman J, Hall N, Caler E. Genome wide survey, discovery and evolution of repetitive elements in three Entamoeba species. BMC Genom 2008;9:595. https://doi.org/10.1186/1471-2164-9-595.

[11] Lorenzi HA, Puiu D, Miller JR, Brinkac LM, Amedeo P, Hall N, et al. New assembly, reannotation and analysis of the *Entamoeba histolytica* genome reveal new genomic features and protein content information. PLoS Negl Trop Dis 2010;4(6):e716. https://doi.org/10.1371/journal.pntd.0000716.

[12] Bhattacharya A, Satish S, Bagchi A, Bhattacharya S. The genome of *Entamoeba histolytica*. Int J Parasitol 2000;30(4):401–10. https://doi.org/10.1016/S0020-7519(99)00189-7.

[13] Hon C-C, Weber C, Sismeiro O, Proux C, Koutero M, Deloger M, et al. Quantification of stochastic noise of splicing and polyadenylation in *Entamoeba histolytica*. Nucleic Acids Res 2013;41(3):1936–52. https://doi.org/10.1093/nar/gks1271.

[14] Baralle FE, Giudice J. Alternative splicing as a regulator of development and tissue identity. Nat Rev Mol Cell Biol 2017;18(7):437–51. https://doi.org/10.1038/nrm.2017.27.

[15] Bracha R, Nuchamowitz Y, Mirelman D. Transcriptional silencing of an amoebapore gene in *Entamoeba histolytica*: molecular analysis and effect on pathogenicity. Eukaryot Cell 2003;2(2):295–305. https://doi.org/10.1128/ec.2.2.295-305.2003.

[16] Anbar M, Bracha R, Nuchamowitz Y, Li Y, Florentin A, Mirelman D. Involvement of a short interspersed element in epigenetic transcriptional silencing of the amoebapore gene in *Entamoeba histolytica*. Eukaryot Cell 2005;4(11):1775–84. https://doi.org/10.1128/EC.4.11.1775-1784.2005.

[17] Bartel DP. Metazoan MicroRNAs. Cell 2018;173(1):20–51. https://doi.org/10.1016/j.cell.2018.03.006.

[18] Ghildiyal M, Zamore PD. Small silencing RNAs: an expanding universe. Nat Rev Genet 2009;10(2): 94–108. https://doi.org/10.1038/nrg2504.

[19] Zeng J, Gupta VK, Jiang Y, Yang B, Gong L, Zhu H. Cross-kingdom small RNAs among animals, plants and microbes. Cells 2019;8(4). https://doi.org/10.3390/cells8040371.

[20] Zhang H, Alramini H, Tran V, Singh U. Nucleus-localized antisense small RNAs with 5′-polyphosphate termini regulate long term transcriptional gene silencing in *Entamoeba histolytica* G3 strain. J Biol Chem 2011;286(52):44467–79. https://doi.org/10.1074/jbc.M111.278184.

[21] Zhang H, Ehrenkaufer GM, Hall N, Singh U. Identification of oligo-adenylated small RNAs in the parasite Entamoeba and a potential role for small RNA control. BMC Genom 2020;21(1):879. https://doi.org/10.1186/s12864-020-07275-6.

[22] Zhang H, Ehrenkaufer GM, Pompey JM, Hackney JA, Singh U. Small RNAs with 5′-polyphosphate termini associate with a Piwi-related protein and regulate gene expression in the single-celled eukaryote *Entamoeba histolytica*. PLoS Pathogens 2008;4(11):e1000219. https://doi.org/10.1371/journal.ppat.1000219.

[23] Morf L, Pearson RJ, Wang AS, Singh U. Robust gene silencing mediated by antisense small RNAs in the pathogenic protist *Entamoeba histolytica*. Nucleic Acids Res 2013;41(20):9424–37. https://doi.org/10.1093/nar/gkt717.

[24] Pak J, Fire A. Distinct populations of primary and secondary effectors during RNAi in *C. elegans*. Science 2007;315(5809):241–4. https://doi.org/10.1126/science.1132839.

[25] Pak J, Maniar JM, Mello CC, Fire A. Protection from feed-forward amplification in an amplified RNAi mechanism. Cell 2012;151(4):885–99. https://doi.org/10.1016/j.cell.2012.10.022.

[26] Foda BM, Singh U. Dimethylated H3K27 is a repressive epigenetic histone mark in the protist *Entamoeba histolytica* and is significantly enriched in genes silenced via the RNAi pathway. J Biol Chem 2015;290(34): 21114–30. https://doi.org/10.1074/jbc.M115.647263.

[27] Wilson RC, Doudna JA. Molecular mechanisms of RNA interference. Annu Rev Biophys 2013;42:217–39. https://doi.org/10.1146/annurev-biophys-083012-130404.

[28] Chendrimada TP, Gregory RI, Kumaraswamy E, Norman J, Cooch N, Nishikura K, et al. TRBP recruits the Dicer complex to Ago2 for microRNA processing and gene silencing. Nature 2005;436(7051):740–4. https://doi.org/10.1038/nature03868.

[29] Motamedi MR, Verdel A, Colmenares SU, Gerber SA, Gygi SP, Moazed D. Two RNAi complexes, RITS and RDRC, physically interact and localize to noncoding centromeric RNAs. Cell 2004;119(6):789–802. https://doi.org/10.1016/j.cell.2004.11.034.

[30] Pompey JM, Foda B, Singh U. A single RNaseIII domain protein from *Entamoeba histolytica* has dsRNA cleavage activity and can help mediate RNAi gene silencing in a heterologous system. PLoS One 2015; 10(7):e0133740. https://doi.org/10.1371/journal.pone.0133740.

[31] Yu X, Li X, Zheng L, Ma J, Gan J. Structural and functional studies of a noncanonical Dicer from *Entamoeba histolytica*. Sci Rep 2017;7:44832. https://doi.org/10.1038/srep44832.

[32] Zhang H, Veira J, Bauer ST, Yip C, Singh U. RISC in *Entamoeba histolytica*: identification of a protein-protein interaction network for the RNA interference pathway in a deep-branching eukaryote. mBio 2021:e0154021. https://doi.org/10.1128/mBio.01540-21.

[33] Swarts DC, Makarova K, Wang Y, Nakanishi K, Ketting RF, Koonin EV, et al. The evolutionary journey of Argonaute proteins. Nat Struct Mol Biol 2014;21(9):743–53. https://doi.org/10.1038/nsmb.2879.

[34] Aravin AA, Hannon GJ, Brennecke J. The Piwi-piRNA pathway provides an adaptive defense in the transposon arms race. Science 2007;318(5851):761–4. https://doi.org/10.1126/science.1146484.

[35] Brennecke J, Aravin AA, Stark A, Dus M, Kellis M, Sachidanandam R, et al. Discrete small RNA-generating loci as master regulators of transposon activity in *Drosophila*. Cell 2007;128(6):1089–103. https://doi.org/10.1016/j.cell.2007.01.043.

[36] Chuma S, Nakano T. piRNA and spermatogenesis in mice. Philos Trans R Soc Lond Ser B Biol Sci 2013; 368(1609):20110338. https://doi.org/10.1098/rstb.2011.0338.

[37] Iwasaki YW, Siomi MC, Siomi H. PIWI-interacting RNA: its biogenesis and functions. Annu Rev Biochem 2015;84:405–33. https://doi.org/10.1146/annurev-biochem-060614-034258.

[38] Inoue K, Ichiyanagi K, Fukuda K, Glinka M, Sasaki H. Switching of dominant retrotransposon silencing strategies from post-transcriptional to transcriptional mechanisms during male germ-cell development in mice. PLoS Genet 2017;13(7):e1006926. https://doi.org/10.1371/journal.pgen.1006926.

[39] Kaur D, Agrahari M, Singh SS, Mandal PK, Bhattacharya A, Bhattacharya S. Transcriptomic analysis of *Entamoeba histolytica* reveals domain-specific sense strand expression of LINE-encoded ORFs with massive antisense expression of RT domain. Plasmid 2021;114:102560. https://doi.org/10.1016/j.plasmid.2021.102560.

[40] Yadav VP, Mandal PK, Bhattacharya A, Bhattacharya S. Recombinant SINEs are formed at high frequency during induced retrotransposition in vivo. Nat Commun 2012;3:854. https://doi.org/10.1038/ncomms1855.

[41] Djebali S, Davis CA, Merkel A, Dobin A, Lassmann T, Mortazavi A, et al. Landscape of transcription in human cells. Nature 2012;489(7414):101–8. https://doi.org/10.1038/nature11233.

[42] Statello L, Guo C-J, Chen L-L, Huarte M. Gene regulation by long non-coding RNAs and its biological functions. Nat Rev Mol Cell Biol 2021;22(2):96–118. https://doi.org/10.1038/s41580-020-00315-9.

[43] Geisler S, Coller J. RNA in unexpected places: long non-coding RNA functions in diverse cellular contexts. Nat Rev Mol Cell Biol 2013;14(11):699–712. https://doi.org/10.1038/nrm3679.

[44] Rinn JL, Chang HY. Genome regulation by long noncoding RNAs. Annu Rev Biochem 2012;81:145–66. https://doi.org/10.1146/annurev-biochem-051410-092902.

[45] Schmitz SU, Grote P, Herrmann BG. Mechanisms of long noncoding RNA function in development and disease. Cell Mol Life Sci: CM 2016;73(13):2491–509. https://doi.org/10.1007/s00018-016-2174-5.

[46] Lakhotia SC. Long non-coding RNAs coordinate cellular responses to stress. Wiley Interdisc Rev RNA 2012;3(6):779–96. https://doi.org/10.1002/wrna.1135.

[47] Shrimal S, Bhattacharya S, Bhattacharya A. Serum-dependent selective expression of EhTMKB1-9, a member of *Entamoeba histolytica* B1 family of transmembrane kinases. PLoS Pathog 2010;6(6):e1000929. https://doi.org/10.1371/journal.ppat.1000929.

[48] Saha A, Bhattacharya S, Bhattacharya A. Serum stress responsive gene EhslncRNA of *Entamoeba histolytica* is a novel long noncoding RNA. Sci Rep 2016;6:27476. https://doi.org/10.1038/srep27476.

[49] Audas TE, Lee S. Stressing out over long noncoding RNA. Biochim Biophys Acta 2016;1859(1):184–91. https://doi.org/10.1016/j.bbagrm.2015.06.010.

[50] Kino T, Hurt DE, Ichijo T, Nader N, Chrousos GP. Noncoding RNA gas5 is a growth arrest- and starvation-associated repressor of the glucocorticoid receptor. Sci Signal 2010;3(107):ra8. https://doi.org/10.1126/scisignal.2000568.

[51] Bazzini AA, Johnstone TG, Christiano R, Mackowiak SD, Obermayer B, Fleming ES, et al. Identification of small ORFs in vertebrates using ribosome footprinting and evolutionary conservation. EMBO J 2014;33(9): 981–93. https://doi.org/10.1002/embj.201488411.

[52] Guttman M, Russell P, Ingolia NT, Weissman JS, Lander ES. Ribosome profiling provides evidence that large noncoding RNAs do not encode proteins. Cell 2013;154(1):240–51. https://doi.org/10.1016/j.cell.2013.06.009.

[53] van Heesch S, van Iterson M, Jacobi J, Boymans S, Essers PB, de Bruijn E, et al. Extensive localization of long noncoding RNAs to the cytosol and mono- and polyribosomal complexes. Genome Biol 2014;15(1):R6. https://doi.org/10.1186/gb-2014-15-1-r6.
[54] Carrieri C, Cimatti L, Biagioli M, Beugnet A, Zucchelli S, Fedele S, et al. Long non-coding antisense RNA controls Uchl1 translation through an embedded SINEB2 repeat. Nature 2012;491(7424):454–7. https://doi.org/10.1038/nature11508.
[55] Speek M. Antisense promoter of human L1 retrotransposon drives transcription of adjacent cellular genes. Mol Cell Biol 2001;21(6):1973–85. https://doi.org/10.1128/MCB.21.6.1973-1985.2001.
[56] Denli AM, Narvaiza I, Kerman BE, Pena M, Benner C, Marchetto MCN, et al. Primate-specific ORF0 contributes to retrotransposon-mediated diversity. Cell 2015;163(3):583–93. https://doi.org/10.1016/j.cell.2015.09.025.
[57] Criscione SW, Theodosakis N, Micevic G, Cornish TC, Burns KH, Neretti N, et al. Genome-wide characterization of human L1 antisense promoter-driven transcripts. BMC Genom 2016;17:463. https://doi.org/10.1186/s12864-016-2800-5.
[58] Kuwabara T, Hsieh J, Muotri A, Yeo G, Warashina M, Lie DC, et al. Wnt-mediated activation of NeuroD1 and retro-elements during adult neurogenesis. Nat Neurosci 2009;12(9):1097–105. https://doi.org/10.1038/nn.2360.
[59] Wesel J, Shuman J, Bastuzel I, Dickerson J, Ingram-Smith C. Encystation of *Entamoeba histolytica* in axenic culture. Microorganisms 2021;9(4). https://doi.org/10.3390/microorganisms9040873.
[60] Ojha S, Singh N, Bhattacharya A, Bhattacharya S. The ribosomal RNA transcription unit of Entamoeba invadens: accumulation of unprocessed pre-rRNA and a long non coding RNA during encystation. Mol Biochem Parasitol 2013;192(1–2):30–8. https://doi.org/10.1016/j.molbiopara.2013.10.002.
[61] Mayer C, Grummt I. Ribosome biogenesis and cell growth: mTOR coordinates transcription by all three classes of nuclear RNA polymerases. Oncogene 2006;25(48):6384–91. https://doi.org/10.1038/sj.onc.1209883.
[62] Moss T. At the crossroads of growth control; making ribosomal RNA. Curr Opin Genet Dev 2004;14(2):210–7. https://doi.org/10.1016/j.gde.2004.02.005.
[63] Vazquezdelara-Cisneros LG, Arroyo-Begovich A. Induction of encystation of Entamoeba invadens by removal of glucose from the culture medium. J Parasitol 1984;70(5):629. https://doi.org/10.2307/3281741.
[64] Vydzhak O, Luke B, Schindler N. Non-coding RNAs at the eukaryotic rDNA locus: RNA-DNA hybrids and beyond. J Mol Biol 2020;432(15):4287–304. https://doi.org/10.1016/j.jmb.2020.05.011.
[65] Pirogov SA, Gvozdev VA, Klenov MS. Long noncoding RNAs and stress response in the nucleolus. Cells 2019;8(7). https://doi.org/10.3390/cells8070668.
[66] Grummt I. The nucleolus—guardian of cellular homeostasis and genome integrity. Chromosoma 2013;122(6):487–97. https://doi.org/10.1007/s00412-013-0430-0.
[67] McStay B, Grummt I. The epigenetics of rRNA genes: from molecular to chromosome biology. Annu Rev Cell Dev Biol 2008;24:131–57. https://doi.org/10.1146/annurev.cellbio.24.110707.175259.
[68] Santoro R, Grummt I. Epigenetic mechanism of rRNA gene silencing: temporal order of NoRC-mediated histone modification, chromatin remodeling, and DNA methylation. Mol Cell Biol 2005;25(7):2539–46. https://doi.org/10.1128/MCB.25.7.2539-2546.2005.
[69] Ojha S, Ahamad J, Bhattacharya A, Bhattacharya S. Ribosomal RNA and protein transcripts persist in the cysts of Entamoeba invadens. Mol Biochem Parasitol 2014;195(1):6–9. https://doi.org/10.1016/j.molbiopara.2014.05.003.
[70] Morgan RS, Uzman BG. Nature of the packing of ribosomes within chromatoid bodies. Science 1966;152(3719):214–6. https://doi.org/10.1126/science.152.3719.214.
[71] Lake JA, Slayter HS. Three dimensional structure of the chromatoid body of Entamoeba invadens. Nature 1970;227(5262):1032–7. https://doi.org/10.1038/2271032a0.

[72] Gupta AK, Panigrahi SK, Bhattacharya A, Bhattacharya S. Self-circularizing 5′-ETS RNAs accumulate along with unprocessed pre ribosomal RNAs in growth-stressed *Entamoeba histolytica*. Sci Rep 2012;2:303. https://doi.org/10.1038/srep00303.

[73] Ahamad J, Ojha S, Srivastava A, Bhattacharya A, Bhattacharya S. Post-transcriptional regulation of ribosomal protein genes during serum starvation in *Entamoeba histolytica*. Mol Biochem Parasitol 2015;201(2): 146−52. https://doi.org/10.1016/j.molbiopara.2015.07.006.

[74] Moshitch-Moshkovitch S, Petter R, Levitan A, Stolarsky T, Mirelman D. Regulation of expression of ribosomal protein L-21 genes of *Entamoeba histolytica* and E. dispar is at the post-transcriptional level. Mol Microbiol 1998;27(4):677−85. https://doi.org/10.1046/j.1365-2958.1998.00686.x.

[75] López-Camarillo C, Luna-Arias JP, Marchat LA, Orozco E. EhPgp5 mRNA stability is a regulatory event in the *Entamoeba histolytica* multidrug resistance phenotype. J Biol Chem 2003;278(13):11273−80. https://doi.org/10.1074/jbc.M211757200.

[76] Sachs A. The role of poly(A) in the translation and stability of mRNA. Curr Opin Cell Biol 1990;2(6): 1092−8. https://doi.org/10.1016/0955-0674(90)90161-7.

[77] Singh SS, Naiyer S, Bharadwaj R, Kumar A, Singh YP, Ray AK, et al. Stress-induced nuclear depletion of *Entamoeba histolytica* 3′−5′ exoribonuclease EhRrp6 and its role in growth and erythrophagocytosis. J Biol Chem 2018;293(42):16242−60. https://doi.org/10.1074/jbc.RA118.004632.

[78] Makino DL, Baumgärtner M, Conti E. Crystal structure of an RNA-bound 11-subunit eukaryotic exosome complex. Nature 2013;495(7439):70−5. https://doi.org/10.1038/nature11870.

[79] Mitchell P, Petfalski E, Shevchenko A, Mann M, Tollervey D. The exosome: a conserved eukaryotic RNA processing complex containing multiple $3' \rightarrow 5'$ exoribonucleases. Cell 1997;91(4):457−66. https://doi.org/10.1016/S0092-8674(00)80432-8.

[80] Parker R, Song H. The enzymes and control of eukaryotic mRNA turnover. Nat Struct Mol Biol 2004;11(2): 121−7. https://doi.org/10.1038/nsmb724.

[81] van Hoof A, Frischmeyer PA, Dietz HC, Parker R. Exosome-mediated recognition and degradation of mRNAs lacking a termination codon. Science 2002;295(5563):2262−4. https://doi.org/10.1126/science.1067272.

[82] Kadaba S, Krueger A, Trice T, Krecic AM, Hinnebusch AG, Anderson J. Nuclear surveillance and degradation of hypomodified initiator tRNAMet in *S. cerevisiae*. Gene Dev 2004;18(11):1227−40. https://doi.org/10.1101/gad.1183804.

[83] Isken O, Maquat LE. Quality control of eukaryotic mRNA: safeguarding cells from abnormal mRNA function. Gene Dev 2007;21(15):1833−56. https://doi.org/10.1101/gad.1566807.

[84] Maquat LE, Carmichael GG. Quality control of mRNA function. Cell 2001;104(2):173−6. https://doi.org/10.1016/S0092-8674(01)00202-1.

[85] Allmang C, Kufel J, Chanfreau G, Mitchell P, Petfalski E, Tollervey D. Functions of the exosome in rRNA, snoRNA and snRNA synthesis. EMBO J 1999;18(19):5399−410. https://doi.org/10.1093/emboj/18.19.5399.

[86] Lardenois A, Liu Y, Walther T, Chalmel F, Evrard B, Granovskaia M, et al. Execution of the meiotic noncoding RNA expression program and the onset of gametogenesis in yeast require the conserved exosome subunit Rrp6. Proc Natl Acad Sci USA 2011;108(3):1058−63. https://doi.org/10.1073/pnas.1016459108.

CHAPTER

Post-transcriptional regulation of gene expression in human malaria parasite *Plasmodium falciparum*

17

Karina Simantov and Manish Goyal

Department of Microbiology & Molecular Genetics, The Kuvin Center for the Study of Infectious and Tropical Diseases, IMRIC, The Hebrew University-Hadassah Medical School, Jerusalem, Israel

Introduction

Malaria is a mosquito-borne infectious disease that represents one of the world's most devastating human parasitic infections. Malaria accounts for ~1.5–2.7 million deaths annually and therefore imposes a huge socioeconomic burden [1,2]. Approximately 40% of the world's population is at risk due to malaria. Malaria is highly endemic throughout the tropical and subtropical areas, mostly occurring in the Sub-Saharan African region [3,4]. The disease is caused by the unicellular eukaryotic protozoan parasite *Plasmodium* belonging to the phylum "Apicomplexa" (family Plasmodiidae, suborder Haemosporia, order Eucoccida) [5]. As per the reports, ~156 species of *Plasmodium* have been identified that infect various species of vertebrates and invertebrates. In humans, five species namely, *P. falciparum*, *P. vivax*, *P. ovale*, *P. malariae*, and *P. knowlesi* cause infection [6]. Out of them, *P. falciparum* is the deadliest, accounting alone for about 90% of deaths from malaria worldwide, although *P. vivax* is responsible for the largest number of malaria infections [7].

Plasmodium has a complex life cycle where it undergoes several morphological and developmental stages in both the mosquito vector and the human host (Fig. 17.1) [8–11]. In humans, the malarial infection begins with infectious sporozoites injected during a blood meal by an infected female *Anopheles* mosquito [9]. The sporozoites then traverse through the circulatory system and reach the liver to infect liver cells (i.e., hepatocytes), where they undergo replication (exo-erythrocytic cycle) and produce numerous merozoites [12]. Following this initial silent asexual multiplication, the merozoites are released into the bloodstream to invade the erythrocytes (erythrocytic cycle) [9]. In the erythrocytes, the parasite undergoes several rounds of multiplication known as intraerythrocytic cycles, where they grow and develop into morphologically different intraerythrocytic blood stages (Fig. 17.1) [11].

Each intraerythrocytic cycle takes 48 h to complete and releases a large number of merozoites. These merozoites reinitiate another round of the blood-stage replicative cycle, thus ensuing parasite population explosions [11]. This results in severe anemia resulting in pathological manifestations of

Post-Transcriptional Gene Regulation in Human Disease, Volume 32. https://doi.org/10.1016/B978-0-323-91305-8.00006-5

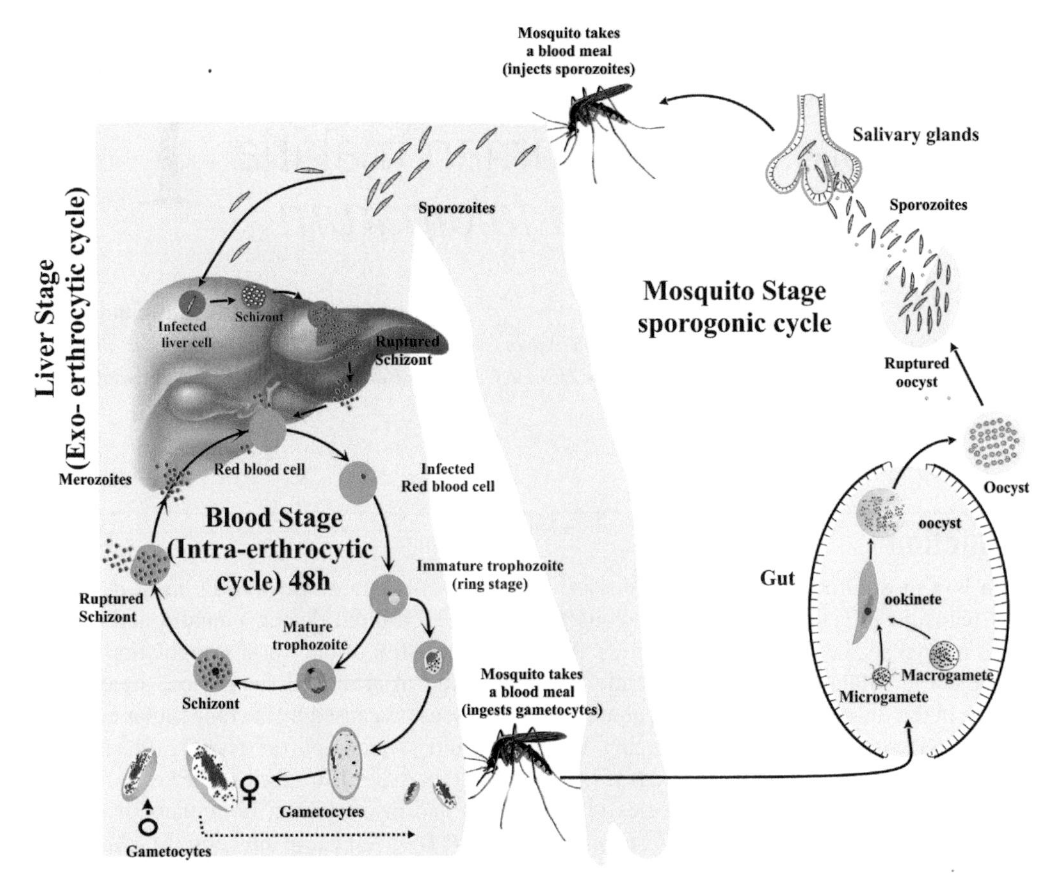

FIGURE 17.1 The life cycle of the malaria parasite *Plasmodium falciparum*.

Plasmodium has a complex life cycle where it undergoes several morphological and developmental stages in both the mosquito vector and the human host. The parasite life cycle can be broadly divided into three distinct stages. First, the pre-erythrocytic stages include the sporozoites (introduce by the mosquito into the skin), and the liver-stage parasites, which differentiate inside liver cells into merozoites. Second, the erythrocytic (red blood cell) stages comprise merozoites, ring, trophozoite, and schizont blood stages. Finally, the mosquito's stages in the lumen of the mosquito midgut and sexual stages (gametes and zygotes), and ookinetes.

malaria i.e., intermittent fever outburst or paroxysm fever, shaking chills, vomiting, and myalgia [13–15]. During the high blood-stage infection, some of the parasites undergo a developmental switch and commit into sexually transmissive stages, i.e., male and female gametocytes (Fig. 17.1) [8,16]. The male and female gametocytes are ingested by a susceptible female anopheline mosquito during a blood meal. Inside the mosquito's midgut, gametocytes produce male and female gametes and undergo a series of distinct morphological developmental phases known as the sporogonic cycle, which includes zygote, motile ookinete, oocyst, and sporozoites [8,10]. Inoculation of the sporozoites into a new human host by an infected female mosquito perpetuates again the *Plasmodium* life cycle [9].

During the mosquito stages, the parasite is exposed to different cellular milieus, such as the mosquito midgut lumen, gut epithelial cells, basal lamina, body cavity, and salivary glands (Fig. 17.1). Overall, the *Plasmodium* life cycle highlights its diversity and exceptional adaptability to invade, survive and propagate under diverse host intracellular and extracellular niches.

Surprisingly, malaria parasites achieve this intricacy (i.e., rapid developmental switching, change in cellular and molecular makeup, diverse morphological stages, primed survival, and escape from the host's defense systems) with a very small genome that lacks well-defined transcription factors, and encodes only ~5700 genes [17–20]. It has been observed that the fine-tuning of gene expression primed the parasite's survival, differentiation, pathogenicity, and immune evasion in its human and mosquito hosts [20,21]. Stage-dependent transcriptional and translational profiles of *P. falciparum* have suggested that dynamic gene expression, where genes are expressed in a "just-in-time" manner is very crucial during the parasite life cycle [22–26]. Gene expression is regulated by several layers of mechanisms that can operate either at the mRNA level, by affecting mRNA transcription (transcriptional), mRNA processing, nuclear export, and degradation (post-transcriptional), or by translation and protein stability (post-translational) (Fig. 17.2). At each of these levels, there are several factors and mechanisms that play important regulatory roles in apicomplexan parasites [27–31]. Recent advents in high-throughput technologies such as next-generation sequencing and transcriptomics advance our understanding of gene regulatory networks in the malaria parasite *Plasmodium*. These studies underline the significance of specific proteins, regulatory elements, and mechanisms to carry out post-transcriptional regulation of gene expression during the parasite life cycle.

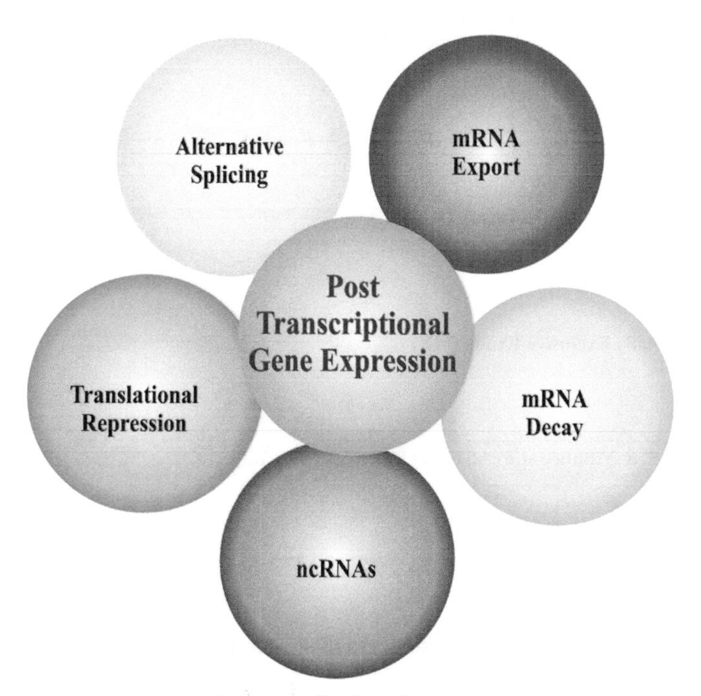

FIGURE 17.2 Key steps in the post-transcriptional regulation of gene expression in malaria parasite *P. falciparum*.

Here, we will discuss and review the insights obtained from these efforts including the involvement of various mechanisms that operate post-transcriptionally, such as mRNA processing (alternative splicing), mRNA export and degradation, noncoding RNAs, and translational repression (Fig. 17.2), and how they might contribute to further characterization of this complex gene regulatory network in the human malaria parasite *P. falciparum*.

Alternative splicing and gene expression in *Plasmodium*

The malaria parasite *Plasmodium* harbors a very complex life that alternates between human and mosquito hosts [10,18,32–34]. This requires a large number of genes that can be encoded only by a big genome. However, the parasite genome is very small (~24 MB) and encodes a relatively small number of genes (~5700), as compared to yeast which has a much simpler life cycle [19]. One way by which eukaryotic organisms expand their proteome is by the alternative splicing (AS) of pre-mRNA. AS of a pre-mRNA transcript is a post-transcriptional process that produces functionally divergent gene products (proteins) from a single coding sequence [35,36]. Protein coding genes in eukaryotes contain both protein-coding sequences (Exons) and noncoding sequences (Introns). The Introns are removed while exons are reassembled in a pre-mRNA to form a mature mRNA by a multiprotein complex known as the spliceosome, and the process is called splicing [35,36]. However, alternative variability while processing and arranging the pre-mRNA (selecting and removing the introns and exons) can produce multiple mRNA isoforms that in turn, if translated, could lead to multiple proteins (Fig. 17.3) [35,36]. It has been observed that AS seems to modulate gene function by the addition or

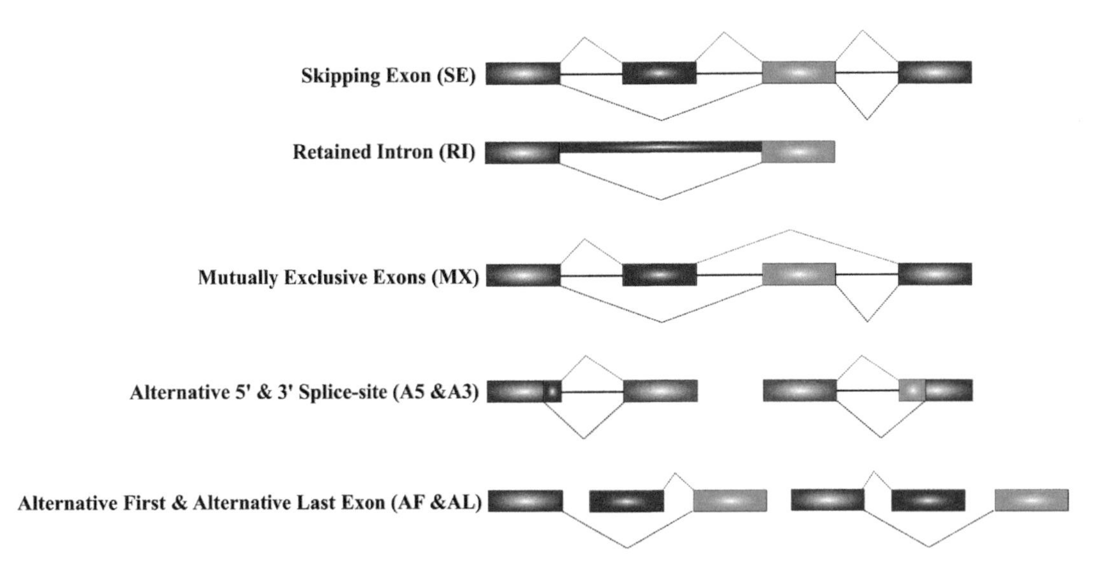

FIGURE 17.3 Schematic diagram showing different types of alternative splicing events reported in *Plasmodium falciparum* as depicted.

(1) SE (skipping exon), (2) RI (retained intron), (3) MX (mutually exclusive exons), (4) A3 (alternative 3′-splice-site), (5) A5 (alternative 5′-splice-site), (6) AF (alternative first exon), and (7) AL (alternative last exon).

removal of protein domains, thereby affecting protein activity [37]. Thus, AS can allow diversity and complexity from a relatively small genome under different stress or developmental conditions, as well as in different cells or host niches.

AS not only produces functionally divergent gene products but also affects the localization and translation efficiency of the mature mRNA [38]. Changes in the splicing pattern can result in mRNA with alternative stop codons or premature termination codons, which in turn leads to its degradation by nonsense-mediated decay (Fig. 17.4) [39]. Besides this, AS also generates different protein isoforms with altered cellular localization, protein-protein interactions, and biological functions with respect to their primary transcript [27,35,36,39,40]. In this view, AS could explain the discrepancy between the genome size and the expanded repertoire of proteins in an organism (Fig. 17.3). The genome sequence in *P. falciparum* suggested that >50% of genes contain predicted introns [18,19]. Transcriptomics and RNA-seq studies during the intraerythrocytic blood stages have revealed the presence of mounting AS events that are comparable with those of other eukaryotes [41–47]. RNA seq analysis in *P. falciparum* has identified that almost ~5% (254 genes) of total genes undergo AS, uncovering the presence of several new splice junctions (5′ GU-AG 3′ junctions), and splicing of antisense transcripts [45]. Recent transcriptomic analysis using single-molecule long-read sequencing during the blood stage further adds new AS events in *P. falciparum*, with a total of 393 AS events, 1555 alternative polyadenylations events, and 1721 fusion transcripts [48]. In *Plasmodium*, several different types of AS events have been reported, including A3 (alternative 3′ splice-site), A5 (alternative 5′ splice-site), skipping exon, mutually exclusive exons, alternative first exon, alternative last exon, and retained intron (RI)

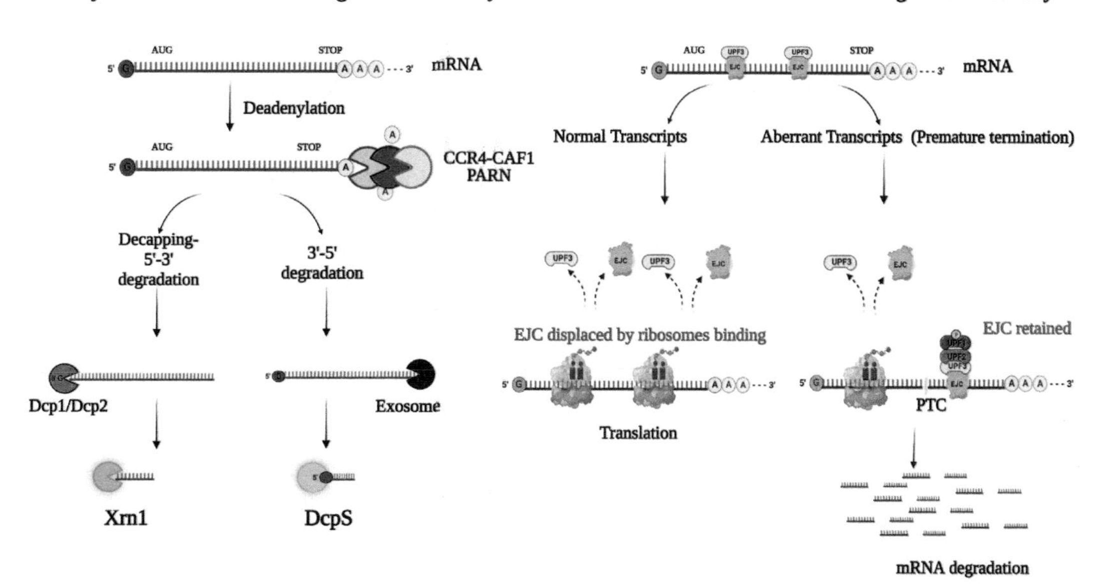

FIGURE 17.4 Messenger RNA (mRNA) degradation pathways and their components in eukaryotes.

The major pathway for cytoplasmic mRNA degradation is triggered by either deadenylation or nonsense-mediated mRNA decay.

(Fig. 17.3) [45,47,48]. However, the majority of the dominant AS events fall into three main categories: A3 (24%), A5 (33%), and RI (31%) (Fig. 17.3) [48]. Single-target high-throughput transcription analyses during the intraerythrocytic cycle reveal stage-specific AS of farnesyl diphosphate synthase/geranyl diphosphate synthase (FPPS/GGPPS) in *P. falciparum* [41]. Comparative RNA-Seq analysis in *P. berghei*, using sexual (either female and/or male) and asexual intraerythrocytic stage parasites also showed that ~90% of the changes in AS events were unique to either male or female gametocytes. Furthermore, perturbation of this sex-specific AS by targeted disruption of an AS factor (*Pb*SR-MG) decreases the parasite's ability to differentiate into male gametes and oocysts [46]. Overall, these studies suggest a widespread presence of AS in *P. falciparum*, where it might provide an additional layer of regulation (post-transcriptional), by modulating the gene functions and by generating a stage/developmental specific diversity of mRNA and protein isoforms.

mRNA degradation and gene expression

mRNA stability and its degradation both play a very important role in the post-transcriptional modulation of gene expression. In any cell, the half-life of an mRNA depends on the cellular need of the protein it encodes. To regulate the amount of proteins translated from each mRNA molecule, cells either degrade or modify mRNA molecules to increase their stability [49,50]. Eukaryotic cells use several different mechanisms for mRNA degradation under specific conditions at certain times, in response to environmental, metabolic, and developmental signals (Fig. 17.4) [49,50].

Deadenylation mediated mRNA degradation is one of the main mRNA decay pathways that is involved in deadenylation of target mRNA, followed by decapping, 5′-3′ and or 3′-5′ decay (Fig. 17.4) [50–53]. In most organisms, the shortening of the poly(A)-tail, i.e., deadenylation, is the rate-limiting step that opens the body of the mRNA and signals it for degradation [51,53]. The poly(A)-tail is one of the intrinsic features of a transcript that ensures its stability in the cytoplasm by facilitating the binding of poly(A) binding proteins [54]. While there are many deadenylases encoded in an organism, the predominant enzymes involved in cytoplasmic deadenylation are components of the CCR4-NOT and Pan2-Pan3 complexes such as CAF1, CCR4, and PARN [50,51,53]. Following deadenylation, decapping enzymes (Dcp1 and Dcp2) removed the mRNA cap, and then the body of the mRNA is degraded by 5′-3′ exonuclease activity (5′-3′ exonuclease, Xrn1) (Fig. 17.4). Alternatively, the mRNA can also be degraded by 3′-5′-end by a large multimeric complex of exonucleases called "the exosome," and subsequent decapping by scavenger decapping protein DcpS (Fig. 17.4) [50,51,53]. Genome-wide studies in *P. falciparum* reported that the rate of mRNA decay increases dramatically during the asexual intraerythrocytic developmental cycle [55]. Furthermore, the rate of mRNA decay (mRNA half-life) was significantly extended in the late schizont stage as compared to ring-stage parasites. The change in mRNA decay rate in respect to the stage of the intraerythrocytic development is a very unique phenomenon to *Plasmodium*, thus indicating the imperative role of post-transcriptional regulation through mRNA decay during asexual blood-stage development [55]. Previous studies in *Plasmodium* have identified several components of deadenylation mediated mRNA decay, including the mRNA decapping enzymes (DCP1 and DCP2), and members of the CCR4-NOT complex [55–57]. Targeted disruption of the deadenylase enzyme Caf1 in *P. falciparum* causes a proliferation defect due to altered gene expression (most notably invasion and egress genes) during asexual development. Apart from this, both CCR4-1 and CAF1 deadenylase play crucial roles in RNA metabolism during the sexual stage development and transmission to mosquitoes [58]. Besides this, various components of the 3–5′ degradation pathway (i.e., Exosome), such as the core proteins and exoribonucleases (exoRNAses) DIS3/RRP44 and RRP6, are also encoded by the *Plasmodium* genome [55]. However, a detailed characterization of these components and the exact composition of the complex remains elusive in *P. falciparum*.

The second mechanism of mRNA degradation is known as nonsense-mediated decay (NMD), which prevents the formation of toxic or undesired proteins by triggering the degradation of aberrant transcripts (Fig. 17.4) [49,50,52,59–61]. The NMD pathway targets mRNAs transcripts harboring "premature" termination codons (PTC). PTCs are generated by either point-nonsense mutations, or by aberrant splicing events, and as a consequence of targeted gene regulation through AS [59,60,62]. The conserved core of the NMD pathway consists of three *trans*-acting up-frameshift (UPF) proteins, namely UPF1, UPF2, and UPF3. UPF2 and UPF3 are integral components of the exon–exon junction complex (EJC) (Fig. 17.4). EJC is primarily bound to mRNA after splicing along with other proteins, and needs to be displaced for efficient translation [59,60,62]. In the case of an aberrant transcript with PTC, the EJCs are retained on the mRNA and results in termination of translation. This leads to the assembly of a complex, composed of UPF1, SMG1, and the release factors (eRF1 and eRF3) on the mRNA. Following this, the interaction of UPF1 with UPF2 and UPF3 triggers the phosphorylation of UPF1 and the subsequent degradation or downregulation of the mRNA by NMD (Fig. 17.4) [49,52,59]. In *Plasmodium*, several core components of the NMD pathway, including UPF1 and UPF2 are well conserved. CRISPR-Cas9 based targeted disruption and co-immunoprecipitation of PfUPF1 (PF3D7_1005500) and PfUPF2 (PF3D7_0925800) suggest the role of additional novel accessory proteins to elicit the mRNA decay [61]. However, the NMD process in *P. falciparum* is quite different from the classical NMD, and requires UPF2 but not UPF1 for mRNA degradation [61]. Nevertheless, it is still unknown how the NMD core complex couples prematurely stop codons containing transcripts to the mRNA decay machinery in *P. falciparum*. Further work is needed to identify specific NMD substrates and NMD accessory proteins to understand how NMD has evolved in the malaria parasite.

mRNA export and gene regulation

Eukaryotic gene expression can be broadly divided into three steps, namely (1) synthesis of mRNA (transcription of pre-mRNA, and pre-mRNA processing in the nucleus), (2) mRNA transport (packaging into messenger ribonucleoprotein (mRNP) complexes, and translocation through nuclear pore complexes [NPCs]), and (3) mRNA translation (directional release of mRNA into the cytoplasm and ribosomes binding). In this view, the movement of mRNA transcripts through NPCs by specific mRNA-binding export factors is a very important step in the post-transcriptional regulation of gene expression [63–68]. Despite the complexity in the export process, the vast majority of cargos are trafficked between the nucleus and the cytoplasm through GTP-dependent karyopherin-based receptors, mediated by GTPase Ran protein [69]. However, in the case of bulk mRNAs export, the cells use a nonkaryopherin-based principle export receptor, namely the heterodimeric Mex67-Mtr2 (budding yeast) or NXF1-NXT1 (TAP-p15 in Human) (Fig. 17.5) [65,66,68,70]. Structural examination of Mex67 and NXF1 suggested that they both have a modular architecture that comprises N-terminus RNA recognition motif (RRM domain), leucine-rich region (LRR), NTF2-like domain (NTF2L), and ubiquitin-associated domains [65,70]. The RRM, LRR, and NTF2L domains all bind

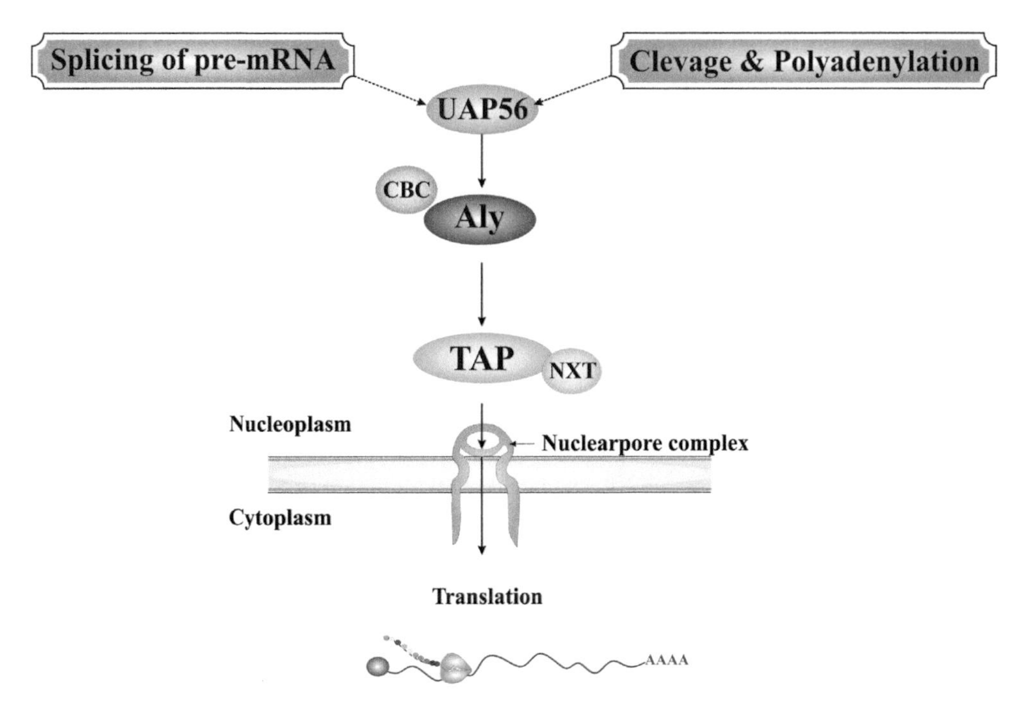

FIGURE 17.5 Illustration of bulk mRNA export pathway in eukaryotes.

Following maturation, the pre-mRNA interacts with different proteins, mRNA export factors [Mex67-Mtr2 (NXF1-NXT1 or TAP-p15 in human)] and protein complexes (TREX) that export mRNAs as mRNPs from the nucleus to the site of translation, that is, cytoplasm.

RNA without any sequence specificity and are involved in capturing the target RNA. Similarly, Mtr2/NXT1 also exhibits an NTF2-like fold that supports their interaction/binding with the NTF2L domain of Mex67/NXF1. Mex67-Mtr2 or NXF1-NXT1 heterodimer does not rely on the RanGTP gradient and is recruited to the mRNA via the TREX component ALY (Fig. 17.5) [70]. TREX (transcription/export complex) is a multisubunit protein complex that contains THO (suppressor of the transcriptional defect of Hpr1 by overexpression) subcomplex, Tex1, UAP56 (Sub2/UAP56), and Yra1p/ALY/REF adaptor proteins (Fig. 17.5) [71,72]. Furthermore, serine arginine-rich proteins (SR proteins) like Gbp2, Hrb1, and Npl3 (recruited to mRNA by the TREX complex), are also involved in mRNA export [73,74]. Overall, mRNA export involves three different steps: (a) recruitment of export factors to assemble the mRNA into RNA-protein (mRNP) complexes marked with Mex67-Mtr2/NXF1-NXT1, mediated by the TREX complex in the nucleus, (b) targeting of mRNA and translocation through the NPC, and (c) release of mRNAs into the cytoplasm, by disassembly of the mRNP export complex by the DEAD-box protein Dbp5/DDX19 ATPase-NAB2 at the cytoplasmic face of the NPC (Fig. 17.5) [63–68].

In silico studies in *P. falciparum* identify several components of the mRNA export pathway [67]. In *P. falciparum*, orthologs of the TREX components, namely UAP56 (PfU52), Tho2, and REF are identified and well characterized. PFUAP56 (PfU52) has been shown as a bonafide DEAD-box helicase that is involved in RNA binding and splicing reactions. Similarly, THO complex subunit 4

(THO4, homolog of yeast YRA1), NAB2, and Dbp5/DDX19 have also been characterized in detail in *Plasmodium* [63,67,74]. Furthermore, recent studies in malaria parasites showed the involvement of nuclear poly(A) binding protein 2 (NAB2), THO complex subunit 4 (THO4), nucleolar protein 3 (NPL3), and G-strand binding protein 2 (GBP2) in nuclear mRNA export. However, the key components, such as orthologs of Mex67-Mtr2/NXF1-NXT1 and TREX complex are completely missing in the *Plasmodium* genome [63,67,74]. Overall, these studies suggest that during evolution, the mRNA export pathway in the malaria parasites underwent a divergence and is somewhat different from that of other higher eukaryotes.

Noncoding RNAs roles in *Plasmodium*

Studies during the recent past have shown an emerging role of noncoding RNAs (ncRNAs) in gene expression, which are orchestrated at multiple levels of transcriptional and post-transcriptional regulation [75–78]. ncRNAs are transcribed similar to other RNAs and undergo processing (alternatively spliced and/or processed into smaller products), but do not translate into a functional protein (Fig. 17.6). ncRNAs can interact with either DNA/RNA or proteins and therefore modulate gene function by altering the chromatin structure (epigenetic memory, chromatin architecture), transcription, and mRNA processing (mRNA splicing, editing, and stability) as well as translation/turnover [75–78]. Based on the length, ncRNAs can be divided into two major categories, small ncRNAs (17 nt-200 nt; S ncRNAs) and long ncRNAs (>200 nt; lncRNAs) [75–78]. Small ncRNAs consist mainly of regulatory RNAs that act as negative regulators of the target gene expression, such as Small nucleolar RNAs (snoRNAs), microRNAs (miRNAs), small interfering RNAs (siRNAs), and PIWI-interacting RNAs (piRNAs) [75]. In comparison, lncRNAs are spliced products of RNA polymerase II (RNAPII) and can act similarly to mRNAs, that is, 5′-capped, 3′-polyadenylated, and translated into stable functional micropeptides [77,78]. Based on the relative position to adjacent encoding genes, lncRNAs can be grouped into different categories, namely sense, antisense, bidirectional, intronic, and intergenic. While some lncRNAs form circular structures (circRNAs) and act as transcript effectors, others can form RNA hybrids (complementary to the endogenous transcript) and can interfere with translation initiation (both inhibitory and stimulatory effects) [77,78]. Altogether lncRNAs have more diversity in terms of their structure, expression (environmental/developmental specific pattern), and regulatory functions [76,78].

Almost 80% of the genome in the human malaria parasite *P. falciparum* is pervasively transcribed where the majority of the mRNAs are translated into proteins, however, many ncRNA transcripts still exist [19,20,23]. There is growing evidence that ncRNA plays an important role in diverse cellular processes in *P. falciparum*, including parasite development and gene expression regulation [79,80]. With the advent of innovative genome-wide sequencing technologies, such as serial analysis of gene expression, high-resolution RNA-Seq, and single-molecule long-read sequencing, a large repertoire of antisense transcripts, lncRNAs, and other ncRNAs have been identified in *P. falciparum* [79,81–83]. The molecular functions of most ncRNAs in *P. falciparum* still remain elusive. Studies during the recent past have revealed the functions of only a small number of lncRNAs that are likely to share with model organism species. In the *Plasmodium* genome, components of the canonical dicer-dependent pathway that generates small ncRNAs (such as siRNAs and miRNAs) are missing [19,84]. To complement the lack of RNA-interference mechanism (RNAi), *Plasmodium* uses natural antisense transcripts and lncRNAs for RNA-induced post-transcriptional gene silencing [82,83].

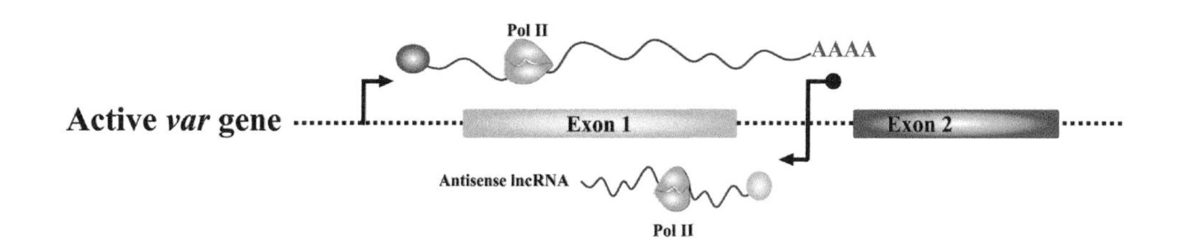

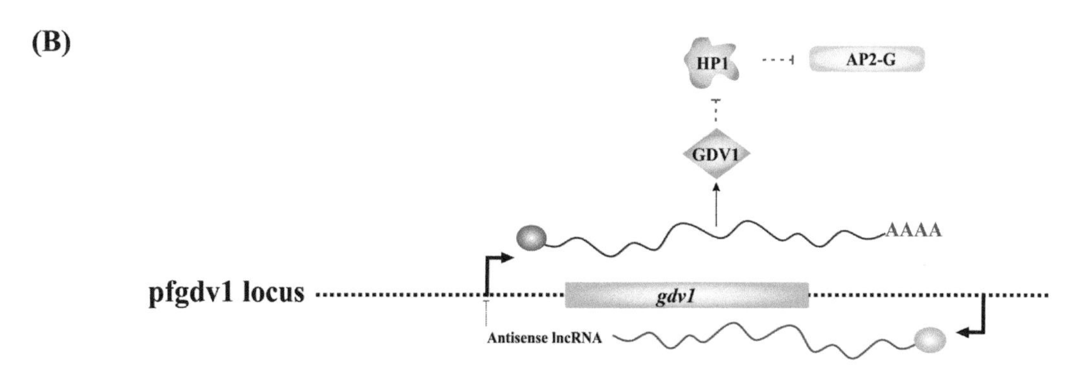

FIGURE 17.6 Long noncoding RNAs (lncRNAs)-mediated regulation of gene expression in *Plasmodium*.

In the human malaria parasite *Plasmodium falciparum* (*P. falciparum*) noncoding RNAs play an important role in gene expression regulation. (A) Antisense lncRNAs transcribed from *var* intron were shown as key regulators of the mutually exclusive expression of virulence genes. (B) Similarly, a antisense lncRNA transcribed from the 3′-region of Pfgdv1 regulates the Pfgdv1 gene expression. PfGDV1 plays an important role in sexual differentiation by displacing PfHP1 mediated AP2-G silencing, thus inducing sexual commitment in *P. falciparum*.

In *P. falciparum*, lncRNAs were shown to play a prominent role in mutually exclusive expression (MEE) of *var* genes, where one *var* gene is activated at a time while the rest are silenced [85,86]. Studies of MEE in *P. falciparum* have revealed that *var* genes specific lncRNAs are transcribed by Pol II from a bidirectional lncRNA promoter placed within the highly conserved *var* gene intron (Fig. 17.6A) [79,87–89]. This promoter can transcribe in both directions and generates either the antisense lncRNA that is complementary to the first exon or sense lncRNA that expands into exon 2 (Fig. 17.6A) [79,87–89].

lncRNA was also shown to be implicated in sexual differentiation in *P. falciparum*. Differentiation into gametocytes in *Plasmodium* is regulated by transcriptional regulator PfAP2-G that triggers sexual commitment [90]. The *P. falciparum* gametocyte development 1 protein (PfGDV1) regulates AP2-G-mediated sexual commitment by displacing heterochromatin protein 1 (PfHP1) locus, thus permitting sexual differentiation [90–92]. It has been reported that PfGDV1 expression is regulated by its antisense lncRNA transcript that negatively regulates its expression (Fig. 17.6B) [90]. The other class of lncRNAs identified in *Plasmodium* is telomeric- and subtelomeric-associated lncRNAs, which are located at the nuclear periphery and implicated in telomere maintenance [81]. It is hypothesized that they are predicted to form a stable and repetitive hairpin structure that allows them to bind histones and engage in assembly and/or disassembly of subtelomeric heterochromatin (Fig. 17.6B) [93]. The other kind of lncRNA found in *Plasmodium* is circRNAs. The majority of circRNAs in *P. falciparum* are short (<200bp), contain human miRNA binding sites, and were predicted to function in host-parasite interaction [81]. Altogether, these studies suggest the emerging role of the ncRNAs in the regulation of gene expression probably via mRNA transcription, stability, or translation in *P. falciparum*. However, their experimental functional validation is still missing requiring additional studies targeting the functional and mechanical roles of lncRNAs in development-specific gene expression regulation in *P. falciparum*.

Translational repression

Translational inhibition or repression of mRNAs is the ultimate layer in post-transcriptional gene regulation [94,95]. It enables the cell to decrease its protein repertoire irrespective of the overall decrease in mRNA level. In this way, the cell can facilitate the temporary storage of transcripts for subsequent use by the translation machinery after the release of repression [94,95]. In eukaryotes, translational repression is mediated by RNA binding translational repressor proteins that condense the transcript as messenger ribonucleoprotein particles (mRNPs) into cytosolic granules. These cytosolic granules either store the mRNA for later protein synthesis as stress granules or final decay/ degradation as P-bodies [94,95]. The 3′-UTR of mRNA plays a pivotal role in translational repression where the regulatory proteins bind and trigger repression or decay. Deadenylation of the poly-A tail of the mRNA is one of the processes that can cause translational repression where CCR4/ Caf1 proteins play a decisive role [96]. Intriguingly, the *P. falciparum* genome encodes a large number of putative RNA-binding proteins (RBPs) that are involved in translational repression. Past studies on *Plasmodium* have suggested the role of several RBPs in transcriptional repression particularly during the transmission stages [97–111]. The probable reason behind this phenomenon is to be proactive and get ready for the change of definite host [57]. In this way, parasites can store the essential protein coding transcripts in mRNPs (needed after transmission) in prior life cycle stages for their translation after transmission. Translational repression was first depicted for surface antigens P25 and P28 in rodent malaria parasite *P. berghei* and subsequently in *P. falciparum*, as well. It has been observed that transcripts of P25 and P28 were abundant during the gametocyte stages; however, proteins were synthesized after the transmission to mosquito stages (zygotes and ookinetes) [20,98,102,105]. The translational repression of P25 and P28 expressions were regulated by the Puf RNA-binding protein family [101,105]. PuF proteins bind to conserved Puf-binding elements (PBEs) located at 3′- UTR of target mRNA [112]. In *P. falciparum*, three members of the Puf family have been reported; namely Puf1-3 [57,101,102,113,114]. Targeted gene deletion studies in *P. falciparum* have further shown that Puf2 is the main regulator of translation, which bind with both 3′ and 5′-UTRs of P25 and P28 transcripts, and the deletion of Puf2 leads to the loss of translational regulation of these transcripts and affects mosquito-to-human transmission [102,108,114].

In addition, other proteins such as development of zygote inhibited (DOZI), and CITH (homolog of worm CAR-I and fly Trailer Hitch) are also involved in the repression of *p25* and *p28* transcripts [103,104]. Loss of either DOZI or CITH caused a significant reduction of several mRNA transcripts suggesting a direct role in transcript stability/repression [103,104]. Identification and characterization of DOZI- and CITH connected proteins revealed many factors; orthologue of poly-A binding protein, *Drosophila* BRUNO/CUGBP Elav-like family member, HoMu (homolog of Musashi), and DNA/RNA-binding proteins ALBA1 and ALBA4 [98,100,103,104,106–108,111,115,116]. ALBA1 and ALBA4 were shown to bind a large number of transcripts in *Plasmodium*, and ALBA1 is supposed to be involved in maintaining mRNA homeostasis, translation, and stage-specific development [107,110]. Besides RBPs, other proteins involved in translational regulation in *Plasmodium* comprise 7-Helix-1 (constitutes stress granules in female gametocytes) and SAP1 (also known as SLARP; sporozoite and liver asparagine-rich protein 1) [97–99,109]. Loss of 7-Helix-1 and SAP1 in *P. falciparum* produced the same phenotype as DOZI/CITH/Puf2 and (i.e., reduced transcripts and incompetent mosquito stages [97,99,109]. Overall, these studies suggest that translation repression is a major phenomenon in *Plasmodium*, where it plays an important role in response to external stimuli and influences developmental and stage-specific gene expression.

Concluding remarks

Gene expression in *Plasmodium* is a coordinated process that comprises different linked steps such as transcription, mRNA processing, and the translation of mRNAs. The journey from an RNA molecule to protein is accompanied by several regulatory layers that act at transcriptional and post-transcriptional levels. This includes but is not limited to AS of nascent pre-mRNA transcripts, mRNA export, and degradation, ncRNA-mediated regulation, and translational repression of mRNA transcripts. Our current understanding of the components and mechanism of these processes and their experimental functional validation is insufficient in *Plasmodium*. Therefore, substantial efforts are needed to find out the important elements and regulators of these processes and their biological function in stage-specific development and host-parasite interactions. Similarly, identified ncRNA candidates also require greater mechanistic and functional characterization to discern their function in parasite biology. A growing body of evidence has suggested the role of a large repertoire of RBPs in these coordinated processes in *Plasmodium*. The specific proteins described earlier represent only a very small fraction of the RBPs that mediate critical steps in post-transcriptional regulation of gene expression in *Plasmodium*. Future efforts are needed using high throughput sequencing approaches, comprehensive demarcation of the parasite's translatome, and CRISPR-Cas9-based genome-editing tool to understand the parasite's complex gene regulatory mechanisms.

References

[1] Greenwood BM, Fidock DA, Kyle DE, Kappe SH, Alonso PL, Collins FH, et al. Malaria: progress, perils, and prospects for eradication. J Clin Invest April 2008;118(4):1266–76. PubMed PMID: 18382739. Pubmed Central PMCID: 2276780. Epub 2008/04/03. eng.

[2] Keusch GT, Kilama WL, Moon S, Szlezak NA, Michaud CM. The global health system: linking knowledge with action–learning from malaria. PLoS Med January 19, 2010;7(1):e1000179. PubMed PMID: 20087415. Pubmed Central PMCID: 2799678.

[3] Nasir SMI, Amarasekara S, Wickremasinghe R, Fernando D, Udagama P. Prevention of re-establishment of malaria: historical perspective and future prospects. Malar J December 7, 2020;19(1):452. PubMed PMID: 33287809. Pubmed Central PMCID: 7720033.

[4] Organization WH. Malaria: WHO. 2021 [15 April 2021]. Available from: https://www.who.int/news-room/fact-sheets/detail/malaria.

[5] Sato S. Plasmodium-a brief introduction to the parasites causing human malaria and their basic biology. J Physiol Anthropol January 7, 2021;40(1):1. PubMed PMID: 33413683. Pubmed Central PMCID: 7792015.

[6] Sharp PM, Plenderleith LJ, Hahn BH. Ape origins of human malaria. Annu Rev Microbiol September 8, 2020;74:39–63. PubMed PMID: 32905751. Pubmed Central PMCID: 7643433.

[7] Zhong D, Lo E, Wang X, Yewhalaw D, Zhou G, Atieli HE, et al. Multiplicity and molecular epidemiology of Plasmodium vivax and Plasmodium falciparum infections in East Africa. Malar J May 2, 2018;17(1): 185. PubMed PMID: 29720181. Pubmed Central PMCID: 5932820.

[8] Alano P. *Plasmodium falciparum* gametocytes: still many secrets of a hidden life. Mol Microbiol October 2007;66(2):291–302. PubMed PMID: 17784927.

[9] Frischknecht F, Matuschewski K. Plasmodium sporozoite biology. Cold Spring Harb Perspect Med May 1, 2017;7(5). PubMed PMID: 28108531. Pubmed Central PMCID: 5411682.

[10] Meibalan E, Marti M. Biology of malaria transmission. Cold Spring Harb Perspect Med March 1, 2017; 7(3). PubMed PMID: 27836912. Pubmed Central PMCID: 5334247.

[11] Venugopal K, Hentzschel F, Valkiunas G, Marti M. Plasmodium asexual growth and sexual development in the haematopoietic niche of the host. Nat Rev Microbiol March 2020;18(3):177–89. PubMed PMID: 31919479. Pubmed Central PMCID: 7223625.

[12] Kori LD, Valecha N, Anvikar AR. Insights into the early liver stage biology of Plasmodium. J Vector Borne Dis 2018 ;55(1):9–13. PubMed PMID: 29916442.

[13] Chen Q, Schlichtherle M, Wahlgren M. Molecular aspects of severe malaria. Clin Microbiol Rev July 2000; 13(3):439–50. PubMed PMID: 10885986. Pubmed Central PMCID: 88942. Epub 2000/07/25. eng.

[14] Dey S, Guha M, Alam A, Goyal M, Bindu S, Pal C, et al. Malarial infection develops mitochondrial pathology and mitochondrial oxidative stress to promote hepatocyte apoptosis. Free Radic Biol Med January 15, 2009;46(2):271–81. PubMed PMID: 19015023. Epub 2008/11/19. eng.

[15] Guha M, Kumar S, Choubey V, Maity P, Bandyopadhyay U. Apoptosis in liver during malaria: role of oxidative stress and implication of mitochondrial pathway. Faseb J June 2006;20(8):1224–6. PubMed PMID: 16603602. Epub 2006/04/11. eng.

[16] Talman AM, Domarle O, McKenzie FE, Ariey F, Robert V. Gametocytogenesis: the puberty of *Plasmodium falciparum*. Malar J July 14, 2004;3:24. PubMed PMID: 15253774.

[17] Bischoff E, Vaquero C. In silico and biological survey of transcription-associated proteins implicated in the transcriptional machinery during the erythrocytic development of *Plasmodium falciparum*. BMC Genom 11:34. PubMed PMID: 20078850. Pubmed Central PMCID: 2821373. Epub 2010/01/19. eng.

[18] Carlton JM, Escalante AA, Neafsey D, Volkman SK. Comparative evolutionary genomics of human malaria parasites. Trends Parasitol December 2008;24(12):545–50. PubMed PMID: 18938107. Epub 2008/10/22. eng.

[19] Gardner MJ, Hall N, Fung E, White O, Berriman M, Hyman RW, et al. Genome sequence of the human malaria parasite *Plasmodium falciparum*. Nature October 3, 2002;419(6906):498–511. PubMed PMID: 12368864. Epub 2002/10/09. eng.

[20] Hall N, Karras M, Raine JD, Carlton JM, Kooij TW, Berriman M, et al. A comprehensive survey of the Plasmodium life cycle by genomic, transcriptomic, and proteomic analyses. Science January 7, 2005; 307(5706):82–6. PubMed PMID: 15637271. Epub 2005/01/08. eng.

[21] Llinas M, Bozdech Z, Wong ED, Adai AT, DeRisi JL. Comparative whole genome transcriptome analysis of three *Plasmodium falciparum* strains. Nucleic Acids Res 2006;34(4):1166–73. PubMed PMID: 16493140. Pubmed Central PMCID: 1380255. Epub 2006/02/24. eng.

[22] Ben Mamoun C, Gluzman IY, Hott C, MacMillan SK, Amarakone AS, Anderson DL, et al. Co-ordinated programme of gene expression during asexual intraerythrocytic development of the human malaria parasite *Plasmodium falciparum* revealed by microarray analysis. Mol Microbiol January 2001;39(1):26–36. PubMed PMID: 11123685. Epub 2000/12/21. eng.

[23] Coleman BI, Duraisingh MT. Transcriptional control and gene silencing in *Plasmodium falciparum*. Cell Microbiol October 2008;10(10):1935–46. PubMed PMID: 18637022. Epub 2008/07/19. eng.

[24] Cui L, Miao J. Chromatin-mediated epigenetic regulation in the malaria parasite *Plasmodium falciparum*. Eukaryot Cell. Aug;9(8):1138–1149. PubMed PMID: 20453074. Pubmed Central PMCID: 2918932. Epub 2010/05/11. eng.

[25] Foth BJ, Zhang N, Mok S, Preiser PR, Bozdech Z. Quantitative protein expression profiling reveals extensive post-transcriptional regulation and post-translational modifications in schizont-stage malaria parasites. Genome Biol 2008;9(12):R177. PubMed PMID: 19091060. Pubmed Central PMCID: 2646281. Epub 2008/12/19. eng.

[26] Hakimi MA, Deitsch KW. Epigenetics in Apicomplexa: control of gene expression during cell cycle progression, differentiation and antigenic variation. Curr Opin Microbiol August 2007;10(4):357–62. PubMed PMID: 17719264. Epub 2007/08/28. eng.

[27] de Klerk E, t Hoen PA. Alternative mRNA transcription, processing, and translation: insights from RNA sequencing. Trends Genet March 2015;31(3):128–39. PubMed PMID: 25648499.

[28] Gil N, Ulitsky I. Regulation of gene expression by cis-acting long non-coding RNAs. Nat Rev Genet February 2020;21(2):102–17. PubMed PMID: 31729473.

[29] Jaenisch R, Bird A. Epigenetic regulation of gene expression: how the genome integrates intrinsic and environmental signals. Nat Genet March 2003;33(Suppl. l):245–54. PubMed PMID: 12610534.

[30] Morita T, Mochizuki Y, Aiba H. Translational repression is sufficient for gene silencing by bacterial small noncoding RNAs in the absence of mRNA destruction. Proc Natl Acad Sci USA March 28, 2006;103(13): 4858–63. PubMed PMID: 16549791. Pubmed Central PMCID: 1458760.

[31] Vembar SS, Droll D, Scherf A. Translational regulation in blood stages of the malaria parasite Plasmodium spp.: systems-wide studies pave the way. Wiley Interdiscip Rev RNA November 2016;7(6):772–92. PubMed PMID: 27230797. Pubmed Central PMCID: 5111744. Epub 2016/05/28. eng.

[32] Le Roch KG, Zhou Y, Blair PL, Grainger M, Moch JK, Haynes JD, et al. Discovery of gene function by expression profiling of the malaria parasite life cycle. Science September 12, 2003;301(5639):1503–8. PubMed PMID: 12893887. Epub 2003/08/02. eng.

[33] Mu J, Seydel KB, Bates A, Su XZ. Recent progress in functional genomic research in *Plasmodium falciparum*. Curr Genom June 2010;11(4):279–86. PubMed PMID: 21119892. Pubmed Central PMCID: 2930667. Epub 2010/12/02. eng.

[34] Zekar L, Sharman T. Plasmodium falciparum malaria. Treasure Island (FL): StatPearls; 2021.

[35] Ule J, Blencowe BJ. Alternative splicing regulatory networks: functions, mechanisms, and evolution. Mol Cell October 17, 2019;76(2):329–45. PubMed PMID: 31626751.

[36] Wang Y, Liu J, Huang BO, Xu YM, Li J, Huang LF, et al. Mechanism of alternative splicing and its regulation. Biomed Rep March 2015;3(2):152–8. PubMed PMID: 25798239. Pubmed Central PMCID: 4360811.

[37] Lareau LF, Brooks AN, Soergel DA, Meng Q, Brenner SE. The coupling of alternative splicing and nonsense-mediated mRNA decay. Adv Exp Med Biol 2007;623:190–211. PubMed PMID: 18380348.
[38] Zhiguo E, Wang L, Zhou J. Splicing and alternative splicing in rice and humans. BMB Rep September 2013;46(9):439–47. PubMed PMID: 24064058. Pubmed Central PMCID: 4133877.
[39] Kalyna M, Simpson CG, Syed NH, Lewandowska D, Marquez Y, Kusenda B, et al. Alternative splicing and nonsense-mediated decay modulate expression of important regulatory genes in Arabidopsis. Nucleic Acids Res March 2012;40(6):2454–69. PubMed PMID: 22127866. Pubmed Central PMCID: 3315328.
[40] Yang X, Coulombe-Huntington J, Kang S, Sheynkman GM, Hao T, Richardson A, et al. Widespread expansion of protein interaction capabilities by alternative splicing. Cell February 11, 2016;164(4):805–17. PubMed PMID: 26871637. Pubmed Central PMCID: 4882190.
[41] Gabriel HB, de Azevedo MF, Palmisano G, Wunderlich G, Kimura EA, Katzin AM, et al. Single-target high-throughput transcription analyses reveal high levels of alternative splicing present in the FPPS/GGPPS from *Plasmodium falciparum*. Sci Rep December 21, 2015;5:18429. PubMed PMID: 26688062. Pubmed Central PMCID: 4685265.
[42] Hull R, Dlamini Z. The role played by alternative splicing in antigenic variability in human endo-parasites. Parasit Vectors January 28, 2014;7:53. PubMed PMID: 24472559. Pubmed Central PMCID: 4015677.
[43] Iriko H, Jin L, Kaneko O, Takeo S, Han ET, Tachibana M, et al. A small-scale systematic analysis of alternative splicing in *Plasmodium falciparum*. Parasitol Int June 2009;58(2):196–9. PubMed PMID: 19268714.
[44] Lunghi M, Spano F, Magini A, Emiliani C, Carruthers VB, Di Cristina M. Alternative splicing mechanisms orchestrating post-transcriptional gene expression: intron retention and the intron-rich genome of apicomplexan parasites. Curr Genet February 2016;62(1):31–8. PubMed PMID: 26194054.
[45] Sorber K, Dimon MT, DeRisi JL. RNA-Seq analysis of splicing in *Plasmodium falciparum* uncovers new splice junctions, alternative splicing and splicing of antisense transcripts. Nucleic Acids Res May 2011;39(9):3820–35. PubMed PMID: 21245033. Pubmed Central PMCID: 3089446.
[46] Yeoh LM, Goodman CD, Mollard V, McHugh E, Lee VV, Sturm A, et al. Alternative splicing is required for stage differentiation in malaria parasites. Genome Biol August 1, 2019;20(1):151. PubMed PMID: 31370870. Pubmed Central PMCID: 6669979.
[47] Yeoh LM, Lee VV, McFadden GI, Ralph SA. Alternative splicing in apicomplexan parasites. mBio February 19, 2019;10(1). PubMed PMID: 30782661. Pubmed Central PMCID: 6381282.
[48] Yang M, Shang X, Zhou Y, Wang C, Wei G, Tang J, et al. Full-length transcriptome analysis of *Plasmodium falciparum* by single-molecule long-read sequencing. Front Cell Infect Microbiol 2021;11:631545. PubMed PMID: 33708645. Pubmed Central PMCID: 7942025.
[49] Siwaszek A, Ukleja M, Dziembowski A. Proteins involved in the degradation of cytoplasmic mRNA in the major eukaryotic model systems. RNA Biol 2014;11(9):1122–36. PubMed PMID: 25483043. Pubmed Central PMCID: 4615280. Epub 2014/12/09. eng.
[50] Tourriere H, Chebli K, Tazi J. mRNA degradation machines in eukaryotic cells. Biochimie August 2002;84(8):821–37. PubMed PMID: 12457569. Epub 2002/11/30. eng.
[51] Funakoshi Y, Doi Y, Hosoda N, Uchida N, Osawa M, Shimada I, et al. Mechanism of mRNA deadenylation: evidence for a molecular interplay between translation termination factor eRF3 and mRNA deadenylases. Genes Dev December 1, 2007;21(23):3135–48. PubMed PMID: 18056425. Pubmed Central PMCID: 2081979. Epub 2007/12/07. eng.
[52] Labno A, Tomecki R, Dziembowski A. Cytoplasmic RNA decay pathways-enzymes and mechanisms. Biochim Biophys Acta December 2016;1863(12):3125–47. PubMed PMID: 27713097. Epub 2016/11/05. eng.
[53] Wahle E, Winkler GS. RNA decay machines: deadenylation by the Ccr4-not and Pan2-Pan3 complexes. Biochim Biophys Acta 2013 ;1829(6–7):561–70. PubMed PMID: 23337855. Epub 2013/01/23. eng.

[54] Bernstein P, Peltz SW, Ross J. The poly(A)-poly(A)-binding protein complex is a major determinant of mRNA stability in vitro. Mol Cell Biol February 1989;9(2):659–70. PubMed PMID: 2565532. Pubmed Central PMCID: 362643. Epub 1989/02/01. eng.
[55] Shock JL, Fischer KF, DeRisi JL. Whole-genome analysis of mRNA decay in *Plasmodium falciparum* reveals a global lengthening of mRNA half-life during the intra-erythrocytic development cycle. Genome Biol 2007;8(7):R134. PubMed PMID: 17612404. Pubmed Central PMCID: 2323219. Epub 2007/07/07. eng.
[56] Hughes KR, Philip N, Starnes GL, Taylor S, Waters AP. From cradle to grave: RNA biology in malaria parasites. Wiley Interdiscip Rev RNA 2010 ;1(2):287–303. PubMed PMID: 21935891.
[57] Reddy BP, Shrestha S, Hart KJ, Liang X, Kemirembe K, Cui L, et al. A bioinformatic survey of RNA-binding proteins in Plasmodium. BMC Genom November 2, 2015;16:890. PubMed PMID: 26525978. Pubmed Central PMCID: 4630921. Epub 2015/11/04. eng.
[58] Hart KJ, Oberstaller J, Walker MP, Minns AM, Kennedy MF, Padykula I, et al. Plasmodium male gametocyte development and transmission are critically regulated by the two putative deadenylases of the CAF1/CCR4/NOT complex. PLoS Pathog January 2019;15(1):e1007164. PubMed PMID: 30703164. Pubmed Central PMCID: 6355032.
[59] Baker KE, Parker R. Nonsense-mediated mRNA decay: terminating erroneous gene expression. Curr Opin Cell Biol June 2004;16(3):293–9. PubMed PMID: 15145354. Epub 2004/05/18. eng.
[60] Hug N, Longman D, Caceres JF. Mechanism and regulation of the nonsense-mediated decay pathway. Nucleic Acids Res February 29, 2016;44(4):1483–95. PubMed PMID: 26773057. Pubmed Central PMCID: 4770240. Epub 2016/01/17. eng.
[61] McHugh E, Bulloch MS, Batinovic S, Sarna DK, Ralph SA. A divergent nonsense-mediated decay machinery in *Plasmodium falciparum* is inefficient and non-essential. bioRxiv 2021. 2021.04.14.439394.
[62] Atkinson GC, Baldauf SL, Hauryliuk V. Evolution of nonstop, no-go and nonsense-mediated mRNA decay and their termination factor-derived components. BMC Evol Biol October 23, 2008;8:290. PubMed PMID: 18947425. Pubmed Central PMCID: 2613156. Epub 2008/10/25. eng.
[63] Avila AR, Cabezas-Cruz A, Gissot M. mRNA export in the apicomplexan parasite Toxoplasma gondii: emerging divergent components of a crucial pathway. Parasit Vectors January 25, 2018;11(1):62. PubMed PMID: 29370868. Pubmed Central PMCID: 5785795.
[64] Carmody SR, Wente SR. mRNA nuclear export at a glance. J Cell Sci June 15, 2009;122(Pt 12):1933–7. PubMed PMID: 19494120. Pubmed Central PMCID: 2723150.
[65] Segref A, Sharma K, Doye V, Hellwig A, Huber J, Luhrmann R, et al. Mex67p, a novel factor for nuclear mRNA export, binds to both poly(A)+ RNA and nuclear pores. EMBO J June 2, 1997;16(11):3256–71. PubMed PMID: 9214641. Pubmed Central PMCID: 1169942.
[66] Serpeloni M, Vidal NM, Goldenberg S, Avila AR, Hoffmann FG. Comparative genomics of proteins involved in RNA nucleocytoplasmic export. BMC Evol Biol January 11, 2011;11:7. PubMed PMID: 21223572. Pubmed Central PMCID: 3032688.
[67] Tuteja R, Mehta J. A genomic glance at the components of the mRNA export machinery in *Plasmodium falciparum*. Commun Integr Biol July 2010;3(4):318–26. PubMed PMID: 20798816. Pubmed Central PMCID: 2928308.
[68] Xie Y, Ren Y. Mechanisms of nuclear mRNA export: a structural perspective. Traffic November 2019; 20(11):829–40. PubMed PMID: 31513326. Pubmed Central PMCID: 7074880.
[69] Madrid AS, Weis K. Nuclear transport is becoming crystal clear. Chromosoma April 2006;115(2):98–109. PubMed PMID: 16421734.
[70] Herold A, Suyama M, Rodrigues JP, Braun IC, Kutay U, Carmo-Fonseca M, et al. TAP (NXF1) belongs to a multigene family of putative RNA export factors with a conserved modular architecture. Mol Cell Biol December 2000;20(23):8996–9008. PubMed PMID: 11073998. Pubmed Central PMCID: 86553.

[71] Heath CG, Viphakone N, Wilson SA. The role of TREX in gene expression and disease. Biochem J October 1, 2016;473(19):2911–35. PubMed PMID: 27679854. Pubmed Central PMCID: 5095910.

[72] Pan H, Liu S, Tang D. The THO/TREX complex functions in disease resistance in Arabidopsis. Plant Signal Behav March 2012;7(3):422–4. PubMed PMID: 22499202. Pubmed Central PMCID: 3443925.

[73] Hurt E, Luo MJ, Rother S, Reed R, Strasser K. Cotranscriptional recruitment of the serine-arginine-rich (SR)-like proteins Gbp2 and Hrb1 to nascent mRNA via the TREX complex. Proc Natl Acad Sci USA February 17, 2004;101(7):1858–62. PubMed PMID: 14769921. Pubmed Central PMCID: 357017.

[74] Niikura M, Fukutomi T, Mitobe J, Kobayashi F. Roles and cellular localization of GBP2 and NAB2 during the blood stage of malaria parasites. Front Cell Infect Microbiol 2021;11:737457. PubMed PMID: 34604117. Pubmed Central PMCID: 8479154.

[75] He L, Hannon GJ. MicroRNAs: small RNAs with a big role in gene regulation. Nat Rev Genet July 2004; 5(7):522–31. PubMed PMID: 15211354.

[76] Hung T, Chang HY. Long noncoding RNA in genome regulation: prospects and mechanisms. RNA Biol 2010 ;7(5):582–5. PubMed PMID: 20930520. Pubmed Central PMCID: 3073254.

[77] Rinn JL, Chang HY. Genome regulation by long noncoding RNAs. Annu Rev Biochem 2012;81:145–66. PubMed PMID: 22663078. Pubmed Central PMCID: 3858397.

[78] Wilusz JE, Sunwoo H, Spector DL. Long noncoding RNAs: functional surprises from the RNA world. Genes Dev July 1, 2009;23(13):1494–504. PubMed PMID: 19571179. Pubmed Central PMCID: 3152381.

[79] Gunasekera AM, Patankar S, Schug J, Eisen G, Kissinger J, Roos D, et al. Widespread distribution of antisense transcripts in the *Plasmodium falciparum* genome. Mol Biochem Parasitol July 2004;136(1): 35–42. PubMed PMID: 15138065.

[80] Patankar S, Munasinghe A, Shoaibi A, Cummings LM, Wirth DF. Serial analysis of gene expression in *Plasmodium falciparum* reveals the global expression profile of erythrocytic stages and the presence of anti-sense transcripts in the malarial parasite. Mol Biol Cell October 2001;12(10):3114–25. PubMed PMID: 11598196. Pubmed Central PMCID: 60160.

[81] Broadbent KM, Broadbent JC, Ribacke U, Wirth D, Rinn JL, Sabeti PC. Strand-specific RNA sequencing in *Plasmodium falciparum* malaria identifies developmentally regulated long non-coding RNA and circular RNA. BMC Genom June 13, 2015;16:454. PubMed PMID: 26070627. Pubmed Central PMCID: 4465157.

[82] Li Y, Baptista RP, Kissinger JC. Noncoding RNAs in apicomplexan parasites: an update. Trends Parasitol October 2020;36(10):835–49. PubMed PMID: 32828659.

[83] Vembar SS, Scherf A, Siegel TN. Noncoding RNAs as emerging regulators of *Plasmodium falciparum* virulence gene expression. Curr Opin Microbiol August 2014;20:153–61. PubMed PMID: 25022240. Pubmed Central PMCID: 4157322.

[84] Baum J, Papenfuss AT, Mair GR, Janse CJ, Vlachou D, Waters AP, et al. Molecular genetics and comparative genomics reveal RNAi is not functional in malaria parasites. Nucleic Acids Res June 2009; 37(11):3788–98. PubMed PMID: 19380379. Pubmed Central PMCID: 2699523.

[85] Deitsch KW, Dzikowski R. Variant gene expression and antigenic variation by malaria parasites. Annu Rev Microbiol September 8, 2017;71:625–41. PubMed PMID: 28697665.

[86] Duraisingh MT, Skillman KM. Epigenetic variation and regulation in malaria parasites. Annu Rev Microbiol September 8, 2018;72:355–75. PubMed PMID: 29927705.

[87] Amit-Avraham I, Pozner G, Eshar S, Fastman Y, Kolevzon N, Yavin E, et al. Antisense long noncoding RNAs regulate var gene activation in the malaria parasite *Plasmodium falciparum*. Proc Natl Acad Sci USA March 3, 2015;112(9):E982–91. PubMed PMID: 25691743. Pubmed Central PMCID: 4352787.

[88] Epp C, Li F, Howitt CA, Chookajorn T, Deitsch KW. Chromatin associated sense and antisense noncoding RNAs are transcribed from the *var* gene family of virulence genes of the malaria parasite *Plasmodium falciparum*. RNA January 2009;15(1):116–27. PubMed PMID: 19037012. Pubmed Central PMCID: 2612763.

[89] Jing Q, Cao L, Zhang L, Cheng X, Gilbert N, Dai X, et al. *Plasmodium falciparum var* gene is activated by its antisense long noncoding RNA. Front Microbiol 2018;9:3117. PubMed PMID: 30619191. Pubmed Central PMCID: 6305453.
[90] Filarsky M, Fraschka SA, Niederwieser I, Brancucci NMB, Carrington E, Carrio E, et al. GDV1 induces sexual commitment of malaria parasites by antagonizing HP1-dependent gene silencing. Science March 16, 2018;359(6381):1259–63. PubMed PMID: 29590075. Pubmed Central PMCID: 6219702.
[91] Eksi S, Morahan BJ, Haile Y, Furuya T, Jiang H, Ali O, et al. *Plasmodium falciparum* gametocyte development 1 (*Pfgdv1*) and gametocytogenesis early gene identification and commitment to sexual development. PLoS Pathog 2012;8(10):e1002964. PubMed PMID: 23093935. Pubmed Central PMCID: 3475683.
[92] Usui M, Prajapati SK, Ayanful-Torgby R, Acquah FK, Cudjoe E, Kakaney C, et al. Plasmodium falciparum sexual differentiation in malaria patients is associated with host factors and GDV1-dependent genes. Nat Commun May 13, 2019;10(1):2140. PubMed PMID: 31086187. Pubmed Central PMCID: 6514009.
[93] Sierra-Miranda M, Delgadillo DM, Mancio-Silva L, Vargas M, Villegas-Sepulveda N, Martinez-Calvillo S, et al. Two long non-coding RNAs generated from subtelomeric regions accumulate in a novel perinuclear compartment in *Plasmodium falciparum*. Mol Biochem Parasitol September 2012;185(1):36–47. PubMed PMID: 22721695. Pubmed Central PMCID: 7116675.
[94] Decker CJ, Parker R. P-bodies and stress granules: possible roles in the control of translation and mRNA degradation. Cold Spring Harb Perspect Biol September 1, 2012;4(9):a012286. PubMed PMID: 22763747. Pubmed Central PMCID: 3428773.
[95] Hu W, Coller J. What comes first: translational repression or mRNA degradation? The deepening mystery of microRNA function. Cell Res September 2012;22(9):1322–4. PubMed PMID: 22613951. Pubmed Central PMCID: 3434348.
[96] Cooke A, Prigge A, Wickens M. Translational repression by deadenylases. J Biol Chem September 10, 2010;285(37):28506–13. PubMed PMID: 20634287. Pubmed Central PMCID: 2937876.
[97] Aly AS, Lindner SE, MacKellar DC, Peng X, Kappe SH. SAP1 is a critical post-transcriptional regulator of infectivity in malaria parasite sporozoite stages. Mol Microbiol February 2011;79(4):929–39. PubMed PMID: 21299648.
[98] Bennink S, Pradel G. The molecular machinery of translational control in malaria parasites. Mol Microbiol December 2019;112(6):1658–73. PubMed PMID: 31531994.
[99] Bennink S, von Bohl A, Ngwa CJ, Henschel L, Kuehn A, Pilch N, et al. A seven-helix protein constitutes stress granules crucial for regulating translation during human-to-mosquito transmission of *Plasmodium falciparum*. PLoS Pathog August 2018;14(8):e1007249. PubMed PMID: 30133543. Pubmed Central PMCID: 6122839.
[100] Goyal M, Banerjee C, Nag S, Bandyopadhyay U. The Alba protein family: structure and function. Biochim Biophys Acta May 2016;1864(5):570–83. PubMed PMID: 26900088.
[101] Liang X, Hart KJ, Dong G, Siddiqui FA, Sebastian A, Li X, et al. Puf3 participates in ribosomal biogenesis in malaria parasites. J Cell Sci March 26, 2018;(6):131. PubMed PMID: 29487181. Pubmed Central PMCID: 5897713.
[102] Lindner SE, Mikolajczak SA, Vaughan AM, Moon W, Joyce BR, Sullivan Jr WJ, et al. Perturbations of Plasmodium Puf2 expression and RNA-seq of Puf2-deficient sporozoites reveal a critical role in maintaining RNA homeostasis and parasite transmissibility. Cell Microbiol July 2013;15(7):1266–83. PubMed PMID: 23356439. Pubmed Central PMCID: 3815636.
[103] Mair GR, Braks JA, Garver LS, Wiegant JC, Hall N, Dirks RW, et al. Regulation of sexual development of Plasmodium by translational repression. Science August 4, 2006;313(5787):667–9. PubMed PMID: 16888139. Pubmed Central PMCID: 1609190.

[104] Mair GR, Lasonder E, Garver LS, Franke-Fayard BM, Carret CK, Wiegant JC, et al. Universal features of post-transcriptional gene regulation are critical for Plasmodium zygote development. PLoS Pathog February 12, 2010;6(2):e1000767. PubMed PMID: 20169188. Pubmed Central PMCID: 2820534.

[105] Miao J, Fan Q, Parker D, Li X, Li J, Cui L. Puf mediates translation repression of transmission-blocking vaccine candidates in malaria parasites. PLoS Pathog 2013;9(4):e1003268. PubMed PMID: 23637595. Pubmed Central PMCID: 3630172.

[106] Minns AM, Hart KJ, Subramanian S, Hafenstein S, Lindner SE. Nuclear, cytosolic, and surface-localized poly(A)-Binding proteins of *Plasmodium yoelii*. mSphere 2018 ;3(1). PubMed PMID: 29359180. Pubmed Central PMCID: 5760745.

[107] Munoz EE, Hart KJ, Walker MP, Kennedy MF, Shipley MM, Lindner SE. ALBA4 modulates its stage-specific interactions and specific mRNA fates during *Plasmodium yoelii* growth and transmission. Mol Microbiol October 2017;106(2):266–84. PubMed PMID: 28787542. Pubmed Central PMCID: 5688949.

[108] Silva PA, Guerreiro A, Santos JM, Braks JA, Janse CJ, Mair GR. Translational control of UIS4 protein of the host-parasite interface is mediated by the RNA binding protein Puf2 in *Plasmodium berghei* sporozoites. PLoS One 2016;11(1):e0147940. PubMed PMID: 26808677. Pubmed Central PMCID: 4726560.

[109] Silvie O, Goetz K, Matuschewski K. A sporozoite asparagine-rich protein controls initiation of Plasmodium liver stage development. PLoS Pathog June 13, 2008;4(6):e1000086. PubMed PMID: 18551171. Pubmed Central PMCID: 2398788.

[110] Vembar SS, Macpherson CR, Sismeiro O, Coppee JY, Scherf A. The PfAlba1 RNA-binding protein is an important regulator of translational timing in *Plasmodium falciparum* blood stages. Genome Biol September 28, 2015;16:212. PubMed PMID: 26415947. Pubmed Central PMCID: 4587749.

[111] Wongsombat C, Aroonsri A, Kamchonwongpaisan S, Morgan HP, Walkinshaw MD, Yuthavong Y, et al. Molecular characterization of *Plasmodium falciparum* Bruno/CELF RNA binding proteins. Mol Biochem Parasitol November 2014;198(1):1–10. PubMed PMID: 25447287.

[112] Miller MA, Olivas WM. Roles of Puf proteins in mRNA degradation and translation. Wiley Interdiscip Rev RNA 2011 ;2(4):471–92. PubMed PMID: 21957038.

[113] Cui L, Fan Q, Li J. The malaria parasite *Plasmodium falciparum* encodes members of the Puf RNA-binding protein family with conserved RNA binding activity. Nucleic Acids Res November 1, 2002;30(21): 4607–17. PubMed PMID: 12409450. Pubmed Central PMCID: 135818.

[114] Miao J, Li J, Fan Q, Li X, Li X, Cui L. The Puf-family RNA-binding protein PfPuf2 regulates sexual development and sex differentiation in the malaria parasite *Plasmodium falciparum*. J Cell Sci April 1, 2010;123(Pt 7):1039–49. PubMed PMID: 20197405. Pubmed Central PMCID: 2844316.

[115] Chene A, Vembar SS, Riviere L, Lopez-Rubio JJ, Claes A, Siegel TN, et al. PfAlbas constitute a new eukaryotic DNA/RNA-binding protein family in malaria parasites. Nucleic Acids Res April 2012;40(7): 3066–77. PubMed PMID: 22167473. Pubmed Central PMCID: 3326326.

[116] Goyal M, Alam A, Iqbal MS, Dey S, Bindu S, Pal C, et al. Identification and molecular characterization of an Alba-family protein from human malaria parasite *Plasmodium falciparum*. Nucleic Acids Res February 2012;40(3):1174–90. PubMed PMID: 22006844. Pubmed Central PMCID: 3273813.

CHAPTER

18

Hepcidin-induced degradation of iron exporter ferroportin determines anemia of chronic diseases

Chinmay K. Mukhopadhyay, Pragya Mishra, Ayushi Aggarwal and Sameeksha Yadav

Special Centre for Molecular Medicine, Jawaharlal Nehru University, New Delhi, Delhi, India

Introduction

Iron is an essential nutritional element for almost all organisms due to its redox-active nature. However, the same redox capacity of this element may cause toxicity both at cellular and tissue levels leading to a multitude of diseases. While a deficiency causes anemia, an excess of iron leads to a variety of pathological conditions from infections to secondary aging. Thus, iron homeostasis is tightly regulated mostly at the post-transcriptional level to maintain the delicate balance between deficiency and excess.

As per the current report of the World Health Organization about 25% of the world population is affected by anemia [64]. Among them, about 50% are suffering due to iron deficiency. The second highest common anemia results from chronic diseases such as infection (viral, parasitic, or bacterial), cancer, chronic kidney disease, autoimmune diseases, rejection of the transplanted organ, heart failure, and obesity [63]. Recent estimates suggest that about 40% of the total patients suffering from anemia worldwide are considered as anemia of chronic diseases (ACD) [63]. The pathophysiology of ACD includes hypoferremia leading to iron-restricted erythropoiesis [38]. A combined effort from many research groups has revealed the post-translational degradation of iron exporter ferroportin by an innate immune peptide hormone namely hepcidin as a mechanism involved in ACD. The current chapter will focus on the mechanism of inflammation-induced ACD.

Requirement of iron

Hemoglobin (Hb) carries oxygen all through the body. This oxygen-carrying capacity depends on the iron-bound to the heme of Hb in erythrocytes. Humans normally synthesize about two million erythrocytes per second. Each of the mature erythrocytes contains about 280 million Hb molecules. Hb is a tetrameric protein consisting of four globin protein subunits. Every globin subunit contains one iron atom in heme. Thus, $2-3 \times 10^{15}$ iron atoms per second are fluxed to maintain erythropoiesis in an adult human [15]. Like Hb, myoglobin is an iron-binding protein and carries oxygen mainly to cardiac

Post-Transcriptional Gene Regulation in Human Disease, Volume 32. https://doi.org/10.1016/B978-0-323-91305-8.00016-8

and skeletal muscle cells. Thus, the activity of these muscle cells depends on iron availability. Iron is also needed for numerous biological functions like replication, electron transport in mitochondria, antioxidant defense, energy homeostasis, neurotransmitter synthesis, and myelin synthesis. Anemia leads to depleted availability of iron decreasing all these functions.

Recycling of iron

About 30 mg of iron is recycled in humans per day of which about 28 mg is handled by macrophages while 1–2 mg enter through duodenal enterocytes to compensate for the similar amount of iron losses [62]. The majority of recycled iron is required for Hb synthesis. These processes are dependent on the level of ferroportin. Thus, when ferroportin is degraded by hepcidin in enterocytes and macrophages, the major iron handling capacity of the body is affected resulting in anemia (that is refractory to iron supplementation). Macrophages are mainly responsible for maintaining adequate levels of plasma iron as only 1–2 mg of iron enters through intestinal absorption that is even less than 10% of the daily iron needs. Rest is covered by macrophages by the recycling process.

Macrophages phagocytose aged as well as damaged RBCs and degrade heme by heme oxygenase-1 to release iron [34]. This released iron is exported out via ferroportin that is regulated at multiple levels. Erythrophagocytosis and heme iron promote its transcription; high iron regulates translation, and its protein stability is regulated by hepcidin.

Iron homeostasis: cellular and systemic

Mammalian iron homeostasis is maintained both at the cellular level as well as a systemic level to coordinate iron uptake, utilization, and storage for preventing toxicity and to assure availability during requirement. Remarkably, these two levels of iron homeostasis use entirely different components and mechanisms except iron exporter ferroportin. In almost all cell types transferrin receptor 1 (TfR1) functions as the major iron uptake component to utilize holo-transferrin (holo-Tf) as an iron source. TfR1 binds holo-Tf and the complex is internalized by clathrin-dependent endocytosis [22]. Iron is freed from the complex in the acidic endosomal compartment. STEAP family of metalloreductases [42] keep the iron in a reduced form to facilitate iron transport into the cytosol by divalent metal transporter 1 (DMT1). In the cytosol, iron initially exists as a "labile iron pool" (LIP) and then gets utilized and released as per the cellular demand. The mechanism of this distribution remains uncertain. Unutilized iron from LIP is stored within the nanocavity of ferritin that consists of 24 subunits of heavy (Ft-H) and light (Ft-L) chains [4]. Both these subunits are ubiquitously expressed with varying ratios in different cell types. Excess iron is released out by unique exporter ferroportin (Fpn, SLC40A1) that is present in all cell types. The ferroportin level in macrophages and duodenal enterocytes is particularly important for maintaining plasma iron.

Cellular iron homeostasis is maintained by coordinated regulation of iron uptake (by TfR1 and DMT1), iron storage (by Ft-H and Ft-L), and iron release (by Fpn) components by a post-transcriptional mechanism mediated by iron regulatory protein 1 (IRP1) and iron regulatory protein 2 (IRP2). These two RNA-binding orthologous proteins interact with iron-responsive elements (IREs) present in the 5′- or 3′-untranslated regions of iron uptake, iron storage, and iron release components to regulate them according to cellular iron level. IRPs inhibit translation initiation by binding to the single IRE present in the 5′-untranslated regions (UTRs) of Ft-H, Ft-L, and Fpn when the iron is scarce but when the iron is excess IRPs are released from IREs to increase translation of ferritin subunits and Fpn. This regulation ensures excess iron is stored in ferritin and/or released by Fpn to avoid iron-induced damage of cells. During iron depletion, IRPs bind to multiple IREs present in the 3′-UTR of TfR1 and in the single IRE present in the 3′-UTR of DMT1 to stabilize respective mRNAs. As a result, expressions of TfR1 and DMT1 are increased to ensure higher iron uptake to maintain an appropriate level of iron. Cellular iron level alters IRP1 activity and expression of IRP2. In iron-replete condition IRP1 converts to cytosolic aconitase and IRP2 is degraded by proteasomal mechanism while during iron depletion the cytosolic aconitase converts to IRP1 form and IRP2 is stabilized to bind respective IREs to maintain cellular iron homeostasis [22].

The plasma iron level is maintained by systemic iron homeostasis and controlled by hepcidin. Hepcidin is a member of the defensin family of innate immune peptides and is regulated both by inflammation and infection. Higher iron level increases hepcidin that binds to ferroportin resulting in internalization, ubiquitination, and subsequent lysosomal degradation of the latter [36]. Thereby iron intake through enterocytes and release from macrophages are highly affected resulting in limitation of plasma iron level. In contrast, conditions like iron deficiency and hypoxia inhibit hepcidin synthesis resulting in an increased level of ferroportin. This ensures a sufficient level of plasma iron to avoid anemia and to increase erythropoiesis.

Ferroportin is the unique iron exporter in mammalian cells and also acts at the crossroad of cellular and systemic iron homeostasis. While systemic iron homeostasis is maintained by hepcidin-ferroportin, the cellular iron homeostasis is maintained by the IRE/IRP system in which ferroportin is one of the targets. Thus, it is important to understand the role and regulation of ferroportin during inflammation-associated anemia.

Ferroportin—the unique iron exporter

The mechanism of entry of iron into plasma from absorptive enterocytes as well as from macrophages was unknown till the year 2000. Three different groups discovered ferroportin independently as MTP1 (metal transporter protein 1), ferroportin1, and IREG1 (iron-regulated 1), respectively [1,13,31]. To identify new genes that are important in the regulation of iron metabolism, Abboud and Haile used a library of mRNA sequences enriched for IRP1 binding and eventually identified a new gene important for iron metabolism and named it MTP1. Overexpression of MTP1 in mammalian cells resulted in intracellular iron depletion. The group also identified an active IRE in the 5′-UTR of MTP1. Donovan and group [13] identified ferroportin in Zebrafish by positional cloning and discovered its function as an iron exporter by using radioactive iron isotopes in *Xenopus laevis* oocytes injected with ferroportin cRNA. McKie et al. isolated and characterized IREG1 cDNA encoding a duodenal protein that was found to be localized at the basolateral membrane of polarized epithelial cells [31]. They also found that IREG1 could stimulate iron efflux when expressed in *Xenopus* oocytes. Early evidence revealed a high expression level of ferroportin in cells and tissues including duodenal enterocytes, splenic red pulp macrophages, Kupffer cells, periportal hepatocytes, and the placental syncytiotrophoblast, which are associated with iron transport.

Fpn null mice did not complete embryonic development [14]. This happened due to the requirement of Fpn for transferring iron to the embryos from the extraembryonic visceral endoderm (exVE).

Global deletion led to a defect in embryonic growth and consequent death due to lack of iron. Deletion of Fpn solely in the embryo proper and not in exVE and placenta resulted in the survival of the mice. After the birth, these animals became anemic and were detected with marked iron accumulation in enterocytes, splenic macrophages, Kupffer cells, and hepatocytes. These observations established the key role of Fpn in iron export in vivo.

Ferroportin is about 62 kDa transmembrane protein and is systematically named SLC40A1. The variation in molecular mass is likely due to differential glycosylation in different tissues through the functional consequences of differential glycosylation are not understood so far [8]. To date, it is the solitary cellular iron exporter protein reported in all mammals. It is an ancient and well-conserved protein in mammals as found by the alignment of human, rat, and mouse protein sequences [31]. They also identified ferroportin-related proteins in *Arabidopsis thaliana* (accession number AAC28758) and *Caenorhabditis*. This indicates the presence of orthologs of Fpn in plants and worms. Fpn is not closely related to any other known mammalian transporter although there may be some distant homology to bacterial transport proteins [15]. Ferroportin plays a critical role in different tissues involved in mammalian iron homeostasis, including duodenal enterocytes (iron uptake and export into circulation), hepatocytes (storage), syncytiotrophoblasts (transfer to the embryo), and reticuloendothelial macrophages (iron recycling from senescent red blood cells) [10].

Structure of ferroportin

Fpn gene encodes for the protein-containing 571 amino acids [10]. Its protein structure was a matter of debate among the researchers. Several studies proposed it as 9–12 *trans*-membrane spanning protein. However, the current understanding of the Fpn structure is that it consists of two six-transmembrane-helix bundles (N-lobe and C-lobe), organized into two six-helix halves connected by a large cytoplasmic loop between the sixth and the seventh domain. These two helix bundles form a cavity that is thought to be involved in an iron release from the cell. The exported iron is found as ferrous form [33]; however, the energetics of the exporting iron is not clear yet [12]. Analysis of Fpn structure indicates the presence of two divalent metal-binding sites, one in each lobe. However, how these sites mediate iron export is not understood so far. How cytoplasmic iron reaches Fpn for export is a matter of interest. It has been suggested that ferrous iron–glutathione complex might be formed to be delivered to Fpn by iron chaperone PCBP2 [45,66].

The oligomeric status of Fpn has been proposed but remains controversial. Fpn is proposed to be monomeric by some and dimeric/multimeric by other groups. Biophysical characterization of purified detergent-solubilized Fpn was performed, and evidence showed that only monomer form could bind to hepcidin [50]. It was reported that both the N- and C-termini are cytosolic, indicating an even number of transmembrane regions [7,10].

Ferroportin is encoded by a gene on chromosome 2q in humans, which is larger than 20 kb in length and is composed of eight exons [28]. The 5′-UTR of Fpn mRNA comprises a putative IRE, which is translationally regulated by iron regulatory proteins (IRPs) like other 5′-UTR-IRE-regulated genes like ferritin (both H and L subunit), erythroid δ-aminolevulinate synthase (ALAS2), and mitochondrial aconitase [10,22].

It is encoded by two tissue-specific differentially spliced transcripts, Fpn1A and Fpn1B. They differ by the presence (in Fpn1A) or absence (in Fpn1B) of 5′-IRE, respectively, to translationally repress its synthesis when cellular iron is low [19,68]. Fpn1B is highly expressed in the duodenum and erythroid precursors.

There are various mechanisms involved in ferroportin regulation. It is controlled at the transcriptional, post-transcriptional, post-translational, and cell-lineage levels.

Regulation of ferroportin by inflammation

Inflammation decreases ferroportin expression by transcription and posts translational protein degradation mechanisms. Liver and splenic Fpn mRNA are down-regulated in vivo due to the administration of bacterial lipopolysaccharide (LPS), which is a ligand for the innate immune sensor Toll-like receptor 4 (TLR 4) [15,18].

A similar trend was detected in human monocytic cells in response to LPS and interferon-γ [15]. LPS injection into mice or rats also decreased the Fpn transcript in the spleen and intestine [61]. TLR2 stimulation also downregulates ferroportin expression. TLR4 stimulation causes slight downregulation of ferroportin, and the effect is intensified after priming with proinflammatory cytokines like IL-1β and TNF-α [2]. Iron demand distinctly expands in the case of hemolytic anemia that is fulfilled by increased expression of Fpn. Hepcidin levels drop in case of experimentally induced anemia, facilitating iron absorption through increased Fpn membrane expression across the small intestine [11]. In iron-restricted erythropoiesis, synchronized upregulation of Fpn takes place in enterocytes and macrophages [8].

Post-translational degradation of ferroportin by hepcidin during inflammation and chronic diseases is the key event in causing ACD and shall be described later.

Ferroportin regulation by iron

Iron is a critical factor in regulating the expression of ferroportin. Several studies have provided ample information that heme-induced Fpn transcription requires the release of iron from heme [61]. Iron released from heme and erythrocytes results in increased expression of ferroportin. The presence of a 5′ IRE in ferroportin mRNA (isoform 1A, which is abundant in macrophages) is key for the regulation of this gene at the translational level. IRP proteins bind to IRE in case of insufficiency of iron in the cell, leading to translational repression of ferroportin. Moreover, the 3′-untranslated region of ferroportin is targeted by miR-485-3p microRNA, which is induced by cellular iron deficiency [52]. An increase in the cellular iron level causes downregulation of miRNA ultimately contributing to the increased Fpn expression [15]. An increased expression during iron surplus and decreased Fpn mRNA expression in iron-deficient conditions was found in osteoblasts [69]. Excess iron causes oxidative stress, resulting in nuclear accumulation of Nrf2 promoting increased Fpn mRNA expression. This observation provided evidence that alteration of intracellular iron level can promote Fpn expression [69]. Bronchial epithelial cells, macrophages, cardiomyocytes, and hepatocytes also control ferroportin by iron. In contrast, a rise in iron level decreases the expression of Fpn in enterocytes and placental syncytiotrophoblast cells [21,30].

Regulation of ferroportin by hypoxia

Hypoxia-inducible factors (HIF1α and HIF2α) are mainly involved in mediating cellular response to hypoxia [53]. These transcription factors act as the regulators of cellular adaptation during hypoxic stress. In normoxia, the regulatable subunits of HIFs (HIF1α and HIF2α) are degraded by proteasomes. They consist of defined proline residues, those are hydroxylated in an iron, oxoglutarate, and oxygen-dependent reaction by prolyl hydroxylases under normoxia. This leads to the process of poly-ubiquitination by von Hippel-Lindau tumor suppressor E3 ligase complex and ultimately causes degradation of the HIFα subunits via 26S proteasome [58]. In contrast, low oxygen conditions do not hydroxylate HIFs by prolyl hydroxylases, so they get stabilized. Then, HIFα subunits translocate to the nucleus, dimerize with their partner ARNT/HIFβ, and recruit coactivators such as CBP/p300. This active transcription factor eventually causes the transcription of hypoxia target genes [55]. The iron-mediated induction of Fpn was found to be reduced in the ARNT-deleted intestine. Additionally, there is decreased transcription of Fpn in the intestine with deleted HIF2α but not in deleted HIF1α [57]. The Fpn promoter contains hypoxia-response element (HRE) and HRE-containing reporter construct responded to low oxygen. Mutation of the HREs prevented that response. Finally, chromatin immunoprecipitation studies in mouse duodenum showed that HIF2 could bind to the Fpn promoter region. Thus, HIF2 has a demonstrable role in increasing Fpn transcription during hypoxia and iron deficiency [15,61]. Furthermore, HIF activation in enterocytes caused induction of the basolateral iron export machinery Fpn [15].

Hepcidin—the master controller of iron homeostasis

Hepcidin is a peptide hormone originally reported to be produced in the liver (hepatocytes). The peptide contains 25 amino acids and 4 disulfide bonds. It was originally found as a new defensin-like innate immune component [19]. Currently, it is established as the master regulator of systemic iron homeostasis by its ability to bind with Fpn and promote degradation of the latter involving ubiquitination and subsequent proteasomal degradation. The molecule is a 2.8 kDa monomer and orchestrates the interplay of molecules involved in inflammation, iron homeostasis, and erythropoiesis.

It was discovered by several groups almost at the same time. It was detected in the plasma ultrafiltrate and named LEAP-1 (liver expressed antimicrobial peptide) [27]. The sequence of the peptide was deposited to the Swiss Prot database in the year 1998 and was named as hepcidin due to the high expression of the mRNA in the liver [43]. The peptide was detected with weak antimicrobial activity in vitro. The same report revealed a more than 100-fold increase of hepcidin in the urine of a septic donor that suggested a link of the peptide to inflammation and innate immunity. Another group also identified hepcidin mRNA as an iron-induced transcript that was also found to be increased during inflammation [47].

A study in mice highlighted the role of hepcidin in iron homeostasis when USF2 (upstream stimulatory factor 2) knockout mouse showed signs of iron overload [40]. It was later concluded that the mutation had caused suppression of the adjacent gene-HAMP responsible for producing hepcidin. It was inferred that hepcidin could regulate iron absorption in the duodenum crypt cells, iron release from the macrophages into the blood, and be involved in placental iron exchange [40]. Hepcidin was found to be regulated by inflammation [47], body iron stores [47], hypoxia, and anemia [41]. At the systemic level, mammals have no physiological excretory mechanism for iron; it exits the body by sloughing of the mucosal layer and blood loss. Therefore, iron levels are maintained by tight regulation of the absorption or release of free iron in blood from duodenal crypt cells, macrophages, and hepatocytes.

The major production site of hepcidin is the liver or hepatocytes. Higher iron level increases hepcidin production to curb the iron absorption from the diet into the blood through duodenal cells while lower iron causes suppression of hepcidin expression so that more iron could be released into the blood from the macrophages that recycles iron from senescent RBCs.

Structure of hepcidin

The human hepcidin gene locus is at chromosome 19q13.1, and the gene is named HAMP [49]. The gene encodes an 84 amino acids precursor peptide, preprohepcidin that undergoes enzymatic cleavage to become 64 amino acids containing prohepcidin peptide in the lumen of the endoplasmic reticulum [60]. By the time the peptide is secreted out of the cells, it is modified into a bioactive peptide of 25 amino acids called hepcidin by the action of a furin-like proprotein convertase [43]. More variants of hepcidin with N-terminal deletions resulting in shorter peptides of 22 and 20 amino acids were identified in urine. Those are N-terminally truncated isoforms of hepcidin-25 [43]. Hepcidin-25 is almost exclusive to show the iron regulating bioactivity. It is cysteine-rich and a member of the defensin family of proteins. It contains eight cysteine residues to form four disulfide bonds in a ladder-like fashion resulting in a β-sheet structure with a hairpin loop of antiparallel strands providing extra stability to the molecule [24]. The high cysteine content of hepcidin is well conserved among species [43].

Other than hepatocytic hepcidin synthesis it is also produced by macrophages [44], fat cells [5], and cardiomyocytes [32]. These findings suggest the involvement of hepcidin in local tissue iron homeostasis.

Regulation of hepcidin

The abundance of hepcidin in hepatocytes is regulated at the transcriptional level [17,39]. While many genes involved in iron homeostasis are regulated through IRE-IRP interaction at the cellular level, the regulation of hepcidin has no direct feedback from the IRE-IRP system. Studies involving human mutations and transgenic mice revealed identities and mechanisms of molecules involved in hepcidin regulation [19]. Interestingly, hepcidin level is feedback-regulated by iron. A number of proteins orchestrate to regulate the HAMP gene by iron. Human homeostatic iron regulator protein or High Fe^{2+} (HFE), transferrin receptor 1 and 2 (TfR1, TfR2), hemojuvelin (HJV), bone morphogenetic protein 6 (BMP6), and the transcription factor Smad4 are involved in hepcidin regulation. The BMP receptor and its signaling components regulate hepcidin transcription. HJV increases the sensitivity of BMPs as an iron-specific adaptor ligand. The information about higher iron levels is communicated by induced BMP6 [56,59]. Increased extracellular iron level results in more holo-transferrin formation from apo-transferrin [19]. Holotransferrin in turn binds to TfR1 and TfR2 in which HFE protein acts as an intermediary. This complex causes a conformational change in membrane receptor HJV, neogenin, and transmembrane protease serine 6 (TMPRSS6) to favor the binding of BMP6 to HJV [39,49].

The binding of BMP6 causes further intracellular signaling to induce hepcidin transcription. TMPRSS6 is stabilized by iron deficiency to cleave membrane HJV and inactivate the latter to affect hepcidin transcription.

Inflammation augments hepcidin synthesis by IL-6

Invading pathogens and other inflammatory signals increase proinflammatory cytokines by activating T cells (CD3+) and monocytes. IFN-γ, IL-6, IL-10, IL-1β, and TNF-α are released in the system. Among them, IL-6 is the main regulatory cytokine that acts on hepatocytes to increase hepcidin production. It also causes the induction of ferritin in macrophages. Ferritin induction is also aided by IFN-γ, TNF-α, and IL-10. These cytokines also cause increased absorption of iron from the blood by destructing RBCs in macrophages. IL-1β and IFN-γ also act directly to inhibit erythropoiesis to cause anemia [63].

Inflammation decreases serum iron by reducing intestinal iron absorption and by sequestering iron in macrophages. Research from the last 2 decades established hepcidin as the main mediator of this physiological effect. In response to inflammation, hepcidin mRNA was increased with the development of hypoferremia in wild-type mice but the hypoferremia response was lost in HAMP knockout mice [41]. This response is mediated by IL-6. It could induce hepcidin in human hepatocytes and hepatic cell lines [35]. Infusion of IL-6 into human volunteers resulted in a more than 7-fold increase in urinary hepcidin level and coincided with hypoferremia [26,37]. This effect of IL-6 on hepcidin transcription is promoted by STAT3 binding to STAT3-binding element in the hepcidin promoter [16,46,65].

Inflammation-induced IL-6 mediated hepcidin increase causes hypoferremia that is developed during infections and with diseases related to chronic inflammation. In this way, hepcidin forms a part of the innate immune system by protecting the body against invading microbes. The decrease in circulating plasma iron levels cuts down the supply of iron, which is essential for the replication and survival of almost all invading microbes. The inflammation-induced hepcidin synthesis thus results in iron sequestration and hypoferremia to limit the iron availability for erythropoiesis.

Hepcidin promotes ferroportin degradation

In the year 2004, it was reported for the first time that the distribution of ectopically expressed Fpn-GFP changed strikingly from the cell surface to intracellular vesicles upon the addition of hepcidin [36]. It was also observed that hepcidin prevents ferroportin from exporting iron by dislocating it from the cell surface that ultimately led to iron retention in the cell. It was then discovered that hepcidin actually could bind to Fpn to induce its endocytosis and lysosomal degradation [36]. Even a low concentration of hepcidin is effective in Fpn endocytosis and subsequent degradation. Evidence was presented for the requirement of hepcidin-mediated conformational change of Fpn to induce ubiquitination of the lysine-rich cytoplasmic segment that connects two six-helix domains [48,51]. Ubiquitinated Fpn is then subjected to proteasomal degradation in lysosomes [39]. Involvement of Rnf217, an E3 ubiquitin ligase has been reported for degradation of Fpn in response to hepcidin [25]. Targeted mutagenesis of hepcidin and Fpn [6] and nanodisc membrane model [39] suggest the binding of hepcidin to the C-lobe of Fpn. It further reveals that C326 thiol cysteine is critical for the interaction and utilizes an iron atom for coordinating the hepcidin binding [6].

This hepcidin-Fpn interaction is specific and unique in vertebrates. Fpn is present in hydra, plants, and multicellular organisms, while the presence of hepcidin is not detected before the evolution of fish (vertebrate). Remarkably, the hepcidin binding site for Fpn C326 is strictly conserved in all vertebrates. Isosteric mutation C326S in mice and human Fpn causes a complete loss of binding of hepcidin to Fpn [3].

Hepcidin-induced ferroportin degradation results in anemia

Hepcidin induction is multifactorial. Infection, inflammation, or in rare instances, adenomas of the liver can cause hepcidin surge. Increased urinary hepcidin was found with lipopolysaccharide treatment in both humans and mice [19,63]. Induction of hepcidin impacts Fpn degradation causing retention of iron in the macrophages, thus recycling of iron is hampered. Iron is slowly accumulated eventually resulting in decreased serum iron (<60 μg/dL) and total iron-binding capacity and most of the iron is diverted to increased ferritin [63]. Elevated hepcidin inhibits the release of stored and recycled iron as well as intestinal iron absorption, leading to low plasma iron concentrations and microcytic anemia (Fig. 18.1).

Anemia is detected in chronic kidney diseases (CKD) wherein erythropoietin production is hampered [20]. Less erythropoietin results in anemia that induces hepcidin expression either by hypoxia, iron shortage signals, or inflammation related to the underlying disease process. Induced hepcidin causes retention of iron inside the macrophages, thus restricting iron availability for erythropoiesis and contributing to the anemic condition (Fig. 18.1). The high circulating hepcidin levels in CKD patients also could be due to decreased renal clearance of the hepcidin [67].

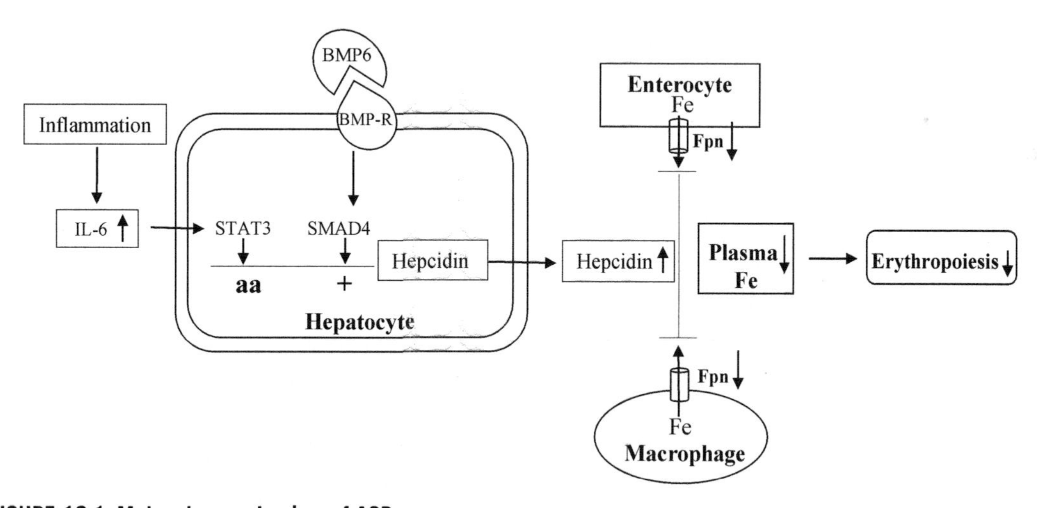

FIGURE 18.1 Molecular mechanism of ACD

Invading pathogens and other inflammatory conditions promote generation of the proinflammatory cytokine IL-6. IL-6 elevates hepcidin transcription mediated by JAK-STAT3 pathway. BMP receptor binds to its ligand BMP6 leading to increased hepcidin transcription by SMAD 4 on sensing extracellular and intracellular iron. Increased level of hepcidin binds and degrades ferroportin (Fpn) in macrophages and enterocytes to affect iron recycling and entry, respectively, resulting in lower plasma iron levels and decreased erythropoiesis.

Anemia is also detected during in initial stage of malignancies that worsens with the progression of the disease. The anemia during malignancies, including Hodgkin's disease [23] and multiple myeloma [29,54], is accompanied by increased hepcidin levels resulting from inflammatory cytokines [20].

In general, chronic diseases are accompanied by increased hepcidin production stimulated by inflammatory cytokines. The increased hepcidin level then promotes Fpn degradation both in enterocytes and macrophages resulting in less availability of iron in plasma. Thus, identification of the mechanism of hepcidin-induced post-translational ferroportin degradation helped in understanding one of the key reasons for anemia related to chronic diseases.

Summary

ACD is reported for more than 50 years [9]; however, research from the last 2 decades revealed a detailed understanding of the molecular mechanism of this pathophysiological condition, in which the critical role of hepcidin in promoting degradation of ferroportin by binding to the C326 thiol cysteine has been established. This knowledge is helping to develop therapeutics for the treatment of ACD that is regarded as the most frequent anemia in hospitalized and chronically ill patients. The current efforts of treatment include antagonists of hepcidin function and mobilization of iron from macrophages for promoting erythropoiesis. These strategies have been worked out in animal models and have now reached human clinical trials.

Acknowledgments

CKM acknowledges the financial support received from the Department of Biotechnology and SERB-Department of Science and Technology, India. AA and SY are supported by fellowships from the Council of Scientific and Industrial Research, India.

References

[1] Abboud S, Haile DJ. A novel mammalian iron-regulated protein involved in intracellular iron metabolism. J Biol Chem 2000;275:19906–12.

[2] Agoro R, Mura C. Inflammation-induced up-regulation of hepcidin and down-regulation of ferroportin transcription are dependent on macrophage polarization. Blood Cells Mol Dis 2016;61:16–25. https://doi.org/10.1016/j.bcmd.2016.07.006.

[3] Altamura S, Kessler R, Groene HJ, Gretz N, Hentze MW, Galy B, Muckenthaler MU. Resistance of ferroportin to hepcidin binding causes exocrine pancreatic failure and fatal iron overload. Cell Metab 2014;20:359–67.

[4] Arosio P, Levi S. Cytosolic and mitochondrial ferritins in the regulation of cellular iron homeostasis and oxidative damage. Biochim Biophys Acta 2010;1800(8):783–92. https://doi.org/10.1016/j.bbagen.2010.02.005.

[5] Bekri S, Gaul P, Anty R, Luciani N, Dahman M, Ramesh B, Iannelli A, Staccini-Myx A, Casanova D, Ben Amor I, Saint-Paul MC, Huet PM, Sadoul JL, Gugenheim J, Srai SK, Tran A, Le Marchand-Brustel Y. Increased adipose tissue expression of hepcidin in severe obesity is independent from diabetes and NASH. Gastroenterology 2006;131:788–96.

[6] Billesbolle CB, Azumaya CM, Kretsch RC, Powers AS, Gonen S, Schneider S, Arvedson T, Dror RO, Cheng Y, Manglik A. Structure of hepcidin-bound ferroportin reveals iron homeostatic mechanisms. Nature 2020;586:807–11.

[7] Bolotta A, Abruzzo PM, Baldassarro VA, Ghezzo A, Scotlandi K, Marini M, Zucchini C. New insights into the hepcidin-ferroportin axis and iron homeostasis in iPSC-derived cardiomyocytes from Friedreich's ataxia patient. Oxid Med Cell Longev 2019;2019:7623023. https://doi.org/10.1155/2019/7623023. eCollection 2019.

[8] Canonne-Hergaux F, Donovan A, Delaby C, Wang HJ, Gros P. Comparative studies of duodenal and macrophage ferroportin proteins. Am J Physiol Gastrointest Liver Physiol 2006;290:G156–63.

[9] Cartwright GE, Lee GR. The anaemia of chronic disorders. Br J Haematol 1971;21:147–52.

[10] Cianetti L, Gabbianelli M, Sposi NM. Ferroportin and erythroid cells: an update. Adv Hematol 2010;2010: 404173. https://doi.org/10.1155/2010/404173.

[11] D'Anna MC, Veuthey TV, Roque ME. Immunolocalization of ferroportin in healthy and anemic mice. J Histochem Cytochem 2009;57:9–16. https://doi.org/10.1369/jhc.2008.951616.

[12] Deshpande CN, Ruwe TA, Shawki A, Xin V, Vieth KR, Valore EV, Qiao B, Ganz T, Nemeth E, Mackenzie B, Jormakka M. Calcium is an essential cofactor for metal efflux by the ferroportin transporter family. Nat Commun 2018;9:3075.

[13] Donovan A, Brownlie A, Zhou Y, Shepard J, Pratt SJ, Moynihan J, Paw BH, Drejer A, Barut B, Zapata A, Law TC, Brugnara C, Lux SE, Pinkus GS, Pinkus JL, Kingsley PD, Palis J, Fleming MD, Andrews NC, Zon LI. Positional cloning of zebrafish ferroportin1 identifies a conserved vertebrate iron exporter. Nature 2000;403:776–81.

[14] Donovan A, Lima CA, Pinkus JL, Pinkus GS, Zon LI, Robine S, Andrews NC. The iron exporter ferroportin/Slc40a1 is essential for iron homeostasis. Cell Metab 2005;1:191–200.

[15] Drakesmith H, Nemeth E, Ganz T. Ironing out ferroportin. Cell Metab 2015;22:777–87. https://doi.org/10.1016/j.cmet.2015.09.006.

[16] Falzacappa MVV, Vujic SM, Kessler R, Stolte J, Hentze MW, Muckenthaler MU. STAT3 mediates hepatic hepcidin expression and its inflammatory stimulation. Blood 2007;109:353–8.

[17] Flanagan JM, Truksa J, Peng H, Lee P, Beutler E. In vivo imaging of hepcidin promoter stimulation by iron and inflammation. Blood Cells Mol Dis 2007;38:253–7.

[18] Ganz T, Nemeth E. Iron sequestration and anemia of inflammation. Semin Hematol 2009;46:387–93. https://doi.org/10.1053/j.seminhematol.2009.06.001.

[19] Ganz T. Hepcidin and iron regulation, 10 years later. Blood 2011;117:4425–33.

[20] Ganz T, Nemeth E. Hepcidin and iron homeostasis. Biochim Biophys Acta 2012;1823:1434–43.

[21] Gulec S, Anderson GJ, Collins JF. Mechanistic and regulatory aspects of intestinal iron absorption. Am J Physiol Gastrointest Liver Physiol 2014;307(4):G397–G409. doi:10.1152/ajpgi.00348.2013.

[22] Hentze MW, Muckenthaler MU, Galy B, Camaschella C. Two to tango: regulation of mammalian iron metabolism. Cell 2010;142:24–38. https://doi.org/10.1016/j.cell.2010.06.028.

[23] Hohaus S, Massini G, Giachelia M, Vannata B, Bozzoli V, Cuccaro A, D'Alo F, Larocca LM, Raymakers RA, Swinkels DW, Voso MT, Leone G. Anemia in Hodgkin's lymphoma: the role of interleukin-6 and hepcidin. J Clin Oncol 2010;28:2538–43.

[24] Hunter HN, Fulton DB, Ganz T, Vogel HJ. The solution structure of human hepcidin, a peptide hormone with antimicrobial activity that is involved in iron uptake and hereditary hemochromatosis. J Biol Chem 2002; 277:37597–603.

[25] Jiang L, Wang J, Wang K, Wang H, Wu Q, Yang C, Yu Y, Ni P, Zhong Y, Song Z, Xie E, Hu R, Min J, Wang F. RNF217 regulates iron homeostasis through its E3 ubiquitin ligase activity by modulating ferroportin degradation. Blood 2021;138:689–705. https://doi.org/10.1182/blood.2020008986.

[26] Kemna E, Pickkers P, Nemeth E, van der Hoeven H, Swinkels D. Time-course analysis of hepcidin, serum iron, and plasma cytokine levels in humans injected with LPS. Blood 2005;106:1864–6.
[27] Krause A, Neitz S, Magert HJ, Schulz A, Forssmann WG, Schulz-Knappe P, Adermann K. LEAP-1, a novel highly disulfide-bonded human peptide, exhibits antimicrobial activity. FEBS Lett 2000;480(2–3):147–50. https://doi.org/10.1016/s0014-5793(00)01920-7.
[28] MacKenzie EL, Iwasaki K, Tsuji Y. Intracellular iron transport and storage: from molecular mechanisms to health implications. Antioxidants Redox Signal 2008;10:997–1030. https://doi.org/10.1089/ars.2007.
[29] Maes K, Nemeth E, Roodman GD, Huston A, Esteve F, Freytes C, Callander N, Katodritou E, Tussing-Humphreys L, Rivera S, Vanderkerken K, Lichtenstein A, Ganz T. In anemia of multiple myeloma, hepcidin is induced by increased bone morphogenetic protein 2. Blood 2010;116:3635–44.
[30] McDonald EA, Gundogan F, Olveda RM, Bartnikas TB, Kurtis JD, Friedman JF. Iron transport across the human placenta is regulated by hepcidin. Pediatr Res 2020. https://doi.org/10.1038/s41390-020-01201-y.
[31] McKie AT, Marciani P, Rolfs A, Brennan K, Wehr K, Barrow D, Miret S, Bomford A, Peters TJ, Farzaneh F, Hediger MA, Hentze MW, Simpson RJ. A novel duodenal iron-regulated transporter, IREG1, implicated in the basolateral transfer of iron to the circulation. Mol Cell 2000;5:299–309.
[32] Merle U, Fein E, Gehrke SG, Stremmel W, Kulaksiz H. The iron regulatory peptide hepcidin is expressed in the heart and regulated by hypoxia and inflammation. Endocrinology 2007;148:2663–8.
[33] Mitchell CJ, Shawki A, Ganz T, Nemeth E, Mackenzie B. Functional properties of human ferroportin, a cellular iron exporter reactive also with cobalt and zinc. Am J Physiol Cell Physiol 2013;306:C450–9.
[34] Naito Y, Takagi T, Higashimura Y. Heme oxygenase-1 and anti-inflammatory M2 macrophages. Arch Biochem Biophys 2014;564:83–8. https://doi.org/10.1016/j.abb.2014.09.005.
[35] Nemeth E, Valore EV, Territo M, Schiller G, Lichtenstein A, Ganz T. Hepcidin, a putative mediator of anemia of inflammation, is a type II acute-phase protein. Blood 2003;101:2461–3.
[36] Nemeth E, Tuttle MS, Powelson J, Vaughn MB, Donovan A, Ward DM, Ganz T, Kaplan J. Hepcidin regulates cellular iron efflux by binding to ferroportin and inducing its internalization. Science 2004;306:2090–3.
[37] Nemeth E, Rivera S, Gabayan V, Keller C, Taudorf S, Pedersen BK, Ganz T. IL-6 mediates hypoferremia of inflammation by inducing the synthesis of the iron regulatory hormone hepcidin. J Clin Invest 2004;113:1271–6.
[38] Nemeth E, Ganz T. Anemia of inflammation. Hematol Oncol Clin N Am 2014;28:671–81.
[39] Nemeth E, Ganz T. Hepcidin-ferroportin interaction controls systemic iron homeostasis. Int J Mol Sci 2021;22:6493.
[40] Nicolas G, Bennoun M, Devaux I, Beaumont C, Grandchamp B, Kahn A, Vaulont S. Lack of hepcidin gene expression and severe tissue iron overload in upstream stimulatory factor 2 (USF2) knockout mice. Proc Natl Acad Sci USA 2001;98:8780–5.
[41] Nicolas G, Chauvet C, Viatte L, Danan JL, Bigard X, Devaux I, Beaumont C, Kahn A, Vaulont S. The gene encoding the iron regulatory peptide hepcidin is regulated by anemia, hypoxia, and inflammation. J Clin Invest 2002;110:1037–44. https://doi.org/10.1172/JCI15686.
[42] Ohgami RS, Campagna DR, McDonald A, Fleming MD. The Steap proteins are metalloreductases. Blood 2006;108:1388–94.
[43] Park CH, Valore EV, Waring AJ, Ganz T. Hepcidin, a urinary antimicrobial peptide synthesized in the liver. J Biol Chem 2001;276:7806–10.
[44] Peyssonnaux C, Zinkernagel AS, Datta V, Lauth X, Johnson RS, Nizet V. TLR4-dependent hepcidin expression by myeloid cells in response to bacterial pathogens. Blood 2006;107:3727–32.
[45] Philpott CC, Patel SJ, Protchenko O. Management versus miscues in the cytosolic labile iron pool: the varied functions of iron chaperones. Biochim Biophys Acta Mol Cell Res 2020;1867:118830.

[46] Pietrangelo A, Dierssen U, Valli L, Garuti C, Rump A, Corradini E, Ernst M, Klein C, Trautwein C. STAT3 is required for IL-6-gp130-dependent activation of hepcidin in vivo. Gastroenterology 2007;132:294–300.
[47] Pigeon C, Ilyin G, Courselaud B, Leroyer P, Turlin B, Brissot P, Loréal O. A new mouse liver-specific gene, encoding a protein homologous to human antimicrobial peptide hepcidin, is overexpressed during iron overload. J Biol Chem 2001;276:7811–9. https://doi.org/10.1074/jbc.M008923200.
[48] Qiao B, Sugianto P, Fung E, del-Castillo-Rueda A, Moran-Jimenez MJ, Ganz T, Nemeth E. Hepcidin-induced endocytosis of ferroportin is dependent on ferroportin ubiquitination. Cell Metab 2012;15:918–24.
[49] Reichert CO, da Cunha J, Levy D, Maselli LMF, Bydlowski SP, Spada C. Hepcidin: homeostasis and diseases related to iron metabolism. Acta Haematol 2017;137:220–36. https://doi.org/10.1159/000471838.
[50] Rice AE, Mendez MJ, Hokanson CA, Rees DC, Björkman PJ. Investigation of the biophysical and cell biological properties of ferroportin, a multipass integral membrane protein iron exporter. J Mol Biol 2009; 386:717–32.
[51] Ross SL, Tran L, Winters A, Lee KJ, Plewa C, Foltz I, King C, Miranda LP, Allen J, Beckman H, Cooke KS, Moody G, Sasu BJ, Nemeth E, Ganz T, Molineux G, Arvedson TL. Molecular mechanism of hepcidin-mediated ferroportin internalization requires ferroportin lysines, not tyrosines or JAK-STAT. Cell Metab 2012;15:905–17.
[52] Sangokoya C, Doss JF, Chi J-T. Iron-responsive miR-485-3p regulates cellular iron homeostasis by targeting ferroportin. PLoS Genet 2013;9(4):e1003408. https://doi.org/10.1371/journal.pgen.1003408.
[53] Semenza GL. HIF-1 and mechanisms of hypoxia sensing. Curr Opin Cell Biol 2001;13:167–71. https://doi.org/10.1016/s0955-0674(00)00194-0.
[54] Sharma S, Nemeth E, Chen YH, Goodnough J, Huston A, Roodman GD, Ganz T, Lichtenstein A. Involvement of hepcidin in the anemia of multiple myeloma. Clin Cancer Res 2008;14:3262–7.
[55] Shay JE, Celeste Simon M. Hypoxia-inducible factors: crosstalk between inflammation and metabolism. Semin Cell Dev Biol 2012;23:389–94. https://doi.org/10.1016/j.semcdb.2012.04.004.
[56] Silvestri L, Nai A, Dulja A, Pagani A. Hepcidin and the BMP-SMAD pathway: an unexpected liaison. Vitam Horm 2019;110:71–99. https://doi.org/10.1016/bs.vh.2019.01.004.
[57] Singhal R, Shah YM. Oxygen battle in the gut: hypoxia and hypoxia-inducible factors in metabolic and inflammatory responses in the intestine. J Biol Chem 2020;295:10493–505. https://doi.org/10.1074/jbc.REV120.011188.
[58] Tanimoto K, Makino Y, Pereira T, Poellinger L. Mechanism of regulation of the hypoxia- inducible factor-1 alpha by the von Hippel-Lindau tumor suppressor protein. EMBO J 2000;19:4298–309. https://doi.org/10.1093/emboj/19.16.4298.
[59] Viatte L, Vaulont S. Hepcidin, the iron watcher. Biochimie 2009;91:1223–8. https://doi.org/10.1016/j.biochi.2009.06.012.
[60] Wallace DF, Jones MD, Pedersen P, Rivas L, Sly LI, Subramaniam VN. Purification and partial characterization of recombinant human hepcidin. Biochimie 2006;88:31–7.
[61] Ward DM, Kaplan J. Ferroportin-mediated iron transport: expression and regulation. Biochim Biophys Acta 2012;1823(9):1426–33. https://doi.org/10.1016/j.bbamcr.2012.03.004.
[62] Weiss G, Goodnough LT. Anemia of chronic disease. N Engl J Med 2005;352:1011–23.
[63] Weiss G, Ganz T, Goodnough LT. Anemia of inflammation. Blood 2019;133:40–50.
[64] WHO. The global prevalence of anaemia in 2011. Geneva: World Health Organization; 2015.
[65] Wrighting DM, Andrews NC. Interleukin-6 induces hepcidin expression through STAT3. Blood 2006;108: 3204–9.
[66] Yanatori I, Richardson DR, Imada K, Kishi F. Iron export through the transporter ferroportin 1 is modulated by the iron chaperone PCBP2. J Biol Chem 2016;291:17303–18.

[67] Zaritsky J, Young B, Wang HJ, Westerman M, Olbina G, Nemeth E, Ganz T, Rivera S, Nissenson AR, Salusky IB. Hepcidin- a potential novel biomarker for iron status in chronic kidney disease. Clin J Am Soc Nephrol 2009;4:1051–6.
[68] Zhang DL, Hughes RM, Ollivierre-Wilson H, Ghosh MC, Rouault TA. A ferroportin transcript that lacks an iron-responsive element enables duodenal and erythroid precursor cells to evade translational repression. Cell Metab 2009;9:461–73. https://doi.org/10.1016/j.cmet.2009.03.006.
[69] Zhao G-Y, Di D-H, Wang B, Zhang P, Xu Y-J. Iron regulates the expression of ferroportin 1 in the cultured hFOB 1.19 osteoblast cell line. Exp Ther Med 2014;8(3):826–30. https://doi.org/10.3892/etm.2014.

CHAPTER

Post-transcriptional regulation of genes and mitochondrial disorder

19

Ankit Sabharwal[a] and Bibekananda Kar[a]

Department of Biochemistry and Molecular Biology, Mayo Clinic, Rochester, MN, United States

Introduction

A mitochondrion is a semiautonomous organelle that has evolved over the years to be an important player in cellular homeostasis and disease progression. The mitochondrion has evolved from an alphaproteobacterial ancestor, *Rickettsia prowazekii* [1]. The evolution of mitochondria has been a curious gesture of independence, during which the genetic material of mitochondria was transferred or lost in a retrograde manner to the nuclear genome [1–3]. Mitochondria as the site of energy production have unraveled a way for the scientific community to dissect out and understand the significance of this organelle in cellular dynamics. Mitochondria perform diverse cellular functions including ATP generation and act as reservoirs for calcium homeostasis, generators of signaling molecules and metabolites such as heme, apart from playing an essential role in thermogenesis, apoptosis, growth, development, and maintenance of other vital processes [3–7]. Given the increasingly evident role of mitochondria in cell maintenance and homeostasis, it is not surprising that genetic crosstalk exists between the nucleus and mitochondria. This crosstalk involves anterograde import of proteins and nucleic acids from the nucleus or cytosol to mitochondria. The myriad functions that mitochondria can execute, apart from oxidative phosphorylation, are not orchestrated just by the subset of 13 proteins that it encodes. Comprehensive proteomic studies in the past have cataloged 1136 nuclear-encoded mitochondrial resident human proteins [8]. These proteins are instrumental in maintaining mitochondrial biogenesis, homeostasis, and function [3]. Mitochondria is completely reliant on the nuclear genome to carry out its genome maintenance as almost all the genetic circuitry required for replication, transcription, post-transcription, and translation is imported from the nucleus [9]. Therefore, mutations in any of these genes can lead to diverse clinical phenotypes affecting mitochondrial function. Whole-genome and transcriptome studies have identified novel nuclear-encoded genes influencing mitochondrial gene expression. In this chapter, we summarize the crosstalk involved at the post-transcriptional level between the nuclear and mitochondrial genomes. Here we also aim to highlight the significance of these nuclear-encoded players and their implications in mitochondrial functions and disorders.

[a]Contributed equally.

Post-Transcriptional Gene Regulation in Human Disease, Volume 32. https://doi.org/10.1016/B978-0-323-91305-8.00008-9

Mitochondrial genome

The mitochondrial genome is a 16.5 kb circular double-stranded DNA molecule. The circular genome consists of heavy and light strands that are classified based on their nuclear content, where heavy strands have a higher proportion of guanine residues and light strands are rich in cytosines [10,11]. Mitochondrial genome codes for 37 genes, out of which 2 are rRNAs (12S and 16S), 13 are protein-coding genes that comprise the respiratory chain complex and 22 are tRNAs, important for the translation of these protein-coding genes. Besides that, a noncoding region D-loop is also present in the mitochondrial genome that serves as a regulatory locus for replication, transcription, and translation of mtDNA [12]. Unlike its nuclear counterpart, multiple copies of mtDNA varying between a few hundred to thousands, exist within each cell. Negligible intramolecular recombination has been observed in the mitochondrial genome at the population level [13]. The mitochondrial genome displays a uniparental mode of inheritance, that is, maternal inheritance. Around 100–100,000 copies of circular, double-stranded mtDNA molecules are present in a prototypical cell [14]. When a cell possesses identical copies of the mitochondrial genome, it is said to be homoplasmic and when different copies of wild-type and mutant mtDNA coexist in the cell, it is called heteroplasmic. Copy number of the mtDNA molecules is governed by genetic factors, epigenetic and environmental modifications [15–19]. Transmission of mutant mtDNA does not follow a Mendelian pattern of inheritance and usually displays random segregation that can occur within and across generations, a phenomenon known as heteroplasmy shift or mitochondrial genetic bottleneck [20].

Mitochondrial transcription

Transcription in mammalian mitochondria is carried out by a multicomponent system consisting of the single subunit mtRNA polymerase (POLRMT) and several accessory transcription factors. This machinery catalyzes all the major transcription steps that include promoter recognition, transcription initiation, transcription elongation, and transcription termination. A noncoding region with two promoters, L-strand (LSP) and divergent H-strand (HSP) promoters, is the initiation site for mitochondrial transcription. Both these promoters are closely located [21] and regulate the generation of polycistronic transcripts [22,23]. These long transcripts are processed in multiple ways to yield individual mature tRNAs and mRNAs. The LSP directs the transcription of the *MT-ND6* gene and eight of the total 22 tRNAs; whereas, the rest of the remaining mitochondrial genes are transcribed under the control of HSP [24].

Transcription initiation

The DNA-dependent RNA polymerase (POLRMT) drives the promotor-specific initiation in mammalian mitochondria. POLRMT requires the mitochondrial transcription factor A (TFAM) and the mitochondrial transcription factor B2 (TFB2M) to initiate the transcription. TFAM is a DNA binding protein and has an established function in promoter recruitment [25,26]. Recent structural studies have shown that TFAM recruits POLRMT to promoter DNA through its "tether" helix present in the N-terminal extension. The TFB2M is required to melt the duplex DNA and induce some conformational changes in POLRMT to stabilize the open DNA during the initiation of transcription [25–27].

Transcription elongation

The replication and transcription events of human mitochondria are tied to each other. Studies have demonstrated that POLRMT requires transcriptional elongation factor (TEFM) protein for the elongation process [28,29]. Most of the transcription initiation events from LSP are prematurely terminated at conserved sequence block 2 (CSB2), which is around 200 bases downstream from the promoter. These shorter RNA species have an important role to play in priming DNA replication. TEFM functions as the switch from replication to transcription by stimulating POLRMT to abolish the premature transcription termination by preventing the formation of G-quadruplexes at CSB2, therefore, inhibiting the progression of the elongation complex [30,31].

Transcription termination

The exact mechanism of termination of HSP-initiated transcription is still poorly understood. The mitochondrial transcription terminator factor (mTERF) is a 39 kDa protein and is believed to play an essential role in the termination process initiated by HSP by binding to mtDNA located within the $tRNA^{Leu}$ gene. This proposed model explains the higher abundance of mitochondrial rRNAs, but recent studies have revealed that this abundance might be due to the high stability of rRNAs rather than the high synthesis rate [32,33]. Recent knockdown and knockout studies support the fact that mTERF is responsible for the termination of LSP-initiated transcription rather than that of HSP-initiated transcription [34,35]. Evidence from the structural studies also suggests that mTERF binds to the DNA through base flipping and helix unwinding that enables mTERF to terminate the mitochondrial transcription [32,36].

Processing of mitochondrial primary transcript

Both HSP and LSP produce long polycistronic transcripts and within the polycistronic transcripts, mitochondrial tRNAs demarcate the mt-rRNAs and mt-mRNAs. The post-transcriptional processing involves endonucleolytic cleavage, polyadenylation of mRNAs, aminoacylation of tRNAs, and other modifications. According to the "tRNA punctuation model" described by Ojala et al. the endonucleolytic cleavage of the polycistronic transcripts occurs at junctions between mt-mRNAs or rRNAs and tRNAs to release mature and functional mtRNAs [37]. Cleavage of the primary polycistronic transcripts is performed by two specific enzymes, RNase P and RNase Z at the 5′- and 3′-end, respectively, releasing individual or bicistronic RNA molecules that undergo further processing. Mitochondrial RNase P is a heterotrimeric endonuclease composed of MRPP1 (tRNA m^1R9 methyltransferase), MRPP2 (short-chain dehydrogenase), and MRRP3 (PIN domain-like nucleases that hydrolyze the phosphodiester bond) and does not have any RNA component unlike cytoplasmic and bacterial RNase P enzymes [38]. RNase Z (ELAC2) is an endonuclease that performs the cleavage at mt-tRNA 3′-ends from the primary transcript and is known to prefer the substrates already processed by RNase P [39–41]. There are four genes in mitochondrial primary polycistronic transcripts that are not flanked by tRNAs on both sides; therefore, these four do not comply with the "tRNA punctuation model." These transcript junctions are noncoding RNA (ncRNA)-*MT-CO1*, *MT-ND6*-ncRNA, *MT-ND5*-CYB, and *MT-ATP6/8-CO3*. The ncRNA at the 5′-end of the *MT-CO1* gene is believed to adopt a tRNA-like structure and is processed by RNaseP [41,42]. The other three junctions are thought to be not processed by either RNase P or Z, and their exact processing mechanism remains elusive. Brzezniak et al. observed no significant effect on the levels of these noncanonical cleavage sites by knocking down both RNase P and RNase Z [39]. Recent studies have demonstrated that proteins belonging to the FASTK family (Fas-activated serine/threonine kinase domain-containing protein) and pentatricopeptide repeat domain protein 2 (PTCD2) proteins are required for the processing of these noncanonical cleavage sites [43]. Several knockdown and CLIP-based studies revealed that FASTKD1, FASTKD2, and FASTKD4 are responsible for the processing of *MT-ND3*, *MT-ND6*, and *MT-ND5*, and PTCD2 is essential for the cleavage of the *MT-ND5-CYB* junction [43–46]. More studies in the future will be required to know the detailed mechanism of this pathway. The cleavage of mitochondrial polycistronic transcripts yields both individual and bicistronic transcripts. Subsequently, they undergo further post-transcriptional modifications as described in the following.

Mitochondrial mRNA polyadenylation and stability

Mitochondrial mRNA processing is different from that of its nuclear counterpart as the precursor mRNAs neither have introns nor undergo splicing and do not require 5′-cap modification. Like nuclear mRNAs, 3′-ends of most of the mt-mRNAs are polyadenylated but with much shorter tails (~40–50 adenine bases). Temperley et al. reported two exceptions to the polyadenylation process: (1) no Poly(A) tails have been detected for ND6 mt-mRNA; (2) ND5 mt-mRNA transcripts are either oligoadenylated or not adenylated at all [47]. The mitochondrial Poly(A) tails are synthesized by a dimeric human mitochondrial polyA polymerase enzyme (mtPAP) [48,49]. Nagaike et al. observed that knock-down of the mtPAP is associated with shortening of poly(A) tails and interference with the respiratory function [48]. Another study by Bai et al. showed that mtPAP mostly prefers ATP as a substrate although it can use all four NTPs [50]. The exact role of mitochondrial poly(A) tails is not very clear and depends on the specific mt-mRNA. Polyadenylation appears to destabilize the subunits of complex IV (MT-CO1, MT-CO2, and MT-CO3) and stabilize subunits of complex I (MT-ND1 and MT-ND2). Polyadenylation also seems to stabilize MT-ND3, MT-ND4, MT-ND5, and MT-CYB but to a lesser extent [48,51–53]. As the by-product of punctuation, 7 mtDNA-derived ORFs out of the 13 do not have a complete stop codon to terminate the translation process, and polyadenylation seems to generate a complete stop codon by adding the adenine bases [10,37].

RNA stability is maintained by important genetic players such as leucine-rich pentatricopeptide repeats containing (LRPPRC), stem-loop interacting RNA-binding protein (SLIRP), and HuR. The RNA-binding proteins, LRPPRC and SLIRP, play key roles in the stability of RNAs after transcription especially in maintaining the stability of HSP-derived mitochondrial transcripts [54,55]. It has been shown that deficiency of LRPPRC protein reduces the stability and polyadenylation of mt-mRNAs, leading to decreased activity of respiratory complexes [55–58]. Evidence by Sirra et al. underscored that LRPPRC acts like an RNA chaperone that stabilizes the RNA structures [59]. SLIRP forms a high molecular weight complex with LRPPRC and is vital for mt-RNA maintenance [60]. Although SLIRP is dispensable for the polyadenylation process, it protects LRPPRC from degradation and stabilizes its interacting partner protein [61]. Chujo and coworkers reported that the LRPPRC-SLIRP complex suppresses 3′-exonucleolytic mRNA degradation mediated by polynucleotide phosphorylase (PNPase) and SUV3 nucleases [62].

HuR belongs to the ELAV type family of RNA-binding proteins. HuR, a key regulator of mitochondrial biogenesis, has been reported to stabilize TFAM transcripts. This interaction is facilitated by the ataxia-telangiectasia mutated kinase/p38 DNA damage response [63]. HuR is also reported to play another role apart from transcript stability, where it suppresses the translation of protein-coding genes such as TRF1-interacting nuclear protein 2 (*TIN2*) and B-cell lymphoma-extra-large *BCL-xL* by binding to the processed transcript and destabilizing it. TIN2 when localized to mitochondria increases the reactive oxygen species (ROS) levels leading to induction and maintenance of cellular senescence [64]. HuR inhibits the translation of BCLxL, a dominant player in the programmed cell death, by binding to the internal ribosome entry site, thereby affecting the rearrangement of the mitochondrial network [65]. In humans, a protein complex containing Suv3 protein (hSuv3, encoded by suv3 like RNA helicase, *SUPV3L1* gene) and PNPase (encoded by *PNPT1* gene) mediates the decay of human mitochondrial RNA. hSuv3 is an NTP-dependent helicase that can unwind multiple DNA and RNA substrates [66,67]. It has been reported that the lack of active hSUV3 leads to the accumulation of aberrant mitochondrial RNA [68]. Knockdown of PNPase leads to an increase in the life span of mitochondrial RNA transcripts and the accumulation of RNA decay intermediates. Recent studies showed that PNPase copurifies with hSuv3 and hSuv3-PNPase complexes facilitate RNA degradation [68].

G-rich RNA sequence binding factor 1 (GRSF1) plays an important role in the clearing of mitochondrial RNA transcripts. This mitochondrial protein forms a complex with PNPT1 in the RNA granules. These granules bind to newly synthesized transcripts such as *MT-ND6* mRNA, long noncoding RNA *MT-CYTB*, *MT-ND5* mRNA. These noncoding RNAs play a crucial role in apoptosis, mitochondrial bioenergetics, and biosynthesis [69,70]. Loss or aberrant function of GRSF1 protein leads to loss of mature mtRNA transcripts, thus affecting mtRNA stability and abnormal loading of transcripts on mitoribosomes [69,71]. RNA turnover rate in mitochondria is regulated by RNA exonuclease 2 (REXO2). REXO2 is an oligoribonuclease that mediates the degradation of tetra- and penta-nucleotides generated from the activity of the mitochondrial degradosome complex. It also degrades the short mitochondrial transcripts that are the byproducts of processing of primary mitochondrial RNA [72]. Silencing of this gene has been associated with impaired cell growth, mitochondrial DNA depletion, reduced mitochondrial protein synthesis, and decreased respiratory activity [73]. This protein is responsible for maintaining the steady-state levels of double-stranded RNA and mitochondrial antisense transcripts [72]. Pumilo and FBF (PuF) are other families of RNA-binding proteins involved in mitochondrial biogenesis and mitophagy. Studies have demonstrated that yeast ortholog Puf3p associates with nuclear-encoded mitochondrial RNAs to the cytosolic face of mitochondrial outer membrane assisting in the organelle localized translation [74–76]. It is also involved in maintaining the nuclear-encoded mtRNA turnover rate by regulating the steady-state levels of specific transcripts involved in Coq biosynthesis [77]. Foat and colleagues showed that the post-transcriptional regulation carried out by Puf3 is somewhat carbon-source dependent. Puf3 is known to bind to the 3′-UTR site of the nuclear-encoded mtRNA transcripts and facilitates the rapid mtRNA deadenylation and decay when sugars such as glucose and fructose are present [78]. Loss of function studies carried out in *Saccharomyces cerevisiae* revealed accumulation and mislocalization of nuclear-encoded mtRNA transcripts. Respiratory dysfunction, abnormal mitochondrial morphology, and motility were some of the phenotypes observed upon Puf3 overexpression in yeast strain [75,79].

Another RNA-binding protein that has been reported to regulate the expression of the nuclear-encoded mitochondrial *OXPHOS* gene at the post-transcriptional level is YB-1. YB-1 was shown to be associated with nuclear mRNAs, wherein it mediates the recruitment of OXPHOS mRNAs from the messenger ribonucleoprotein in the translation-off state to active polysomes. It also acts as a translational inhibitor or activator depending on the concentration of the endogenous transcripts. Matsumoto and colleagues demonstrated that overexpression of YB-1 protein led to the decrease of expression of OXPHOS subunit mRNAs such as NDUFA9, NDUFB8, SDHB, and UQCRFS1 [80]. Clustered mitochondria homolog (CLUH) is a cytoplasmic RNA-binding protein that exhibits a post-transcriptional role in mitochondrial aiding in nuclear-encoded mRNA stability. This protein has tetratricopeptide RNA-binding domains that have been shown to bind to nuclear-encoded mRNAs involved in mitochondrial bioenergetics. Knockdown of CLUH in HeLa cells leads to clustering of mitochondria and reduced levels of proteins involved in the previously mentioned bioenergetic pathways [81,82].

Recent studies have highlighted the incidence of nuclear-mitochondrial crosstalk at the transcript level. Few examples of nuclear-encoded transcripts imported to mitochondria and involved in the mitochondrial post-transcriptional modifications are RNA component of mitochondrial RNA processing endoribonuclease (*MRP RNA*) [83], RNA component of Ribonuclease P (*RNase P RNA*) [84–86], and microRNAs [87,88]. *MRP RNA* is an extensively studied nuclear-encoded transcript located to mammalian mitochondria where it is part of RNA-processing endoribonuclease complex (RNase MRP) [83]. The nuclear role of MRP RNA involves facilitating the processing and maturation of small ribosomal RNAs in the nucleus [89]. Inside mitochondria, MRP RNA is involved in the processing of mtRNA to form RNA primers for the leading strand replication of the mitochondrial genome [83,90]. PNPase protein is involved in the import of *MRP RNA*, where it recognizes and binds to the stem-loop structure of *MRP RNA*, and this nucleus to cytosol translocation is mediated via HuR and CRM1 (chromosome region maintenance 1)-dependent nuclear export pathway. Upon its recruitment to the mitochondrial matrix, it interacts with GRSF1, an RNA-binding protein, to form RNP complexes, thus influencing mitochondrial respiration and genome maintenance [91,92].

In recent years, emerging evidence of the presence of microRNAs (miRNAs) has been shown to regulate mitochondrial function, also popularly known as MitomiRs. MicroRNAs (miRNAs) are small noncoding RNAs, about 22 nucleotides long, expressed in a wide range of eukaryotic organisms. MitomiRs have been shown to bind to the 5′- and 3′-UTR of mRNAs, thereby regulating their translation and life span [93,94]. Recent studies indicate that miRNA localizes in some subcellular organelles such as mitochondria, endoplasmic reticulum, and exosomes. Mitochondrial miRNA (mitomiR) is a group of miRNAs that appears to tightly orchestrate mitochondrial gene regulation. MitomiRs are broadly divided into two categories based on their mode of action: (1) mitomiRs that bind to cytoplasmic RNA and inhibits the gene expression in the cytoplasm to regulate the mitochondrial function and (2) mitomiRs that target mt-mRNAs and are imported into mitochondria as part of RNA-induced silencing complex [95]. Possible roles of such mitochondrial small noncoding RNAs have been proposed in mediating cell survival, apoptosis, mitochondrial transcription, energy production, cell division, and disease-like cancer. For instance, miR-181c negatively regulates the expression of mitochondrial gene cytochrome *c* oxidase I (*MT-COX1*) [96]. miR-338 is a miRNA specifically expressed in neuronal tissue and has been shown to regulate the expression of *COXIV* by binding to the COXIV 3′-UTR [97,98]. A functional study by Aschrafi and coworkers revealed that overexpression of miR-338 reduced ATP production, metabolic activity, and mitochondrial oxygen consumption significantly [99]. Similarly, miR-210 is reported to repress mitochondrial respiration [100]. In mouse β-cells, miR-15a inhibits the expression of endogenous uncoupling protein-2 (UCP-2), thereby promoting insulin biosynthesis [101]. Bandiera and colleagues found 57 miRNAs that are differentially expressed in HeLa mitochondria and cytosol. Out of these 57, about 13 nuclear-encoded miRNAs were observed to be enriched in mitochondria. Some of the miRNAs such as miR-328, miR-494, miR-513, and miR-638 are involved in mitochondrial homeostasis [102]. Organellar transcriptomics studies have enabled the identification of a compendium of mitochondrial encoded and localized transcripts, also known as mitotranscriptome [87,103,104].

Mitochondrial rRNA maturation

The two subunits of mitochondrial ribosomes are composed of 12S, 16S mt-rRNAs, and ribosomal proteins. Five and four nucleotide modifications have been identified to be associated with mammalian 12S and 16S mt-rRNAs, respectively. In 12S mt-rRNA, single methylation occurs at m^5U429 (mt-DNA position-1076), m^4C839 (mt-DNA position-1486), and m^5C841 (mt-DNA position-1488) and double methylation occurs at $m^6{}_2A936$ and $m^6{}_2A937$ (mt-DNA position 1583 and 1584) [105]. Only two of the proteins, transcription factor B1, mitochondrial (TFB1M), and NOP2/Sun RNA methyltransferase 4 (NSUN4) responsible for these 12S mt-rRNAs modifications are characterized to date [106,107]. In 16S mt-rRNA, ribose methylation occurs at G_m1145, G_m1370 and U_m1369 (mt-DNA position 2815, 3040, and 3039, respectively), and one pseudo-uridylated base at Psi1397 position (mt-DNA position 3067) [108,109]. Mitochondrial rRNA methyltransferases (MRM1, MRM2, and MRM3) are responsible for 2′-O-ribose methylation of 16S mt-rRNA at positions G_m1145, U_m1369, and G_m1370, respectively [110,111]. The enzyme for pseudo-uridylation at the 1397th position of 16S mt-rRNA is unknown to date. A recent study by Baar-Yaacov et al. observed the RNA-DNA differences at the 947th position of 16S mt-rRNA. They showed that this is a post-transcriptional modification and resulted from a 1-methyladenosine (m^1A) modification introduced by tRNA methyltransferase 61B (TRMT61B) enzyme. In addition, knockdown experiments and in vitro methylation assays have revealed that TRMT61B is the first enzyme that methylates both tRNA and rRNAs (Fig. 19.1) [112].

Mitochondrial tRNA maturation

After the release of mt-tRNAs from the polycistronic transcripts, numerous post-transcriptional modifications (CAA 3′-tail addition, isomerization, methylation, formylation, thiolation, ribosylation, etc.) occur. These modifications bring correct folding, stability, and appropriate functions to mt-tRNAs. The anticodon of the tRNAs "wobble" base (position 34) is one of the important sites in tRNA for modifications. This modification is important for expanding the codon recognition process during mitochondrial translation. Some of the proteins involved in this modification are NSUN3, ABH1, MTO1, GTPBP3, and MTU1 (Fig. 19.1) [113–116]. In addition to modifications at the 34th position, modifications at the 37th position are critical for maintaining translation fidelity and accuracy. Enzymes that are responsible for this modification include tRNA isopentenyltransferase 1 (TRIT1) and tRNA methyltransferase 5 (TRMT5), which introduce an isopentenyl group onto N^6 of 37th adenine and methyl group to the N^1 of the 37th guanosine, respectively [117,118]. Pseudouridine (Psi) is a structural isomer of the uridine base and provides stabilization to the RNA molecules through structural rigidity and has been detected in several mt-tRNAs. Some of the enzymes that are characterized for this modification include RPUSD4, RPUSD3, PUS1 (pseudouridine synthase 1), and TruB pseudouridine synthase family member 2 TRUB2 [119–121].

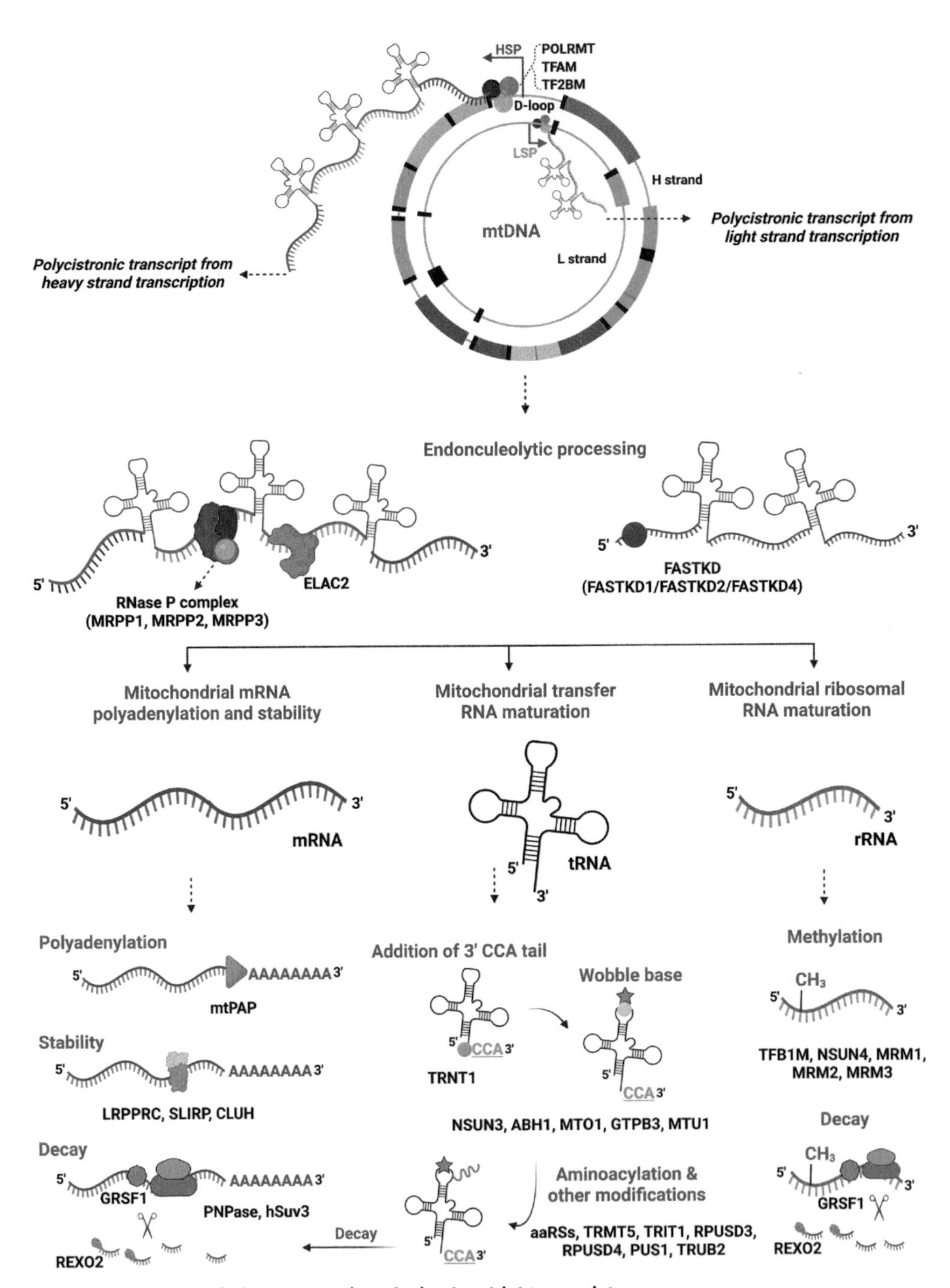

FIGURE 19.1 Post-transcriptional regulation of mitochondrial transcripts.

Two polycistronic transcripts are transcribed from each strand of the mitochondrial genome (green ribosomal RNA; black transfer RNA; blue messenger RNA). Proteins involved in the endonucleolytic processing are recruited to the polycistronic transcripts aiding in the generation of mRNA, tRNA, and rRNA precursors.

The 3′-CCA tail of the mt-tRNAs is not transcribed from mtDNA but synthesized by tRNA-nucleotidyltransferase 1 (TRNT1) as a part of post-transcriptional modification (Fig. 19.1) [122]. This addition of the CCA tail facilitates the attachment of the amino acid group required for translation. In addition, unstable tRNA species can be targeted for degradation by the addition of a double CCA tail can be added to Ref. [123]. Although the irregular polyadenylation of 16S mt-rRNA does not affect the integrity of the mitoribosome, reduced levels of an aminoacylated pool of certain mt-tRNAs can be caused by spurious additions of poly(A) to mt-tRNA. Poly(A)-specific 3′−5′ exoribonuclease, phosphodiesterase 12 (PDE12), removes the spurious poly(A) tails thus helping in quality control [124].

Post-transcriptional modification and mitochondrial disorders

Mitochondrial disorders are clinical manifestations that are the result of either spontaneous or inherited mutations in the mtDNA or nuclear DNA leading to the impaired functions of biomolecules that normally reside in mitochondria [3,125]. Mitochondrial disorders are usually associated with defects in oxidative phosphorylation and may be accompanied by the accumulation of metabolites such as lactate, pyruvate, and the generation of reactive oxygen species. They can occur at any stage during development either in the neonatal period, childhood, or adulthood, and can present with diverse clinical phenotypes involving a single tissue or multiple organ systems [126−128]. The spectrum of clinical and genetic heterogeneity of mitochondrial disorders has broadened over the years. To date, mutations in more than 245 nuclear-encoded genes have been associated with mitochondrial clinical manifestations, presenting with varying onsets and severity [14,129]. The major component involved in the pathology of these disorders is the nuclear-encoded mitochondrial proteins that account for 90% of the compendium of disease-associated genes. The consequences of genetic alterations in the nuclear genes can lead to catastrophic consequences concerning mitochondrial functions such as impaired bioenergetics, defects in mtDNA maintenance, and mtDNA copy number loss. Over the last few decades, studies on patients and families of those suffering from mitochondrial

RNase P complex and ELAC2 (cleaves at mt-tRNA 3′-ends from the primary transcript) are shown to bind to the transcripts that obey the tRNA punctuation model, where protein-coding and rRNA genes are interspersed by tRNA genes. FASTKD family of proteins is known to bind to the transcripts that do not follow this rule. mRNA, tRNA, and rRNA precursors undergo a sequential process of maturation and subsequent decay. mRNAs are polyadenylated by the mtPAP enzyme, and the complex of LRPPRC and SLIRP maintains the stability of these protein-coding transcripts. tRNAs precursors are modified by the addition of the 3′-CCA tail by TRNT1 protein. They undergo a series of modifications such as at the wobble base pair position (mediated by NSUN3, ABH1, MTO1, GTPB3, MTU1) and subsequent aminoacylation (aaRSs—aminoacyl-tRNA synthetases). Other modifications include the addition of isopentyl and methyl groups by TRIT1 and TRMT5 proteins, respectively. Pseudouridine modification is catalyzed by proteins RPUSD3, RPUSD4, PUS1, and TRUB2. rRNAs undergo maturation by the addition of the methyl group. This process is carried out by proteins such as MRM1, MRM2, and MRM3. RNA degradation or decay is modulated by proteins such as PNPase complex, hSuv3, REXO2, and GRSF1.

Figure was created with BioRender.com.

disorders have provided unique insights into understanding the genomic architecture of such mitochondrial disorders. Animal models harboring pathogenic clinical variations and patient-derived cell lines serve as ideal substrates to quench scientific myths and mysteries regarding mitochondrial disorders. In this chapter, we will focus on the post-transcriptional mitochondrial disorders caused by mutations in nuclear-encoded mitochondrial genes. Post-transcriptional mitochondrial disorders (PTMD) [130–133] have been studied in much detail in various model organisms such as mice, zebrafish, yeast, and drosophila and have helped in envisaging our understanding of the pathophysiology of mitochondrial disorders. We have classified the diseases and disease model studies according to the post-transcriptional pathway.

Defects in mitochondrial mRNA polyadenylation and stability

Mutations in the *LRPPRC* gene are associated with a rare pediatric mitochondrial disorder, Leigh Syndrome French-Canadian (LSFC) type. It was first discovered in the Saguenay-Lac-Saint-Jean region of Quebec in Canada where it was reported that 1 out of 23 individuals carries the allele harboring a pathogenic variant of this gene, predominant of which is p.Ala354Val transition. Patients with mutations in this gene display an early onset of chronic lactic acidosis, hepatic microvesicular steatosis, and cytochrome *c* oxidase deficiency. The lifespan of these patients is significantly reduced to early childhood that is marked by episodes of neurological and metabolic crisis [57,134–136]. Liver-specific knockout mice mimic the symptoms observed in LSFC patients such as growth delay, mitochondrial hepatopathy, reduced fatty acid oxidation along with a decrease in the activity of cytochrome *c* oxidase and ATP synthase [137]. A gene-trap revertible zebrafish *lrpprc* mutant phenocopies the hallmarks of LSFC such as early lethality, muscle defects, and decrease in endogenous transcript levels. Zebrafish mutants also show defects in lipid homeostasis marked by the accumulation of triglycerides. Reversion of the mutagenicity cassette specifically in the liver leads to the reversion of severe phenotypes such as larval lethality and lipid accumulation [138]. A very recent study by Guo et al. identified two compound heterozygous mutations in *SLIRP* in patients affected by congenital hypotonia, cerebral atrophy, complex I, and IV deficiency. It was observed that in the patient-derived fibroblasts, expression of *MT-ND1*, *MT-ND6*, and *MT-CO1* was reduced. Pathogenicity of this variation was confirmed by complementation studies carried out using wild-type *SLIRP* cDNA in fibroblasts [139]. Interestingly, insights from knockout studies suggest that *Slirp*-deficient mice are healthy, despite a reduction in the steady-state levels of mitochondrial encoded proteins [61]. Mutations in *ELAC2* are associated with combined OXPHOS deficiencies. Individuals harboring p.Phe154Leu substitution in the Metallo β-lactamase domain display clinical phenotypes such as lactic acidosis, cardiomyopathy, infant mortality accompanied by accumulation of mt-tRNA intermediates [140]. *ELAC2* variants have been correlated with infantile-onset hypertrophic cardiomyopathy and respiratory chain deficiency [141]. Apart from these predominant phenotypes, intellectual disability with minimal cardiac involvement was also observed in a patient carrying homozygous splicing mutation at the *ELAC2* genetic locus [142]. Defects caused by deficiency of *Elac2* have been successfully modeled in mice. Knockout models harboring loss of function-specific for the heart and skeletal muscle tissue display dilated cardiomyopathy and premature death [143]. They recapitulate the in vitro findings wherein accumulation of unprocessed mt-tRNAs is observed due to the loss of endonuclease activity [39]. Affected members of a large consanguineous family were shown to carry homozygous sequence variants in the *mtPAP* gene. In this case, individuals presented with cerebellar ataxia, spastic paraparesis, and optic atrophy. An increased percentage of oligoadenylated transcripts as compared to polyadenylated was also observed in patients [144]. In another study, loss of protein encoded by this gene has been involved in the disruption of ROS homeostasis recognizing its role in radiosensitivity and DNA double-strand break repair [145].

PNPT1 plays an important role in RNA processing and degradation, wherein it forms a degradosome complex with mitochondrial helicase, SUPV3L1 to mediate in RNA processing via its 3′-5′-endonuclease activity. Genetic investigations of a consanguineous Moroccan family revealed that homozygous carriers were affected by hearing impairment [146]. Other reports that have documented the pathogenic variants in this gene describe patients presenting with clinical phenotypes such as Leigh syndrome, myoclonic epilepsy, seizures, hypotonia, developmental delay, optic atrophy, and sensorineural deafness [147−149]. Patients harboring mutations in this gene often present themselves with symptoms such as neurodevelopmental defects, microcephaly, hypotonia, and in some cases with hearing loss [146,150]. Whole-body knockout mice for *Pnpt1* were embryonic lethal. However, liver-specific knockout mice were viable and showed mitochondria with altered cristae and decreased respiration [91]. Shimada and colleagues studied the PNPase knockout specific to the inner ear hair cells of mice and showed the knockout of this protein leads to progressive hearing loss [151]. Loss of function studies for the SUPV3L1 ortholog in *Drosophila melanogaster* demonstrated that mutants exhibited pupal lethal phenotype accompanied by a reduced mitochondrial translation and severe respiratory chain complex deficiency [152]. In mice models, impaired function for this gene has been associated with loss of adipose tissue muscle mass, where it also plays a role in the maintenance of skin barrier and normal growth aging [153]. In another study, knockdown of Grsf1 in mice caused a series of phenotypes such as increased lipid peroxidation, mental retardation, and apoptotic alterations [154]. Tissue-specific ablation of Grsf1, involved in mtRNA degradation, is linked to weak muscle endurance in the mice model system. Mice knockout models have been established to elucidate the role of REXO2, an RNA exonuclease. Knockout of this gene results in embryonic lethality, highlighting its role in embryonic development [155]. Mouse *Cluh* knockout models exhibit normal embryonic development but neonatal lethality. As expected, deficiency of Cluh protein in mice leads to down-regulation of proteins involved in amino acid metabolism, tricarboxylic acid cycle, fatty acid degradation, and oxidative phosphorylation. $Cluh^{-/-}$ mice also develop hypoglycemia after birth accompanied by respiratory deficiency, reduced steady-state levels of nuclear-encoded mitochondrial resident protein mRNAs, and abnormal mitochondrial morphology with a reduction in the mitochondrial matrix density. Liver-specific knockout mice have an impeded ability to adapt their metabolism to high energy requirements that affect glucose homeostasis and production of ketone bodies during starvation, highlighting the role of this protein in specific metabolism niches rather than general mitochondrial function overall [82]. Deficiency of PTCD2 in mice leads to defects in the myocardium with the reduction in the activity of ubiquinol-cytochrome c reductase complex in energy-demanding tissues such as heart, liver, and skeletal muscle, indicating its role in the processing of transcripts from the cytochrome *b* complex [46].

Combined oxidative phosphorylation deficiency caused by mutations in *FASTKD2* has been successfully modeled in zebrafish. Zebrafish morphants display multiple OXPHOS deficiencies with slow heart rates [156]. These phenotypes partially overlap with the clinical manifestations associated with this gene such as sinus tachycardia and hypertrophic cardiomyopathy. They did not exhibit neurological phenotypes such as visual activity and movement ability unlike what was documented in patients harboring mutations in these genes [157]. Whole-exome sequencing carried out on patients presenting with symptoms such as lactic acidosis, feeding difficulties and deafness revealed pathogenic variation in the *MRPP1* gene. These phenotypes were associated with infant mortality as observed in the patients. It was also observed that there was an increase in the levels of mtRNA precursors indicative of impaired mtRNA processing and thereby affecting mitochondrial protein synthesis [158]. Studies shedding light on the clinical significance of *MRPP2* mention that mutations in this genetic locus present with symptoms such as seizures loss of visual activity coupled with mitochondrial dysfunction [159,160]. Knockout mice models established to study loss of function for *MRPP3* gene reveal that this protein is essential for embryonic development, loss of which leads to embryonic lethality. Heart and skeletal muscle-specific mice knockouts displayed severe cardiomyopathy. Accumulation of necrotic foci was seen in the myocardium and reduced muscle fiber appearance was observed in the tissue-specific knockout animals. The altered proteomic signature was also seen in the tissues with a significant reduction in the protein synthesis of mitochondrial encoded proteins affecting the biogenesis of the electron transport chain [161].

Defects in mitochondrial rRNA maturation

Genetic players involved in post-transcriptional maturation of mt-rRNA have been associated with the risk of progression of diseases such as diabetes, roles of which have been elucidated using studies carried out in animal model systems. Genome-wide association studies (GWAS) have identified a risk variant (rs950994) in the *TFB1M* gene that has been associated with reduced insulin secretion and developing type 2 diabetes in the future [162]. These findings have been corroborated with β-cell-specific knockout mice models that develop diabetes due to impaired insulin secretion. Aberrant mitochondrial architecture, reduction in *12S rRNA* methylation and expression of mtDNA encoded proteins, impaired ATP production and elevation of ROS are some of the other phenotypes observed in *Tfb1m*$^{-/-}$ mice [163]. Heart-specific knockout for this gene leads to progressive cardiomyopathy accompanied by reduced 12s rRNA methylation in the cardiac tissue [164]. The deficiency of Nsun4 involved in the methylation of *12S rRNA* and mitoribosome biogenesis is reported to be embryonic lethal in mice [165]. Targeted exome sequencing of nuclear genes encoding for mitochondrial resident proteins identified a homozygous mutation resulting in p.Gly189Arg substitution. Individuals carrying this mutation presented with symptoms resembling MELAS such as progressive encephalopathy and stroke-like acidosis. OXPHOS defects and mtDNA depletion were observed in the muscle homogenate. Functional validation of this variant was carried out in yeast, where it was shown that knockout strains exhibited respiratory defects and reduced 2′-O-methyl modification at the U2791 position of yeast ortholog of *21S rRNA* [166].

Defects in mitochondrial tRNA maturation

Mutations in the *TRIT1* gene are associated with combined oxidative phosphorylation deficiency accompanied by defective isopentenyl transferase modification of mitochondrial and cytosolic tRNAs [117]. Deletion of *Trit1* in mice leads to embryonic lethality indicating its role in embryonic development [167]. Patients harboring pathogenic variations in the *TRNT5* gene displayed phenotype OXPHOS defects [168] similar to that observed with other genetic loci involved in the mitochondria tRNA modification. TRNT1 is another player for which clinical phenotypes have been reported. Cytotoxicity and apoptosis were observed upon silencing the expression of the *TRNT1* gene in fibroblasts. Pathogenic mutations in this gene lead to manifestations ranging from sideroblastic anemia, B-cell immunodeficiency, fever, developmental delay, and retinitis pigmentosa [169,170]. *trnt1* zebrafish morphants exhibited cardiovascular defects, anemia with reduced visual activity [170]. Patients with myopathy, lactic acidosis, and sideroblastic anemia have been identified as carrying pathogenic mutations in the *PUS1* gene [171,172]. Whole-exome sequencing studies have established the pathogenicity of the *GTPBP3* gene in individuals presenting with hypertrophic cardiomyopathy [173]. These clinical observations have been recapitulated in animal models such as zebrafish that phenocopy these dysfunctions displaying embryonic heart defects in specific hypertrophic cardiomyopathy marked by aberrant mitochondrial tRNA metabolism [174].

Summary

Here, in this review, we have summarized the post-transcriptional modification of mitochondrial RNA and how it is regulated by the nuclear genome. Genetic alterations in this regulatory process had been associated with different clinical phenotypes that we have described in this chapter. The holistic clinical investigation, biohistochemical studies, and advancements in genomics and proteomics technologies have enabled our understanding of the crosstalk between the nuclear and mitochondrial genomes. These advancements have not only led to the identification and characterization of novel mitochondrial candidates in post-transcriptional regulation but have also enhanced our knowledge in understanding the pathophysiology of the diseases caused by pathogenic variations in these genes. More mechanistic studies are required to understand the regulation of post-transcriptional regulation of mitochondrial transcripts. The recent discovery of noncoding RNAs inside the mitochondrial has added an extra dimension wherein more investigations are required. It will be interesting and scientifically rewarding to understand the role of such candidates in the mitochondrial genome by carrying out mechanistic studies in the cellular and animal model systems. It is important to identify such regulatory molecules that can, in the future, be validated as potential therapeutic candidates for the management and treatment of PTMD disorders.

References

[1] Muller M, Martin W. The genome of *Rickettsia prowazekii* and some thoughts on the origin of mitochondria and hydrogenosomes. Bioessays 1999;21(5):377–81.

[2] Andersson SG, Zomorodipour A, Andersson JO, Sicheritz-Ponten T, Alsmark UC, Podowski RM, et al. The genome sequence of *Rickettsia prowazekii* and the origin of mitochondria. Nature 1998;396(6707):133–40.

[3] Vafai SB, Mootha VK. Mitochondrial disorders as windows into an ancient organelle. Nature 2012; 491(7424):374–83.

[4] Rizzuto R, Marchi S, Bonora M, Aguiari P, Bononi A, De Stefani D, et al. Ca(2+) transfer from the ER to mitochondria: when, how and why. Biochim Biophys Acta 2009;1787(11):1342–51.

[5] Valero T. Mitochondrial biogenesis: pharmacological approaches. Curr Pharmaceut Des 2014;20(35): 5507–9.

[6] Green DR. Apoptotic pathways: the roads to ruin. Cell 1998;94(6):695–8.

[7] Adhya S, Mahato B, Jash S, Koley S, Dhar G, Chowdhury T. Mitochondrial gene therapy: the tortuous path from bench to bedside. Mitochondrion 2011;11(6):839–44.

[8] Rath S, Sharma R, Gupta R, Ast T, Chan C, Durham TJ, et al. MitoCarta3.0: an updated mitochondrial proteome now with sub-organelle localization and pathway annotations. Nucleic Acids Res 2021;49(D1): D1541–7.
[9] Kotrys AV, Szczesny RJ. Mitochondrial gene expression and beyond-novel aspects of cellular physiology. Cells 2019;9(1).
[10] Anderson S, Bankier AT, Barrell BG, de Bruijn MH, Coulson AR, Drouin J, et al. Sequence and organization of the human mitochondrial genome. Nature 1981;290(5806):457–65.
[11] Andrews RM, Kubacka I, Chinnery PF, Lightowlers RN, Turnbull DM, Howell N. Reanalysis and revision of the Cambridge reference sequence for human mitochondrial DNA. Nat Genet 1999;23(2):147.
[12] Taanman JW. The mitochondrial genome: structure, transcription, translation and replication. Biochim Biophys Acta 1999;1410(2):103–23.
[13] Elson JL, Andrews RM, Chinnery PF, Lightowlers RN, Turnbull DM, Howell N. Analysis of European mtDNAs for recombination. Am J Hum Genet 2001;68(1):145–53.
[14] Craven L, Alston CL, Taylor RW, Turnbull DM. Recent advances in mitochondrial disease. Annu Rev Genom Hum Genet 2017;18:257–75.
[15] Moraes CT. What regulates mitochondrial DNA copy number in animal cells? Trends Genet 2001;17(4): 199–205.
[16] Trinei M, Berniakovich I, Pelicci PG, Giorgio M. Mitochondrial DNA copy number is regulated by cellular proliferation: a role for Ras and p66(Shc). Biochim Biophys Acta 2006;1757(5–6):624–30.
[17] Hori A, Yoshida M, Shibata T, Ling F. Reactive oxygen species regulate DNA copy number in isolated yeast mitochondria by triggering recombination-mediated replication. Nucleic Acids Res 2009;37(3):749–61.
[18] Clay Montier LL, Deng JJ, Bai Y. Number matters: control of mammalian mitochondrial DNA copy number. J Genet Genomics 2009;36(3):125–31.
[19] Kelly RD, Mahmud A, McKenzie M, Trounce IA, St John JC. Mitochondrial DNA copy number is regulated in a tissue specific manner by DNA methylation of the nuclear-encoded DNA polymerase gamma A. Nucleic Acids Res 2012;40(20):10124–38.
[20] Taylor RW, Turnbull DM. Mitochondrial DNA mutations in human disease. Nat Rev Genet 2005;6(5): 389–402.
[21] Chang DD, Clayton DA. Precise identification of individual promoters for transcription of each strand of human mitochondrial DNA. Cell 1984;36(3):635–43.
[22] Aloni Y, Attardi G. Expression of the mitochondrial genome in HeLa cells. II. Evidence for complete transcription of mitochondrial DNA. J Mol Biol 1971;55(2):251–67.
[23] Aloni Y, Attardi G. Symmetrical in vivo transcription of mitochondrial DNA in HeLa cells. Proc Natl Acad Sci U S A 1971;68(8):1757–61.
[24] Montoya J, Christianson T, Levens D, Rabinowitz M, Attardi G. Identification of initiation sites for heavy-strand and light-strand transcription in human mitochondrial DNA. Proc Natl Acad Sci U S A 1982;79(23): 7195–9.
[25] Morozov YI, Agaronyan K, Cheung AC, Anikin M, Cramer P, Temiakov D. A novel intermediate in transcription initiation by human mitochondrial RNA polymerase. Nucleic Acids Res 2014;42(6):3884–93.
[26] Hillen HS, Temiakov D, Cramer P. Structural basis of mitochondrial transcription. Nat Struct Mol Biol 2018;25(9):754–65.
[27] Hillen HS, Morozov YI, Sarfallah A, Temiakov D, Cramer P. Structural basis of mitochondrial transcription initiation. Cell 2017;171(5):1072–1081 e10.
[28] Minczuk M, He J, Duch AM, Ettema TJ, Chlebowski A, Dzionek K, et al. TEFM (c17orf42) is necessary for transcription of human mtDNA. Nucleic Acids Res 2011;39(10):4284–99.
[29] Posse V, Shahzad S, Falkenberg M, Hallberg BM, Gustafsson CM. TEFM is a potent stimulator of mitochondrial transcription elongation in vitro. Nucleic Acids Res 2015;43(5):2615–24.

[30] Pham XH, Farge G, Shi Y, Gaspari M, Gustafsson CM, Falkenberg M. Conserved sequence box II directs transcription termination and primer formation in mitochondria. J Biol Chem 2006;281(34):24647–52.
[31] Wanrooij PH, Uhler JP, Simonsson T, Falkenberg M, Gustafsson CM. G-quadruplex structures in RNA stimulate mitochondrial transcription termination and primer formation. Proc Natl Acad Sci U S A 2010; 107(37):16072–7.
[32] Jimenez-Menendez N, Fernandez-Millan P, Rubio-Cosials A, Arnan C, Montoya J, Jacobs HT, et al. Human mitochondrial mTERF wraps around DNA through a left-handed superhelical tandem repeat. Nat Struct Mol Biol 2010;17(7):891–3.
[33] Martin M, Cho J, Cesare AJ, Griffith JD, Attardi G. Termination factor-mediated DNA loop between termination and initiation sites drives mitochondrial rRNA synthesis. Cell 2005;123(7):1227–40.
[34] Terzioglu M, Ruzzenente B, Harmel J, Mourier A, Jemt E, Lopez MD, et al. MTERF1 binds mtDNA to prevent transcriptional interference at the light-strand promoter but is dispensable for rRNA gene transcription regulation. Cell Metabol 2013;17(4):618–26.
[35] Hyvarinen AK, Kumanto MK, Marjavaara SK, Jacobs HT. Effects on mitochondrial transcription of manipulating mTERF protein levels in cultured human HEK293 cells. BMC Mol Biol 2010;11:72.
[36] Yakubovskaya E, Mejia E, Byrnes J, Hambardjieva E, Garcia-Diaz M. Helix unwinding and base flipping enable human MTERF1 to terminate mitochondrial transcription. Cell 2010;141(6):982–93.
[37] Ojala D, Montoya J, Attardi G. tRNA punctuation model of RNA processing in human mitochondria. Nature 1981;290(5806):470–4.
[38] Holzmann J, Frank P, Loffler E, Bennett KL, Gerner C, Rossmanith W. RNase P without RNA: identification and functional reconstitution of the human mitochondrial tRNA processing enzyme. Cell 2008; 135(3):462–74.
[39] Brzezniak LK, Bijata M, Szczesny RJ, Stepien PP. Involvement of human ELAC2 gene product in 3′end processing of mitochondrial tRNAs. RNA Biol 2011;8(4):616–26.
[40] Rossmanith W. Localization of human RNase Z isoforms: dual nuclear/mitochondrial targeting of the ELAC2 gene product by alternative translation initiation. PLoS One 2011;6(4):e19152.
[41] Sanchez MI, Mercer TR, Davies SM, Shearwood AM, Nygard KK, Richman TR, et al. RNA processing in human mitochondria. Cell Cycle 2011;10(17):2904–16.
[42] Mercer TR, Gerhardt DJ, Dinger ME, Crawford J, Trapnell C, Jeddeloh JA, et al. Targeted RNA sequencing reveals the deep complexity of the human transcriptome. Nat Biotechnol 2011;30(1):99–104.
[43] Boehm E, Zaganelli S, Maundrell K, Jourdain AA, Thore S, Martinou JC. FASTKD1 and FASTKD4 have opposite effects on expression of specific mitochondrial RNAs, depending upon their endonuclease-like RAP domain. Nucleic Acids Res 2017;45(10):6135–46.
[44] Popow J, Alleaume AM, Curk T, Schwarzl T, Sauer S, Hentze MW. FASTKD2 is an RNA-binding protein required for mitochondrial RNA processing and translation. RNA 2015;21(11):1873–84.
[45] Antonicka H, Shoubridge EA. Mitochondrial RNA granules are centers for post-transcriptional RNA processing and ribosome biogenesis. Cell Rep 2015;10(6):920–32.
[46] Xu F, Ackerley C, Maj MC, Addis JB, Levandovskiy V, Lee J, et al. Disruption of a mitochondrial RNA-binding protein gene results in decreased cytochrome b expression and a marked reduction in ubiquinol-cytochrome c reductase activity in mouse heart mitochondria. Biochem J 2008;416(1):15–26.
[47] Temperley RJ, Wydro M, Lightowlers RN, Chrzanowska-Lightowlers ZM. Human mitochondrial mRNAs–like members of all families, similar but different. Biochim Biophys Acta 2010;1797(6–7):1081–5.
[48] Nagaike T, Suzuki T, Katoh T, Ueda T. Human mitochondrial mRNAs are stabilized with polyadenylation regulated by mitochondria-specific poly(A) polymerase and polynucleotide phosphorylase. J Biol Chem 2005;280(20):19721–7.

[49] Lapkouski M, Hallberg BM. Structure of mitochondrial poly(A) RNA polymerase reveals the structural basis for dimerization, ATP selectivity and the SPAX4 disease phenotype. Nucleic Acids Res 2015;43(18): 9065–75.
[50] Bai Y, Srivastava SK, Chang JH, Manley JL, Tong L. Structural basis for dimerization and activity of human PAPD1, a noncanonical poly(A) polymerase. Mol Cell 2011;41(3):311–20.
[51] Tomecki R, Dmochowska A, Gewartowski K, Dziembowski A, Stepien PP. Identification of a novel human nuclear-encoded mitochondrial poly(A) polymerase. Nucleic Acids Res 2004;32(20):6001–14.
[52] Rorbach J, Nicholls TJ, Minczuk M. PDE12 removes mitochondrial RNA poly(A) tails and controls translation in human mitochondria. Nucleic Acids Res 2011;39(17):7750–63.
[53] Wydro M, Bobrowicz A, Temperley RJ, Lightowlers RN, Chrzanowska-Lightowlers ZM. Targeting of the cytosolic poly(A) binding protein PABPC1 to mitochondria causes mitochondrial translation inhibition. Nucleic Acids Res 2010;38(11):3732–42.
[54] Sterky FH, Ruzzenente B, Gustafsson CM, Samuelsson T, Larsson NG. LRPPRC is a mitochondrial matrix protein that is conserved in metazoans. Biochem Biophys Res Commun 2010;398(4):759–64.
[55] Ruzzenente B, Metodiev MD, Wredenberg A, Bratic A, Park CB, Camara Y, et al. LRPPRC is necessary for polyadenylation and coordination of translation of mitochondrial mRNAs. EMBO J 2012;31(2):443–56.
[56] Sasarman F, Brunel-Guitton C, Antonicka H, Wai T, Shoubridge EA, Consortium L. LRPPRC and SLIRP interact in a ribonucleoprotein complex that regulates post-transcriptional gene expression in mitochondria. Mol Biol Cell 2010;21(8):1315–23.
[57] Gohil VM, Nilsson R, Belcher-Timme CA, Luo B, Root DE, Mootha VK. Mitochondrial and nuclear genomic responses to loss of LRPPRC expression. J Biol Chem 2010;285(18):13742–7.
[58] Wilson WC, Hornig-Do HT, Bruni F, Chang JH, Jourdain AA, Martinou JC, et al. A human mitochondrial poly(A) polymerase mutation reveals the complexities of post-transcriptional mitochondrial gene expression. Hum Mol Genet 2014;23(23):6345–55.
[59] Siira SJ, Spahr H, Shearwood AJ, Ruzzenente B, Larsson NG, Rackham O, et al. LRPPRC-mediated folding of the mitochondrial transcriptome. Nat Commun 2017;8(1):1532.
[60] Baughman JM, Nilsson R, Gohil VM, Arlow DH, Gauhar Z, Mootha VK. A computational screen for regulators of oxidative phosphorylation implicates SLIRP in mitochondrial RNA homeostasis. PLoS Genet 2009;5(8):e1000590.
[61] Lagouge M, Mourier A, Lee HJ, Spahr H, Wai T, Kukat C, et al. SLIRP regulates the rate of mitochondrial protein synthesis and protects LRPPRC from degradation. PLoS Genet 2015;11(8):e1005423.
[62] Chujo T, Ohira T, Sakaguchi Y, Goshima N, Nomura N, Nagao A, et al. LRPPRC/SLIRP suppresses PNPase-mediated mRNA decay and promotes polyadenylation in human mitochondria. Nucleic Acids Res 2012;40(16):8033–47.
[63] Zhang R, Wang J. HuR stabilizes TFAM mRNA in an ATM/p38-dependent manner in ionizing irradiated cancer cells. Cancer Sci 2018;109(8):2446–57.
[64] Lee JH, Jung M, Hong J, Kim MK, Chung IK. Loss of RNA-binding protein HuR facilitates cellular senescence through post-transcriptional regulation of TIN2 mRNA. Nucleic Acids Res 2018;46(8): 4271–85.
[65] Durie D, Hatzoglou M, Chakraborty P, Holcik M. HuR controls mitochondrial morphology through the regulation of BclxL translation. Translation 2013;1(1).
[66] Minczuk M, Piwowarski J, Papworth MA, Awiszus K, Schalinski S, Dziembowski A, et al. Localisation of the human hSuv3p helicase in the mitochondrial matrix and its preferential unwinding of dsDNA. Nucleic Acids Res 2002;30(23):5074–86.
[67] Shu Z, Vijayakumar S, Chen CF, Chen PL, Lee WH. Purified human SUV3p exhibits multiple-substrate unwinding activity upon conformational change. Biochemistry 2004;43(16):4781–90.

[68] Szczesny RJ, Borowski LS, Brzezniak LK, Dmochowska A, Gewartowski K, Bartnik E, et al. Human mitochondrial RNA turnover caught in flagranti: involvement of hSuv3p helicase in RNA surveillance. Nucleic Acids Res 2010;38(1):279–98.
[69] Antonicka H, Sasarman F, Nishimura T, Paupe V, Shoubridge EA. The mitochondrial RNA-binding protein GRSF1 localizes to RNA granules and is required for post-transcriptional mitochondrial gene expression. Cell Metabol 2013;17(3):386–98.
[70] Noh JH, Kim KM, Pandey PR, Noren Hooten N, Munk R, Kundu G, et al. Loss of RNA-binding protein GRSF1 activates mTOR to elicit a proinflammatory transcriptional program. Nucleic Acids Res 2019; 47(5):2472–86.
[71] Jourdain AA, Koppen M, Wydro M, Rodley CD, Lightowlers RN, Chrzanowska-Lightowlers ZM, et al. GRSF1 regulates RNA processing in mitochondrial RNA granules. Cell Metabol 2013;17(3):399–410.
[72] Szewczyk M, Malik D, Borowski LS, Czarnomska SD, Kotrys AV, Klosowska-Kosicka K, et al. Human REXO2 controls short mitochondrial RNAs generated by mtRNA processing and decay machinery to prevent accumulation of double-stranded RNA. Nucleic Acids Res 2020;48(10):5572–90.
[73] Bruni F, Gramegna P, Oliveira JM, Lightowlers RN, Chrzanowska-Lightowlers ZM. REXO2 is an oligoribonuclease active in human mitochondria. PLoS One 2013;8(5):e64670.
[74] Garcia-Rodriguez LJ, Gay AC, Pon LA. Puf3p, a Pumilio family RNA-binding protein, localizes to mitochondria and regulates mitochondrial biogenesis and motility in budding yeast. J Cell Biol 2007; 176(2):197–207.
[75] Gerber AP, Herschlag D, Brown PO. Extensive association of functionally and cytotopically related mRNAs with Puf family RNA-binding proteins in yeast. PLoS Biol 2004;2(3):E79.
[76] Lesnik C, Golani-Armon A, Arava Y. Localized translation near the mitochondrial outer membrane: an update. RNA Biol 2015;12(8):801–9.
[77] Lapointe CP, Stefely JA, Jochem A, Hutchins PD, Wilson GM, Kwiecien NW, et al. Multi-omics reveal specific targets of the RNA-binding protein Puf3p and its orchestration of mitochondrial biogenesis. Cell Syst 2018;6(1):125–135 e6.
[78] Foat BC, Houshmandi SS, Olivas WM, Bussemaker HJ. Profiling condition-specific, genome-wide regulation of mRNA stability in yeast. Proc Natl Acad Sci U S A 2005;102(49):17675–80.
[79] Ravanidis S, Doxakis E. RNA-binding proteins implicated in mitochondrial damage and mitophagy. Front Cell Dev Biol 2020;8:372.
[80] Matsumoto S, Uchiumi T, Tanamachi H, Saito T, Yagi M, Takazaki S, et al. Ribonucleoprotein Y-box-binding protein-1 regulates mitochondrial oxidative phosphorylation (OXPHOS) protein expression after serum stimulation through binding to OXPHOS mRNA. Biochem J 2012;443(2):573–84.
[81] Gao J, Schatton D, Martinelli P, Hansen H, Pla-Martin D, Barth E, et al. CLUH regulates mitochondrial biogenesis by binding mRNAs of nuclear-encoded mitochondrial proteins. J Cell Biol 2014;207(2): 213–23.
[82] Schatton D, Pla-Martin D, Marx MC, Hansen H, Mourier A, Nemazanyy I, et al. CLUH regulates mitochondrial metabolism by controlling translation and decay of target mRNAs. J Cell Biol 2017;216(3): 675–93.
[83] Chang DD, Clayton DA. A mammalian mitochondrial RNA processing activity contains nucleus-encoded RNA. Science 1987;235(4793):1178–84.
[84] Doersen CJ, Guerrier-Takada C, Altman S, Attardi G. Characterization of an RNase P activity from HeLa cell mitochondria. Comparison with the cytosol RNase P activity. J Biol Chem 1985;260(10):5942–9.
[85] Puranam RS, Attardi G. The RNase P associated with HeLa cell mitochondria contains an essential RNA component identical in sequence to that of the nuclear RNase P. Mol Cell Biol 2001;21(2):548–61.
[86] Evans D, Marquez SM, Pace NR. RNase P. Interface of the RNA and protein worlds. Trends Biochem Sci 2006;31(6):333–41.

[87] Bian Z, Li LM, Tang R, Hou DX, Chen X, Zhang CY, et al. Identification of mouse liver mitochondria-associated miRNAs and their potential biological functions. Cell Res 2010;20(9):1076–8.
[88] Srinivasan H, Das S. Mitochondrial miRNA (MitomiR): a new player in cardiovascular health. Can J Physiol Pharmacol 2015;93(10):855–61.
[89] Chu S, Archer RH, Zengel JM, Lindahl L. The RNA of RNase MRP is required for normal processing of ribosomal RNA. Proc Natl Acad Sci U S A 1994;91(2):659–63.
[90] Chang DD, Clayton DA. Mouse RNAase MRP RNA is encoded by a nuclear gene and contains a decamer sequence complementary to a conserved region of mitochondrial RNA substrate. Cell 1989;56(1):131–9.
[91] Wang G, Chen HW, Oktay Y, Zhang J, Allen EL, Smith GM, et al. PNPASE regulates RNA import into mitochondria. Cell 2010;142(3):456–67.
[92] Noh JH, Kim KM, Abdelmohsen K, Yoon JH, Panda AC, Munk R, et al. HuR and GRSF1 modulate the nuclear export and mitochondrial localization of the lncRNA RMRP. Genes Dev 2016;30(10):1224–39.
[93] Lee Y, Ahn C, Han J, Choi H, Kim J, Yim J, et al. The nuclear RNase III Drosha initiates microRNA processing. Nature 2003;425(6956):415–9.
[94] Moretti F, Thermann R, Hentze MW. Mechanism of translational regulation by miR-2 from sites in the 5' untranslated region or the open reading frame. RNA 2010;16(12):2493–502.
[95] Purohit PK, Saini N. Mitochondrial microRNA (MitomiRs) in cancer and complex mitochondrial diseases: current status and future perspectives. Cell Mol Life Sci 2021;78(4):1405–21.
[96] Das S, Bedja D, Campbell N, Dunkerly B, Chenna V, Maitra A, et al. miR-181c regulates the mitochondrial genome, bioenergetics, and propensity for heart failure in vivo. PLoS One 2014;9(5):e96820.
[97] Kim J, Krichevsky A, Grad Y, Hayes GD, Kosik KS, Church GM, et al. Identification of many microRNAs that copurify with polyribosomes in mammalian neurons. Proc Natl Acad Sci U S A 2004;101(1):360–5.
[98] Wienholds E, Kloosterman WP, Miska E, Alvarez-Saavedra E, Berezikov E, de Bruijn E, et al. MicroRNA expression in zebrafish embryonic development. Science 2005;309(5732):310–1.
[99] Aschrafi A, Schwechter AD, Mameza MG, Natera-Naranjo O, Gioio AE, Kaplan BB. MicroRNA-338 regulates local cytochrome c oxidase IV mRNA levels and oxidative phosphorylation in the axons of sympathetic neurons. J Neurosci 2008;28(47):12581–90.
[100] Chan SY, Zhang YY, Hemann C, Mahoney CE, Zweier JL, Loscalzo J. MicroRNA-210 controls mitochondrial metabolism during hypoxia by repressing the iron-sulfur cluster assembly proteins ISCU1/2. Cell Metabol 2009;10(4):273–84.
[101] Bordone L, Motta MC, Picard F, Robinson A, Jhala US, Apfeld J, et al. Sirt1 regulates insulin secretion by repressing UCP2 in pancreatic beta cells. PLoS Biol 2006;4(2):e31.
[102] Bandiera S, Ruberg S, Girard M, Cagnard N, Hanein S, Chretien D, et al. Nuclear outsourcing of RNA interference components to human mitochondria. PLoS One 2011;6(6):e20746.
[103] Sabharwal A, Sharma D, Vellarikkal SK, Jayarajan R, Verma A, Senthivel V, et al. Organellar transcriptome sequencing reveals mitochondrial localization of nuclear-encoded transcripts. Mitochondrion 2019;46:59–68.
[104] Barrey E, Saint-Auret G, Bonnamy B, Damas D, Boyer O, Gidrol X. Pre-microRNA and mature microRNA in human mitochondria. PLoS One 2011;6(5):e20220.
[105] Baer RJ, Dubin DT. Methylated regions of hamster mitochondrial ribosomal RNA: structural and functional correlates. Nucleic Acids Res 1981;9(2):323–37.
[106] Metodiev MD, Lesko N, Park CB, Camara Y, Shi Y, Wibom R, et al. Methylation of 12S rRNA is necessary for in vivo stability of the small subunit of the mammalian mitochondrial ribosome. Cell Metabol 2009;9(4):386–97.
[107] Spahr H, Habermann B, Gustafsson CM, Larsson NG, Hallberg BM. Structure of the human MTERF4-NSUN4 protein complex that regulates mitochondrial ribosome biogenesis. Proc Natl Acad Sci U S A 2012;109(38):15253–8.

[108] Dubin DT, Taylor RH. Modification of mitochondrial ribosomal RNA from hamster cells: the presence of GmG and late-methylated UmGmU in the large subunit (17S) RNA. J Mol Biol 1978;121(4):523–40.

[109] Ofengand J, Bakin A. Mapping to nucleotide resolution of pseudouridine residues in large subunit ribosomal RNAs from representative eukaryotes, prokaryotes, archaebacteria, mitochondria and chloroplasts. J Mol Biol 1997;266(2):246–68.

[110] Lee KW, Okot-Kotber C, LaComb JF, Bogenhagen DF. Mitochondrial ribosomal RNA (rRNA) methyltransferase family members are positioned to modify nascent rRNA in foci near the mitochondrial DNA nucleoid. J Biol Chem 2013;288(43):31386–99.

[111] Lee KW, Bogenhagen DF. Assignment of 2'-O-methyltransferases to modification sites on the mammalian mitochondrial large subunit 16 S ribosomal RNA (rRNA). J Biol Chem 2014;289(36):24936–42.

[112] Bar-Yaacov D, Frumkin I, Yashiro Y, Chujo T, Ishigami Y, Chemla Y, et al. Mitochondrial 16S rRNA is methylated by tRNA methyltransferase TRMT61B in all vertebrates. PLoS Biol 2016;14(9):e1002557.

[113] Haag S, Sloan KE, Ranjan N, Warda AS, Kretschmer J, Blessing C, et al. NSUN3 and ABH1 modify the wobble position of mt-tRNAMet to expand codon recognition in mitochondrial translation. EMBO J 2016; 35(19):2104–19.

[114] Nakano S, Suzuki T, Kawarada L, Iwata H, Asano K, Suzuki T. NSUN3 methylase initiates 5-formylcytidine biogenesis in human mitochondrial tRNA(Met). Nat Chem Biol 2016;12(7):546–51.

[115] Asano K, Suzuki T, Saito A, Wei FY, Ikeuchi Y, Numata T, et al. Metabolic and chemical regulation of tRNA modification associated with taurine deficiency and human disease. Nucleic Acids Res 2018;46(4): 1565–83.

[116] Sasarman F, Antonicka H, Horvath R, Shoubridge EA. The 2-thiouridylase function of the human MTU1 (TRMU) enzyme is dispensable for mitochondrial translation. Hum Mol Genet 2011;20(23):4634–43.

[117] Yarham JW, Lamichhane TN, Pyle A, Mattijssen S, Baruffini E, Bruni F, et al. Defective i6A37 modification of mitochondrial and cytosolic tRNAs results from pathogenic mutations in TRIT1 and its substrate tRNA. PLoS Genet 2014;10(6):e1004424.

[118] Powell CA, Kopajtich R, D'Souza AR, Rorbach J, Kremer LS, Husain RA, et al. TRMT5 mutations cause a defect in post-transcriptional modification of mitochondrial tRNA associated with multiple respiratory-chain deficiencies. Am J Hum Genet 2015;97(2):319–28.

[119] Patton JR, Bykhovskaya Y, Mengesha E, Bertolotto C, Fischel-Ghodsian N. Mitochondrial myopathy and sideroblastic anemia (MLASA): missense mutation in the pseudouridine synthase 1 (PUS1) gene is associated with the loss of tRNA pseudouridylation. J Biol Chem 2005;280(20):19823–8.

[120] Zaganelli S, Rebelo-Guiomar P, Maundrell K, Rozanska A, Pierredon S, Powell CA, et al. The pseudouridine synthase RPUSD4 is an essential component of mitochondrial RNA granules. J Biol Chem 2017; 292(11):4519–32.

[121] Antonicka H, Choquet K, Lin ZY, Gingras AC, Kleinman CL, Shoubridge EA. A pseudouridine synthase module is essential for mitochondrial protein synthesis and cell viability. EMBO Rep 2017;18(1):28–38.

[122] Nagaike T, Suzuki T, Tomari Y, Takemoto-Hori C, Negayama F, Watanabe K, et al. Identification and characterization of mammalian mitochondrial tRNA nucleotidyltransferases. J Biol Chem 2001;276(43): 40041–9.

[123] Betat H, Morl M. The CCA-adding enzyme: a central scrutinizer in tRNA quality control. Bioessays 2015; 37(9):975–82.

[124] Pearce SF, Rorbach J, Van Haute L, D'Souza AR, Rebelo-Guiomar P, Powell CA, et al. Maturation of selected human mitochondrial tRNAs requires deadenylation. Elife 2017;6.

[125] Chinnery PF. Primary mitochondrial disorders overview. In: Adam MP, Ardinger HH, Pagon RA, Wallace SE, Bean LJH, Mirzaa G, et al., editors. GeneReviews((R)). Seattle (WA); 1993.

[126] McFarland R, Taylor RW, Turnbull DM. A neurological perspective on mitochondrial disease. Lancet Neurol 2010;9(8):829–40.

[127] Gorman GS, Schaefer AM, Ng Y, Gomez N, Blakely EL, Alston CL, et al. Prevalence of nuclear and mitochondrial DNA mutations related to adult mitochondrial disease. Ann Neurol 2015;77(5):753–9.
[128] Gorman GS, Chinnery PF, DiMauro S, Hirano M, Koga Y, McFarland R, et al. Mitochondrial diseases. Nat Rev Dis Primers 2016;2:16080.
[129] Calvo SE, Clauser KR, Mootha VK. MitoCarta2.0: an updated inventory of mammalian mitochondrial proteins. Nucleic Acids Res 2016;44(D1):D1251–7.
[130] Van Haute L, Pearce SF, Powell CA, D'Souza AR, Nicholls TJ, Minczuk M. Mitochondrial transcript maturation and its disorders. J Inherit Metab Dis 2015;38(4):655–80.
[131] Pearce SF, Rebelo-Guiomar P, D'Souza AR, Powell CA, Van Haute L, Minczuk M. Regulation of mammalian mitochondrial gene expression: recent advances. Trends Biochem Sci 2017;42(8):625–39.
[132] Schatton D, Rugarli EI. Post-transcriptional regulation of mitochondrial function. Curr Opin Physiol 2018; 3:6–15.
[133] Jedynak-Slyvka M, Jabczynska A, Szczesny RJ. Human mitochondrial RNA processing and modifications: overview. Int J Mol Sci 2021;22(15).
[134] Morin C, Mitchell G, Larochelle J, Lambert M, Ogier H, Robinson BH, et al. Clinical, metabolic, and genetic aspects of cytochrome C oxidase deficiency in Saguenay-Lac-Saint-Jean. Am J Hum Genet 1993; 53(2):488–96.
[135] Mootha VK, Lepage P, Miller K, Bunkenborg J, Reich M, Hjerrild M, et al. Identification of a gene causing human cytochrome c oxidase deficiency by integrative genomics. Proc Natl Acad Sci U S A 2003;100(2): 605–10.
[136] Debray FG, Morin C, Janvier A, Villeneuve J, Maranda B, Laframboise R, et al. LRPPRC mutations cause a phenotypically distinct form of Leigh syndrome with cytochrome c oxidase deficiency. J Med Genet 2011; 48(3):183–9.
[137] Cuillerier A, Honarmand S, Cadete VJJ, Ruiz M, Forest A, Deschenes S, et al. Loss of hepatic LRPPRC alters mitochondrial bioenergetics, regulation of permeability transition and trans-membrane ROS diffusion. Hum Mol Genet 2017;26(16):3186–201.
[138] Sabharwal A, Wishman MD, Cervera RL, Serres MR, Anderson JL, Treichel AJ, et al. A genetic model therapy proposes a critical role for liver dysfunction in mitochondrial biology and disease. bioRxiv 2020. 2020.05.08.084681.
[139] Guo L, Engelen BPH, Hemel I, de Coo IFM, Vreeburg M, Sallevelt S, et al. Pathogenic SLIRP variants as a novel cause of autosomal recessive mitochondrial encephalomyopathy with complex I and IV deficiency. Eur J Hum Genet 2021;29(12):1789–95.
[140] Haack TB, Kopajtich R, Freisinger P, Wieland T, Rorbach J, Nicholls TJ, et al. ELAC2 mutations cause a mitochondrial RNA processing defect associated with hypertrophic cardiomyopathy. Am J Hum Genet 2013;93(2):211–23.
[141] Saoura M, Powell CA, Kopajtich R, Alahmad A, Al-Balool HH, Albash B, et al. Mutations in ELAC2 associated with hypertrophic cardiomyopathy impair mitochondrial tRNA 3'-end processing. Hum Mutat 2019;40(10):1731–48.
[142] Akawi NA, Ben-Salem S, Hertecant J, John A, Pramathan T, Kizhakkedath P, et al. A homozygous splicing mutation in ELAC2 suggests phenotypic variability including intellectual disability with minimal cardiac involvement. Orphanet J Rare Dis 2016;11(1):139.
[143] Siira SJ, Rossetti G, Richman TR, Perks K, Ermer JA, Kuznetsova I, et al. Concerted regulation of mitochondrial and nuclear non-coding RNAs by a dual-targeted RNase Z. EMBO Rep 2018;19(10).
[144] Crosby AH, Patel H, Chioza BA, Proukakis C, Gurtz K, Patton MA, et al. Defective mitochondrial mRNA maturation is associated with spastic ataxia. Am J Hum Genet 2010;87(5):655–60.
[145] Martin NT, Nakamura K, Paila U, Woo J, Brown C, Wright JA, et al. Homozygous mutation of MTPAP causes cellular radiosensitivity and persistent DNA double-strand breaks. Cell Death Dis 2014;5:e1130.

[146] von Ameln S, Wang G, Boulouiz R, Rutherford MA, Smith GM, Li Y, et al. A mutation in PNPT1, encoding mitochondrial-RNA-import protein PNPase, causes hereditary hearing loss. Am J Hum Genet 2012;91(5): 919−27.
[147] Slavotinek AM, Garcia ST, Chandratillake G, Bardakjian T, Ullah E, Wu D, et al. Exome sequencing in 32 patients with anophthalmia/microphthalmia and developmental eye defects. Clin Genet 2015;88(5): 468−73.
[148] Matilainen S, Carroll CJ, Richter U, Euro L, Pohjanpelto M, Paetau A, et al. Defective mitochondrial RNA processing due to PNPT1 variants causes Leigh syndrome. Hum Mol Genet 2017;26(17):3352−61.
[149] Dhir A, Dhir S, Borowski LS, Jimenez L, Teitell M, Rotig A, et al. Mitochondrial double-stranded RNA triggers antiviral signalling in humans. Nature 2018;560(7717):238−42.
[150] Bamborschke D, Kreutzer M, Koy A, Koerber F, Lucas N, Huenseler C, et al. PNPT1 mutations may cause Aicardi-Goutieres-Syndrome. Brain Dev 2021;43(2):320−4.
[151] Shimada E, Ahsan FM, Nili M, Huang D, Atamdede S, TeSlaa T, et al. PNPase knockout results in mtDNA loss and an altered metabolic gene expression program. PLoS One 2018;13(7):e0200925.
[152] Clemente P, Pajak A, Laine I, Wibom R, Wedell A, Freyer C, et al. SUV3 helicase is required for correct processing of mitochondrial transcripts. Nucleic Acids Res 2015;43(15):7398−413.
[153] Paul E, Cronan R, Weston PJ, Boekelheide K, Sedivy JM, Lee SY, et al. Disruption of Supv3L1 damages the skin and causes sarcopenia, loss of fat, and death. Mamm Genome 2009;20(2):92−108.
[154] Ufer C, Wang CC, Fahling M, Schiebel H, Thiele BJ, Billett EE, et al. Translational regulation of glutathione peroxidase 4 expression through guanine-rich sequence-binding factor 1 is essential for embryonic brain development. Genes Dev 2008;22(13):1838−50.
[155] Nicholls TJ, Spahr H, Jiang S, Siira SJ, Koolmeister C, Sharma S, et al. Dinucleotide degradation by REXO2 maintains promoter specificity in mammalian mitochondria. Mol Cell 2019;76(5):784−796 e6.
[156] Wei X, Du M, Li D, Wen S, Xie J, Li Y, et al. Mutations in FASTKD2 are associated with mitochondrial disease with multi-OXPHOS deficiency. Hum Mutat 2020;41(5):961−72.
[157] Ghezzi D, Saada A, D'Adamo P, Fernandez-Vizarra E, Gasparini P, Tiranti V, et al. FASTKD2 nonsense mutation in an infantile mitochondrial encephalomyopathy associated with cytochrome c oxidase deficiency. Am J Hum Genet 2008;83(3):415−23.
[158] Metodiev MD, Thompson K, Alston CL, Morris AAM, He L, Assouline Z, et al. Recessive mutations in TRMT10C cause defects in mitochondrial RNA processing and multiple respiratory chain deficiencies. Am J Hum Genet 2016;98(5):993−1000.
[159] Yang SY, He XY, Olpin SE, Sutton VR, McMenamin J, Philipp M, et al. Mental retardation linked to mutations in the HSD17B10 gene interfering with neurosteroid and isoleucine metabolism. Proc Natl Acad Sci U S A 2009;106(35):14820−4.
[160] Rauschenberger K, Scholer K, Sass JO, Sauer S, Djuric Z, Rumig C, et al. A non-enzymatic function of 17beta-hydroxysteroid dehydrogenase type 10 is required for mitochondrial integrity and cell survival. EMBO Mol Med 2010;2(2):51−62.
[161] Rackham O, Busch JD, Matic S, Siira SJ, Kuznetsova I, Atanassov I, et al. Hierarchical RNA processing is required for mitochondrial ribosome assembly. Cell Rep 2016;16(7):1874−90.
[162] Koeck T, Olsson AH, Nitert MD, Sharoyko VV, Ladenvall C, Kotova O, et al. A common variant in TFB1M is associated with reduced insulin secretion and increased future risk of type 2 diabetes. Cell Metabol 2011; 13(1):80−91.
[163] Sharoyko VV, Abels M, Sun J, Nicholas LM, Mollet IG, Stamenkovic JA, et al. Loss of TFB1M results in mitochondrial dysfunction that leads to impaired insulin secretion and diabetes. Hum Mol Genet 2014; 23(21):5733−49.

[164] Lee S, Rose S, Metodiev MD, Becker L, Vernaleken A, Klopstock T, et al. Overexpression of the mitochondrial methyltransferase TFB1M in the mouse does not impact mitoribosomal methylation status or hearing. Hum Mol Genet 2015;24(25):7286–94.
[165] Metodiev MD, Spahr H, Loguercio Polosa P, Meharg C, Becker C, Altmueller J, et al. NSUN4 is a dual function mitochondrial protein required for both methylation of 12S rRNA and coordination of mitoribosomal assembly. PLoS Genet 2014;10(2):e1004110.
[166] Garone C, D'Souza AR, Dallabona C, Lodi T, Rebelo-Guiomar P, Rorbach J, et al. Defective mitochondrial rRNA methyltransferase MRM2 causes MELAS-like clinical syndrome. Hum Mol Genet 2017;26(21): 4257–66.
[167] Khalique A, Mattijssen S, Haddad AF, Chaudhry S, Maraia RJ. Targeting mitochondrial and cytosolic substrates of TRIT1 isopentenyltransferase: specificity determinants and tRNA-i6A37 profiles. PLoS Genet 2020;16(4):e1008330.
[168] Haller RG, Lewis SF, Estabrook RW, DiMauro S, Servidei S, Foster DW. Exercise intolerance, lactic acidosis, and abnormal cardiopulmonary regulation in exercise associated with adult skeletal muscle cytochrome c oxidase deficiency. J Clin Invest 1989;84(1):155–61.
[169] Chakraborty PK, Schmitz-Abe K, Kennedy EK, Mamady H, Naas T, Durie D, et al. Mutations in TRNT1 cause congenital sideroblastic anemia with immunodeficiency, fevers, and developmental delay (SIFD). Blood 2014;124(18):2867–71.
[170] DeLuca AP, Whitmore SS, Barnes J, Sharma TP, Westfall TA, Scott CA, et al. Hypomorphic mutations in TRNT1 cause retinitis pigmentosa with erythrocytic microcytosis. Hum Mol Genet 2016;25(1):44–56.
[171] Fernandez-Vizarra E, Berardinelli A, Valente L, Tiranti V, Zeviani M. Nonsense mutation in pseudouridylate synthase 1 (PUS1) in two brothers affected by myopathy, lactic acidosis and sideroblastic anaemia (MLASA). J Med Genet 2007;44(3):173–80.
[172] Metodiev MD, Assouline Z, Landrieu P, Chretien D, Bader-Meunier B, Guitton C, et al. Unusual clinical expression and long survival of a pseudouridylate synthase (PUS1) mutation into adulthood. Eur J Hum Genet 2015;23(6):880–2.
[173] Kopajtich R, Nicholls TJ, Rorbach J, Metodiev MD, Freisinger P, Mandel H, et al. Mutations in GTPBP3 cause a mitochondrial translation defect associated with hypertrophic cardiomyopathy, lactic acidosis, and encephalopathy. Am J Hum Genet 2014;95(6):708–20.
[174] Chen D, Zhang Z, Chen C, Yao S, Yang Q, Li F, et al. Deletion of Gtpbp3 in zebrafish revealed the hypertrophic cardiomyopathy manifested by aberrant mitochondrial tRNA metabolism. Nucleic Acids Res 2019;47(10):5341–55.

Index

Note: 'Page numbers followed by "*f*" indicate figures and "*t*" indicates tables.'

I

J

K

L

M

N

O

P

T

U

9780323913058